AF539932

Encyclopedia of Water Science

Encyclopedia of Water Science

Gaj Raj Singh

2012

SBS Publishers & Distributors Pvt. Ltd.
New Delhi

ISBN: 9789380090207

First Published in 2012

Published by:

SBS PUBLISHERS & DISTRIBUTORS PVT. LTD.
2/9, Ground Floor, Ansari Road, Darya Ganj,
New Delhi - 110002,
INDIA
Tel: 0091.11.23289119 / 41563911 / 32945311
Email: mail@sbspublishers.com
www.sbspublishers.com

Printed in India by Chaman Enterprises, New Delhi.

Contents

Preface

All living things require water. One in five people on this planet does not have access to safe and affordable drinking water, and half do not have access to sanitation. With global population predicted to reach almost 8 billion by 2025, water management and treatment will become critical.

Most of the world's water is salt water, unsuitable for most uses. Fresh water makes up only 2.5% of the water supply and two-thirds of this is in the form of glaciers and permafrost, leaving less than 1% of the world's water available for use. Of this remaining 1%, agriculture is the biggest user of water withdrawals from groundwater and surface water supplies: Agriculture comprises 69% of water use compared with industrial and domestic users, who consume 21% and 10%, respectively.

But as population growth continues and industrialization expands, there will be greater competition among all users and an increasing need for more efficient water use. The information assembled will be a useful tool in helping humanity address and meet water use challenges of the 21st century.

Water science treats the occurrence, circulation, distribution, and properties of the waters of the earth, and their reaction with the environment. Encyclopedia of Water Science covers topics such as Earth Water, Water Basics, Water Cycle, Water Quality and Standards, Oceans and Saltwater, Fresh Water, Estuaries and Wetlands, Ice, Water, Weather and Climate, Pollution, etc. The author examines the impact of human use, misuse, and reuse of freshwater and wastewater on the overall water supply. Authoritative, informative, and up-to-date, the book blends real-world experience with theoretical models. This work provides the valuable insight all water practitioners need and includes important information for policy-makers and anyone else tasked with making decisions concerning water resource utilization.

Author

Chapter 1

Earth Water

AN OVERVIEW

Water is continually moving around, through, and above the Earth as water vapour, liquid water, and ice. In fact, water is continually changing its form. The Earth is pretty much a "closed system," like a terrarium. That means that the Earth neither, as a whole, gains nor loses much matter, including water. Although some matter, such as meteors from outer space, are captured by Earth, very little of Earth's substances escape into outer space. This is certainly true about water. This means that the same water that existed on Earth millions of years ago is still here. Thanks to the water cycle, the same water is continually being recycled all around the globe. It is entirely possible that the water you drank for lunch was once used by Mama Allosaurus to give her baby a bath. By the way, there is a theory that much of Earth's water came from comets hitting the planet over billions of years.

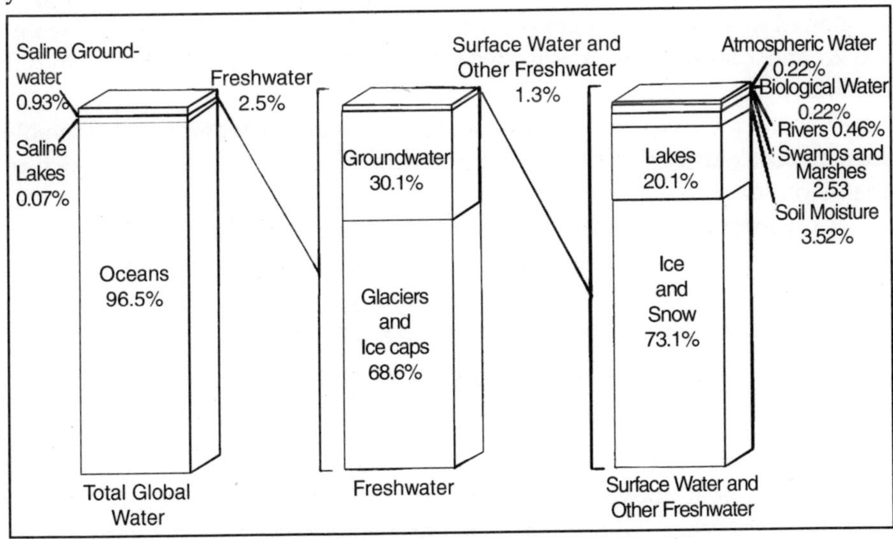

Fig. Distribution of Earth's Water

WATER ON AND IN THE EARTH

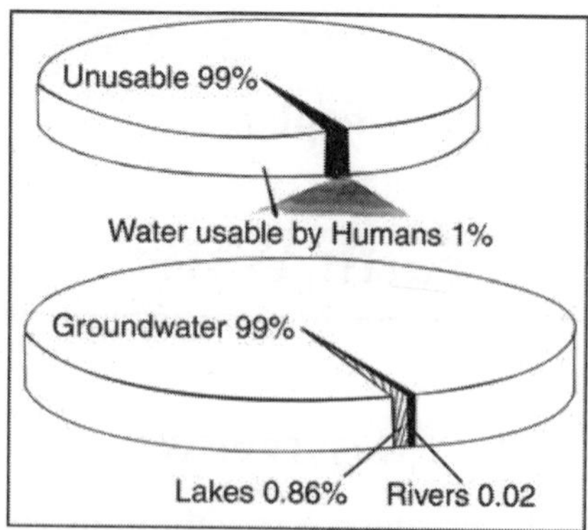

Fig. All on Earth

Where is Earth's water located and in what forms does it exist? You can see how water is distributed by viewing these bar charts. The left-side bar shows where the water on Earth exists; about 97 per cent of all water is in the oceans.

The middle bar shows the distribution of that three per cent of all Earth's water that is freshwater. The majority, about 69 per cent, is locked up in glaciers and icecaps, mainly in Greenland and Antarctica.

You might be surprised that of the remaining freshwater, almost all of it is below your feet, as ground water. No matter where on Earth you are standing, chances are that, at some depth, the ground below you is saturated with water. Of all the freshwater on Earth, only about 0.3 per cent is contained in rivers and lakes—yet rivers and lakes are not only the water we are most familiar with, it is also where most of the water we use in our everyday lives exists.

The top pie chart shows that over 99 per cent of all water is not available for our uses. And even of the remaining fraction of one per cent much of that is out of reach.

Considering that most of the water we use in everyday life comes from rivers, you'll see we generally only make use of a tiny portion of the available water supplies. The bottom pie shows that the vast majority of the fresh water available for our uses is stored in the ground.

HOW MUCH WATER IS THERE ON, IN, AND ABOVE THE EARTH?

As you know, the Earth is a watery place. But just how much water exists on, in, and above our planet? You're probably thinking I missed a decimal point when running my calculator since surely all the water on, in, and above the Earth would fill a ball a lot larger than that "tiny" blue sphere sitting on the United States, reaching from about Salt Lake City, Utah to Topeka, Kansas. But, no, this diagram is indeed correct.

About 70 per cent of the Earth's surface is water–covered, and the oceans hold about 96.5 per cent of all Earth's water. But water also exists in the air as water vapour, in rivers and lakes, in icecaps and glaciers, in the ground as soil moisture and aquifers, and even in you and your dog. Still, all that water would fit into that tiny ball. The ball is actually much larger than it looks like on your computer monitor or printed page because we're talking about volume, a 3-dimensional shape, but trying to show it on a flat, 2-dimensional screen or piece of document. That tiny water bubble has a diameter of about 860 miles, meaning the height would be 860 miles high, too! That is a lot of water.

But, as far as people are concerned, almost all of Earth's water is not usable in everyday life. Water on, in, and above the Earth is never sitting still, and thanks to the water cycle our planet's water supply is constantly moving from one place to another and from one form to another. Things would get pretty stale without the water cycle!

The vast majority of water on the Earth's surface, over 96 per cent, is saline water in the oceans. But it is the freshwater resources, such as the water in streams, rivers, lakes, and ground water that provide people with most of the water they need everyday to live. Water sitting on the surface of the Earth is easy to visualize, and your view of the water cycle might be that rainfall fills up the rivers and lakes. But, the unseen water below our feet is critically important to life, also.

How would you account for the flow in rivers after weeks without rain? In fact, how would you account for the water flowing down this driveway on a day when it didn't rain? The answer is that there is more to our water supply than just surface water, there is also plenty of water beneath our feet. Even though you may only notice water on the Earth's surface, there is much more freshwater stored in the ground than there is in liquid form on the surface. In fact, some of the water you see flowing in rivers comes from seepage of ground water into river beds. Water from precipitation continually seeps into the ground to recharge the aquifers, while at the same time water from underground aquifers continually recharges rivers through seepage.

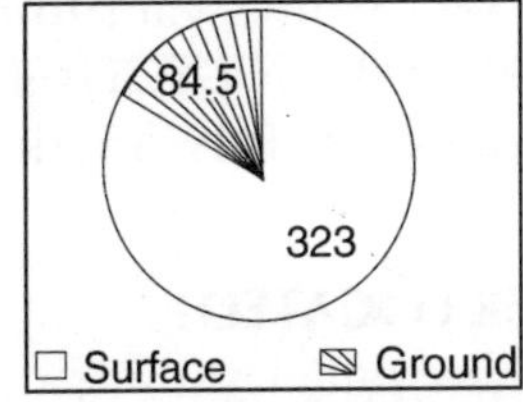

Fig. Total Water Withdrawals, in Billion Gallons Per Day

Humans are happy this happens because people make use of both kinds of water. In the United States in 2000, we used about 323 billion gallons per day of surface water and about 84.5 billion gallons per day of ground water. Although surface water is used more to supply drinking water and to irrigate crops, ground water is vital in that it not only helps to keep rivers and lakes full, it also provides water for people in places where visible water is scarce, such as in the desert towns of the western United States. Without ground water, people would be sand-surfing in Palm Springs, California. instead of playing golf.

Just how much water is there on the Earth? Here are some numbers you can think about:

- If all of Earth's water (oceans, icecaps and glaciers, lakes, rivers, ground water, and water in the atmosphere was put into a sphere, then the diameter of that water ball would be about 860 miles (about 1,385 kilometers) across, a bit more than the distance between Salt Lake City, Utah to Topeka, Kansas. The volume of all water would be about 332.5 million cubic miles (mi^3), or 1,386 million cubic kilometers (km^3). A cubic mile of water equals more than 1.1 trillion gallons. A cubic kilometer of water equals about 264 billion gallons.
- About 3,100 mi^3 (12,900 km^3) of water, mostly in the form of water vapour, is in the atmosphere at any one time. If it all fell as precipitation at once, the Earth would be covered with only about 1 inch of water.
- The 48 contiguous United States receives a total volume of about 4 mi^3 (17.7 km^3) of precipitation each day.
- Each day, 280 mi^3 (1,170 km^3)of water evaporate or transpire into the atmosphere.
- If all of the world's water was poured on the United States, it would cover the land to a depth of 90 miles (145 kilometers).
- Of the freshwater on Earth, much more is stored in the ground than is available in lakes and rivers. More than 2,000,000 mi^3 (8,400,000 km^3)of freshwater is stored in the Earth, most within one-half mile of the surface. Contrast that with the 60,000 mi^3 (250,000 km^3) of water stored as freshwater in lakes, inland seas, and rivers. But, if you really want to find freshwater, the most is stored in the 7,000,000 mi^3 (29,200,000 km^3) of water found in glaciers and icecaps, mainly in the polar regions and in Greenland.

WHERE IS EARTH'S WATER LOCATED?

Explanation of where Earth's water is, look at the data table below. Notice how of the world's total water supply of about 333 million mi^3 of water, over 96 per cent is saline. And, of the total freshwater, over 68 per cent is locked up in ice and glaciers. Another 30 per cent of freshwater is in the ground.

Thus, surface–water sources only constitute about 300 mi^3 (1,250 km^3) (about 1/10,000 th of one per cent of total water), yet rivers are the source of most of the water people use. Note: percentages may not sum to 100% due to rounding.

Table. One Estimate of Global Water Distribution

Water source	Water volume, in cubic miles	Water volume, in cubic kilometers	Per cent of freshwater	Per cent of total water
Oceans, Seas, and Bays	321,000,000	1,338,000,000	—	96.54
Ice caps, Glaciers, and Permanent Snow	5,773,000	24,064,000	68.6	1.74
Ground water	5,614,000	23,400,000	—	1.69
Fresh	2,526,000	10,530,000	30.1	0.76
Saline	3,088,000	12,870,000	—	0.93
Soil Moisture	3,959	16,500	0.05	0.001
Ground Ice and Permafrost	71,970	300,000	0.86	0.022
Lakes	42,320	176,400	—	0.013
Fresh	21,830	91,000	0.26	0.007
Saline	20,490	85,400	—	0.007
Atmosphere	3,095	12,900	0.04	0.001
Swamp Water	2,752	11,470	0.03	0.0008
Rivers	509	2,120	0.006	0.0002
Biological Water	269	1,120	0.003	0.0001

WATER CYCLE

Earth's water is always in movement, and the water cycle, also known as the hydrologic cycle, describes the continuous movement of water on, above, and below the surface of the Earth.

Although the balance of water on Earth remains fairly constant over time, individual water molecules can come and go in a hurry. Since the water cycle is truly a "cycle," there is no beginning or end.

Water can change states among liquid, vapour, and ice at various places in the water cycle, with these processes happening in the blink of an eye and over millions of years.

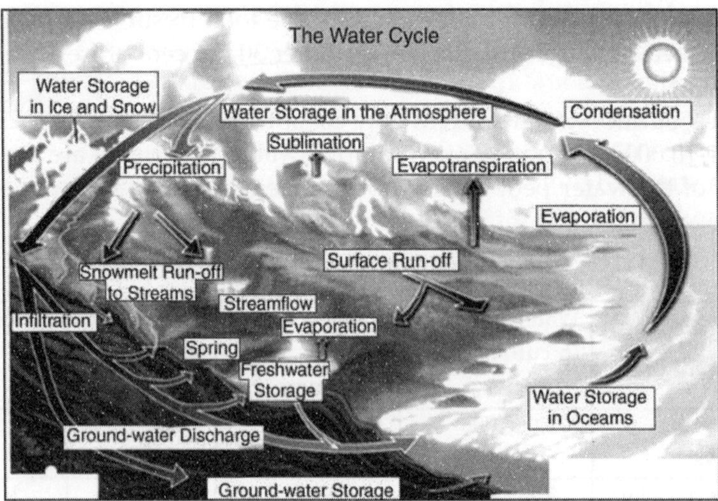

EARTH'S SURFACE WATER

WATERSHED

When looking at the location of rivers and the amount of streamflow in rivers, the key concept is the river's "watershed". What is a watershed? Easy, if you are standing on ground right now, just look down. You're standing, and everyone is standing, in a watershed. A watershed is the area of land where all of the water that falls in it and drains off of it goes into the same place. Watersheds can be as small as a footprint or large enough to encompass all the land that drains water into rivers that drain into Chesapeake Bay, where it enters the Atlantic Ocean. This map shows one set of watersheds in the continental United States; these are known as National 8-digit hydrologic units (watersheds). A watershed is an area of land that drains all the streams and rainfall to a common outlet such as the outflow of a reservoir, mouth of a bay, or any point along a stream channel. The word watershed is sometimes used interchangeably with drainage basin or catchment. Ridges and hills that separate two watersheds are called the drainage divide. The watershed consists of surface water—lakes, streams, reservoirs, and wetlands—and all the underlying ground water. Larger watersheds contain many smaller watersheds. It all depends on the outflow point; all of the land that drains water to the outflow point is the watershed for that outflow location. Watersheds are important because the streamflow and the water quality of a river are affected by things, human–induced or not, happening in the land area "above" the river–outflow point

A WATERSHED IS A PRECIPITATION COLLECTOR

Most of the precipitation that falls within the drainage area of a stream's monitoring site collects in the stream and eventually flows by the monitoring site. Many factors determine how much of the streamflow will flow by the monitoring site. Imagine that the whole basin is covered with a big (and strong) plastic sheet. Then if it rained one inch, all of that rain would fall on the plastic, run downslope into gulleys and small creeks and then drain into main stream. Ignoring evaporation and any other losses, and using a 1-square mile example watershed, then all of the approximately 17,378,560 gallons of water that fell as rainfall would eventually flow by the watershed-outflow point.

NOT ALL PRECIPITATION THAT FALLS IN A WATERSHED FLOWS OUT

To picture a watershed as a plastic-covered area of land that collects precipitation is overly simplistic and not at all like a real–world watershed. A career could be built on trying to model a watershed water budget (correlating water coming into a watershed to water leaving a watershed).

There are many factors that determine how much water flows in a stream:

- *Precipitation*: The greatest factor controlling streamflow, by far, is the amount of precipitiation that falls in the watershed as rain or snow. However, not all precipitation that falls in a watershed flows out, and a stream will often continue to flow where there is no direct run-off from recent precipitation.
- *Infiltration*: When rain falls on dry ground, some of the water soaks in, or infiltrates the soil. Some water that infiltrates will remain in the shallow soil layer, where it will gradually move downhill, through the soil, and eventually enters the stream by seepage into the stream bank. Some of the water may infiltrate much deeper, recharging ground-water aquifers. Water may travel long distances or remain in storage for long periods before returning to the surface. The amount of water that will soak in over time depends on several characteristics of the watershed:
  - *Soil characteristics*: In Georgia, clayey and rockey soils of the northern areas absorb less water at a slower rate than sandy soils, such as in Georgia's Coastal Plain. Soils absorbing less water results in more run-off overland into streams.
  - *Soil saturation*: Like a wet sponge, soil already saturated from previous rainfall can't absorb much more... thus more rainfall will become surface run-off.
  - *Land cover*: Some land covers have a great impact on infiltration and rainfall run-off. Impervious surfaces, such as parking lots, roads, and developments, act as a "fast lane" for rainfall—right into storm drains that drain directly into

streams. Flooding becomes more prevalent as the area of impervious surfaces increase.

– *Slope of the land*: Water falling on steeply–sloped land runs off more quickly than water falling on flat land.

- *Evaporation*: Water from rainfall returns to the atmosphere largely through evaporation. The amount of evaporation depends on temperature, solar radiation, wind, atmospheric pressure, and other factors.
- *Transpiration*: The root systems of plants absorb water from the surrounding soil in various amounts. Most of this water moves through the plant and escapes into the atmosphere through the leaves. Transpiration is controlled by the same factors as evaporation, and by the characteristics and density of the vegetation. Vegetation slows run-off and allows water to seep into the ground.
- *Storage*: Reservoirs store water and increase the amount of water that evaporates and infiltrates. The storage and release of water in reservoirs can have a significant effect on the streamflow patterns of the river below the dam.
- *Water use by people*: Uses of a stream might range from a few homeowners and businesses pumping small amounts of water to irrigate their lawns to large amounts of water withdrawals for irrigation, industries, mining, and to supply populations with drinking water.

SEA LEVEL AND CLIMATE

Global sea level and the Earth's climate are closely linked. The Earth's climate has warmed about 1°C (1.8°F) during the last 100 years. As the climate has warmed following the end of a recent cold period known as the "Little Ice Age" in the 19th century, sea level has been rising about 1 to 2 millimeters per year due to the reduction in volume of ice caps, ice fields, and mountain glaciers in addition to the thermal expansion of ocean water. If present trends continue, including an increase in global temperatures caused by increased greenhouse-gas emissions, many of the world's mountain glaciers will disappear. For example, at the current rate of melting, all glaciers will be gone from Glacier National Park, Montana, by the middle of the next century. In Iceland, about 11 per cent of the island is covered by glaciers. If warming continues, Iceland's glaciers will decrease by 40 per cent by 2100 and virtually disappear by 2200.

The Grinnell Glacier in Glacier National Park, Montana in 1981. The glacier has been retreating rapidly since the early 1900's. The arrows point to the former extent of the glacier in 1850, 1937, and 1968. Mountain glaciers are excellent monitors of climate change; the worldwide shrinkage of mountain glaciers is thought to be caused by a combination of a temperature increase from the Little Ice Age, which ended in the latter half of the 19th century, and increased greenhouse-gas emissions.

Most of the current global land ice mass is located in the Antarctic and Greenland ice sheets. Complete melting of these ice sheets could lead to a sea-level rise of about 80 meters, whereas melting of all other glaciers could lead to a sea-level rise of only one-half meter.

Estimated Potential Maximum Sea-level Rise from the Total Melting of Present-day Glaciers

The table shows estimated potential maximum sea-level rise from the total melting of present-day glaciers.

Location	Volume (km^3)	Potential Sea-level Rise (m)
East Antarctic ice sheet	26,039,200	64.80
West Antaretic ice sheet	3,262,000	8.06
Greenland	2,620,000	6.55
All other ice caps, ice fields and valley glaciers	180.000	.45
Total	**32,328,300**	**80.32**

Glacial-Interglacial Cycles

Climate-related sea-level changes of the last century are very minor compared with the large changes in sea level that occur as climate oscillates between the cold and warm intervals that are part of the Earth's natural cycle of long-term climate change.

During cold–climate intervals, known as glacial epochs or ice ages, sea level falls because of a shift in the global hydrologic cycle: water is evaporated from the oceans and stored on the continents as large ice sheets and expanded ice caps, ice fields, and mountain glaciers. Global sea level was about 125 meters below today's sea level at the last glacial maximum about 20,000 years ago. As the climate warmed, sea level rose because the melting North American, Eurasian, South American, Greenland, and Antarctic ice sheets returned their stored water to the world's oceans.

During the warmest intervals, called interglacial epochs, sea level is at its highest. Today we are living in the most recent interglacial, an interval that started about 10,000 years ago and is called the Holocene Epoch by geologists.

Sea levels during several previous interglacials were about 3 to as much as 20 meters higher than current sea level. The evidence comes from two different but complementary types of studies. One line of evidence is provided by old shoreline features. Wave-cut terraces and beach deposits from regions as separate as the Caribbean and the North Slope of Alaska suggest higher sea levels during past interglacial times. A second line of evidence comes from sediments cored from below the existing Greenland and West Antarctic ice sheets. The fossils and chemical signals in the sediment cores indicate that both major ice sheets were greatly reduced from their current size or even completely melted one or more times in the recent geologic past. The precise timing and details of past sea-level history are still being debated, but there is clear evidence for past sea levels significantly higher than current sea level.

This wave–cut terraces on San Clemente Island, California. Nearly horizontal surfaces, separated by step-like cliffs, were created during former intervals of high sea level; the highest terrace represents the oldest sea–level high stand.

Because San Clemente Island is slowly rising, terraces cut during an interglacial continue to rise with the island during the following glacial interval. When sea level rises during the next interglacial, a new wave-cut terrace is eroded below the previous interglacial terrace. Geologists can calculate the height of the former high sea levels by knowing the tectonic uplift rate of the island.

Potential Sea-level Changes

If Earth's climate continues to warm, then the volume of present–day ice sheets will decrease. Melting of the current Greenland ice sheet would result in a sea–level rise of about 6.5 meters; melting of the West Antarctic ice sheet would result in a sea–level rise of about 8 meters. The West Antarctic ice sheet is especially vulnerable, because much of it is grounded below sea level. Small changes in global sea level or a rise in ocean temperatures could cause a breakup of the two buttressing ice shelves. The resulting surge of the West Antarctic ice sheet would lead to a rapid rise in global sea level.

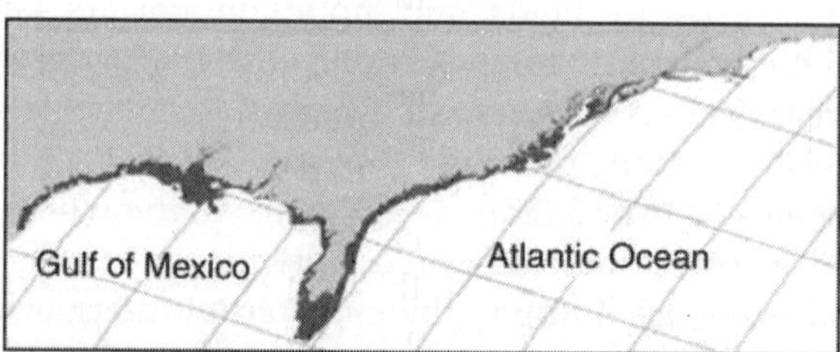

Reduction of the West Antarctic and Greenland ice sheets similar to past reductions would cause sea level to rise 10 or more meters. A sea–level rise of 10 meters would flood about 25 per cent of the U.S. population, with the major impact being mostly on the people and infrastructures in the Gulf and East Coast States. Red shows areas along the Gulf Coast and East Coast of the United States that would be flooded by a 10-meter rise in sea level. Population figures for 1996 indicate that a 10-meter rise in sea level would flood approximately 25 per cent of the Nation's population. Researchers at the U.S. Geological Survey and elsewhere are investigating the magnitude and timing of sea–level changes during previous interglacial intervals. Better documentation and understanding of these past changes will improve our ability to estimate the potential for future large-scale changes in sea level.

LAKES AND RESERVOIRS

If people had to pick their favourite water body, they'd probably choose a crystal–clear lake nestled in the mountains. Not all lakes are clear or are near mountains, though. The world is full of lakes of all types and sizes. A lake really is just another component of Earth's surface water. A lake is where surface-water run-off have accumulated in a low spot, relative to the surrounding countryside. It's not that the water that forms lakes get trapped, but that the water entering a lake comes in faster than it can escape, either via outflow in a river, seepage into the ground, or by evaporation. A reservoir is the same thing as a lake in many people's minds. But, in fact, a reservoir is a manmade lake that is created when a dam is built on a river. River water backs up behind the dam creating a reservoir.

The Earth has a tremendous variety of freshwater lakes, from fishing ponds to Lake Superior to many reservoirs. Most lakes contain fresh water, but some, especially those where water cannot escape via a river, can be salty. In fact, some lakes, such as the Great Salt Lake, are saltier than the oceans. Most lakes support a lot of aquatic life, but the Dead Sea isn't called "Dead" for nothing—it is too salty for aquatic life! Lakes formed by the erosive force of ancient glaciers, such as the Great Lakes, can be thousands of feet deep. Some very large lakes may be only a few dozen feet deep—Lake Pontchartrain in Louisiana has a maximum depth of only about 15 feet.

Some of the salty lakes were formed in ancient times when they were connected to seas and when rainfall may have been heavier. These lakes have been shrinking since the last ice age. The ancient Lake Bonneville in the United States was once as big as Lake Michigan, and the Great Salt Lake was once about 14 times as large as it is now.

Lakes are highly valued for their recreational, aesthetic and scenic qualities, and the water they contain is one of the most treasured of our natural resources. Lakes constitute important habitats and food resources for a diverse array of fish, aquatic life, and wildlife, but lake ecosystems are fragile. Lake ecosystems can undergo rapid environmental changes, often leading to significant declines in their aesthetic, recreational, and aquatic ecosystem functions. Exposed to external effects from the atmosphere, their watersheds, and ground water, lakes are subject to change through time. Human activities can further accelerate the rates of change. If the causes of the changes are known, however, human intervention sometimes can control, or even reverse, detrimental changes.

Limnology–The Study of Lakes

Limnology is the science that can provide improved understanding of lake ecosystem dynamics and information that can lead to sound management policies. As more studies are conducted on a variety of lake systems, the accumulated information leads to the development of general concepts about how lakes function and respond to environmental changes. The condition of a lake at a given time is the result of the interaction of many factors—its watershed, climate, geology, human influence, and characteristics of the lake itself. With constantly expanding databases and increased knowledge, limnologists and hydrologists are able to better understand problems that develop in particular lakes, and further develop comprehensive models that can be used to predict how lakes might change in the future.

While the development of a limnological database and knowledge is important, no amount of generalization can provide a full understanding or predict conditions of any particular lake. Each lake system is unique, and its dynamics can be understood only to a limited degree based on information from other lakes. Just as a physician would not diagnose an individual's medical condition or prescribe treatment without a personal medical examination, a limnologist or hydrologist cannot accurately assess a lake system or suggest a management strategy without data and analysis from that particular lake and its environment.

Characteristics of Lakes

The following are some of the most important basic factors that give unique character to each lake ecosystem:

- *Climate*: Temperature, wind, precipitation, and solar radiation all critically affect the lake's hydrologic and chemical characteristics, and indirectly affect the composition of the biological community. Precipitation is the main factor affecting run-off and the delivery of nutrients and sediments. Temperature, wind, and energy from the sun affect lake stratification and mixing, plant growth, and evaporation.
- *Atmospheric inputs*: The surface of a lake is directly exposed to atmospheric inputs. Not only wet precipitation, but also dry particles, can be major sources of certain contaminants to a lake. Each lake also receives indirect atmospheric inputs by way of the run-off from its watershed.
- *Geologic substrate and soils in the basin*: The soil type affects the

potential for run-off and erosion. The physical characteristics of the substrate determine the extent, nature, and quality of ground-water inflows and outflows. These are primary factors affecting the lake's chemistry, because of transfers between water and sediments, and input of sediment, minerals, and nutrients from the watershed by run-off water flowing into the lake.

- *Physiography*: The area, surface topography, existence of upstream lakes and wetlands, altitude, and land slope of the lake's watershed affect surface–water run-off and the amount and nature of chemicals and sediments entering the lake. The physiography of the region affects the size of a lake's watershed and ground–water contributing area. The boundaries of a lake watershed and ground–water contributing area may not necessarily coincide. Interactions with land use by people can appreciably change how these factors affect run-off and the export of nutrients and sediment.
- *Land use:* The type, location, extent, and history of land cover/ land use can greatly affect the quantity of surfacewater and ground-water inflows and outflows, as well as the amounts and types of sediment, nutrients and chemicals that are transported into the lake from the watershed.
- *Lake morphometry*: Size, shape, and depth characteristics of a lake are critical in determining currents and mixing of the lake, as well as its thermal and chemical stratification characteristics.

Common Environmental Problems in Lakes and Probable Causes

Eutrophication is the natural process of physical, chemical, and biological changes associated with nutrient, organic matter, and silt enrichment of a lake. If the natural process is accelerated by human influences, it is termed "cultural" eutrophication. Lakes are subject to a variety of physical, chemical, and biological problems that can diminish their aesthetic beauty, recreational value, water quality, and habitat suitability.

Among the most common lake problems, and the conditions that often occur with eutrophication are the following.

- Algal blooms Extensive and rapid growth of planktonic algae, caused by an increased input of nutrients is a common problem in lakes. Lakes normally undergo aging over timescales of centuries or thousands of years, but the process can be accelerated rapidly to only decades by human activities that cause increases in sedimentation and nutrient inflow to the lake. Accelerated eutrophication and excessive algal growth reduces water clarity, inhibits growth of other plants, and can lead to extensive oxygen depletion, accumulation of unsightly and decaying organic matter, unpleasant odours, and fish kills.

- Sedimentation/turbidity Increases in accumulation and/or resuspension of sediments can be a detriment to water quality and habitat for many aquatic species. Such events usually are caused by heavy rains that produce erosion and intense run-off, carrying heavy sediment loads into lakes. High winds, boating activity, and bottom-feeding fish, such as carp, may also resuspend bottom sediments and increase turbidity.
- Oxygen depletion Decreases in dissolved oxygen to less than 3 mg/L in the water can be harmful or lethal to many desirable species of aquatic life. The primary mechanism of oxygen loss is consumption by high rates of respiration and organic decomposition. Ideally, such consumption is offset by oxygen inputs from the atmosphere and from photosynthesis by aquatic plants. However, in stratified lakes, the atmospheric source is cut off from the hypolimnion and oxygen concentrations in the hypolimnion may decline to zero until the lake mixes again. Under anoxic conditions, phosphorus may be released from the bottom sediments into the overlying water. This "internal loading" may be considerable with phosphorus–enriched sediments and prolonged anoxia. Prolonged low oxygen concentrations in the summer or under ice in the winter can lead to fish kills.
- Growth of aquatic plants Normal macrophytic growth generally is beneficial for the lake ecosystem; among other benefits, the plants provide refuge for fish and other organisms. However, in some lakes, the growth of aquatic plants can become excessive and create a serious nuisance for lake users, interfering with swimming, boating, and other recreational activities. Excessive macrophytes commonly are caused by increased nutrients, invasion of exotic species, or accumulation of organic sediment. The improvement of water clarity resulting from management actions designed to control algal production can provide better conditions for growth of rooted plants.
- Water–level changes Wide fluctuations in stage can create major hardships for lakeside residences, marinas, and businesses, and they also may impair the habitat suitability for nearshore biota. These changes most commonly are linked to weather anomalies but also may be associated with human activities such as withdrawals for water use.
- Species shifts Populations of desirable animal and plant species might decline sharply or disappear, to be replaced by other species. Usually, the new dominant species will become a nuisance

and degrade some or all desirable qualities of the lake. Species shifts can be caused by introduction of invasive species that may have little or no natural controls on their population growth, or are stimulated by changes in environmental conditions.

RIVERS AND STREAMS

Rivers? Streams? Creeks? They are all names for water flowing on the Earth's surface. As far as the Water Science site is concerned, they are pretty much interchangeable. I tend to think of creeks as the smallest of the three, with streams being in the middle, and rivers being the largest.

Most of the water you see flowing in rivers comes from precipitation run-off from the land surface alongside the river. Of course, not all run-off ends up in rivers.

Some of it evaporates on the journey downslope, can be diverted and used by people for their uses, and can even be lapped up by thirsty animals. Rivers flow through valleys in the landscape with ridges of higher land separating the valleys. The area of land between ridges that collects precipitation is a watershed or drainage basin. Most, but not all, precipitation that falls in a watershed runs off directly into rivers-part of it soaks into the ground to recharge groundwater aquifers, some of which can then seep back into riverbeds.

What is a River?

A river is nothing more than surface water finding its way over land from a higher altitude to a lower altitude, all due to gravity. When rain falls on the land, it either seeps into the ground or becomes run-off, which flows downhill into rivers and lakes, on its journey towards the seas. In most landscapes the land is not perfectly flat—it slopes downhill in some direction.

Flowing water finds its way downhill initially as small creeks. As small creeks flow downhill they merge to form larger streams and rivers. Rivers eventually end up flowing into the oceans. If water flows to a place that is surrounded by higher land on all sides, a lake will form. If people have built a dam to hinder a river's flow, the lake that forms is a reservoir.

Where does the Water Come from?

The water in a river doesn't all come from surface run-off. Rain falling on the land also seeps into the Earth to form ground water. At a certain depth below the land surface, called the water table, the ground becomes saturated with water. If a river bank happens to cut into this saturated layer, as most rivers do, then water will seep out of the ground into the river. Groundwater seepage can sometimes be seen when a road is built through water-bearing layers, and even on a driveway!

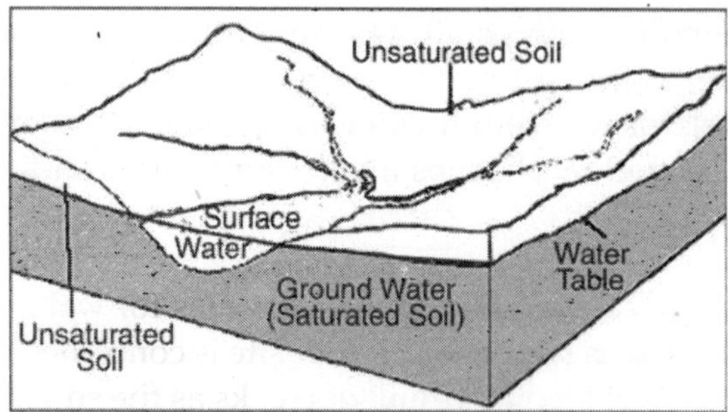

Look at the diagram. The ground below the water table, the aquifer, is saturated, whereas the ground above is not. The top layer is usually wet, but not totally saturated.

Saturated, water-bearing materials often exist in horizontal layers beneath the land surface. Since rivers, in time, may cut vertically into the ground as they flow, the water-bearing layers of rock can become exposed on the river banks. Thus, some of the water in rivers is attributed to flow coming out of the banks. This is why even during droughts there is usually some water in streams.

Rivers Serve Many Uses

The phrase "river of life" is not just a random set of words. Rivers have been essential not only to humans, but to all life on earth, ever since life began. Plants and animals grow and congregate around rivers simply because water is so essential to all life.

It might seem that rivers happen to run through many cities in the world, but it is not that the rivers go through the city, rather, the city was built and grew up around the river. For humans, rivers are diverted for flood control, irrigation, power generation, municipal uses, and even waste disposal.

And, if you ask any of these people recreating on the Chattahoochee River in Atlanta, Georgia what the best use of a river is, they might just say "to have fun."

Facts

To get a better perspective of the relative importance of large and small rivers in maintaining continental water balance, consider some statistics on the amounts of water that flow or discharge out of rivers. The Mississippi, North America's largest river, has a drainage area of 1,243,000 square miles and discharges at an average rate of 620,000 cubic feet per second. This amounts to some 133 cubic miles per year and about 34 per cent of the total discharge from all rivers of the United States.

The Columbia, nearest competitor of the Mississippi, discharges less than 75 cubic miles per year. Relatively speaking, the great Colorado River is a watery dwarf, discharging only about 5 cubic miles annually. On the other hand, the Amazon, the largest river in the world, is nearly 10 times the size of the Mississippi, discharging about 4 cubic miles each day or some 1,300 cubic miles per year—about 3 times the flow of all U.S. rivers. Africa's great Congo River, with a discharge of about 340 cubic miles per year, is the world's second largest.

The estimated annual discharge of all African rivers is about 510 cubic miles. It has been estimated that the total amount of water physically present in stream channels throughout the world at a given moment is about 500 cubic miles, The estimated total discharge from all rivers, large and small, measured and unmeasured, is about 8,430 cubic miles yearly. The estimated total discharge from all rivers about 23 cubic miles daily, with about 4 cubic miles coming from the Amazon River and about 1 cubic mile from the Congo River in Africa.

MONTHLY AND YEARLY STREAMFLOW PATTERNS: AN EXAMPLE

The U.S. Geological Survey has been measuring streamflow at thousands of streams for over a century. When extensive records of past streamflows exist, it is possible to see a pattern of streamflow variation by month and season.

Of course, every stream exhibits its own unique patterns, each stream "resides" in its own spot on the Earth's landscape, and each reacts differently to weather conditions, such as precipitation, seasonal differences, and evaporation.

This is just meant as an example of streamflow patterns, in this case for Peachtree Creek in Atlanta, Ga., USA. The chart shows the mean of monthly streamflow at Peachtree Creek for each month. The January value of 174 cubic feet per second (ft^3/s) was computed by averaging the 44 mean January streamflows. Atlantans would not be surprised that highest mean streamflows occurred in March. The chart also shows that while summer is indeed normally dry in Atlanta, October has been the month when the lowest mean streamflows have been recorded.

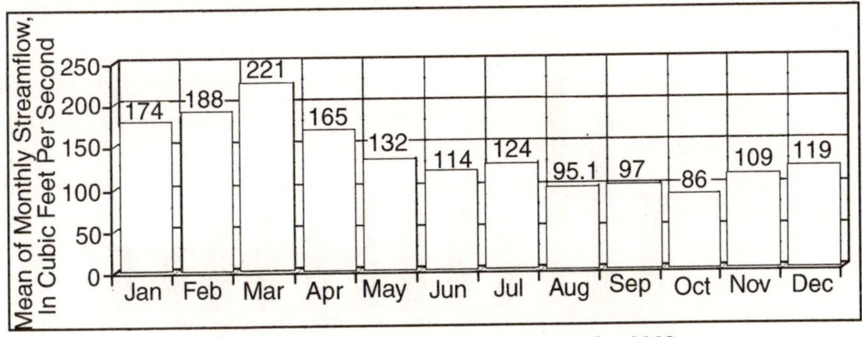

Fig. Streamflow by Months, Jul 1958-Sep2002

Seasonal Streamflow Patterns

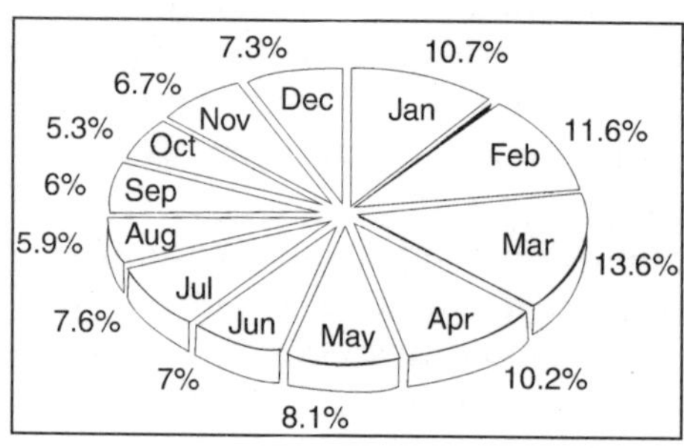

Fig. Streamflow by Months 1958–2001

To view the seasonal pattern of streamflow at Peachtree Creek is to look at a pie chart showing the percentage of yearly streamflow occuring in each month. Naturally, the streamflow patterns closely follow rainfall patterns. In Georgia, winter frontal storms result in moderate rainfall over multiple days; spring storms result in heavy rainfall; summer produces brief thunderstorms; and autumn is generally dry. This chart shows that 35.9 per cent of the yearly streamflow occurred in the months of January to March, whereas only 17.2 per cent of the yearly streamflow occurred during the typically drier months of August to October.

Yearly Patterns: Dry Years Versus Wet Years

Streamflow at Peachtree Creek can vary greatly both in the short term and long term.

Streamflows Vary by Month

The streamflow for 1975 show a more typical pattern of streamflow in Georgia, with relatively high streamflow occuring during the spring and lower streamflow during the autumn. The typical springtime streamflow pattern did not occur during 1999, with mean streamflow for March being only 80 cubic feet per second (ft^3/s) as compared to 487 ft^3/s during 1975. This makes the March 1975 streamflow about six times greater than streamflow in March of 1999.

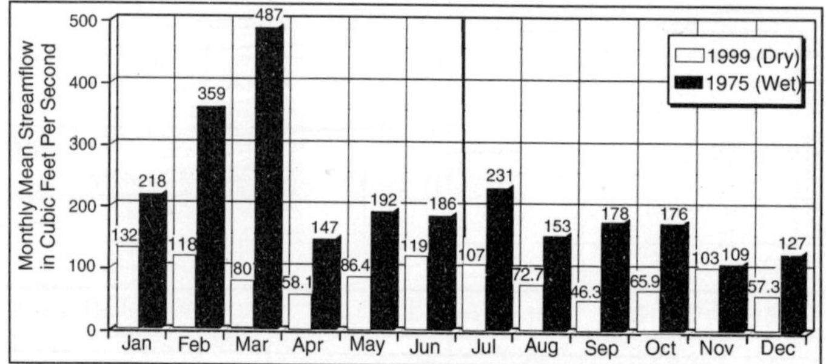

Fig. Monthly Mean Streamflow During a Dry Year and a Wet Year

Streamflow Varies by Year

The table shows the annual mean streamflow for the years 1959–2001. Annual mean streamflow is the average of all daily streamflows for the year.

The long–term annual mean streamflow for the period of record is about 136 ft^3/s. Variability in annual mean streamflow is shown by contrasting the wet year of 1975, when annual mean streamflow was 213 ft^3/s to the dry year of 1999, when mean streamflow was 86.9 ft^3/s for the dry year of 1999.

Thus, streamflow during 1975 was about 56 per cent above the long-term average, and streamflow during 1999 was only about 63 per cent of the long-term annual average.

Table. Annual Mean Streamflow at Peachtree Creek at Atlanta, 1959–2001

Year	Annual mean streamflow, in ft^3/s	Year	Annual mean streamflow, in ft^3/s
1959	78.4	1960	87.9
1961	138	1962	112
1963	137	1964	188
1965	102	1966	155
1967	172	1968	134
1969	145	1970	119
1971	137	1972	134
1973	179	1974	143
1975	213	1976	148
1977	118	1978	97.7
1979	163	1980	149
1981	70.0	1982	145
1983	165	1984	172
1985	125	1986	82.4
1987	116	1988	81.2
1989	166	1990	175
1991	175	1992	171
1993	125	1994	150
1995	157	1996	153
1997	155	1998	139
1999	86.9	2000	91.4
2001	99.8		

RUN-OFF

When rain or snow falls onto the earth, it just doesn't sit there—it starts moving just as to the laws of gravity. A portion of the precipitation seeps into the ground to replenish Earth's groundwater. Most of it flows downhill as run-off. Run-off is extremely important in that not only does it keep rivers and lakes full of water, but it also changes the landscape by the action of erosion. Flowing water has tremendous power—it can move boulders and carve out canyons. Run-off of course occurs during storms, and much more water flows in rivers during storms.

Some definitions of run-off:

- That part of the precipitation, snow melt, or irrigation water that appears in uncontrolled surface streams, rivers, drains or sewers. Run-off may be classified just as to speed of appearance after rainfall or melting snow as direct run-off or base run-off, and just as to source as surface run-off, storm interflow, or ground-water run-off.
- The sum of total discharges during a specified period of time.
- The depth to which a watershed would be covered if all of the run-off for a given period of time were uniformly distributed over it.

Meteorological factors affecting run-off:

- Type of precipitation
- Rainfall intensity
- Rainfall amount
- Rainfall duration
- Distribution of rainfall over the watershedS
- Direction of storm movement
- Antecedent precipitation and resulting soil moisture
- Other meteorological and climatic conditions that affect evapotranspiration, such as temperature, wind, relative humidity, and season.

Physical characteristics affecting run-off:

- Land use
- Vegetation
- Soil type
- Drainage area
- Basin shape
- Elevation
- Slope
- Topography
- Direction of orientation
- Drainage network patterns

- Ponds, lakes, reservoirs, sinks, etc. in the basin, which prevent or alter run-off from continuing downstream

Run-off and Water Quality

A significant portion of rainfall in forested watersheds is absorbed into soils is stored as ground water, and is slowly discharged to streams through seeps and springs. Flooding is less significant in these conditions because some of the run-off during a storm is absorbed into the ground, thus lessening the amount of run-off into a stream during the storm.

As watersheds are urbanized, much of the vegetation is replaced by impervious surfaces, thus reducing the area where infiltration to ground water can occur. Thus, more stormwater run-off occurs-run-off that must be collected by extensive drainage systems that combine curbs, storm sewers and ditches to carry stormwater run-off directly to streams. More simply, in a developed watershed, much more water arrives into a stream much more quickly, resulting in an increased likelihood of more frequent and more severe flooding.

Drainage ditches to carry stormwater run-off to storage ponds are often built to hold run-off and collect excess sediment in order to keep it out of streams. Run-off from agricultural land can carry excess nutrients, such as nitrogen and phosphorus into streams, lakes, and ground-water supplies. These excess nutrients have the potential to degrade water quality.

GLACIERS AND ICECAPS: STOREHOUSES OF FRESHWATER

Even though you've probably never seen a glacier, they are a big item when we talk about the world's water supply. Almost 10 per cent of the world's land mass is currently covered with glaciers, mostly in places like Greenland and Antarctica. Glaciers are important features in the hydrologic cycle and affect the volume, variability, and water quality of run-off in areas where they occur.

In a way, glaciers are just frozen rivers of ice flowing downhill. Glaciers begin life as snowflakes. When the snowfall in an area far exceeds the melting that occurs during summer, glaciers start to form. The weight of the accumulated snow compresses the fallen snow into ice. These "rivers" of ice are tremendously heavy, and if they are on land that has a downhill slope the whole ice patch starts to slowly grind its way downhill. These glaciers can vary greatly in size, from a football–field sized patch to a river a hundred miles long.

Glaciers Affect the Landscape

Glaciers have had a profound effect on the topography in some areas, as in the northern U.S. You can imagine how a billion-ton ice cube can rearrange the landscape as it slowly grinds its way overland. In this image you can see the bowl–shaped valley in a glacial valley in Wyoming where an ancient glacier forced its way through the landscape. Many lakes, such as the Great Lakes, and valleys have been carved out by ancient glaciers. A massive icecap can be found in Greenland, where practically the whole country is covered with ice?

The ice on Greenland approaches two miles in thickness in some places and is so heavy that some of the land has been compressed so much that it is way below sea level.

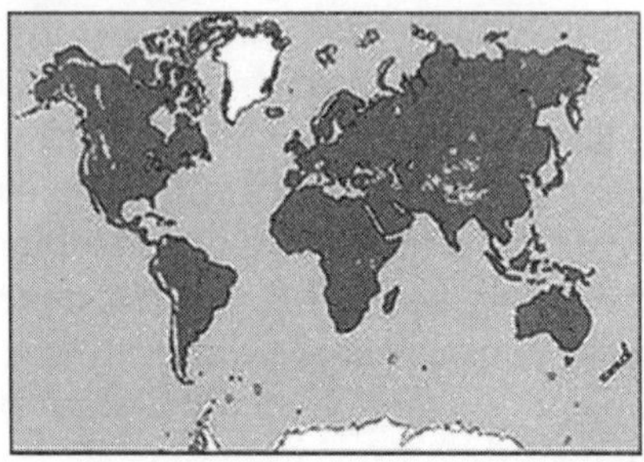

Fig. Here's a map of where glaciers and icecaps exist in the world.

White areas show glaciers and ice sheets around the world. The white spots in the oceans are islands where glaciers are found.

Ice and Glaciers Come and Go

There are many long–term weather patterns that the Earth goes through. The climate, on a global scale, is always changing, although usually not at a rate fast enough for people to notice. There have been many warm periods, such as when the dinosaurs lived and many cold periods, such as the last ice age of about 20,000 years ago.

During the last ice age much of the northern hemisphere was covered in ice and glaciers, and, as this map from the University of Arizona shows, they covered nearly all of Canada, much of northern Asia and Europe, and extended well into the United States.

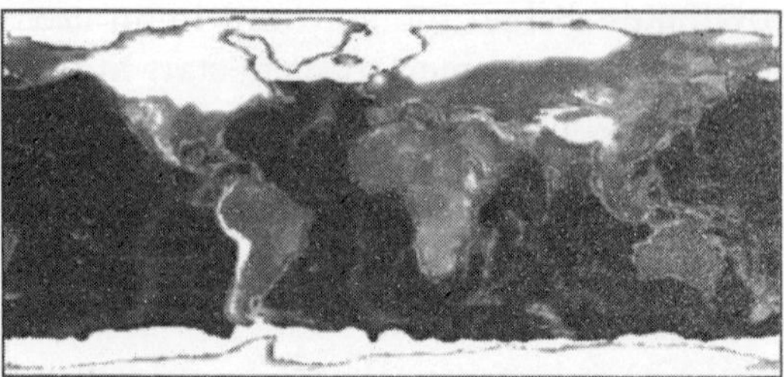

Glaciers are still around today; tens of thousands of them are in Alaska. Climatic factors still affect them today and during the current warmer climate today, they can retreat in size at a rate easily measured on a yearly scale.

Some glacier and icecap facts:

- Glaciers store about 69% of the world's freshwater, and if all land ice melted the seas would rise about 70 meters.
- During the last ice age the sea level was about 400 feet lower than it is today. At that time, glaciers covered almost one-third of the land.
- During the last warm spell, 125,000 years ago, the seas were about 18 feet higher than they are today. About three million years ago the seas could have been up to 165 feet higher.
- North America's longest glacier is the Bering Glacier in Alaska, measuring 204 kilometers long.
- Glacial ice can be very old—in some Canadian Arctic icecaps, ice at the base is over 100 000 years old.
- The land underneath parts of the West Antarctic Ice Sheet may be up to 2.5 kilometers below sea level, due to the weight of the ice.
- Antarctic ice shelves may calve icebergs that are over 80 kilometers long.
- The Kutiah Glacier in Pakistan holds the record for the fastest glacial surge. In 1953, it raced more than 12 kilometers in 3 months, averaging about 112 meters per day.
- Glacial ice often appears blue when it has become very dense. Years of compression gradually make the ice denser over time, forcing out the tiny air pockets between crystals. When glacier ice becomes extremely dense, the ice absorbs all other colours in the spectrum and reflects primarily blue, which is what we see. When glacier ice is white, that usually means that there are many tiny air bubbles still in the ice.

SURFACE WATER USE

The Nation's surface-water resources-the water in the nation's rivers, streams, creeks, lakes, and reservoirs—are vitally important to our everyday life. The main uses of surface water include drinking–water and other public uses, irrigation uses, and for use by the thermoelectric–power industry to cool electricity–generating equipment. The majority of water used for thermoelectric power, public supply, irrigation, mining, and industrial purposes came from surface–water sources.

Of all the water used in the United States in 2005, about 80 per cent came from surface–water sources. Water from groundwater sources accounted for the remaining 20 per cent.

Over 85 per cent of all water used in 2005 was freshwater, although saline water was heavily used in the thermoelectric-power industry, and, to a lesser extent, for industrial and mining purposes.

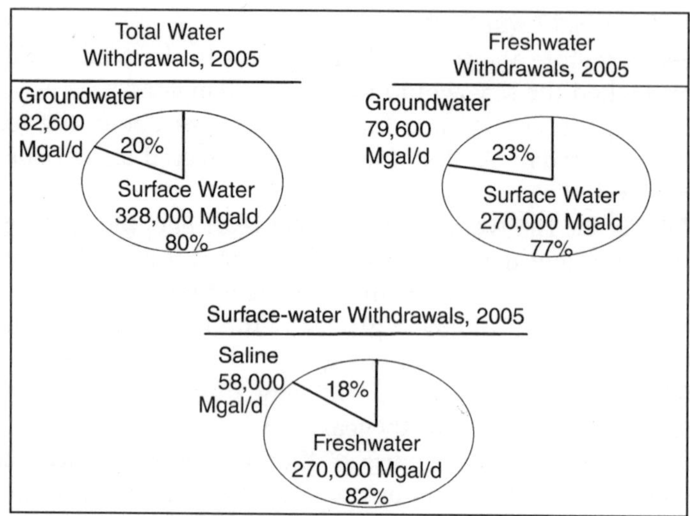

About 77 per cent of the freshwater used in the United States in 2005 came from surface–water sources.

The other 23 per cent came from groundwater. Surface water is an important natural resource used for many purposes, especially irrigation and public supply.

Surface–water Use, by Category of Use, 2005

For 2005, most of the fresh surface–water withdrawals, 53 per cent, was used in the thermoelectric-power industry to cool electricity–generating equipment. Water used in this manner is most often returned to its source.

That is why the more significant use of surface water is irrigation, which used about 28 per cent of all fresh surface water, but, ignoring thermoelectric–power withdrawals, irrigation accounted for about 58 per cent of the Nation's surface water withdrawals. Public supply and industrial were the next largest users of surface water.

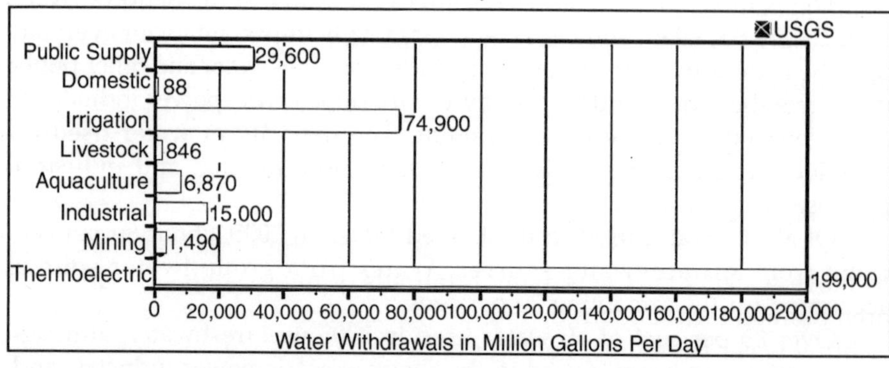

Fig. Total Surface-Water Withdrawals in the United States, 2005

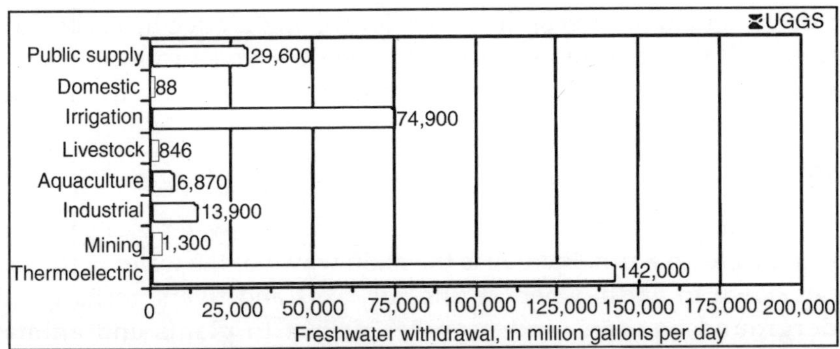

Fig. Total Surface Freshwater Withdrawals in the United States, 2005

The pie charts below show the percentage of total water that was used in 2005 for various water-use categories, broken out by surface water and groundwater.

For most categories, surface water is used more than groundwater, although this pattern varies geographically across the United States. Domestic water use is almost exclusively groundwater, whereas the water used to produce electricity comes totally from surface water.

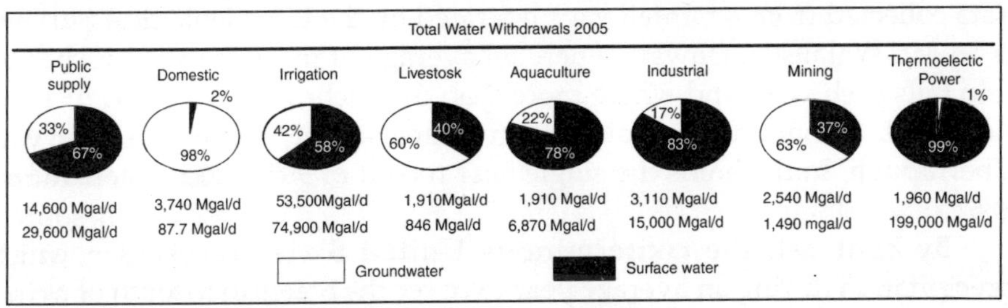

AQUEDUCTS MOVE WATER

If you live in an area where ample rain falls all year, you won't see many aqueducts like the ones pictured here. But there are many areas of the world, such as the western United States, where much less rainfall occurs and it may only occur during certain times of the year. Large cities and communities in the dry areas need lots of water, and nature doesn't always supply it to them.

Some parts of the western U.S. do have ample water supplies, though. So, some states have developed ways of moving water from the place of ample supply to the thirsty areas. Engineers have built aqueducts, or canals, to move water, sometimes many hundreds of miles. Actually, aqueducts aren't a high-tech modern invention—the ancient Romans had aqueducts to bring water from the mountains above Rome, Italy to the city.

Can you see something about the aqueduct in these depictions that causes some water to be lost in transit? In places where the climate is hot and dry, a certain portion of the water flowing in the aqueduct is bound to evaporate. It would be more efficient to cover the aqueduct to stop loss by evaporation, but the cost of covering it must be weighed against the value of the evaporated water.

RAIN

Rain and snow are key elements in the Earth's water cycle, which is vital to all life on Earth. Rainfall is the main way that the water in the skies comes down to Earth, where it fills our lakes and rivers, recharges the underground aquifers, and provides drinks to plants and animals. Fortunately for everyone, water is a renewable resource that moves in a cycle with neither beginning nor end. Water vapour condenses and returns to Earth as precipitation, once again replenishing reservoirs, lakes, rivers, and other sources of water and providing the moisture required by plants and animals's

The amount of precipitation that falls around the world may range from less than 0.1 inch per year in some deserts to more than 900 inches per year in the tropics. One of the driest spots on Earth is Lquique, Chile, where no rain fell for a period of 14 years. The world's wettest spot, as shown by data collected from a rainfall gage operated by the U.S. Geological Survey, is on Mt. Waialeale, Hawaii, where an average of more than 451 inches of rain falls each year, and where more than 642 inches fell from July 1947 to July 1948. Although Mt. Waialeale averages slightly more rain per year, Cherrapunji, India, holds the single year record of 905 inches measured in 1861.

By contrast, the conterminous United States receives enough precipitation during an average year to cover the States to a depth of about 30 inches. This is equivalent to about 1,430 cubic miles of water each year and would weigh about 6.6 billion tons. What happens to the water after it reaches the ground depends upon many factors such as the rate of rainfall, topography, soil condition, density of vegetation, temperature, and the extent of urbanization.

For example, the direct run-off in a highly urbanized area is relatively great, not only because of the density of roofs and impermeable pavements permits less rain to infiltrate the ground, but also because storm-sewer systems carry more water directly to the streams and lakes. In a more natural or undeveloped area, the direct run-off would be considerable less.

In the United States, an average of some 70 per cent of the annual precipitation returns to the atmosphere by evaporation from land and water surfaces and by transpiration from vegetation. The remaining 30 per cent eventually reaches a stream, lake, or ocean, partly by overland run-off during and immediately after rain, and partly by a much slower route through the natural ground–water reservoir.

Much of the rain that enters the ground filters down into subsurface water-bearing rocks and eventually reaches lakes, streams, and rivers where these surface-water bodies intercept the aquifers.

The portion of the precipitation that reaches the streams produces an average annual streamflow in the United States of approximately 1,200 billion gallons a day. By comparison, the Nations's homes, farms, and factories withdraw and use about 400 billion gallons a day.

How MuchWateralls During a Rainstorm?

Have you ever wondered how much water falls onto your yard during a rainstorm? Using a 1-inch rainstorm as an example, the table below gives example of how much water falls during your storm for various land areas.

Table. Amount of Water Received When an inch of Rain Occurs

Area	Area (Square miles)	Area (Square kilometers)	Amount of water (gallons)	Amount of water (liters)
1 acre	.00156	.004	27,154 gallons	102,789 liters
1 square mile	1	2.6	17.38 million gallons	65.78 million liters
Washington,	61.4	159	1.07 billion gallons	4.04 billion litersDC
United States	3,537,438	9,161,922	61,474 billion gallons	232,700 billion liters

There are 640 acres in a square mile. Once on the land, rainfall either seeps into the ground or becomes run-off, which flows into rivers and lakes.

What happens to the rain after it falls depends on many factors such as:

- *The rate of rainfall*: A lot of rain in a short period tends to run off the land into streams rather than soak into the ground.
- *The topography of the land*: Topography is the lay of the land—the hills, valleys, mountains, and canyons. Water falling on unlevel land drains downhill until it becomes part of a stream, finds a hollow place to accumulate, like a lake, or soaks into the ground.
- *Soil conditions*: There is a lot of dense clay in the southeastern United States that rain has a hard time soaking into. Contrast that to the sandy soils in more desert areas, which allow water to quickly be absorbed, at least initially.
- *Density of vegetation*: It has long been known that plant growth helps decrease erosion caused by flowing water. If you look at hills without vegetation you'll see gullies dug out by running water. Land with plant cover slows the speed of the water flowing on it and thus helps to keep soil from eroding.
- *Amount of urbanization*: As a city is being built, a lot of money and construction goes into moving water out of built-up areas. Roads, pavement, and parking lots create impervious areas where water can no longer seep into the ground. Rather, water is

funneled into creeks and streams that were never meant by nature to handle so much run-off. This can cause problems in urban areas.

The table below gives example of how much water falls within the city limits of selected cities when one inch of rainfall occurs.

Table. Amount of Water Received When an inch of Rain Occurs

City	Area (Square miles)	Amount of water (Million gallons)
Atlanta, GA	131.7	2,289
Baltimore, MD	80.8	1,404
Chicago, IL	227.1	3,947
Cincinnati, OH	78.0	1,356
Denver, CO	153.4	2,666
Detroit, MI	138.8	2,412
Honolulu, HI	85.7	1,489
Houston, TX	579.4	10,069
Jacksonville, FL	757.7	13,168
Louisville, KY	62.1	1,079
Milwaukee, WI	96.1	1,670
New Orleans, LA	180.6	3,139
New York, NY	303.3	5,271
Philadelphia, PA	135.1	2,348
Salt Lake City, UT	109.1	1,906
Seattle, WA	83.9	1,458
Washington, DC	61.4	1,067

Consider for a moment how much rainwater some cities may receive during a year. for example, Atlanta, Ga. averages about 45 inches of precipitation per year; multiplying this by the 2.36 billion gallons shown in the table as the number of gallons in 1 inch reveals that some 106 billion gallons of water fall on Atlanta in an average year. In a city the size of Atlanta, the per capita water use is about 110 gallons per day or 40,150 gallons per year. Thus, the water from a year's precipitation, if it could be collected and stored without evaporation loss, would supply the needs of about 2,640,000.

Water Equivalents (Approximate)

The following equivalents show the relationship between the volume and weight of water and between the volume and speed of flowing water.

Volume and weight:

- One inch of rain falling on 1 acre of ground is equal to about 27,154 gallons and weighs about 113 tons.
- An inch of snow falling evenly on 1 acre of ground is equivalent to about 2,715 gallons of water. However, based upon the "rule-of-thumb" that 10 inches of snow is equal to 1 inch of water, can vary considerable, depending on whether the snow is heavy and wet, or powdery and dry. Heavy, wet snow has a very high water content—4 or 5 inches of this kind of snow contains about 1 inch of water. Thus, an inch of very wet snow over an acre might amount to more than 5,400 gallons of water, while an inch of powdery snow might yield only about 1,300 gallons.
- One acre–foot of water equals 326,000 gallons or 43,560 cubic feet of water, and weighs 2.7 million pounds.
- One cubic mile of water equals 1.1 trillion gallons, 147.2 billion cubic feet, or 3.38 million acre-feet, and weighs 9.2 trillion pounds.

Rate of flow (in a stream):

- *Water flowing at the steady rate of 1 gallon per minute is equivalent to*: 1,440 gallons per day; 0.00223 cubic foot per second; 192.7 cubic feet per day; or 0.00442 acre-foot of water per day.
- *Water flowing at the steady rate of 1 cubic foot per second is equivalent to*: 449 gallons per minute; 646,000 gallons per day; 86,400 cubic feet per day; or 1.98 acre-feet of water per day.
- *Water flowing at the steady rate of 1 acre-foot per day is equivalent to*: 226 gallons per minute; 326,000 gallons per day; 0.504 cubic foot per second; or 43, 560 cubic feet of water per day.
- *Water flowing at the steady rate of 1 cubic mile per day is equivalent to*: 764.6 million gallons per minute; 1.1 trillion gallons per day; 1.7 million cubic feet per day; or 3.38 million acre-feet of water per day.

FOLLOW A DROP THROUGH THE WATER CYCLE

You may be familiar with how water is always cycling around, through, and above the Earth, continually changing from liquid water to water vapour to ice. One way to envision the water cycle is to follow a drop of water around as it moves on its way. I could really begin this story anywhere along the cycle, but I think the ocean is the best place to start, since that is where most of Earth's water is. If the drop wanted to stay in the ocean then it shouldn't have been sunbathing on the surface of the sea. The heat from the sun found the drop, warmed it, and evaporated it into water vapour. It rose into the air and continued rising until strong winds aloft grabbed it and took it hundreds of miles until it was over land. There, warm updrafts coming from the heated land surface took the dropettes up even higher, where the air is quite cold. When the vapour got cold it changed back into it a liquid. If it was cold enough, it would have turned into tiny ice crystals, such as those that make up cirrus clouds. The vapour condenses on tiny particles of dust, smoke, and salt crystals to become part of a cloud.

After a while our drop combined with other drops to form a bigger drop and fell to the earth as precipitation. Earth's gravity helped to pull it down to the surface. Once it starts falling there are many places for water drops to go. Maybe it would land on a leaf in a tree, in which case it would probably evaporate and begin its process of heading for the clouds again. If it misses a leaf there are still plenty of places to go.

The drop could land on a patch of dry dirt in a flat field. In this case it might sink into the ground to begin its journey down into an underground aquifer as groundwater. The drop will continue moving as groundwater, but the journey might end up taking tens of thousands of years until it finds its way back out of the ground. Then again, the drop could be pumped out of the ground via a water well and be sprayed on crops. Or the well water containing the drop could end up in a baby's drinking bottle or be sent to wash a car or a dog. From these places, it is back again either into the air, down sewers into rivers and eventually into the ocean, or back into the ground.

But our drop may be a land–lover. Plenty of precipitation ends up staying on the earth's surface to become a component of surface water. If the drop lands in an urban area it might hit your house's roof, go down the gutter and your driveway to the curb. If a dog or squirrel doesn't lap it up it will run down the curb into a storm sewer and end up in a small creek. It is likely the creek will flow into a larger river and the drop will begin its journey back towards the ocean.

If no one interferes, the trip will be fast back to the ocean, or at least to a lake where evaporation could again take over. But, with billions of people worldwide needing water for most everything, there is a good chance that our drop will get picked up and used before it gets back to the sea. A lot of surface water is used for irrigation. Even more is used by power-production facilities to cool their electrical equipment. From there it might go into the cooling tower to be evaporated. Talk about a quick trip back into the atmosphere as water vapour—this is it. But maybe a town pumped the drop out of the river and into a water tank. From here the drop could go on to help wash your dishes, fight a fire, water the tomatoes, or flush your toilet. Maybe the local steel mill will grab the drop, or it might end up at a fancy restaurant mopping the floor. The possibilities are endless—but it doesn't matter to the drop, because eventually it will get back into the environment. From there it will again continue its cycle into and then out of the clouds, this time maybe to end up in the water glass of the President of the United States.

GROUNDWATER

Groundwater is water located beneath the ground surface in soil pore spaces and in the fractures of rock formations. A unit of rock or an unconsolidated deposit is called an aquifer when it can yield a usable quantity of water. The depth at which soil pore spaces or fractures and voids in rock become completely saturated with water is called the water table. Groundwater is recharged from, and eventually flows to, the surface naturally; natural discharge often occurs at springs and seeps, and can form oases or wetlands. Groundwater is also often withdrawn for agricultural, municipal and industrial use by constructing and operating extraction wells. The study of the distribution and movement of groundwater is hydrogeology, also called groundwater hydrology. Typically, groundwater is thought of as liquid water flowing through shallow aquifers, but technically it can also include soil moisture, permafrost, immobile water in very low permeability bedrock, and deep geothermal or oil formation water. Groundwater is hypothesized to provide lubrication that can possibly influence the movement of faults. It is likely that much of the Earth's subsurface contains some water, which may be mixed with other fluids in some instances. Groundwater may not be confined only to the Earth. The formation of some of the landforms observed on Mars may have been influenced by groundwater. There is also evidence that liquid water may also exist in the subsurface of Jupiter's moon Europa.

AQUIFERS

An aquifer is a layer of porous substrate that contains and transmits groundwater. When water can flow directly between the surface and the saturated zone of an aquifer, the aquifer is unconfined. The deeper parts of unconfined aquifers are usually more saturated since gravity causes water to flow downward. The upper level of this saturated layer of an unconfined aquifer is called the water table or phreatic surface. Generally all pore spaces are saturated with water is the phreatic zone. Substrate with low porosity that permits limited transmission of groundwater is known as an aquitard. An aquiclude is a substrate with porosity that is so low it is virtually impermeable to groundwater.

A confined aquifer is an aquifer that is overlain by a relatively impermeable layer of rock or substrate such as an aquiclude or aquitard. If a confined aquifer follows a downward grade from its recharge zone, groundwater can become pressurized as it flows. This can create artesian wells that flow freely without the need of a pump and rise to a higher elevation than the static water table at the unconfined, aquifer.

The characteristics of aquifers vary with the geology and structure of the substrate and topography in which they occur. Generally, the more productive aquifers occur in sedimentary geologic formations. By comparison, weathered and fractured crystalline rocks yield smaller quantities of groundwater in many environments. Unconsolidated to poorly cemented alluvial materials that have accumulated as valley-filling sediments in major river valleys and geologically subsiding structural basins are included among the most productive sources of groundwater.

The high specific heat capacity of water and the insulating effect of soil and rock can mitigate the effects of climate and maintain groundwater at a relatively steady temperature. In some places where groundwater temperatures are maintained by this effect at about 50°F/10°C, groundwater can be used for controlling the temperature inside structures at the surface. For example, during hot weather relatively cool groundwater can be pumped through radiators in a home and then returned to the ground in another well. During cold seasons, because it is relatively warm, the water can be used in the same way as a source of heat for heat pumps that is much more efficient than using air.

WATER CYCLE

Groundwater makes up about twenty per cent of the world's fresh water supply, which is about 0.61% of the entire world's water, including oceans and permanent ice. Global groundwater storage is roughly equal to the total amount of freshwater stored in the snow and ice pack, including the north and south poles. This makes it an important resource which can act as a natural storage that can buffer against shortages of surface water, as in during times of drought. Groundwater is naturally replenished by surface water from precipitation, streams, and rivers when this recharge reaches the water table. Groundwater can be a long-term 'reservoir' of the natural water cycle as opposed to short-term water reservoirs like the atmosphere and fresh surface water.

The Great Artesian Basin in central and eastern Australia is one of the largest confined aquifer systems in the world, extending for almost 2 million km^2. By analysing the trace elements in water sourced from deep underground, hydrogeologists have been able to determine that water extracted from these aquifers can be more than 1 million years old.

By comparing the age of groundwater obtained from different parts of the Great Artesian Basin, hydrogeologists have found it increases in age across the basin. Where water recharges the aquifers along the Eastern Divide, ages are young. As groundwater flows westward across the continent, it increases in age, with the oldest groundwater occurring in the western parts. This means that in order to have travelled almost 1000 km from the source of recharge in 1 million years, the groundwater flowing through the Great Artesian Basin travels at an average rate of about 1 metre per year.

ISSUES

Certain problems have beset the use of groundwater around the world. Just as river waters have been over–used and polluted in many parts of the world, so too have aquifers. The big difference is that aquifers are out of sight. The other major problem is that water management agencies, when calculating the 'sustainable yield' of aquifer and river water, have often counted the same water twice, once in the aquifer, and once in its connected river. This problem, although understood for centuries, has persisted, partly through inertia within government agencies. In Australia, for example, prior to the statutory reforms initiated by the Council of Australian Governments water reform framework in the 1990s, many Australian States managed groundwater and surface water through separate government agencies, an approach beset by rivalry and poor communication.

The time lags inherent in the dynamic response of groundwater to development have generally been ignored by water management agencies, decades after scientific understanding of the issue was consolidated. In brief, the effects of groundwater overdraft may take decades or centuries to manifest themselves. In a classic study in 1982, Bredehoeft and colleagues modelled a situation where groundwater extraction in an intermontane basin withdrew the entire annual recharge, leaving 'nothing' for the natural groundwater—dependent vegetation community. Even when the borefield was situated close to the vegetation, 30% of the original vegetation demand could still be met by the lag inherent in the system after 100 years. By year 500 this had reduced to 0%, signalling complete death of the groundwater-dependent vegetation.

The science has been available to make these calculations for decades; however water management agencies have generally ignored effects which will appear outside the rough timeframe of political elections. Marios Sophocleous argued strongly that management agencies must define and use appropriate timeframes in groundwater planning. This will mean calculating groundwater withdrawal permits based on predicted effects decades, sometimes centuries in the future.

As water moves through the landscape it collects soluble salts, mainly sodium chloride. Where such water enters the atmosphere through evapotranspiration, these salts are left behind. In irrigation districts, poor drainage of soils and surface aquifers can result in water tables coming to the surface in low-lying areas. Major land degradation problems of soil salinity and waterlogging result, combined with increasing levels of salt in surface waters. As a consequence, major damage has occurred to local economies and environments.

First, flood mitigation schemes, intended to protect infrastructure built on floodplains, have had the unintended consequence of reducing aquifer recharge associated with natural flooding. Second, prolonged depletion of groundwater in extensive aquifers can result in land subsidence, with associated infrastructure damage–as well as saline intrusion. Fourth, draining acid sulphate soils, often found in low-lying coastal plains, can result in acidification and pollution of formerly freshwater and estuarine streams.

Another cause for concern is that groundwater drawdown from over-allocated aquifers has the potential to cause severe damage to both terrestrial and aquatic ecosystems–in some cases very conspicuously but in others quite imperceptibly because of the extended period over which the damage occurs.

Overdraft

Groundwater is a highly useful and often abundant resource. However, over-use, or overdraft, can cause major problems to human users and to the environment. The most evident problem is a lowering of the water table beyond the reach of existing wells. Wells must consequently be deepened to reach the groundwater; in some places the water table has dropped hundreds of feet because of extensive well pumping. In the Punjab region of India, for example, groundwater levels have dropped 10 meters since 1979, and the rate of depletion is accelerating. A lowered water table may, in turn, cause other problems such as groundwater–related subsidence and saltwater intrusion. Groundwater is also ecologically important. The importance of groundwater to ecosystems is often overlooked, even by freshwater biologists and ecologists. Groundwaters sustain rivers, wetlands and lakes, as well as subterranean ecosystems within karst or alluvial aquifers.

Not all ecosystems need groundwater, of course. Some terrestrial ecosystems—for example, those of the open deserts and similar arid environments—exist on irregular rainfall and the moisture it delivers to the soil, supplemented by moisture in the air. While there are other terrestrial ecosystems in more hospitable environments where groundwater plays no central role, groundwater is in fact fundamental to many of the world's major ecosystems.

Water flows between groundwaters and surface waters. Most rivers, lakes and wetlands are fed by, and feed groundwater, to varying degrees. Groundwater feeds soil moisture through percolation, and many terrestrial vegetation communities depend directly on either groundwater or the percolated soil moisture above the aquifer for at least part of each year. Hyporheic zones and riparian zones are examples of ecotones largely or totally dependent on groundwater.

Aquifer drawdown or overdrafting and the pumping of fossil water increases the total amount of water within the hydrosphere subject to transpiration and evaporation processes, thereby causing accretion in water vapour and cloud cover, the primary absorbers of infrared radiation in the Earth's atmosphere. Adding water to the system has a forcing effect on the whole earth system, an accurate estimate of which hydrogeological fact is yet to be quantified.

A Solution to Over-use of Groundwater

Irrigated agriculture is a big business in Asia. It accounts for 70 per cent of the world's irrigated land and 73 per cent of water used each year by the farming industry. As ageing large-scale surface irrigation schemes have become increasingly inefficient, and farmers have begun growing a wider range of crops requiring water on demand, the number of groundwater wells in India has exploded. In 1960, there were fewer than 100,000 such wells; by 2006 the figure had risen to nearly 12 million. In India, a possible solution to over-use of groundwater is emerging, known as 'groundwater recharge'. It involves capturing rainwater that would otherwise run-off, and using it to refill aquifers.

Since 2000, the International Water Management Institute has been working with the Indian authorities to help improve the availability of water for agriculture in India. In 2006, India's finance minister invited IWMI to submit policy recommendations based on its research on groundwater depletion.

One of the key recommendations was to instigate a programme of recharging groundwater across the 65 per cent of India that has hard-rock aquifers. As a result, the Indian government allocated ₹1800 crore to fund dug–well recharge projects in 100 districts within seven states where water stored in hard–rock aquifers has been over-exploited.

These geological formations have a much lower capacity to store rainwater than alluvial areas with porous sand or clay rocks, hence being given priority. The money is sufficient to fund seven million structures to be installed on dug–wells to divert monsoon run-off. The structures include a de-siltation chamber, plus pipes to collect surplus rainwater and divert de-silted water from the chamber to the well. As of the end of November 2009, funds amounting to ₹216.98 crore had been released to the concerned states. Subsidies had been released to 566,637 beneficiaries.

Subsidence

Subsidence occurs when too much water is pumped out from underground, deflating the space below the above–surface, and thus causing the ground to actually collapse. The result can look like craters on plots of land. This occurs because in its natural equilibrium state, the hydraulic pressure of groundwater in the pore spaces of the aquifer and the aquitard supports some of the weight of the overlying sediments. When groundwater is removed from aquifers by excessive pumping, pore pressures in the aquifer drop and compression of the aquifer may occur. This compression may be partially recoverable if pressures rebound, but much of it is not.

When the aquifer gets compressed it may cause land subsidence, a drop in the ground surface. The city of New Orleans, Louisiana, is actually below sea level today, and its subsidence is partly caused by removal of groundwater from the various aquifer/aquitard systems beneath it. In the first half of the 20th century, the city of San Jose, California, dropped 13 feet from land subsidence caused by overpumping; this subsidence has been halted with improved groundwater management.

Seawater Intrusion

Generally, in very humid or undeveloped regions, the shape of the water table mimics the slope of the surface. The recharge zone of an aquifer near the seacoast is likely to be inland, often at considerable distance. In these coastal areas, a lowered water table may induce sea water to reverse the flow towards the land. Sea water moving inland is called a saltwater intrusion. Alternatively, salt from mineral beds may leach into the groundwater of its own accord.

Mining

Sometimes the water movement from the recharge zone to the place where it is withdrawn may take centuries. When the usage of water is greater than the recharge, it is referred to as mining water. Under those circumstances it is not a renewable resource.

Pollution

Water pollution of groundwater, from pollutants released to the ground that can work their way down into groundwater, can create a contaminant plume within an aquifer. Movement of water and dispersion within the aquifer spreads the pollutant over a wider area, its advancing boundary often called a plume edge, which can then intersect with groundwater wells or daylight into surface water such as seeps and springs, making the water supplies unsafe for humans and wildlife. The interaction of groundwater contamination with surface waters is Analysed by use of hydrology transport models.

The stratigraphy of the area plays an important role in the transport of these pollutants. An area can have layers of sandy soil, fractured bedrock, clay, or hardpan. Areas of karst topography on limestone bedrock are sometimes vulnerable to surface pollution from groundwater. Earthquake faults can also be entry routes for downward contaminant entry. Water table conditions are of great importance for drinking water supplies, agricultural irrigation, waste disposal, wildlife habitat, and other ecological issues.

In the US, upon commercial real estate property transactions both groundwater and soil are the subjects of scrutiny, with a Phase I Environmental Site Assessment normally being prepared to investigate and disclose potential pollution issues. In the San Fernando Valley of California, Real estate contracts for property transfer below the Santa Susana Field Laboratory and eastward have clauses releasing the seller from liability for groundwater contamination consequences from existing or future water pollution of the Valley Aquifer.

Love Canal was one of the most widely known examples of groundwater pollution. In 1978, residents of the Love Canal neighbourhood in upstate New York noticed high rates of cancer and an alarming number of birth defects. This was eventually traced to organic solvents and dioxins from an industrial landfill that the neighbourhood had been built over and around, which had then infiltrated into the water supply and evaporated in basements to further contaminate the air. Eight hundred families were reimbursed for their homes and moved, after extensive legal battles and media coverage.

Another example of widespread groundwater pollution is in the Ganges Plain of northern India and Bangladesh where severe contamination of groundwater by naturally occurring arsenic affects 25% of water wells in the shallower of two regional aquifers. The pollution occurs because aquifer sediments contain organic matter that generates anaerobic conditions in the aquifer.

These conditions result in the microbial dissolution of iron oxides in the sediment and thus the release of the arsenic, normally strongly bound to iron oxides, into the water. As a consequence, arsenic-rich groundwater is often iron-rich, although secondary processes often obscure the association of dissolved arsenic and dissolved iron.

GROUNDWATER QUALITY

Just because you have a well that yields plenty of water doesn't mean you can go ahead and just take a drink. Because water is such an excellent solvent it can contain lots of dissolved chemicals. And since groundwater moves through rocks and subsurface soil, it has a lot of opportunity to dissolve substances as it moves. For that reason, groundwater will often have more dissolved substances than surface water will.

Even though the ground is an excellent mechanism for filtering out particulate matter, such as leaves, soil, and bugs, dissolved chemicals and gases can still occur in large enough concentrations in groundwater to cause problems. Undergroundwater can get contaminated from industrial, domestic, and agricultural chemicals from the surface. This includes chemicals such as pesticides and herbicides that many homeowners apply to their lawns.

Contamination of groundwater by road salt is of major concern in northern areas of the United States. Salt is spread on roads to melt ice, and, with salt being so soluble in water, excess sodium and chloride is easily transported into the subsurface groundwater. The most common water–quality problem in rural water supplies is bacterial contamination from septic tanks, which are often used in rural areas that don't have a sewage-treatment system. Effluent from a septic tank can percolate down to the water table and maybe into a homeowner's own well. Just as with urban water supplies, chlorination may be necessary to kill the dangerous bacteria.

The U.S. Geological Survey is involved in monitoring the Nation's ground-water supplies. A national network of observation wells exists to measure regularly the water levels in wells and to investigate water quality.

Contaminants can be Natural or Human-induced

Naturally occurring contaminants are present in the rocks and sediments. As groundwater flows through sediments, metals such as iron and manganese are dissolved and may later be found in high concentrations in the water. Industrial discharges, urban activities, agriculture, ground-water pumpage, and disposal of waste all can affect ground-water quality. Contaminants from leaking fuel tanks or fuel or toxic chemical spills may enter the groundwater and contaminate the aquifer. Pesticides and fertilizers applied to lawns and crops can accumulate and migrate to the water table.

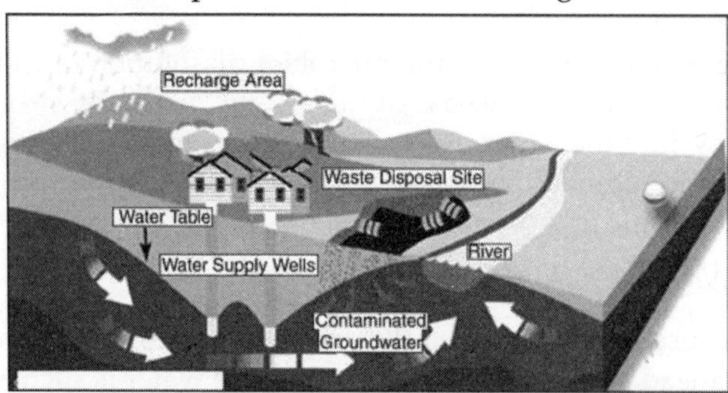

Fig. Groundwater Contamination from a Waste Disposal Site

The physical properties of an aquifer, such as thickness, rock or sediment type, and location, play a large part in determining whether contaminants from the land surface will reach the groundwater. The risk of contamination is greater for unconfined aquifers than for confined aquifers because they usually are nearer to land surface and lack an overlying confining layer to impede the movement of contaminants. Because groundwater moves slowly in the subsurface and many contaminants sorb to the sediments, restoration of a contaminated aquifer is difficult and may require years, decades, centuries, or even millennia.

Table. Inorganic Contaminants Found in Groundwater

Contaminant	Sources to Groundwater	Potential Health and other Effects
Aluminum	Occurs naturally in some rocks and drainage from mines.	Can precipitate out of water after treatment, causing increased turbidity or discoloured water.
Antimony	Enters environment from natural weathering, industrial production, municipal waste disposal, and manufacturing of flame retardants, ceramics, glass, batteries, fireworks, and explosives.	Decreases longevity, alters blood levels of glucose and cholesterol in laboratory animals exposed at high levels over their lifetime.
Arsenic	Enters environment from natural processes, industrial activities, pesticides, and industrial waste, smelting of copper, lead, and zinc ore.	Causes acute and chronic toxicity, liver and kidney damage; decreases blood hemoglobin. Possible carcinogen.
Barium	Occurs naturally in some limestones, sandstones, and soils in the eastern United States.	Can cause a variety of cardiac, gastrointestinal,and neuromuscular effects. Associated with hypertension and cardiotoxicity in animals.
Beryllium	Occurs naturally in soils, groundwater, and surface water. Often used in electrical industry equipment and components, nuclear power and space industry. Enters the environment from mining operations, processing plants, and improper waste disposal. Found in low concentrations in rocks, coal, and petroleum and enters the ground and	Causes acute and chronic toxicity; can cause damage to lungs and bones. Possible carcinogen.
Cadmium	Found in low concentrations in rocks, coal, and petroleum and enters the ground and surface water when dis-	Replaces zinc biochemically in the body and causes high blood pressure, liver and kidney damage, and anemia. Destroys testicular

	solved by acidic waters. May enter the environment from industrial discharge, mining waste, metal plating, water pipes, batteries, paints and pigments, plastic stabilizers, and landfill leachate.	tissue and red blood cells. Toxic to aquatic biota.
Chloride	May be associated with the presence of sodium in drinking water when present in high concentrations. Often from saltwater intrusion, mineral dissolution, industrial and domestic waste.	Deteriorates plumbing, water heaters, and municipal waterworks equipment at high levels. Above secondary maximum contaminant level, taste becomes noticeable.
Chromium	Enters environment from old mining operations run-off and leaching into groundwater, fossil-fuel combustion, cement-plant emissions, mineral leaching, and waste incineration. Used in metal plating and as acooling-tower water additive.	Chromium III is a nutritionally essential element. Chromium VI is much more toxic than Chromium III and causes liver and kidney damage, internal hemorrhaging, respiratory damage, dermatitis, and ulcers on the skin at high concentrations.
Copper	Enters environment from metal plating, industrial and domestic waste, mining, and mineral leaching.	Can cause stomach and intestinal distress, liver and kidney damage, anemia in high doses. Imparts an adverse taste and significant staining to clothes and fixtures. Essential trace element but toxic to plants and algae at moderate levels.
Cyanide	Often used in electroplating, steel processing, plastics, synthetic fabrics, and fertilizer production; also from improper waste disposal.	Poisoning is the result of damage to spleen, brain, and liver.

Dissolved solids	Occur naturally but also enters environment from man-made sources such as landfill leachate, feedlots, or sewage. A measure of the dissolved "salts" or minerals in the water. May also include dissolved organic compounds.	May have an influence on the acceptability of water in general. May be indicative of the presence of excess concentrations of specific substances not included in the Safe Water Drinking Act, which would make water objectionable. High concentrations of dissolved solids shorten the life of hot water heaters.
Fluoride	Occurs naturally or as an additive to municipal water supplies; widely used in industry.	Decreases incidence of tooth decay but high levels can stain or mottle teeth. Causes crippling bone disorder (calcification of the bones and joints) at very high levels.
Hardness	Result of metallic ions dissolved in the water; reported as concentration of calcium carbonate. Calcium carbonate is derived from dissolved limestone or discharges from operating or abandoned mines.	Decreases the lather formation of soap and increases scale formation in hot-water heaters and low-pressure boilers at high levels.
Iron	Occurs naturally as a mineral from sediment and rocks or from mining, industrial waste, and corroding metal.	Imparts a bitter astringent taste to water and a brownish Colour to laundered clothing and plumbing fixtures.
Lead	Enters environment from industry, mining, plumbing, gasoline, coal, and as a water additive.	Affects red blood cell chemistry; delays normal physical and mental development in babies and young children. Causes slight deficits in attention span, hearing, and learning in children. Can cause slight increase in blood pressure in some adults. Probable carcinogen.
Manganese	Occurs naturally as a mineral from sediment and rocks or from mining and industrial waste.	Causes aesthetic and economic damage, and imparts brownish stains to laundry. Affects taste of water, and causes dark brown or

		black stains on plumbing fixtures. Relatively non-toxic to animals but toxic to plants at high levels.
Mercury	Occurs as an inorganic salt and as organic mercury compounds. Enters the environment from industrial waste, mining, pesticides, coal, electrical equipment (batteries, lamps, switches), smelting, and fossil-fuel combustion.	Causes acute and chronic toxicity. Targets the kidneys and can cause nervous system disorders.
Nickel	Occurs naturally in soils, groundwater, and surface water. Often used in electro-plating, stainless steel and alloy products, mining, and refining.	Damages the heart and liver of laboratory animals exposed to large amounts over their lifetime.
Nitrate (as nitrogen)	Occurs naturally in mineral deposits, soils, seawater, freshwater systems, the atmosphere, and biota. More stable form of combined nitrogen in oxygenated water. Found in the highest levels in groundwater under extensively developed areas. Enters the environment from fertilizer, feedlots, and sewage.	Toxicity results from the body's nitrite. Causes "bluebaby disease," or methemoglobinemia, which threatens oxygen-carrying capacity of the blood.
Nitrite (combined nitrate /nitrite)	Enters environment from fertilizer, sewage, and human or farm-animal waste.	Toxicity resul ts from the body's natural breakdown of nitrate to nitrite. Causes "bluebaby disease," or methemoglobinemia, which threatens oxygen-carrying capacity of the blood.
Selenium	Enters environment from naturally occurring geologic sources, sulfur, and coal.	Causes acute and chronic toxic effects in animals—"blind staggers" in cattle. Nutritionally essential element at low doses but toxic at high doses.

Silver	Enters environment from ore mining and processing, product fabrication, and disposal. Often used in photography, electric and elec- equipment, sterling and solder. Because of great economic value of silver, recovery practices are typically used to minimize loss.	Can cause argyria, a blue-gray colouration of the skin, mucous membranes, eyes, and organs in humans and animals with chronic and electroplating, alloy, tronic exposure.
Sodium	Derived geologically from leaching of surface and underground deposits of salt and decomposition of various minerals. Human activities contribute through de-icing and washing products.	Can be a health risk factor for those individuals on a low-sodium diet.
Sulfate	Elevated concentrations may result from saltwater intrusion, mineral dissolution, and domestic or industrial waste.	Forms hard scales on boilers and heat exchangers; can change the taste of water, and has a laxative effect in high doses.
Thallium	Enters environment from soils; used in electronics, pharmaceuticals manufacturing, glass, and alloys.	Damages kidneys, liver, brain, and intestines in laboratory animals when given in high doses over their lifetime.
Zinc	Found naturally in water, most frequently in areas where it is mined. Enters environment from industrial waste, metal plating, and plumbing, and is a major component of sludge.	Aids in the healing of wounds. Causes no ill health effects except in very high doses. Imparts an undesirable taste to water. Toxic to plants at high levels.
Organic Contaminants Found in Groundwater		
Contaminant	**Sources to groundwater**	**Potential health and other effects**
Volatile	Enter environment when	Can cause cancer and liver

organic compounds	used to make plastics, dyes, rubbers, polishes, solvents, crude oil, insecticides, inks, varnishes, paints, disinfectants, gasoline products, pharmaceuticals, preservatives, spot removers, paint removers, degreasers, and many more.	damage, anemia, gastrointestinal disorder, skin irritation, blurred vision, exhaustion, weight loss, damage to the nervous system, and respiratory tract irritation.
Pesticides	Enter environment as herbicides, insecticides, fungicides, rodenticides, and algicides.	Cause poisoning, headaches, dizziness, gastrointestinal disturbance, numbness, weakness, and cancer. Destroys nervous system, thyroid, reproductive system, liver, and kidneys.
Plasticizers, chlorinated solvents, benzo[a]pyrene, and dioxin	Used as sealants, linings, solvents, pesticides, plasticizers, components of gasoline disinfectant, and wood preservative. Enters the environment from improper waste disposal, leaching run-off, leaking storage tank, and industrial run-off.	Cause cancer. Damages nervous and reproductive systems, kidney, stomach, and liver.
Microbiological Contaminants Found in Groundwater		
Coliform bacteria	Occur naturally in the environment from soils and plants and in the intestines of humans and other warm-blooded animals. Used as an indicator for the presence of pathogenic bacteria, viruses, and parasites from domestic sewage, animal waste, or plant or soil material.	Bacteria, viruses, and parasites can cause polio, cholera, typhoid fever, dysentery, and infectious hepatitis.
Radiological Contaminants Found in Groundwater		
Gross alpha-particle activity	A category of radioactive isotopes. Occurs from either natural or man-made	Damages tissues and destroys bone marrow.

	sources including weapons, nuclear reactors, atomic energy for power, medical treatment and diagnosis, mining radioactive material, and naturally occurring radioactive geologic formations. Primary concern is natural sources, which are ubiquitous in the environment; secondary concern is man-made sources.	
Combined radium-226 and radium-228	Enters environment from natural and man-made sources. Historical industrial waste sites are the main man-made source.	Causes cancer by concentrating in the bone and skeletal tissue.
Beta-particle and photon radioactivity	A category of radioactive isotopes from either natural or man-made sources including weapons, nuclear reactors, atomic energy for power, medical treatment and diagnosis, mining radioactive material, and naturally occurring radioactive geologic formations. Primary concern is man-made sources because of widespread use; secondary concern is natural sources.	Damages tissues and destroys bone marrow.
Physical Characteristics of Groundwater		
Turbidity	**Caused by the presence of**	**Objectionable for aesthetic reasons.**
	suspended matter such as clay, silt, and fine particles of organic and inorganic matter, plankton, and other microscopic organisms. A measure how much light can filter through the water sample.	Indicative of clay or other inert suspended particles in drinking water. May not adversely affect health but may cause need for additional treatment. Following rainfall, variations in ground-water turbidity may be an indicator of surface contamination.
Colour	Can be caused by decaying	Suggests that treatment is needed.

	leaves, plants, organic matter, matter, copper, iron, and manganese, which may be objectionable. Indicative of large amounts of organic chemicals, inadequate treatment, and high disinfection demand. Potential for production of excess amounts of disinfection byproducts.	No health concerns. Aesthetically unpleasing.
pH	Indicates, by numerical expr-ession, the degree to which water is alkaline or acidic. Represented on a scale of 0-14 where 0 is the most acidic, 14 is the most alkaline, and 7 is neutral.	High pH causes a bitter taste; water pipes and water-using appliances become encrusted; depresses the effectiveness of the disinfection of chlorine, thereby causing the need for additional chlorine when pH is high. Low-pH water will corrode or dissolve metals and other substances.
Odour	Certain odours may be indicative of organic or non-organic contaminants that originate from municipal or industrial waste discharges or from natural sources.	
Taste	Some substances such as certain organic salts produce a taste without an odour and can be evaluated by a taste test. Many other sensations ascribed to the sense of taste actually are odours, even though the sensation is not noticed until the material is taken into the mouth.	

PESTICIDES IN GROUNDWATER

If you ask your grandparents what life was like when they were kids, the answer will probably be that things were simpler, slower, less automated, and that people did not move so often.

But since your grandparent's time two major things have happened:

1. The population of the United States has increased greatly, and
2. Technology and scientific innovations have come to play a major role in our lives.

Pesticide use has grown because not only must our exploding population be supplied with food, but crops and food are grown for export to other countries.

The United States has become the largest producer of food products in the world, partly owing to our use of modern chemicals to control the insects, weeds, and other organisms that attack food crops. But, as with many things in life, there's a hidden cost to the benefit we get from pesticides. We've learned that pesticides can potentially harm the environment and our own health.

Water plays an important role here because it is one of the main ways that pesticides are transported from the areas where they are applied to other locations, where they may cause health problems.

Pesticides can Contaminate Groundwater

Pesticide contamination of ground water is a subject of national importance because ground water is used for drinking water by about 50 per cent of the Nation's population. This especially concerns people living in the agricultural areas where pesticides are most often used, as about 95 per cent of that population relies upon ground water for drinking water.

Before the mid–1970s, it was thought that soil acted as a protective filter that stopped pesticides from reaching ground water. Studies have now shown that this is not the case.

Pesticides can reach water–bearing aquifers below ground from applications onto crop fields, seepage of contaminated surface water, accidental spills and leaks, improper disposal, and even through injection waste material into wells.

Chemicals can take a Long Time to Appear in Groundwater

The effects of past and present land-use practices may take decades to become apparent in ground water. When weighing management decisions for protection of ground–water quality, it is important to consider the time lag between application of pesticides and fertilizers to the land and arrival of the chemicals at a well.

This time lag generally decreases with increasing aquifer permeability and with decreasing depth to water. In response to reductions in chemical applications to the land, the quality of shallow ground water will improve before the quality of deep ground water, which could take decades.

Pesticides are mostly modern chemicals. There are many hundreds of these compounds, and extensive tests and studies of their effect on humans have not been completed. That leads us to ask just how concerned we should be about their presence in our drinking water. Certainly it would be wise to treat pesticides as potentially dangerous and, thus, to handle them with care. We can say they pose a potential danger if they are consumed in large quantities, but, as any experienced scientist knows, you cannot draw factual conclusions unless scientific tests have been done.

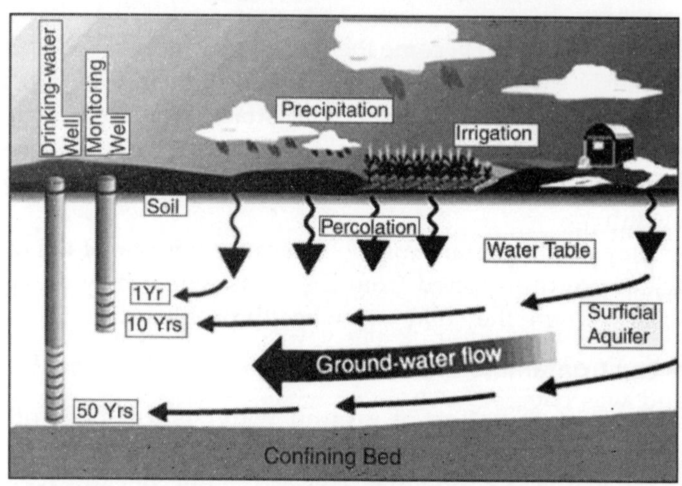

Some pesticides have had a designated Maximum Contaminant Limit in drinking water set by the U.S. Environmental Protection Agency but many have not. Also, the effect of combining more than one pesticide in drinking water might be different than the effects of each individual pesticide alone. It is another situation where we don't have sufficient scientific data to draw reliable conclusions.

GROUNDWATER: WELLS

There's a good chance that the average Joe who had to dig a well in ancient Egypt probably did the work with his hands, a shovel, and a bucket. He would have kept digging until he reached the water table and water filled the bottom of the hole. Some wells are still dug by hand today, but more modern methods are available. It's still a dirty job, though! Wells are extremely important to all societies. In many places wells provide a reliable and ample supply of water for home uses, irrigation, and industries. Where surface water is scarce, such as in deserts,people couldn't survive and thrive without ground water.

Types of Wells

Digging a well by hand is becoming outdated today. Modern wells are more often drilled by a truck-mounted drill rig. Still, there are many ways to put in a well—here are some of the common methods.

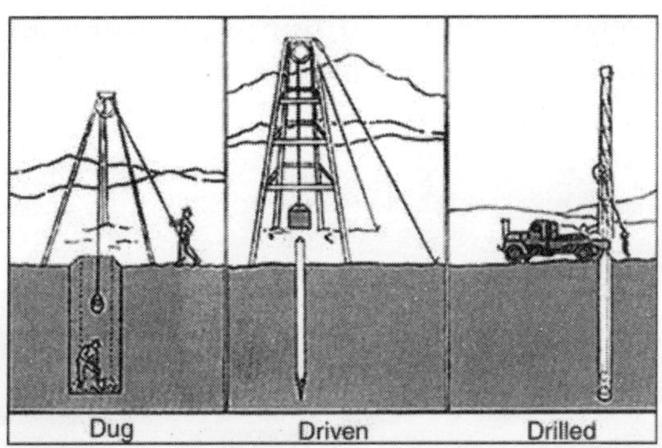

Dug Wells

Hacking at the ground with a pick and shovel is one way to dig a well. If the ground is soft and the water table is shallow,then dug wells can work. Historically, dug wells were excavated by hand shovel to below the water table until incoming water exceeded the digger's bailing rate. The well was lined with stones, brick, tile, or other material to prevent collapse, and was covered with a cap of wood, stone, or concrete. They cannot be dug much deeper than the water table—just as you cannot dig a hole very deep when you are at the beach... it keeps filling up with water!

Driven Wells

Driven wells are still common today. They are built by driving a small-diameter pipe into soft earth, such as sand or gravel. A screen is usually attached to the bottom of the pipe to filter out sand and other particles. Problems? They can only tap shallow water, and because the source of the water is so close to the surface, contamination from surface pollutants can occur.

Drilled Wells

Most modern wells are drilled, which requires a fairly complicated and expensive drill rig. Drill rigs are often mounted on big trucks. They use rotary drill bits that chew away at the rock, percussion bits that smash the rock, or, if the ground is soft,large auger bits. Drilled wells can be drilled more than 1,000 feet deep. Often a pump is placed at the bottom to push water up to the surface.

Water Levels in Wells

Ground-water users would find life easier if the water level in the aquifer that supplied their well always stayed the same. Seasonal variations in rainfall and the occasional drought affect the "height" of the underground water level. If a well is pumped at a faster rate than the aquifer around it is recharged by precipitation or other underground flow, then water levels around the well can be lowered.

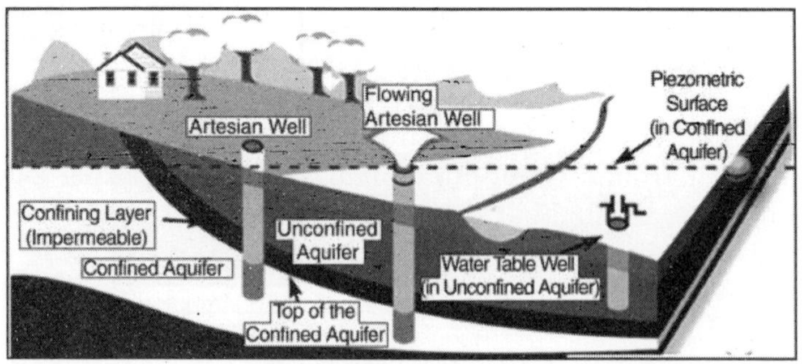

Fig. Aquifers and Wells

The water level in a well can also be lowered if other wells near it are withdrawing too much water. When water levels drop below the levels of the pump intakes, then wells will begin to pump air—they will "go dry."

ARTESIAN WATER AND ARTESIAN WELLS

Maybe you've heard advertisements by water companies wanting to sell you "artesian–well drinking water." Is this water different from other bottled water taken from springs?

The water may not be different, but it comes to the earth's surface a bit differently. Groundwater in aquifers between layers of poorly permeable rock, such as clay or shale, may be confined under pressure. If such a confined aquifer is tapped by a well, water will rise above the top of the aquifer and may even flow from the well onto the land surface. Water confined in this way is said to be under artesian pressure, and the aquifer is called an artesian aquifer. The word artesian comes from the town of Artois in France, the old Roman city of Artesium, where the best known flowing artesian wells were drilled in the Middle Ages. The level to which water will rise in tightly cased wells in artesian aquifers is called the potentiometric surface. Deep wells drilled into rock to intersect the water table and reaching far below it are often called artesian wells in ordinary conversation, but this is not necessarily a correct use of the term. Such deep wells may be just like ordinary, shallower wells; great depth alone does not automatically make them artesian wells. The word artesian, properly used, refers to situations where the water is confined under pressure below layers of relatively impermeable rock.

Example of an Aquifer System with Artesian Wells

The diagram below shows the aquifer system near Brunswick, Georgia, as it was before development of the Floridan aquifer system in the 1880's. The aquifer system was under artesian conditions and the pressure in the aquifer system was great enough that wells flowed at land surface throughout most of the coastal area. In some areas, pressure was high enough to elevate water to multi–story buildings without pumping. The artesian water level was about 65 feet above sea level at Brunswick. Ground water discharged naturally to springs, rivers, ponds, wetlands, and other surface-water bodies and to the Atlantic Ocean.

Nowadays, ground-water pumping has caused the water level in the aquifer to decline throughout the entire coastal area, with the result that some artesian aquifers no longer have enough pressure to cause a well to naturally flow to the land surface.

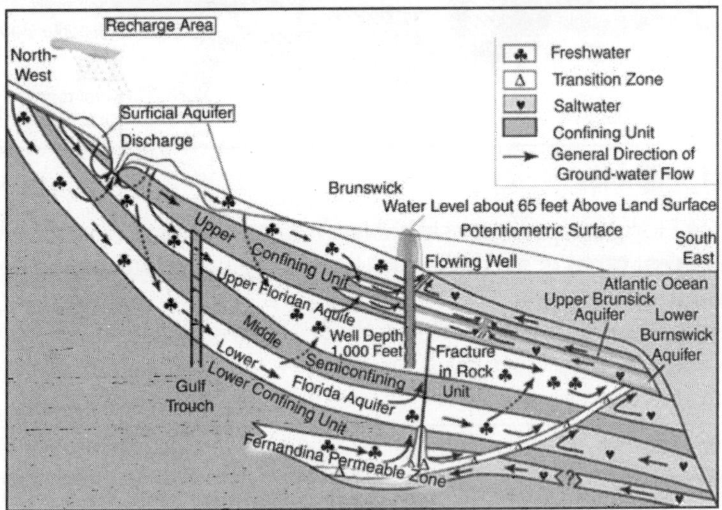

SINKHOLE

A sinkhole is an area of ground that has no natural external surface drainage—when it rains, all of the water stays inside the sinkhole and typically drains into the subsurface. Sinkholes can vary from a few feet to hundreds of acres and from less than 1 to more than 100 feet deep. Some are shaped like shallow bowls or saucers whereas others have vertical walls; some hold water and form natural ponds. Typically, sinkholes form so slowly that little change is seen in one's life-time, but they can form suddenly when a collapse occurs. Such a collapse can have a dramatic effect if it occurs in an urban setting.

Areas Prone to Collapse Sinkholes

The map below shows areas of the United States where certain rock types that are susceptible to dissolution in water occur. In these areas the formation of underground cavities can form and catastrophic sinkholes can happen. These rock types are evaporites and carbonates. Evaporite rocks underlie about 35 to 40 per cent of the United States, though in many areas they are buried at great depths.

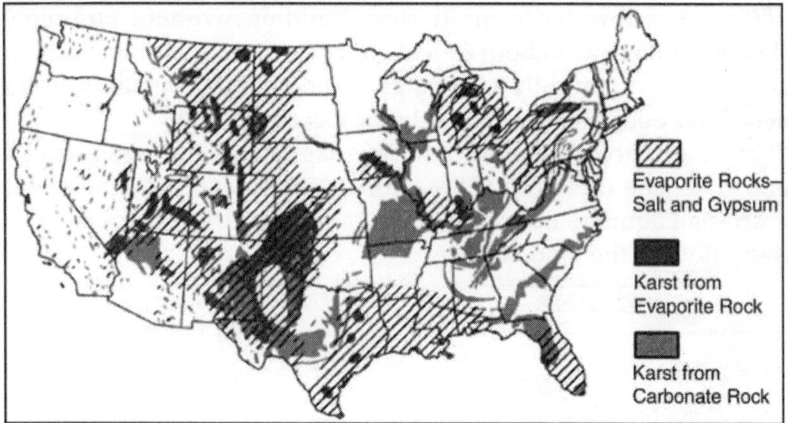

Types of Sinkholes

Since Florida is prone to sinkholes, it is a good place to use to discuss some different types of sinkholes and the geologic and hydrologic processes that form them.

The processes of dissolution, where surface rock that are soluble to weak acids, are dissolved, and suffosion, where cavities form below the land surface, are responsible for virtually all sinkholes in Florida.

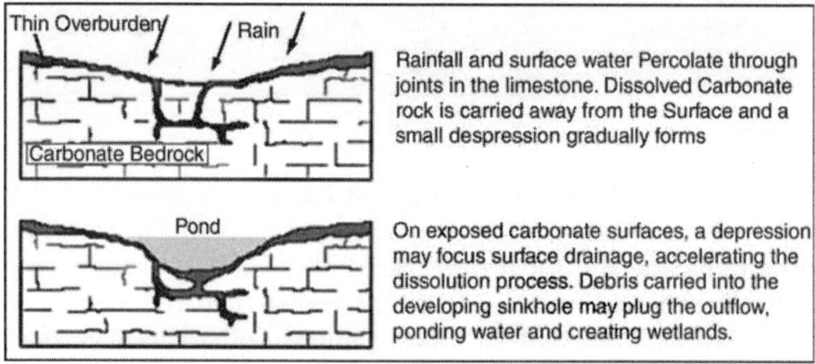

Dissolution Sinkholes

Dissolution of the limestone or dolomite is most intensive where the water first contacts the rock surface. Aggressive dissolution also occurs where flow is focussed in preexisting openings in the rock, such as along joints, fractures, and bedding planes, and in the zone of water-table fluctuation where ground water is in contact with the atmosphere.

Cover-subsidence Sinkholes

Cover-subsidence sinkholes tend to develop gradually where the covering sediments are permeable and contain sand. In areas where cover material is thicker or sediments contain more clay, cover–subsidence sinkholes are relatively uncommon, are smaller, and may go undetected for long periods.

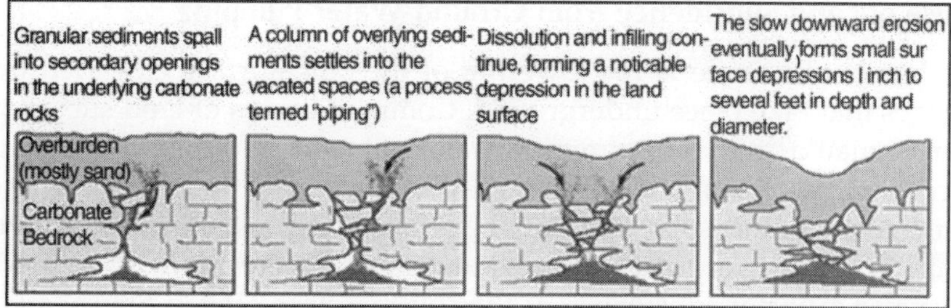

Cover–Collapse Sinkholes

Cover-collapse sinkholes may develop abruptly and cause catastrophic damages. They occur where the covering sediments contain a significant amount of clay. Over time, surface drainage, erosion, and deposition of sinkhole into a shallower bowl-shaped depression.

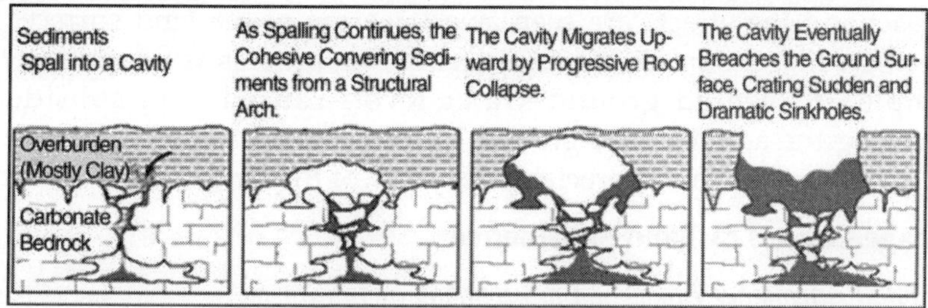

Sinkholes can be Human-Induced

New sinkholes have been correlated to land-use practices, especially from ground-water pumping and from construction and development practices.

Sinkholes can also form when natural water-drainage patterns are changed and new water-diversion systems are developed. Some sinkholes form when the land surface is changed, such as when industrial and run-off-storage ponds are created. The substantial weight of the new material can trigger an underground collapse of supporting material, thus causing a sinkhole.

The overburden sediments that cover buried cavities in the aquifer systems are delicately balanced by ground-water fluid pressure. The water below ground is actually helping to keep the surface soil in place. Ground-water pumping for urban water supply and for irrigation can produce new sinkholes In sinkhole-prone areas. If pumping results in a lowering of ground-water levels, then underground structural failure, and thus, sinkholes, can occur.

LAND SUBSIDENCE

Cause of Land Subsidence from Ground-Water Pumping

Land subsidence is the lowering of the land-surface elevation from changes that take place underground. Common causes of land subsidence from human activity are pumping water, oil, and gas from underground reservoirs; dissolution of limestone aquifers; collapse of underground mines; drainage of organic soils; and initial wetting of dry soils. Land subsidence occurs in nearly every state of the United States. Overdrafting of aquifers is the major cause of subsidence in the southwestern United States, and as ground-water pumping increases, land subsidence also will increase. In many aquifers, ground water is pumped from pore spaces between grains of sand and gravel. If an aquifer has beds of clay or silt within or next to it, the lowered water pressure in the sand and gravel causes slow drainage of water from the clay and silt beds. The reduced water pressure is a loss of support for the clay and silt beds. Because these beds are compressible, they compact and the effects are seen as a lowering of the land surface. The lowering of land surface elevation from this process is permanent. For example, if lowered ground–water levels caused land subsidence, recharging the aquifer until ground water returned to the original levels would not result in an appreciable recovery of the land-surface elevation.

Damage Caused by Land Subsidence

Land subsidence causes many problems including:

- Changes in elevation and slope of streams, canals, and drains;
- Damage to bridges, roads, railroads, storm drains, sanitary sewers, canals, and levees;
- Damage to private and public buildings; and
- Failure of well casings from forces generated by compaction of fine–grained materials in aquifer systems.

In some coastal areas, subsidence has resulted in tides moving into low–lying areas that were previously above high-tide levels. In many areas of the arid Southwest, earth fissures are associated with land subsidence. Earth fissures can be more than 100 feet deep and several hundred feet in length. One extraordinary fissure in central Arizona is 10 miles long. These features start out as narrow cracks, an inch or less in width. They intercept surface drainage and can erode to widths of tens of feet at the surface.

Amounts of Subsidence in Selected areas in the Southwest

After large–scale development of ground-water resources began in the

Southwest after World War II, land subsidence has occurred in many areas. The following table lists approximate maximum subsidence amounts as of 1997 for selected locations in the Southwest:

Arizona		**Nevada**		**California**		**Texas**	
Eloy	15 feet	Las Vegas	6 feet	Lancaster	6 feet	El Paso	1 foot
West of Phoenix	18 feet	New Mexico		Southwest of Mendota	29 feet	Houston	9 feet
Tucson	<1 foot	Albuquerque	"<" 1 foot	Davis	4 feet		
		Mimbres Basin	2 feet	Santa Clara Valley	12 feet		
				Ventura	2 feet		

Relation of Land Subsidence to Climate Change and Population Increase in the Southwest

In areas where climate change results in less precipitation and reduced surface–water supplies, communities will pump more ground water. In the southern part of the United States from states on the Gulf Coast and westward including states of New Mexico, Colorado, Arizona, Utah, Nevada and California, major aquifers include compressible clay and silt that can compact when ground–water is pumped. Also, increased population in the Southwest will increase demands on ground–water supplies, causing more land subsidence in areas already subsiding and new subsidence in areas where subsidence has not yet occurred. In the past, major subsidence areas have been in agricultural settings where ground–water has been pumped for irrigation. In the future, however, increasing population may result in subsidence problems in metropolitan areas where damage from subsidence will be great.

Monitoring Land Subsidence

Several methods are available to monitor land subsidence. The most basic approaches use repeated surveys with conventional or GPS leveling. Another approach is to use permanent compaction recorders, or vertical extensometers. These devices use a pipe or a cable inside a well casing. The pipe inside the casing extends from land surface to some depth through compressible sediments. A table at land surface holds instruments that monitor change in distance between the top of the pipe and the table. If the inner pipe and casing go through the entire thickness of compressible sediments, then the device measures actual land subsidence.

If both ground-water levels and compaction of sediments are measured, then the data can be Analysed to determine properties that can be used to predict future subsidence. About 19 of these installations are operated in Southern Arizona and additional stations are operated in California, Nevada, New Mexico, and Texas. Another subsidence monitoring method under development and testing uses Interferometric Synthetic Aperture Radar. With this method, individual radar images from satellites are compared and interferograms are produced. Under the best conditions, land–surface elevation changes on the order of 1 inch or less can be determined.

Reducing Future Subsidence

In some areas where ground–water pumping has caused subsidence, the subsidence has been stopped by switching from ground-water to surface-water supplies. If surface water is not available, then other means must be taken to reduce subsidence. Possible measures include reducing water use and determining locations for pumping and artificial recharge that will minimize subsidence. Optimization models coupled with ground–water flow models can be used to develop such strategies.

WATER DOWSING

"Water dowsing" refers in general to the practice of using a forked stick, rod, pendulum, or similar device to locate underground water, minerals, or other hidden or lost substances,and has been a subject of discussion and controversy for hundreds, if not thousands, of years. Although tools and methods vary widely, most dowsers probably still use the traditional forked stick, which may come from a variety of trees, including the willow, peach, and witchhazel. Other dowsers may use keys, wire coathangers, pliers, wire rods, pendulums, or various kinds of elaborate boxes and electrical instruments.

In the classic method of using a forked stick, one fork is held in each hand with the palms upward. The bottom or butt end of the "Y" is pointed skyward at an angle of about 45 degrees. The dowser than walks back and forth over the area to be tested. When he passes over a source of water, the butt end of the stick is supposed to rotate or be attracted downward. Water dowsers practice mainly in rural or suburban communities where residents are uncertain as to how to locate the best and cheapest supply of ground water.

Because the drilling and development of a well often costs more than a thousand dollars, homeowners are understandably reluctant to gamble on a dry hole and turn to the water dowser for advice.

How did Water Dowsing begin?

Cave paintings in northwestern Africa that are 6,000–8,000 years old are believed to show a water dowser at work. The exact origin of the divining rod in Europe is not known. The device was introduced into England during the reign of Elizabeth I to locate mineral deposits, and soon afterward it was adopted as a water finder throughout Europe. Water dowsing seems to be a mainly European cultural phenomenon; it was carried across the Atlantic to America by some of the earliest settlers from England and Germany.

What does Science Say About Dowsing?

Case histories and demonstrations of dowsers may seem convincing, but when dowsing is exposed to scientific examination, it presents a very different picture. The natural explanation of "successful" water dowsing is that in many areas underground water is so prevalent close to the land surface that it would be hard to drill a well and not find water. In a region of adequate rainfall and favourable geology, it is difficult not to drill and find water! Some water exists under the Earth's surface almost everywhere. This explains why many dowsers appear to be successful. To locate ground water accurately, however, as to depth, quantity, and quality, a number of techniques must be used. Hydrologic, geologic, and geophysical knowledge is needed to determine the depths and extent of the different water-bearing strata and the quantity and quality of water found in each. The area must be thoroughly tested and studied to determine these facts.

HOW DO HYDROLOGISTS LOCATE GROUNDWATER?

Using Scientific Methods to Locate Water

To locate groundwater accurately and to determine the depth, quantity, and quality of the water, several techniques must be used, and a target area must be thoroughly tested and studied to identify hydrologic and geologic features important to the planning and management of the resource. The landscape may offer clues to the hydrologist about the occurrence of shallow groundwater. Conditions for large quantities of shallow groundwater are more favourable under valleys than under hills.

In some regions—in parts of the arid Southwest, for example—the presence of "water–loving" plants, such as cottonwoods or willows, indicates groundwater at shallow to moderate depth. Areas where water is at the surface as springs, seeps, swamps, or lakes reflect the presence of groundwater, although not necessarily in large quantities or of usable quality.

Geology is the Key

Rocks are the most valuable clues of all. As a first step in locating favourable conditions for ground–water development, the hydrologist prepares geologic maps and cross sections showing the distribution and positions of the different kinds of rocks, both on the surface and underground. Some sedimentary rocks may extend many miles as aquifers of fairly uniform permeability. Other types of rocks may be cracked and broken and contain openings large enough to carry water. Types and orientation of joints or other fractures may be clues to obtaining useful amounts of groundwater. Some rocks may be so folded and displaced that it is difficult to trace them underground.

Existing Wells Provide Clues

Next, a hydrologist obtains information on the wells in the target area. The locations, depth to water, amount of water pumped, and types of rocks penetrated by wells also provide information on groundwater. Wells are tested to determine the amount of water moving through the aquifer, the volume of water that can enter a well, and the effects of pumping on water levels in the area. Chemical analysis of water from wells provides information on quality of water in the aquifer.

How Groundwater Occurs in Rocks

Ground water is simply the subsurface water that fully saturates pores or cracks in soils and rocks. Aquifers are replenished by the seepage of precipitation that falls on the land, although they can be artificially replenished by people, also. There are many geologic, meteorologic, topographic, and human factors that determine the extent and rate to which aquifers are refilled with water.

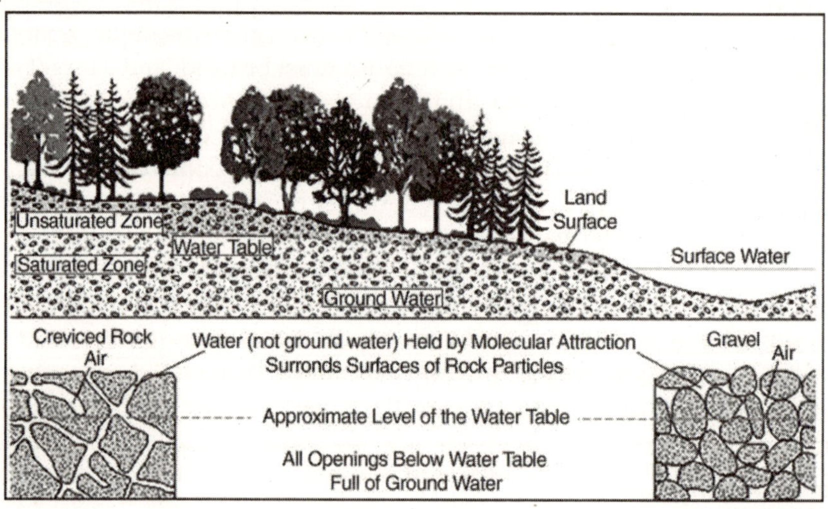

Chapter 2

Water Basics

PROPERTIES OF WATER

Water (H_2O) is the most abundant compound on Earth's surface, covering about 70%. In nature, it exists in liquid, solid, and gaseous states. It is in dynamic equilibrium between the liquid and gas states at standard temperature and pressure. At room temperature, it is a tasteless and odourless liquid, nearly colourless with a hint of blue. Many substances dissolve in water and it is commonly referred to as the universal solvent. Because of this, water in nature and in use is rarely pure and some of its properties may vary slightly from those of the pure substance. However, there are also many compounds that are essentially, if not completely, insoluble in water. Water is the only common substance found naturally in all three common states of matter and it is essential for all life on Earth. Water usually makes up 55% to 78% of the human body.

FORMS OF WATER

Like many substances, water can take numerous forms that are broadly categorized by phase of matter. The liquid phase is the most common among water's phases (within the Earth's atmosphere and surface) and is the form that is generally denoted by the word "water." The solid phase of water is known as ice and commonly takes the structure of hard, amalgamated crystals, such as ice cubes, or loosely accumulated granular crystals, like snow. For a list of the many different crystalline and amorphous forms of solid H_2O. The gaseous phase of water is known as water vapour (or steam), and is characterized by water assuming the configuration of a transparent cloud. (Note that the visible steam and clouds are, in fact, water in the liquid form as minute droplets suspended in the air.) The fourth state of water, that of a supercritical fluid, is much less common than the other three and only rarely occurs in nature, in extremely uninhabitable conditions.

When water achieves a specific critical temperature and a specific critical pressure (647 K and 22.064 MPa), liquid and gas phase merge to one homogeneous fluid phase, with properties of both gas and liquid. One example of naturally occurring supercritical water is found in the hottest parts of deep water hydrothermal vents, in which water is heated to the critical temperature by scalding volcanic plumes and achieves the critical pressure because of the crushing weight of the ocean at the extreme depths at which the vents are located.

Additionally, anywhere there is volcanic activity below a depth of 2.25 km (1.4 miles) can be expected to have water in the supercritical phase. Vienna Standard Mean Ocean Water is the current international standard for water isotopes. Naturally occurring water is almost completely composed of the neutron–less hydrogen isotope protium. Only 155 ppm include deuterium (^{2}H or D), a hydrogen isotope with one neutron, and less than 20 parts per quintillion include tritium (^{3}H or T), which has two.

Heavy water is water with a higher-than-average deuterium content, up to 100%. Chemically, it is similar but not identical to normal water. This is because the nucleus of deuterium is twice as heavy as protium, and this causes noticeable differences in bonding energies. Because water molecules exchange hydrogen atoms with one another, hydrogen deuterium oxide (DOH) is much more common in low-purity heavy water than pure dideuterium monoxide (D_2O). Humans are generally unaware of taste differences, but sometimes report a burning sensation or sweet flavour. Rats, however, are able to avoid heavy water by smell. Toxic to many animals, heavy water is used in the nuclear reactor industry to moderate (slow down) neutrons. Light water reactors are also common, where "light" simply designates normal water. Light water more specifically refers to deuterium-depleted water (DDW), water in which the deuterium content has been reduced below the standard 155ppm level. Light water has been found to be beneficial for improving cancer survival rates in mice and humans undergoing chemotherapy.

PHYSICS AND CHEMISTRY

Water is the chemical substance with chemical formula H_2O: one molecule of water has two hydrogen atoms covalently bonded to a single oxygen atom. Water is a tasteless, odourless liquid at ambient temperature and pressure, and appears colourless in small quantities, although it has its own intrinsic very light blue hue. Ice also appears colourless, and water vapour (steam) is essentially invisible as a gas. Water is primarily a liquid under standard conditions, which is not predicted from its relationship to other analogous hydrides of the oxygen family in the periodic table, which are gases such as hydrogen sulfide.

The elements surrounding oxygen in the periodic table, nitrogen, fluorine, phosphorus, sulfur and chlorine, all combine with hydrogen to produce gases under standard conditions. The reason that water forms a liquid is that oxygen is more electronegative than all of these elements with the exception of fluorine. Oxygen attracts electrons much more strongly than hydrogen, resulting in a net positive charge on the hydrogen atoms, and a net negative charge on the oxygen atom. The presence of a charge on each of these atoms gives each water molecule a net dipole moment.

Electrical attraction between water molecules due to this dipole pulls individual molecules closer together, making it more difficult to separate the molecules and therefore raising the boiling point. This attraction is known as hydrogen bonding. The molecules of water are constantly moving in relation to each other, and the hydrogen bonds are continually breaking and reforming at timescales faster than 200 femtoseconds. However, this bond is sufficiently strong to create many of the peculiar properties of water, such as the those that make it integral to life. Water can be described as a polar liquid that slightly dissociates disproportionately into the hydronium ion ($H_3O^+(aq)$) and an associated hydroxide ion ($OH^-(aq)$).

$$2\,H_2O\,(l) \rightarrow H_3O^+\,(aq) + OH^-(aq)$$

The dissociation constant for this dissociation is commonly symbolized as K_w and has a value of about 10^{-14} at 25°C.

Water, Ice and Vapour

Heat Capacity and Heats of Vapourization and Fusion

Water has the second highest specific heat capacity of all known substances, after ammonia, as well as a high heat of vapourization (40.65 kJ/mol or 2257 kJ/kg), both of which are a result of the extensive hydrogen bonding between its molecules. These two unusual properties allow water to moderate Earth's climate by buffering large fluctuations in temperature. Josh Willis, of NASA's Jet Propulsion Laboratory, the oceans absorb one thousand times more heat than the atmosphere (air) and are holding 80 to 90% of global warming heat. The specific enthalpy of fusion of water is 333.55 kJ/kg at 0°C. Of common substances, only that of ammonia is higher. This property confers resistance to melting on the ice of glaciers and drift ice. Before and since the advent of mechanical refrigeration, ice was and still is in common use for retarding food spoilage.

Temperature (°C)	**Constant–pressure heat capacity C_p (J/(g·K) at 100 kPa)**	**Temperature (°C)**	**Constant–pressure heat capacity C_p (J/(g·K) at 100 kPa)**
0	4.2176	10	4.1921
20	4.1818	30	4.1784
40	4.1785	50	4.1806
60	4.1843	70	4.1895
80	4.1963	90	4.205
100	4.2159		

Note that the specific heat capacity of ice at–10°C is about 2.05 J/(g·K) and that the heat capacity of steam at 100°C is about 2.080 J/(g·K).

Density of Water and Ice

The density of water is approximately one gram per cubic centimeter. More precisely, it is dependent on its temperature, but the relation is not linear and is not even monotonic. When cooled from room temperature liquid water becomes increasingly dense, just like other substances. But at approximately 4°C, pure water reaches its maximum density. As it is cooled further, it expands to become less dense.

This unusual negative thermal expansion is attributed to strong, orientation-dependent, intermolecular interactions and is also observed in molten silica. The solid form of most substances is denser than the liquid phase; thus, a block of most solids will sink in the liquid. However, a block of ice floats in liquid water because ice is less dense. Upon freezing, the density of water decreases by about 9%.

The reason for this is the 'cooling' of intermolecular vibrations allowing the molecules to form steady hydrogen bonds with their neighbours and thereby gradually locking into positions reminiscent of the hexagonal packing achieved upon freezing to ice Ih. Whereas the hydrogen bonds are shorter in the crystal than in the liquid, this locking effect reduces the average coordination number of molecules as the liquid approaches nucleation. Other substances that expand on freezing are silicon, gallium, germanium, antimony, bismuth, plutonium and other compounds that form spacious crystal lattices with tetrahedral coordination.

Only ordinary hexagonal ice is less dense than the liquid. Under increasing pressure, ice undergoes a number of transitions to other allotropic forms with higher density than liquid water, such as high density amorphous ice (HDA) and very high density amorphous ice (VHDA). Water also expands significantly as the temperature increases.

Its density decreases by 4% from its highest value when approaching its boiling point. The melting point of ice is 0°C (32°F, 273 K) at standard pressure, however, pure liquid water can be supercooled well below that temperature without freezing if the liquid is not mechanically disturbed. It can remain in a fluid state down to its homogeneous nucleation point of approximately 231 K (–42°C). The melting point of ordinary hexagonal ice falls slightly under moderately high pressures, but as ice transforms into its allotropes above 209.9 MPa (2,072 atm), the melting point increases markedly with pressure, *i.e.*, reaching 355 K (82°C) at 2.216 GPa (21,870 atm) (triple point of Ice VII).

A significant increase of pressure is required to lower the melting point of ordinary ice—the pressure exerted by an ice skater on the ice only reduces the melting point by approximately 0.09°C (0.16°F). These properties of water have important consequences in its role in the ecosystem of Earth. Water at a temperature of 4°C will always accumulate at the bottom of fresh water lakes, irrespective of the temperature in the atmosphere. Since water and ice are poor conductors of heat (good insulators) it is unlikely that sufficiently deep lakes will freeze completely, unless stirred by strong currents that mix cooler and warmer water and accelerate the cooling. In warming weather, chunks of ice float, rather than sink to the bottom where they might melt extremely slowly. These phenomena thus may help to preserve aquatic life.

Density of Saltwater and Ice

The density of water is dependent on the dissolved salt content as well as the temperature of the water. Ice still floats in the oceans, otherwise they would freeze from the bottom up. However, the salt content of oceans lowers the freezing point by about 2°C and lowers the temperature of the density maximum of water to the freezing point.

This is why, in ocean water, the downward convection of colder water is not blocked by an expansion of water as it becomes colder near the freezing point. The oceans' cold water near the freezing point continues to sink. For this reason, any creature attempting to survive at the bottom of such cold water as the Arctic Ocean generally lives in water that is 4°C colder than the temperature at the bottom of frozen–over fresh water lakes and rivers in the winter.

In cold countries, when the temperature of the ocean reaches 4°C, the layers of water near the top in contact with cold air continue to lose heat energy and their temperature falls below 4°C. On cooling below 4°C, these layers do not sink but may rise up as water has a maximum density at 4°C. (Refer: Polarity and hydrogen bonding) Due to this, the layer of water at 4°C remains at the bottom and above this layers of water 3°C, 2°C, 1°C and 0°C are formed.

Since ice is a poor conductor of heat, it does not allow heat energy from the water beneath the layer of ice which prevents the water freezing. Thus, aquatic creatures survive in such places. As the surface of salt water begins to freeze (at–1.9°C for normal salinity seawater, 3.5%) the ice that forms is essentially salt free with a density approximately equal to that of freshwater ice. This ice floats on the surface and the salt that is "frozen out" adds to the salinity and density of the seawater just below it, in a process known as brine rejection. This denser saltwater sinks by convection and the replacing seawater is subject to the same process. This provides essentially freshwater ice at–1.9°C on the surface.

The increased density of the seawater beneath the forming ice causes it to sink towards the bottom. On a large scale, the process of brine rejection and sinking cold salty water results in ocean currents forming to transport such water away from the Poles. One potential consequence of global warming is that the loss of Arctic and Antarctic ice could result in the loss of these currents as well, which could have unforeseeable consequences on near and distant climates.

Miscibility and Condensation

Water is miscible with many liquids, for example ethanol in all proportions, forming a single homogeneous liquid. On the other hand, water and most oils are immiscible usually forming layers just as to increasing density from the top. As a gas, water vapour is completely miscible with air. On the other hand the maximum water vapour pressure that is thermodynamically stable with the liquid (or solid) at a given temperature is relatively low compared with total atmospheric pressure. For example, if the vapour partial pressure is 2% of atmospheric pressure and the air is cooled from 25 °C, starting at about 22 °C water will start to condense, defining the dew point, and creating fog or dew. The reverse process accounts for the fog burning off in the morning. If the humidity is increased at room temperature, for example, by running a hot shower or a bath, and the temperature stays about the same, the vapour soon reaches the pressure for phase change, and then condenses out as minute water droplets, commonly referred to as steam.

A gas in this context is referred to as saturated or 100% relative humidity, when the vapour pressure of water in the air is at the equilibrium with vapour pressure due to (liquid) water; water (or ice, if cool enough) will fail to lose mass through evaporation when exposed to saturated air. Because the amount of water vapour in air is small, relative humidity, the ratio of the partial pressure due to the water vapour to the saturated partial vapour pressure, is much more useful. Water vapour pressure above 100% relative humidity is called super-saturated and can occur if air is rapidly cooled, for example, by rising suddenly in an updraft.

Vapour Pressusre

Temperature			Pressure				
°C	K	°F	Pa	atm	torr	in Hg	psi
0	273	32	611	0.00603	4.58	0.180	0.0886
5	278	41	872	0.00861	6.54	0.257	0.1265
10	283	50	1,228	0.01212	9.21	0.363	0.1781
12	285	54	1,403	0.01385	10.52	0.414	0.2034
14	287	57	1,599	0.01578	11.99	0.472	0.2318
16	289	61	1,817	0.01793	13.63	0.537	0.2636
17	290	63	1,937	0.01912	14.53	0.572	0.2810
18	291	64	2,064	0.02037	15.48	0.609	0.2993

19	292	66	2,197	0.02168	16.48	0.649	0.3187
20	293	68	2,338	0.02307	17.54	0.691	0.3392
21	294	70	2,486	0.02453	18.65	0.734	0.3606
22	295	72	2,644	0.02609	19.83	0.781	0.3834
23	296	73	2,809	0.02772	21.07	0.830	0.4074
24	297	75	2,984	0.02945	22.38	0.881	0.4328
25	298	77	3,168	0.03127	23.76	0.935	0.4594

Compressibility

The compressibility of water is a function of pressure and temperature. At 0 °C, at the limit of zero pressure, the compressibility is 5.1×10–10 Pa^{-1}. At the zero-pressure limit, the compressibility reaches a minimum of 4.4×10–10 Pa^{-1} around 45°C before increasing again with increasing temperature. As the pressure is increased, the compressibility decreases, being 3.9×10–10 Pa^{-1} at 0°C and 100 MPa.

The bulk modulus of water is 2.2 GPa. The low compressibility of non-gases, and of water in particular, leads to their often being assumed as incompressible. The low compressibility of water means that even in the deep oceans at 4 km depth, where pressures are 40 MPa, there is only a 1.8% decrease in volume.

Triple Point

The temperature and pressure at which solid, liquid, and gaseous water coexist in equilibrium is called the triple point of water. This point is used to define the units of temperature (the kelvin, the SI unit of thermodynamic temperature and, indirectly, the degree Celsius and even the degree Fahrenheit).

As a consequence, water's triple point temperature is a prescribed value rather than a measured quantity. The triple point is at a temperature of 273.16 K (0.01°C) by convention, and at a pressure of 611.73 Pa. This pressure is quite low, about $^{1}/_{166}$ of the normal sea level barometric pressure of 101,325 Pa. The atmospheric surface pressure on planet Mars is remarkably close to the triple point pressure, and the zero-elevation or "sea level" of Mars is defined by the height at which the atmospheric pressure corresponds to the triple point of water.

Although it is commonly named as "the triple point of water", the stable combination of liquid water, ice I, and water vapour is but one of several triple points on the phase diagram of water. Gustav Heinrich Johann Apollon Tammann in GOttingen produced data on several other triple points in the early 20th century. Kamb and others documented further triple points in the 1960s.

Electrical Properties

Electrical Conductivity

Pure water containing no ions is an excellent insulator, but not even "deionized" water is completely free of ions. Water undergoes auto-ionization in the liquid state.

Further, because water is such a good solvent, it almost always has some solute dissolved in it, most frequently a salt. If water has even a tiny amount of such an impurity, then it can conduct electricity readily, as impurities such as salt separate into free ions in aqueous solution by which an electric current can flow.

It is known that the theoretical maximum electrical resistivity for water is approximately 182 kΩm·m at 25°C. This figure agrees well with what is typically seen on reverse osmosis, ultra–filtered and deionized ultrapure water systems used, for instance, in semiconductor manufacturing plants. A salt or acid contaminant level exceeding even 100 parts per trillion (ppt) in ultra-pure water begins to noticeably lower its resistivity level by up to several kOhm·m (or hundreds of nanosiemens per meter).

The low electrical conductivity of water increases significantly upon solvation of a small amount of ionic material, such as hydrogen chloride or any salt. Any electrical conductivity observable in water is the result of ions of mineral salts and carbon dioxide dissolved in it.

Carbon dioxide forms carbonate ions in water. Water self-ionizes, where two water molecules form one hydroxide anion and one hydronium cation, but not enough to carry enough electric current to do any work or harm for most operations.

In pure water, sensitive equipment can detect a very slight electrical conductivity of 0.055 μS/cm at 25°C. Water can also be electrolyzed into oxygen and hydrogen gases but in the absence of dissolved ions this is a very slow process, as very little current is conducted. While electrons are the primary charge carriers in water (and metals), in ice the primary charge carriers are protons.

Electrolysis

Water can be split into its constituent elements, hydrogen and oxygen, by passing an electric current through it. This process is called electrolysis. Water molecules naturally dissociate into H^+ and OH^-ions, which are attracted towards the cathode and anode, respectively. At the cathode, two H^+ ions pick up electrons and form H_2 gas. At the anode, four OH^-ions combine and release O_2 gas, molecular water, and four electrons. The gases produced bubble to the surface, where they can be collected. The standard potential of the water electrolysis cell is 1.23 V at 25°C.

Dielectric Constant

Table. Dielectric Constant of Water

Temperature/°C	ε	Temperature/°C	ε
0	87.9	10	83.95
20	80.18	30	76.58
40	73.18	50	69.88
60	66.76	70	63.78
80	60.93	90	58.2
100	55.58		

Polarity and Hydrogen Bonding

An important feature of water is its polar nature. The water molecule forms an angle, with hydrogen atoms at the tips and oxygen at the vertex. Since oxygen has a higher electronegativity than hydrogen, the side of the molecule with the oxygen atom has a partial negative charge. An object with such a charge difference is called a dipole meaning two poles. The oxygen end is partially negative and the hydrogen end is partially positive, because of this the direction of the dipole moment points towards the oxygen. The charge differences cause water molecules to be attracted to each other (the relatively positive areas being attracted to the relatively negative areas) and to other polar molecules. This attraction contributes to hydrogen bonding, and explains many of the properties of water, such as solvent action.

A water molecule can form a maximum of four hydrogen bonds because it can accept two and donate two hydrogen atoms. Other molecules like hydrogen fluoride, ammonia, methanol form hydrogen bonds but they do not show anomalous behaviour of thermodynamic, kinetic or structural properties like those observed in water. The answer to the apparent difference between water and other hydrogen bonding liquids lies in the fact that apart from water none of the hydrogen bonding molecules can form four hydrogen bonds either due to an inability to donate/accept hydrogens or due to steric effects in bulky residues. In water local tetrahedral order due to the four hydrogen bonds gives rise to an open structure and a 3-dimensional bonding network, resulting in the anomalous decrease of density when cooled below 4°C.

Although hydrogen bonding is a relatively weak attraction compared to the covalent bonds within the water molecule itself, it is responsible for a number of water's physical properties. One such property is its relatively high melting and boiling point temperatures; more energy is required to break the hydrogen bonds between molecules. The similar compound hydrogen sulfide (H_2S), which has much weaker hydrogen bonding, is a gas at room temperature even though it has twice the molecular mass of water. The extra bonding between water molecules also gives liquid water a large specific heat capacity. This high heat capacity makes water a good heat storage medium (coolant) and heat shield.

Cohesion and Adhesion

Water molecules stay close to each other (cohesion), due to the collective action of hydrogen bonds between water molecules. These hydrogen bonds are constantly breaking, with new bonds being formed with different water molecules; but at any given time in a sample of liquid water, a large portion of the molecules are held together by such bonds.

Water also has high adhesion properties because of its polar nature. On extremely clean/smooth glass the water may form a thin film because the molecular forces between glass and water molecules (adhesive forces) are stronger than the cohesive forces. In biological cells and organelles, water is in contact with membrane and protein surfaces that are hydrophilic; that is, surfaces that have a strong attraction to water. Irving Langmuir observed a strong repulsive force between hydrophilic surfaces. To dehydrate hydrophilic surfaces—to remove the strongly held layers of water of hydration—requires doing substantial work against these forces, called hydration forces. These forces are very large but decrease rapidly over a nanometer or less. They are important in biology, particularly when cells are dehydrated by exposure to dry atmospheres or to extracellular freezing.

Surface Tension

Water has a high surface tension of 72.8 mN/m at room temperature, caused by the strong cohesion between water molecules, the highest of the non-metallic liquids. This can be seen when small quantities of water are placed onto a sorption-free (non-adsorbent and non-absorbent) surface, such as polyethylene or Teflon, and the water stays together as drops. Just as significantly, air trapped in surface disturbances forms bubbles, which sometimes last long enough to transfer gas molecules to the water. Another surface tension effect is capillary waves, which are the surface ripples that form around the impacts of drops on water surfaces, and sometimes occur with strong subsurface currents flowing to the water surface. The apparent elasticity caused by surface tension drives the waves.

Capillary Action

Due to an interplay of the forces of adhesion and surface tension, water exhibits capillary action whereby water rises into a narrow tube against the force of gravity. Water adheres to the inside wall of the tube and surface tension tends to straighten the surface causing a surface rise and more water is pulled up through cohesion. The process continues as the water flows up the tube until there is enough water such that gravity balances the adhesive force. Surface tension and capillary action are important in biology. For example, when water is carried through xylem up stems in plants, the strong intermolecular attractions (cohesion) hold the water column together and adhesive properties maintain the water attachment to the xylem and prevent tension rupture caused by transpiration pull.

Water as a Solvent

Water is also a good solvent due to its polarity. Substances that will mix well and dissolve in water (*i.e.,* salts) are known as hydrophilic ("water–loving") substances, while those that do not mix well with water (*e.g.* fats and oils), are known as hydrophobic ("water-fearing") substances. The ability of a substance to dissolve in water is determined by whether or not the substance can match or better the strong attractive forces that water molecules generate between other water molecules. If a substance has properties that do not allow it to overcome these strong intermolecular forces, the molecules are "pushed out" from the water, and do not dissolve. Contrary to the common misconception, water and hydrophobic substances do not "repel", and the hydration of a hydrophobic surface is energetically, but not entropically, favourable.

When an ionic or polar compound enters water, it is surrounded by water molecules (Hydration). The relatively small size of water molecules typically allows many water molecules to surround one molecule of solute. The partially negative dipole ends of the water are attracted to positively charged components of the solute, and vice versa for the positive dipole ends. In general, ionic and polar substances such as acids, alcohols, and salts are relatively soluble in water, and non–polar substances such as fats and oils are not. Non–polar molecules stay together in water because it is energetically more favourable for the water molecules to hydrogen bond to each other than to engage in van der Waals interactions with non-polar molecules. An example of an ionic solute is table salt; the sodium chloride, NaCl, separates into Na^+ cations and Cl^-anions, each being surrounded by water molecules. The ions are then easily transported away from their crystalline lattice into solution. An example of a nonionic solute is table sugar. The water dipoles make hydrogen bonds with the polar regions of the sugar molecule (OH groups) and allow it to be carried away into solution.

Water in Acid-base Reactions

Chemically, water is amphoteric: it can act as either an acid or a base in chemical reactions. The Brønsted-Lowry definition, an acid is defined as a species which donates a proton (a H^+ ion) in a reaction, and a base as one which receives a proton. When reacting with a stronger acid, water acts as a base; when reacting with a stronger base, it acts as an acid.

For instance, water receives an H^+ ion from HCl when hydrochloric acid is formed:

$$HCl\ (acid) + H_2O\ (base) \rightleftharpoons H_3O^+ + Cl^-$$

In the reaction with ammonia, NH_3, water donates a H^+ ion, and is thus acting as an acid:

$$NH_3\ (base) + H_2O\ (acid) \rightleftharpoons NH+$$
$$4 + OH^-$$

Because the oxygen atom in water has two lone pairs, water often acts as a Lewis base, or electron pair donor, in reactions with Lewis acids, although it can also react with Lewis bases, forming hydrogen bonds between the electron pair donors and the hydrogen atoms of water. HSAB theory describes water as both a weak hard acid and a weak hard base, meaning that it reacts preferentially with other hard species:

$$H^+\ (Lewis\ acid) + H_2O\ (Lewis\ base) \rightarrow H_3O^+$$

$$Fe^{3+}\ (Lewis\ acid) + H_2O\ (Lewis\ base) \rightarrow Fe(H_2O)3+6$$

$$Cl^-(Lewis\ base) + H_2O\ (Lewis\ acid) \rightarrow Cl(H_2O)–6$$

When a salt of a weak acid or of a weak base is dissolved in water, water can partially hydrolyze the salt, producing the corresponding base or acid, which gives aqueous solutions of soap and baking soda their basic pH:

$$Na_2CO_3 + H_2O \rightleftharpoons NaOH + NaHCO_3$$

Ligand Chemistry

Water's Lewis base character makes it a common ligand in transition metal complexes, examples of which range from solvated ions, such as $Fe(H_2O)3+6$, to perrhenic acid, which contains two water molecules coordinated to a rhenium atom, to various solid hydrates, such as $CoCl_2 \cdot 6H_2O$. Water is typically a monodentate ligand, it forms only one bond with the central atom.

Organic Chemistry

As a hard base, water reacts readily with organic carbocations, for example in hydration reaction, in which a hydroxyl group (OH^-) and an acidic proton are added to the two carbon atoms bonded together in the carbon-carbon double bond, resulting in an alcohol. When addition of water to an organic molecule cleaves the molecule in two, hydrolysis is said to occur. Notable examples of hydrolysis are saponification of fats and digestion of proteins and polysaccharides. Water can also be a leaving group in S_N2 substitution and E2 elimination reactions, the latter is then known as dehydration reaction.

Acidity in Nature

Pure water has the concentration of hydroxide ions (OH^-) equal to that of the hydronium (H_3O+) or hydrogen ($H+$) ions, which gives pH of 7 at 298 K. In practice, pure water is very difficult to produce. Water left exposed to air for any length of time will dissolve carbon dioxide, forming a dilute solution of carbonic acid, with a limiting pH of about 5.7. As cloud droplets form in the atmosphere and as raindrops fall through the air minor amounts of CO_2 are absorbed and thus most rain is slightly acidic. If high amounts of nitrogen and sulfur oxides are present in the air, they too will dissolve into the cloud and rain drops producing acid rain.

Water in Redox Reactions

Water contains hydrogen in oxidation state +1 and oxygen in oxidation state–2. Because of that, water oxidizes chemicals with reduction potential below the potential of H^+/H_2, such as hydrides, alkali and alkaline earth metals (except for beryllium), etc. Some other reactive metals, such as aluminum, are oxidized by water as well, but their oxides are not soluble, and the reaction stops because of passivation. Note, however, that rusting of iron is a reaction between iron and oxygen, dissolved in water, not between iron and water.

$$2\ Na + 2\ H_2O \rightarrow 2\ NaOH + H_2$$

Water can be oxidized itself, emitting oxygen gas, but very few oxidants react with water even if their reduction potential is greater than the potential of O_2/O^{2-}. Almost all such reactions require a catalyst

$$4\ AgF_2 + 2\ H_2O \rightarrow 4\ AgF + 4\ HF + O_2$$

Geochemistry

Action of water on rock over long periods of time typically leads to weathering and water erosion, physical processes that convert solid rocks and minerals into soil and sediment, but under some conditions chemical reactions with water occur as well, resulting in metasomatism or mineral hydration, a type of chemical alteration of a rock which produces clay minerals in nature and also occurs when Portland cement hardens. Water ice can form clathrate compounds, known as clathrate hydrates, with a variety of small molecules that can be embedded in its spacious crystal lattice. The most notable of these is methane clathrate, $4CH_4{\cdot}23H_2O$, naturally found in large quantities on the ocean floor.

Transparency

Water is relatively transparent to visible light, near ultraviolet light, and far-red light, but it absorbs most ultraviolet light, infrared light, and microwaves. Most photoreceptors and photosynthetic pigments utilize the portion of the light spectrum that is transmitted well through water. Microwave ovens take advantage of water's opacity to microwave radiation to heat the water inside of foods. The very weak onset of absorption in the red end of the visible spectrum lends water its intrinsic blue hue.

Heavy Water and Isotopologues

Several isotopes of both hydrogen and oxygen exist, giving rise to several known isotopologues of water. Hydrogen occurs naturally in three isotopes. The most common (^1H) accounting for more than 99.98% of hydrogen in water, consists of only a single proton in its nucleus. A second, stable isotope, deuterium (chemical symbol D or ^2H), has an additional neutron. Deuterium oxide, D_2O, is also known as heavy water because of its higher density.

It is used in nuclear reactors as a neutron moderator. The third isotope, tritium, has 1 proton and 2 neutrons, and is radioactive, decaying with a half-life of 4500 days. T_2O exists in nature only in minute quantities, being produced primarily via cosmic ray-induced nuclear reactions in the atmosphere. Water with one deuterium atom HDO occurs naturally in ordinary water in low concentrations (~0.03%) and D_2O in far lower amounts (0.000003%).

The most notable physical differences between H_2O and D_2O, other than the simple difference in specific mass, involve properties that are affected by hydrogen bonding, such as freezing and boiling, and other kinetic effects. The difference in boiling points allows the isotopologues to be separated. Consumption of pure isolated D_2O may affect biochemical processes—ingestion of large amounts impairs kidney and central nervous system function. Small quantities can be consumed without any ill-effects, and even very large amounts of heavy water must be consumed for any toxicity to become apparent. Oxygen also has three stable isotopes, with ^{16}O present in 99.76%, ^{17}O in 0.04%, and ^{18}O in 0.2% of water molecules.

Liquid Crystal State in the Exclusion Zone

Near hydrophilic surfaces, water exists in a liquid crystal state.

This liquid crystal state has the following properties:

- The water molecules are constrained in movement (as shown by nuclear magnetic resonance imagery)
- It is more stable (as shown by infrared radiation imagery)
- It has a negative charge (as shown by a test of its electric potential)
- It abosrbs at 270 nm (as shown by light absorption imagery)
- It is more viscous than liquid water (as shown by falling ball viscometry)
- The molecules are aligned (as shown by polarizing microscopy)

Gerald Pollack speculated that this liquid crystal zone remained relatively unexplored recently, despite extensive writing on this topic up through 1949, because of the polywater and water memory debacles.

SYSTEMATIC NAMING

The accepted IUPAC name of water is oxidane or simply water, or its equivalent in different languages, although there are other systematic names which can be used to describe the molecule.

The simplest systematic name of water is hydrogen oxide. This is analogous to related compounds such as hydrogen peroxide, hydrogen sulfide, and deuterium oxide (heavy water). Another systematic name, oxidane, is accepted by IUPAC as a parent name for the systematic naming of oxygen-based substituent groups, although even these commonly have other recommended names. For example, the name hydroxyl is recommended over oxidanyl for the–OH group. The name oxane is explicitly mentioned by the IUPAC as being unsuitable for this purpose, since it is already the name of a cyclic ether also known as tetrahydropyran.

The polarized form of the water molecule, $H^{+}OH^{-}$, is also called hydron hydroxide by IUPAC nomenclature. Dihydrogen monoxide (DHMO) is a rarely used name of water.

This term has been used in various hoaxes that call for this "lethal chemical" to be banned, such as in the dihydrogen monoxide hoax. Other systematic names for water include hydroxic acid, hydroxylic acid, and hydrogen hydroxide. Both acid and alkali names exist for water because it is amphoteric (able to react both as an acid or an alkali). None of these exotic names are used widely.

CAPILLARY ACTION

Capillary action, or capillarity, is the ability of to flow against gravity where liquid spontaneously rises in a narrow space such as a thin tube, or in porous materials such as document or in some non–porous materials such as liquified carbon fibre. This effect can cause liquids to flow against the force of gravity or the magnetic field induction.

It occurs because of inter–molecular attractive forces between the liquid and solid surrounding surfaces; If the diameter of the tube is sufficiently small, then the combination of surface tension (which is caused by cohesion within the liquid) and forces of adhesion between the liquid and container act to lift the liquid.

ETYMOLOGY

The word "capillary," in the non-medical sense, means narrow-tube. The word comes from Latin capillaris ("pertaining to the hair") < capillus ("the hair, prop. of the head")< caput ("head"). In medicine and biology, it usually refers to the smallest blood vessels.

PHENOMENA AND PHYSICS OF CAPILLARY ACTION

Capillary action, capillarity, capillary motion, or wicking refers to two phenomena: A common apparatus used to demonstrate the first phenomenon is the capillary tube. When the lower end of a vertical glass tube is placed in a liquid such as water, a concave meniscus forms. Adhesion pulls the liquid column up until there is a sufficient mass of liquid for gravitational forces to overcome the intermolecular forces. The contact length (around the edge) between the top of the liquid column and the tube is proportional to the diameter of the tube, while the weight of the liquid column is proportional to the square of the tube's diameter, so a narrow tube will draw a liquid column higher than a wide tube.

In hydrology, capillary action describes the attraction of water molecules to soil particles. Capillary action is responsible for moving groundwater from wet areas of the soil to dry areas. Differences in soil potential (ψ_m) drive capillary action in soil.

EXAMPLES

Capillary action is also essential for the drainage of constantly produced tear fluid from the eye. Two canaliculi of tiny diameter are present in the inner corner of the eyelid, also called the lacrimal ducts; their openings can be seen with the naked eye within the lacrymal sacs when the eyelids are everted. Wicking is to absorb something and then drain like a wick. Paper towels absorb liquid through capillary action, allowing a fluid to be transferred from a surface to the towel. The small pores of a sponge act as small capillaries, causing it to absorb a comparatively large amount of fluid.

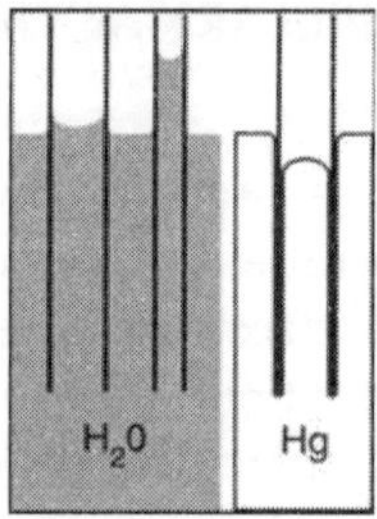

Fig. Capillary Action of Water Compared to Mercury, in Each Case with Respect to a Polar Surface *i.e.*, Glass

Some modern sport and exercise fabrics use capillary action to "wick" sweat away from the skin. These are often referred to as wicking fabrics, after the capillary properties of a candle and lamp wicks. Capillary action is observed in thin layer chromatography, in which a solvent moves vertically up a plate via capillary action. Dissolved solutes travel with the solvent at various speeds depending on their affinity for the solvent (the mobile phase) or the absorbent coating on the plate (the stationary phase).

With some pairs of materials, such as mercury and glass, the intermolecular forces within the liquid exceed those between the solid and the liquid, so a convex meniscus forms and capillary action works in reverse.

HEIGHT OF A MENISCUS

The height h of a liquid column is given by:

$$h = \frac{2\gamma\cos\theta}{\rho g r},$$

where γ is the liquid-air surface tension (force/unit length),θ is the contact angle, ρ is the density of liquid (mass/volume), g is local gravitational field strength (force/unit mass), and r is radius of tube (length).

For a water-filled glass tube in air at standard laboratory conditions,γ = 0.0728 N/m at 20°C, θ = 20° (0.35 rad), ρ is 1000 kg/m^3, and g = 9.8 m/s^2. For these values, the height of the water column is

$$h \approx \frac{1.4\times10^{-5}}{r} m.$$

Thus for a 4 m (13 ft) diameter tube (radius 2 m (6.6 ft), the water would rise an unnoticeable 0.007 mm (0.00028 in). However, for a 4 cm (1.6 in) diameter tube (radius 2 cm (0.79 in), the water would rise 0.7 mm (0.028 in), and for a 0.4 mm (0.016 in) diameter tube (radius 0.2 mm (0.0079 in), the water would rise 70 mm (2.8 in).

LIQUID TRANSPORT IN POROUS MEDIA

When a dry porous medium, such as a brick or a wick, is brought into contact with a liquid, it will start absorbing the liquid at a rate which decreases over time. For a bar of material with cross–sectional area A that is wetted on one end, the cumulative volume V of absorbed liquid after a time t is

$$V = AS\sqrt{t},$$

where S is the sorptivity of the medium, with dimensions m s$^{-1/2}$ or mm min$^{-1/2}$. The quantity

$$i = \frac{V}{A}$$

is called the cumulative liquid intake, with the dimension of length. The wetted length of the bar, that is the distance between the wetted end of the bar and the so-called wet front, is dependent on the fraction *f* of the volume occupied by liquid. This number f is the porosity of the medium; the wetted length is then,

$$x = \frac{i}{f} = \frac{S}{f}\sqrt{t}$$

Some authors use the quantity S/f as the sorptivity. The above description is for the case where gravity and evaporation do not play a role. Sorptivity is a relevant property of building materials, because it affects the amount of rising dampness.

THE WATER IN YOU

Think of what you need to survive, really just survive. Food? Water? Air? MTV? Naturally, I'm going to concentrate on water here. Water is of major importance to all living things; in some organisms, up to 90% of their body weight comes from water. Up to 60% of the human body is water, the brain is composed of 70% water, and the lungs are nearly 90% water. Lean muscle tissue contains about 75% water by weight, as is the brain; body fat contains 10% water and bone has 22% water. About 83% of our blood is water, which helps digest our food, transport waste, and control body temperature. Each day humans must replace 2.4 litres of water, some through drinking and the rest taken by the body from the foods eaten.

Dr. Jeffrey Utz, Neuroscience, pediatrics, Allegheny University, different people have different percentages of their bodies made up of water. Babies have the most, being born at about 78%. By one year of age, that amount drops to about 65%. In adult men, about 60% of their bodies are water. However, fat tissue does not have as much water as lean tissue. In adult women, fat makes up more of the body than men, so they have about 55% of their bodies made of water. Fat men also have less water (as a percentage) than thin men.

Thus:

- Babies and kids have more water (as a percentage) than adults.
- Women have less water than men (as a percentage).
- Fat people have less water than thin people (as a percentage).

There just wouldn't be any you, me, or Fido the dog without the existence of an ample liquid water supply on Earth. The unique qualities and properties of water are what make it so important and basic to life. The cells in our bodies are full of water. The excellent ability of water to dissolve so many substances allows our cells to use valuable nutrients, minerals, and chemicals in biological processes.

Water's "stickiness" (from surface tension) plays a part in our body's ability to transport these materials all through ourselves. The carbohydrates and proteins that our bodies use as food are metabolized and transported by water in the bloodstream. No less important is the ability of water to transport waste material out of our bodies.

HYDRO ELECTRICITY

Hydroelectric power must be one of the oldest methods of producing power. No doubt, Jack the Caveman stuck some sturdy leaves on a pole and put it in a moving stream. The water would spin the pole that crushed grain to make their delicious, low-fat prehistoric bran muffins. People have used moving water to help them in their work throughout history, and modern people make great use of moving water to produce electricity.

HYDROELECTRIC POWER FOR THE NATION

Although most energy in the United States is produced by fossil-fuel and nuclear power plants, hydroelectricity is still important to the Nation, as about 7 per cent of total power is produced by hydroelectric plants. Nowadays, huge power generators are placed inside dams. Water flowing through the dams spin turbine blades (made out of metal instead of leaves) which are connected to generators. Power is produced and is sent to homes and businesses.

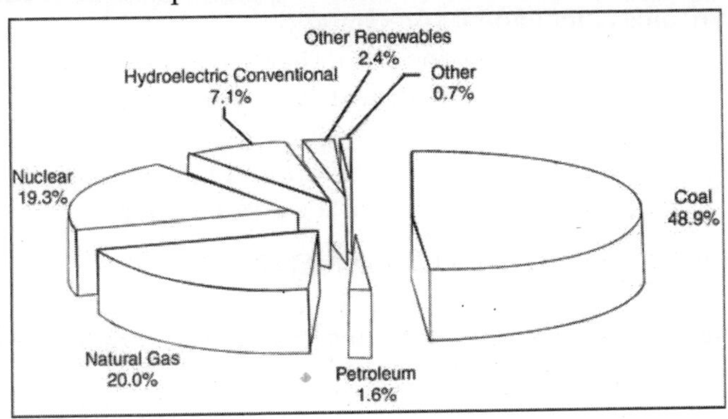

World Distribution of Hydropower

- Hydropower is the most important and widely-used renewable source of energy.
- Hydropower represents 19% of total electricity production.
- China is the largest producer of hydroelectricity, followed by Canada, Brazil, and the United States
- Approximately two-thirds of the economically feasible potential remains to be developed. Untapped hydro resources are still abundant in Latin America, Central Africa, India and China.

Producing electricity using hydroelectric power has some advantages over other power-producing methods.

Advantages to hydroelectric power:

- Fuel is not burned so there is minimal pollution
- Water to run the power plant is provided free by nature
- Hydropower plays a major role in reducing greenhouse gas emissions
- Relatively low operations and maintenance costs
- The technology is reliable and proven over time
- It's renewable—rainfall renews the water in the reservoir, so the fuel is almost always there

Read an expanded list of advantages of hydroelectric power from the Top World Conference on Sustainable Development conference, Johannesburg, South Africa (2002).

Disadvantages to power plants that use coal, oil, and gas fuel:

- They use up valuable and limited natural resources
- They can produce a lot of pollution
- Companies have to dig up the Earth or drill wells to get the coal, oil, and gas
- For nuclear power plants there are waste–disposal problems

Hydroelectric power is not perfect, though, and does have some disadvantages:/p>

- High investment costs
- Hydrology dependent (precipitation)
- In some cases, inundation of land and wildlife habitat
- In some cases, loss or modification of fish habitat
- Fish entrainment or passage restriction
- In some cases, changes in reservoir and stream water quality
- In some cases, displacement of local populations

HYDROPOWER AND THE ENVIRONMENT

Hydropower is Nonpolluting, but does have Environmental Impacts

Hydropower does not pollute the water or the air. However, hydropower facilities can have large environmental impacts by changing the environment and affecting land use, homes, and natural habitats in the dam area. Most hydroelectric power plants have a dam and a reservoir. These structures may obstruct fish migration and affect their populations. Operating a hydroelectric power plant may also change the water temperature and the river's flow. These changes may harm native plants and animals in the river and on land. Reservoirs may cover people's homes, important natural areas, agricultural land, and archeological sites. So building dams can require relocating people. Methane, a strong greenhouse gas, may also form in some reservoirs and be emitted to the atmosphere. (EPA Energy Kids)

RESERVOIR CONSTRUCTION IS "DRYING UP" IN THE UNITED STATES

Gosh, hydroelectric power sounds great—so why don't we use it to produce all of our power? Mainly because you need lots of water and a lot of land where you can build a dam and reservoir, which all takes a LOT of money, time, and construction. In fact, most of the good spots to locate hydro plants have already been taken. In the early part of the century hydroelectric plants supplied a bit less than one-half of the nation's power, but the number is down to about 10 per cent today.

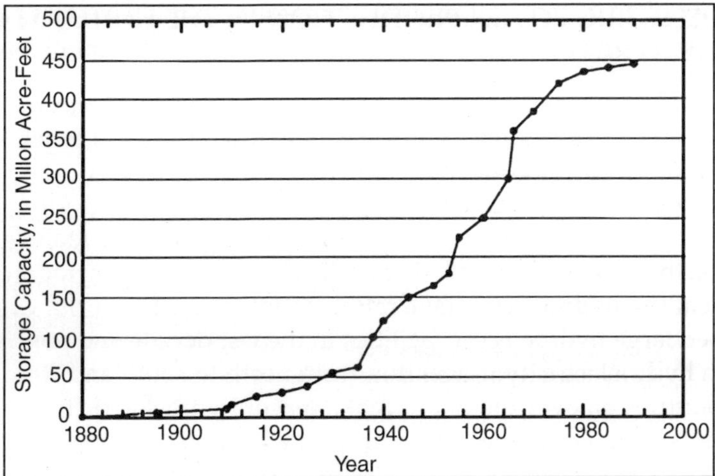

The trend for the future will probably be to build small-scale hydro plants that can generate electricity for a single community. As this chart shows, the construction of surface reservoirs has slowed considerably in recent years. In the middle of the 20th Century, when urbanization was occuring at a rapid rate, many reservoirs were constructed to serve peoples' rising demand for water and power. Since about 1980, the rate of reservoir construction has slowed considerably.

TYPICAL HYDROELECTRIC POWER PLANT

Hydroelectric energy is produced by the force of falling water. The capacity to produce this energy is dependent on both the available flow and the height from which it falls. Building up behind a high dam, water accumulates potential energy. This is transformed into mechanical energy when the water rushes down the sluice and strikes the rotary blades of turbine. The turbine's rotation spins electromagnets which generate current in stationary coils of wire. Finally, the current is put through a transformer where the voltage is increased for long distance transmission over power lines.

Source: Environment Canada

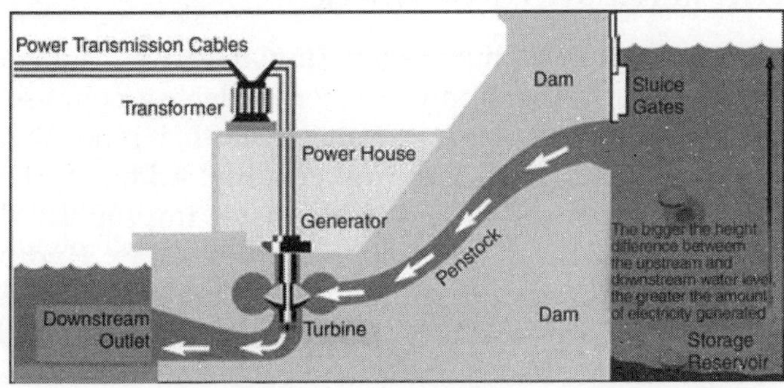

Fig. Hydroelectric Power Generation

HYDROELECTRIC-POWER PRODUCTION IN THE UNITED STATES AND THE WORLD

In the United States, most states make some use of hydroelectric power, although, as you can expect, states with low topographical relief, such as Florida and Kansas, produce very little hydroelectric power.

But some states, such as Idaho, Washington, and Oregon use hydroelectricity as their main power source. in 1995, all of Idaho's power came from hydroelectric plants. The hydroelectric power generation in 2006 for the leading hydroelectric-generating countries in the world. China has developed large hydroelectric facilities in the last decade and now lead the world in hydroelectricity usage. But, from north to south and from east to west, countries all over the world make use of hydroelectricity—the main ingredients are a large river and a drop in elevation.

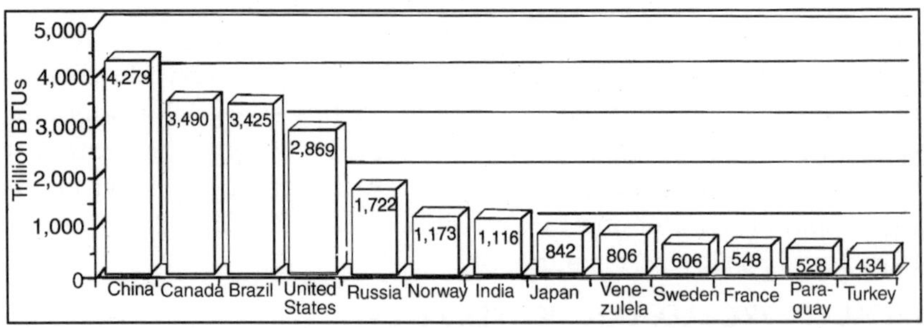

Fig. World Net Hydroelectric Power Generation (Trillion Btu) 2006

HYDROLOGY

Water is one of our most natural resources. Without it, there would be no life on earth. Hydrology has evolved as a science in response to the need to understand the complex water system of the earth and help solve water problems. This hydrology primer gives you information about water on Earth and humans' involvement and use of water.

HYDROLOGY IS THE STUDY OF WATER

Water is one of our most important natural resources. Without it, there would be no life on earth. The supply of water available for our use is limited by nature. Although there is plenty of water on earth, it is not always in the right place, at the right time and of the right quality. Adding to the problem is the increasing evidence that chemical wastes improperly discarded yesterday are showing up in our water supplies today. Hydrology has evolved as a science in response to the need to understand the complex water systems of the Earth and help solve water problems. Hydrologists play a vital role in finding solutions to water problems, and interesting and challenging careers are available to those who choose to study hydrology.

WATER AND PEOPLE

Estimates of water use in the United States indicate that about 408 billion gallons per day (one thousand million gallons per day, abbreviated Bgal/d) were withdrawn for all uses during 2000. This total has varied less than 3 per cent since 1985 as withdrawals have stabilized for the two largest uses—thermoelectric power and irrigation. Fresh ground-water withdrawals (83.3 Bgal/d) during 2000 were 14 per cent more than during 1985. Fresh surface-water withdrawals for 2000 were 262 Bgal/d, varying less than 2 per cent since 1985. Much of our water use is hidden. Think about what you had for lunch.

A hamburger, for example, requires water to raise wheat for the bun, to grow hay and corn to feed the cattle and to process the bread and beef. Together with french fries and a soft drink, this all-American meal uses about 1,500 gallons of water--enough to fill a small swimming pool. How about your clothes? To grow cotton for a pair of jeans takes about 400 gallons. A shirt requires about 400 gallons. How do you get to school or to the store? To produce the amount of finished steel in a car has in the past required about 32,000 gallons of water. Similarly, the steel in a 30-pound bicycle required 480 gallons. This shows that industry must continue to strive to reduce water use through manufacturing processes that use less water, and through recycling of water.

WHAT IS HYDROLOGY?

Hydrology is the science that encompasses the occurrence, distribution, movement and properties of the waters of the earth and their relationship with the environment within each phase of the hydrologic cycle.

The water cycle, or hydrologic cycle, is a continuous process by which water is purified by evaporation and transported from the earth's surface (including the oceans) to the atmosphere and back to the land and oceans.

All of the physical, chemical and biological processes involving water as it travels its various paths in the atmosphere, over and beneath the earth's surface and through growing plants, are of interest to those who study the hydrologic cycle. There are many pathways the water may take in its continuous cycle of falling as rainfall or snowfall and returning to the atmosphere.

It may be captured for millions of years in polar ice caps. It may flow to rivers and finally to the sea. It may soak into the soil to be evaporated directly from the soil surface as it dries or be transpired by growing plants. It may percolate through the soil to ground water reservoirs (aquifers) to be stored or it may flow to wells or springs or back to streams by seepage. They cycle for water may be short, or it may take millions of years. People tap the water cycle for their own uses. Water is diverted temporarily from one part of the cycle by pumping it from the ground or drawing it from a river or lake. It is used for a variety of activities such as households, businesses and industries; for irrigation of farms and parklands; and for production of electric power. After use, water is returned to another part of the cycle: perhaps discharged downstream or allowed to soak into the ground. Used water normally is lower in quality, even after treatment, which often poses a problem for downstream users. The hydrologist studies the fundamental transport processes to be able to describe the quantity and quality of water as it moves through the cycle (evaporation, precipitation, streamflow, infiltration, ground water flow, and other components). The engineering hydrologist, or water resources engineer, is involved in the planning, analysis, design, construction and operation of projects for the control, utilization, and management of water resources. Water resources problems are also the concern of meteorologists, oceanographers, geologists, chemists, physicists, biologists, economists, political scientists, specialists in applied mathematics and computer science, and engineers in several fields.

WHAT HYDROLOGISTS DO?

Hydrologists apply scientific knowledge and mathematical principles to solve water-related problems in society: problems of quantity, quality and availability. They may be concerned with finding water supplies for cities or irrigated farms, or controlling river flooding or soil erosion. Or, they may work in environmental protection: preventing or cleaning up pollution or locating sites for safe disposal of hazardous wastes.

Persons trained in hydrology may have a wide variety of job titles. Scientists and engineers in hydrology may be involved in both field investigations and office work. In the field, they may collect basic data, oversee testing of water quality, direct field crews and work with equipment. Many jobs require travel, some abroad. A hydrologist may spend considerable time doing field work in remote and rugged terrain. In the office, hydrologists do many things such as interpreting hydrologic data and performing analyses for determining possible water supplies. Much of their work relies on computers for organizing, summarizing and analysing masses of data, and for modeling studies such as the prediction of flooding and the consequences of reservoir releases or the effect of leaking underground oil storage tanks. The work of hydrologists is as varied as the uses of water and may range from planning multimillion dollar interstate water projects to advising homeowners about backyard drainage problems.

SURFACE WATER

Most cities meet their needs for water by withdrawing it from the nearest river, lake or reservoir. Hydrologists help cities by collecting and analysing the data needed to predict how much water is available from local supplies and whether it will be sufficient to meet the city's projected future needs. To do this, hydrologists study records of rainfall, snowpack depths and river flows that are collected and compiled by hydrologists in various government agencies. They inventory the extent river flow already is being used by others.

Managing reservoirs can be quite complex, because they generally serve many purposes. Reservoirs increase the reliability of local water supplies. Hydrologists use topographic maps and aerial photographs to determine where the reservoir shorelines will be and to calculate reservoir depths and storage capacity. This work ensures that, even at maximum capacity, no highways, railroads or homes would be flooded.

Deciding how much water to release and how much to store depends upon the time of year, flow predictions for the next several months, and the needs of irrigators and cities as well as downstream water-users that rely on the reservoir. If the reservoir also is used for recreation or for generation of hydroelectric power, those requirements must be considered. Decisions must be coordinated with other reservoir managers along the river. Hydrologists collect the necessary information, enter it into a computer, and run computer models to predict the results under various operating strategies. On the basis of these studies, reservoir managers can make the best decision for those involved.

The availability of surface water for swimming, drinking, industrial or other uses sometimes is restricted because of pollution. Pollution can be merely an unsightly and inconvenient nuisance, or it can be an invisible, but deadly, threat to the health of people, plants and animals.

Hydrologists assist public health officials in monitoring public water supplies to ensure that health standards are met. When pollution is discovered, environmental engineers work with hydrologists in devising the necessary sampling programme. Water quality in estuaries, streams, rivers and lakes must be monitored, and the health of fish, plants and wildlife along their stretches surveyed. Related work concerns acid rain and its effects on aquatic life, and the behaviour of toxic metals and organic chemicals in aquatic environments. Hydrologic and water quality mathematical models are developed and used by hydrologists for planning and management and predicting water quality effects of changed conditions.

Simple analyses such as pH, turbidity, and oxygen content may be done by hydrologists in the field. Other chemical analyses require more sophisticated laboratory equipment. In the past, municipal and industrial sewage was a major source of pollution for streams and lakes. Such wastes often received only minimal treatment, or raw wastes were dumped into rivers. Today, we are more aware of the consequences of such actions, and billions of dollars must be invested in pollution-control equipment to protect the waters of the earth. Other sources of pollution are more difficult to identify and control. These include road deicing salts, storm run-off from urban areas and farmland, and erosion from construction sites.

GROUNDWATER

Groundwater, pumped from beneath the earth's surface, is often cheaper, more convenient and less vulnerable to pollution than surface water. Therefore, it is commonly used for public water supplies. Groundwater provides the largest source of usable water storage in the United States. Underground reservoirs contain far more water than the capacity of all surface reservoirs and lakes, including the Great Lakes. In some areas, ground water may be the only option. Some municipalities survive solely on groundwater.

Hydrologists estimate the volume of water stored underground by measuring water levels in local wells and by examining geologic records from well-drilling to determine the extent, depth and thickness of water-bearing sediments and rocks. Before an investment is made in full-sized wells, hydrologists may supervise the drilling of test wells. They note the depths at which water is encountered and collect samples of soils, rock and water for laboratory analyses. They may run a variety of geophysical tests on the completed hole, keeping and accurate log of their observations and test results. Hydrologists determine the most efficient pumping rate by monitoring the extent that water levels drop in the pumped well and in its nearest neighbours. Pumping the well too fast could cause it to go dry or could interfere with neighbouring wells. Along the coast, overpumping can cause saltwater intrusion. By plotting and analysing these data, hydrologists can estimate the maximum and optimum yields of the well.

Polluted ground water is less visible, but more insidious and difficult to clean up, than pollution in rivers and lakes. Ground water pollution most often results from improper disposal of wastes on land. Major sources include industrial and household chemicals and garbage landfills, industrial waste lagoons, tailings and process wastewater from mines, oil field brine pits, leaking underground oil storage tanks and pipelines, sewage sludge and septic systems. Hydrologists provide guidance in the location of monitoring wells around waste disposal sites and sample them at regular intervals to determine if undesirable leachate—contaminated water containing toxic or hazardous chemicals—is reaching the ground water. In polluted areas, hydrologists may collect soil and water samples to identify the type and extent of contamination.

The chemical data then are plotted on a map to show the size and direction of waste movement. In complex situations, computer modeling of water flow and waste migration provides guidance for a clean-up programme. In extreme cases, remedial actions may require excavation of the polluted soil. Today, most people and industries realise that the amount of money invested in prevention is far less than that of cleanup. Hydrologists often are consulted for selection of proper sites for new waste disposal facilities.

The danger of pollution is minimized by locating wells in areas of deep ground water and impermeable soils. Other practices include lining the bottom of a landfill with watertight materials, collecting any leachate with drains, and keeping the landfill surface covered as much as possible. Careful monitoring is always necessary.

CAREERS IN HYDROLOGY

Students who plan to become hydrologists need a strong emphasis in mathematics, statistics, geology, physics, computer science, chemistry and biology. In addition, sufficient background in other subjects—economics, public finance, environmental law, government policy—is needed to communicate with experts in these fields and to understand the implications of their work on hydrology.

Communicating clearly in writing and speech is a basic requirement essential for any professional person. Hydrologists should be able to work well with people, not only as part of a team with other scientists and engineers, but also in public relations, whether it be advising governmental leaders or informing the general public on water issues. Hydrology offers a variety of interesting and challenging career choices for today and tomorrow. It's a field worth considering.

COMMON WATER MEASUREMENTS

The U.S. Geological Survey has been measuring water for decades. Millions of measurements and analyses have been made.

Some measurements are taken almost every time water is sampled and investigated, no matter where in the U.S. the water is being studied. Even these simple measurements can sometimes reveal something important about the water and the environment around it.

The results of a single measurement of a water's properties are actually less important than looking at how the properties vary over time. For example, if you take the pH of the creek behind your school and find that it is 5.5, you might say "Wow, this water is acidic!" But, a pH of 5.5 might be "normal" for that creek. It is similar to how my normal body temperature (when I'm not sick) is about 97.5 degrees, but my third-grader's normal temperature is "really normal"—right on the 98.6 mark. As with our temperatures, if the pH of your creek begins to change, then you might suspect that something is going on somewhere that is affecting the water, and possibly, the water quality. So, often, the changes in water measurements are more important than the actual measured values.

WATER TEMPERATURE

Water temperature is not only important to swimmers and fisherman, but also to industries and even fish and algae. A lot of water is used for cooling purposes in power plants that generate electricity. They need cool water to start with, and they generally release warmer water back to the environment. The temperature of the released water can affect downstream habitats. Temperature also can affect the ability of water to hold oxygen as well as the ability of organisms to resist certain pollutants.

pH

pH is a measure of how acidic/basic water is. The range goes from 0-14, with 7 being neutral. pHs of less than 7 indicate acidity, whereas a pH of greater than 7 indicates a base. pH is really a measure of the relative amount of free hydrogen and hydroxyl ions in the water. Water that has more free hydrogen ions is acidic, whereas water that has more free hydroxyl ions is basic. Since pH can be affected by chemicals in the water, pH is an important indicator of water that is changing chemically. pH is reported in "logarithmic units," like the Richter scale, which measures earthquakes. Each number represents a 10-fold change in the acidity/basicness of the water. Water with a pH of 5 is ten times more acidic than water having a pH of six.

Pollution can change a water's pH, which in turn can harm animals and plants living in the water. For instance, water coming out of an abandoned coal mine can have a pH of 2, which is very acidic and would definitely affect any fish crazy enough to try to live in it! By using the logarithm scale, this mine-drainage water would be 100,000 times more acidic than neutral water—so stay out of abandoned mines.

SPECIFIC CONDUCTANCE

Specific conductance is a measure of the ability of water to conduct an electrical current. It is highly dependent on the amount of dissolved solids (such as salt) in the water. Pure water, such as distilled water, will have a very low specific conductance, and sea water will have a high specific conductance.

Rainwater often dissolves airborne gasses and airborne dust while it is in the air, and thus often has a higher specific conductance than distilled water. Specific conductance is an important water-quality measurement because it gives a good idea of the amount of dissolved material in the water.

High specific conductance indicates high dissolved-solids concentration; dissolved solids can affect the suitability of water for domestic, industrial, and agricultural uses. At higher levels, drinking water may have an unpleasant taste or odour or may even cause gastrointestinal distress. Additionally, high dissolved-solids concentration can cause deterioration of plumbing fixtures and appliances. Relatively expensive water-treatment processes, such as reverse osmosis, are needed to remove excessive dissolved solids from water.

Agriculture also can be adversely affected by high-specific-conductance water, as crops cannot survive if the water they use is too saline, for instance. Agriculture can also be the cause of increases in the specific conductance of local waters. When water is used for irrigation, part of the water evaporates or is consumed by plants, concentrating the original amount of dissolved solids in less water; thus, the dissolved-solids concentration and the specific conductance in the remaining water is increased. The remaining higher specific-conductance water reenters the river as irrigation-return flow. In a USGS study in Colorado, USA, specific conductance was found to vary during the year as a result of the temporal variability of streamflow. As this chart shows, specific conductance generally was lowest in the Arkansas RIver near Avondale, Colorado, in May to August, when streamflow generally was largest, and increased with decreasing streamflow in the fall, winter, and spring.

Often in school, students do an experiment where they connect a battery to a light bulb and run two wires from the battery into a beaker of water. When the wires are put into a beaker of distilled water, the light will not light. But, the bulb does light up when the beaker contains salt water (saline). In the saline water, the salt has dissolved, releasing free electrons, and the water will conduct an electrical current.

TURBIDITY

Turbidity is the amount of particulate matter that is suspended in water. Turbidity measures the scattering effect that suspended solids have on light: the higher the intensity of scattered light, the higher the turbidity.

Material that causes water to be turbid include:

- Clay
- Silt
- Finely divided organic and inorganic matter
- Soluble coloured organic compounds
- Plankton
- Microscopic organisms

Turbidity makes the water cloudy or opaque. Turbidity is measured by shining a light through the water and is reported in nephelometric turbidity units (NTU). During periods of low flow (base flow), many rivers are a clear green Colour, and turbidities are low, usually less than 10 NTU. During a rainstorm, particles from the surrounding land are washed into the river making the water a muddy brown Colour, indicating water that has higher turbidity values. Also, during high flows, water velocities are faster and water volumes are higher, which can more easily stir up and suspend material from the stream bed, causing higher turbidities.

Turbidity can be measured in the laboratory and also on-site in the river. A handheld turbidity meter measures turbidity of a water sample. The meter is calibrated using standard samples from the meter manufacturer. Once the meter is calibrated to correctly read these standards, the turbidity of a water sample can be taken. State-of-the-art turbidity meters are beginning to be installed in rivers to provide an instantaneous turbidity reading.

The large tube is the turbidity sensor; it reads turbidity in the river by shining a light into the water and reading how much light is reflected back to the sensor. The smaller tube contains a conductivity sensor to measure electrical conductance of the water, which is strongly influenced by dissolved solids (the two holes) and a temperature gauge (the metal rod).

DISSOLVED OXYGEN

You can't tell by looking at water that there is oxygen in it (unless you remember that chemical makeup of a water molecule is hydrogen and oxygen). But, if you look at a closed bottle of a soft drink, you don't see the carbon dioxide dissolved in that-until you shake it up and open the top. The oxygen dissolved in lakes, rivers, and oceans is crucial for the organisms and creatures living in it. As the amount of dissolved oxygen drops below normal levels in water bodies, the water quality is harmed and creatures begin to die off. Indeed, a water body can "die", a process called eutrophication.

Although water molecules contain an oxygen atom, this oxygen is not what is needed by aquatic organisms living in our natural waters. A small amount of oxygen, up to about ten molecules of oxygen per million of water, is actually dissolved in water. This dissolved oxygen is breathed by fish and zooplankton and is needed by them to survive. Rapidly moving water, such as in a mountain stream or large river, tends to contain a lot of dissolved oxygen, while stagnant water contains little. Bacteria in water can consume oxygen as organic matter decays. Thus, excess organic material in our lakes and rivers can cause an oxygen-deficient situation to occur. Aquatic life can have a hard time in stagnant water that has a lot of rotting, organic material in it, especially in summer, when dissolved-oxygen levels are at a seasonal low.

HARDNESS

The amount of dissolved calcium and magnesium in water determines its "hardness." Water hardness varies throughout the United States. If you live in an area where the water is "soft," then you may never have even heard of water hardness. But, if you live in Florida, New Mexico, Arizona,

Utah, Wyoming, Nebraska, South Dakota, Iowa, Wisconsin, or Indiana, where the water is relatively hard, you may notice that it is difficult to get a lather up when washing your hands or clothes. And, industries in your area might have to spend money to soften their water, as hard water can damage equipment. Hard water can even shorten the life of fabrics and clothes! Does this mean that students who live in areas with hard water keep up with the latest fashions since their clothes wear out faster?

SUSPENDED SEDIMENT

Suspended sediment is the amount of soil moving along in a stream. It is highly dependent on the speed of the water flow, as fast-flowing water can pick up and suspend more soil than calm water. During storms, soil is washed from the stream banks into the stream. The amount that washes into a stream depends on the type of land in the river's watershed and the vegetation surrounding the river.

If land is disturbed along a stream and protection measures are not taken, then excess sediment can harm the water quality of a stream. You've probably seen those short, plastic fences that builders put up on the edges of the property they are developing. These silt fences are supposed to trap sediment during a rainstorm and keep it from washing into a stream, as excess sediment can harm the creeks, rivers, lakes, and reservoirs. Sediment coming into a reservoir is always a concern; once it enters it cannot get out-most of it will settle to the bottom. Reservoirs can "silt in" if too much sediment enters them. The volume of the reservoir is reduced, resulting in less area for boating, fishing, and recreation, as well as reducing the power-generation capability of the power plant in the dam.

Chapter 3

Water Cycle

INTRODUCTION

The water cycle, also known as the hydrologic cycle or H_2O cycle, describes the continuous movement of water on, above and below the surface of the Earth. Water can change states among liquid, vapour, and ice at various places in the water cycle. Although the balance of water on Earth remains fairly constant over time, individual water molecules can come and go, in and out of the atmosphere. The water moves from one reservoir to another, such as from river to ocean, or from the ocean to the atmosphere, by the physical processes of evaporation, condensation, precipitation, infiltration, run-off, and subsurface flow.

In so doing, the water goes through different phases: liquid, solid, and gas. The hydrologic cycle involves the exchange of heat energy, which leads to temperature changes. For instance, in the process of evaporation, water takes up energy from the surroundings and cools the environment. Conversely, in the process of condensation, water releases energy to its surroundings, warming the environment.

The water cycle figures significantly in the maintenance of life and ecosystems on Earth. Even as water in each reservoir plays an important role, the water cycle brings added significance to the presence of water on our planet. By transferring water from one reservoir to another, the water cycle purifies water, replenishes the land with freshwater, and transports minerals to different parts of the globe. It is also involved in reshaping the geological features of the Earth, through such processes as erosion and sedimentation. In addition, as the water cycle also involves heat exchange, it exerts an influence on climate as well.

DESCRIPTION

The sun, which drives the water cycle, heats water in oceans and seas. Water evaporates as water vapour into the air. Ice and snow can sublimate directly into water vapour. Evapotranspiration is water transpired from plants and evaporated from the soil. Rising air currents take the vapour up into the atmosphere where cooler temperatures cause it to condense into clouds. Air currents move water vapour around the globe, cloud particles collide, grow, and fall out of the sky as precipitation. Some precipitation falls as snow or hail, and can accumulate as ice caps and glaciers, which can store frozen water for thousands of years.

Snowpacks can thaw and melt, and the melted water flows over land as snowmelt. Most water falls back into the oceans or onto land as rain, where the water flows over the ground as surface run-off. A portion of run-off enters rivers in valleys in the landscape, with streamflow moving water towards the oceans. Run-off and groundwater are stored as freshwater in lakes.

Not all run-off flows into rivers, much of it soaks into the ground as infiltration. Some water infiltrates deep into the ground and replenishes aquifers, which store freshwater for long periods of time. Some infiltration stays close to the land surface and can seep back into surface-water bodies as groundwater discharge. Some groundwater finds openings in the land surface and comes out as freshwater springs. Over time, the water returns to the ocean, where our water cycle started.

PRECIPITATION

In meteorology, precipitation is any product of the condensation of atmospheric water vapour that falls under gravity. The main forms of precipitation include drizzle, rain, sleet, snow, graupel and hail. It occurs when a local portion of the atmosphere becomes saturated with water vapour and the water condenses Two processes, possibly acting together, can lead to air becoming saturated: cooling the air or adding water vapour to the air.

Generally, precipitation will fall to the surface; an exception is Virga which evaporates before reaching the surface. Precipitation forms as smaller droplets coalesce via collision with other rain drops or ice crystals within a cloud. Rain drops range in size from oblate, pancake-like shapes for larger drops, to small spheres for smaller drops. Unlike raindrops, snowflakes grow in a variety of different shapes and patterns, determined by the temperature and humidity characteristics of the air the snowflake moves through on its way to the ground. While snow and ice pellets require temperatures close to the ground to be near or below freezing, hail can occur during much warmer temperature regimes due to the process of its formation.

Moisture overriding associated with weather fronts is an overall major method of precipitation production. If enough moisture and upward motion is present, precipitation falls from convective clouds such as cumulonimbus and can organize into narrow rainbands. Where relatively warm water bodies are present, for example due to water evaporation from lakes, lake-effect snowfall becomes a concern downwind of the warm lakes within the cold cyclonic flow around the backside of extratropical cyclones. Lake-effect snowfall can be locally heavy. Thundersnow is possible within a cyclone's comma head and within lake effect precipitation bands. In mountainous areas, heavy precipitation is possible where upslope flow is maximized within windward sides of the terrain at elevation. On the leeward side of mountains, desert climates can exist due to the dry air caused by compressional heating. The movement of the monsoon trough, or intertropical convergence zone, brings rainy seasons to savannah climes.

Precipitation is a major component of the water cycle, and is responsible for depositing the fresh water on the planet. Approximately 505,000 cubic kilometres of water falls as precipitation each year; 398,000 cubic kilometres of it over the oceans. Given the Earth's surface area, that means the globally averaged annual precipitation is 990 millimetres. Climate classification systems such as the Köppen climate classification system use average annual rainfall to help differentiate between differing climate regimes. The urban heat island effect may lead to increased rainfall, both in amounts and intensity, downwind of cities. Global warming is also causing changes in the precipitation pattern globally. Precipitation may occur on other celestial bodies, *e.g.* when it gets cold, Mars has precipitation which most likely takes the form of ice needles, rather than rain or snow.

Hydrometeor

The term meteor describes an object from outer space which has entered the Earth's atmosphere and produces a light phenomenon. In contrast, any phenomenon which was at some point produced due to condensation or precipitation of moisture within the Earth's atmosphere is known as a hydrometeor. Particles composed of fallen precipitation which fell onto the Earth's surface can become hydrometeors if blown off the landscape by wind. Formations due to condensation such as clouds, haze, fog, and mist are composed of hydrometeors. All precipitation types are hydrometeors by definition, including virga, which is precipitation which evaporates before reaching the ground. Particles removed from the Earth's surface by wind such as blowing snow and blowing sea spray are also hydrometeors.

Types

Precipitation is a major component of the water cycle, and is responsible for depositing most of the fresh water on the planet. Approximately 505,000 km3 of water falls as precipitation each year, 398,000 km3 of it over the oceans. Given the Earth's surface area, that means the globally averaged annual precipitation is 990 millimetres. Mechanisms of producing precipitation include convective, stratiform, and orographic rainfall. Convective processes involve strong vertical motions that can cause the overturning of the atmosphere in that location within an hour and cause heavy precipitation, while stratiform processes involve weaker upward motions and less intense precipitation. Precipitation can be divided into three categories, based on whether it falls as liquid water, liquid water that freezes on contact with the surface, or ice. Mixtures of different types of precipitation, including types in different categories, can fall simultaneously. Liquid forms of precipitation include rain and drizzle. Rain or drizzle that freezes on contact within a subfreezing air mass is called "freezing rain" or "freezing drizzle". Frozen forms of precipitation include snow, ice needles, ice pellets, hail, and graupel.

How the Air becomes Saturated

Cooling Air to its Dew Point

The dew point is the temperature to which a parcel must be cooled in order to become saturated, and condenses to water. Water vapour normally begins to condense on condensation nuclei such as dust, ice, and salt in order to form clouds. An elevated portion of a frontal zone forces broad areas of lift, which form clouds decks such as altostratus or cirrostratus. Stratus is a stable cloud deck which tends to form when a cool, stable air mass is trapped underneath a warm air mass. It can also form due to the lifting of advection fog during breezy conditions.

There are four main mechanisms for cooling the air to its dew point: adiabatic cooling, conductive cooling, radiational cooling, and evaporative cooling. Adiabatic cooling occurs when air rises and expands. The air can rise due to convection, large-scale atmospheric motions, or a physical barrier such as a mountain. Conductive cooling occurs when the air comes into contact with a colder surface, usually by being blown from one surface to another, for example from a liquid water surface to colder land. Radiational cooling occurs due to the emission of infrared radiation, either by the air or by the surface underneath. Evaporative cooling occurs when moisture is added to the air through evaporation, which forces the air temperature to cool to its wet-bulb temperature, or until it reaches saturation.

Adding Moisture to the Air

The main ways water vapour is added to the air are: wind convergence into areas of upward motion, precipitation or virga falling from above, daytime heating evaporating water from the surface of oceans, water bodies or wet land, transpiration from plants, cool or dry air moving over warmer water, and lifting air over mountains.

Formation

Raindrops

Coalescence occurs when water droplets fuse to create larger water droplets, or when water droplets freeze onto an ice crystal, which is known as the Bergeron process. The fall rate of very small droplets is negligible, hence clouds do not fall out of the sky; precipitation will only occur when these coalesce into larger drops. When air turbulence occurs, water droplets collide, producing larger droplets. As these larger water droplets descend, coalescence continues, so that drops become heavy enough to overcome air resistance and fall as rain.

Raindrops have sizes ranging from 0.1 millimetres to 9 millimetres mean diameter, above which they tend to break up. Smaller drops are called cloud droplets, and their shape is spherical. As a raindrop increases in size, its shape becomes more oblate, with its largest cross-section facing the oncoming airflow.

Contrary to the cartoon pictures of raindrops, their shape does not resemble a teardrop. Intensity and duration of rainfall are usually inversely related, *i.e.*, high intensity storms are likely to be of short duration and low intensity storms can have a long duration. Rain drops associated with melting hail tend to be larger than other rain drops. The METAR code for rain is RA, while the coding for rain showers is SHRA.

Ice Pellets

Ice pellets or sleet are a form of precipitation consisting of small, translucent balls of ice. Ice pellets are usually smaller than hailstones. They often bounce when they hit the ground, and generally do not freeze into a solid mass unless mixed with freezing rain. The METAR code for ice pellets is PL.

Ice pellets form when a layer of above-freezing air exists with sub-freezing air both above and below it. This causes the partial or complete melting of any snowflakes falling through the warm layer. As they fall back into the sub-freezing layer closer to the surface, they re-freeze into ice pellets. However, if the sub-freezing layer beneath the warm layer is too small, the precipitation will not have time to re-freeze, and freezing rain will be the result at the surface. A temperature profile showing a warm layer above the ground is most likely to be found in advance of a warm front during the cold season, but can occasionally be found behind a passing cold front.

Hail

Like other precipitation, hail forms in storm clouds when supercooled water droplets freeze on contact with condensation nuclei, such as dust or dirt. The storm's updraft blows the hailstones to the upper part of the cloud. The updraft dissipates and the hailstones fall down, back into the updraft, and are lifted up again. Hail has a diameter of 5 millimetres or more. Within METAR code, GR is used to indicate larger hail, of a diameter of at least 6.4 millimetres. GR is derived from the French word grêle.

Smaller-sized hail, as well as snow pellets, use the coding of GS, which is short for the French word grésil. Stones just larger than golf ball-sized are one of the most frequently reported hail sizes. Hailstones can grow to 15 centimetres and weigh more than.5 kilograms. In large hailstones, latent heat released by further freezing may melt the outer shell of the hailstone. The hailstone then may undergo 'wet growth', where the liquid outer shell collects other smaller hailstones. The hailstone gains an ice layer and grows increasingly larger with each ascent. Once a hailstone becomes too heavy to be supported by the storm's updraft, it falls from the cloud.

Snowflakes

Snow crystals form when tiny supercooled cloud droplets freeze. Once a droplet has frozen, it grows in the supersaturated environment. Because water droplets are more numerous than the ice crystals the crystals are able to grow to hundreds of micrometers or millimeters in size at the expense of the water droplets. This process is known as the Wegner-Bergeron-Findeison process. The corresponding depletion of water vapour causes the droplets to evaporate, meaning that the ice crystals grow at the droplets' expense. These large crystals are an efficient source of precipitation, since they fall through the atmosphere due to their mass, and may collide and stick together in clusters, or aggregates. These aggregates are snowflakes, and are usually the type of ice particle that falls to the ground. Guinness World Records list the world's largest snowflakes as those of January 1887 at Fort Keogh, Montana; allegedly one measured 38 cm wide. The exact details of the sticking mechanism remain a subject of research.

Although the ice is clear, scattering of light by the crystal facets and hollows/imperfections mean that the crystals often appear white in Colour due to diffuse reflection of the whole spectrum of light by the small ice particles. The shape of the snowflake is determined broadly by the temperature and humidity at which it is formed. Rarely, at a temperature of around "2 °C, snowflakes can form in threefold symmetry—triangular snowflakes. The most common snow particles are visibly irregular, although near-perfect snowflakes may be more common in pictures because they are more visually appealing. No two snowflakes are alike, which grow at different rates and in different patterns depending on the changing temperature and humidity within the atmosphere that the snowflake falls through on its way to the ground. The METAR code for snow is SN, while snow showers are coded SHSN.

Diamond Dust

Diamond dust, also known as ice needles or ice crystals, forms at temperatures approaching "40°F due to air with slightly higher moisture from aloft mixing with colder, surface based air. They are made of simple ice crystals that are hexagonal in shape. The METAR identifier for diamond dust within international hourly weather reports is IC.

Causes

Frontal Activity

Stratiform or dynamic precipitation occurs as a consequence of slow ascent of air in synoptic systems such as over surface cold fronts, and over and ahead of warm fronts. Similar ascent is seen around tropical cyclones outside of the eyewall, and in comma-head precipitation patterns around mid-latitude cyclones. A wide variety of weather can be found along an occluded front, with thunderstorms possible, but usually their passage is associated with a drying of the air mass. Occluded fronts usually form around mature low-pressure areas. Precipitation may occur on celestial bodies other than Earth. When it gets cold, Mars has precipitation that most likely takes the form of ice needles, rather than rain or snow.

Convection

Convective rain, or showery precipitation, occurs from convective clouds, *i.e.*, cumulonimbus or cumulus congestus. It falls as showers with rapidly changing intensity. Convective precipitation falls over a certain area for a relatively short time, as convective clouds have limited horizontal extent. Most precipitation in the tropics appears to be convective; however, it has been suggested that stratiform precipitation also occurs. Graupel and hail indicate convection. In mid-latitudes, convective precipitation is intermittent and often associated with baroclinic boundaries such as cold fronts, squall lines, and warm fronts.

Orographic Effects

Orographic precipitation occurs on the windward side of mountains and is caused by the rising air motion of a large-scale flow of moist air across the mountain ridge, resulting in adiabatic cooling and condensation. In mountainous parts of the world subjected to relatively consistent winds a more moist climate usually prevails on the windward side of a mountain than on the leeward or downwind side. Moisture is removed by orographic lift, leaving drier air on the descending and generally warming, leeward side where a rain shadow is observed.

In Hawaii, on the island of Kauai, is notable for its extreme rainfall, as it has the second highest average annual rainfall on Earth, with 460 inches. Storm systems affect the state with heavy rains between October and March. Local climates vary considerably on each island due to their topography, divisible into windward and leeward regions based upon location relative to the higher mountains. Windward sides face the east to northeast trade winds and receive much more rainfall; leeward sides are drier and sunnier, with less rain and less cloud cover.

In South America, the Andes mountain range blocks Pacific moisture that arrives in that continent, resulting in a desertlike climate just downwind across western Argentina. The Sierra Nevada range creates the same effect in North America forming the Great Basin and Mojave Deserts.

Snow

Extratropical cyclones can bring cold and dangerous conditions with heavy rain and snow with winds exceeding 119 km/h. The band of precipitation that is associated with their warm front is often extensive, forced by weak upward vertical motion of air over the frontal boundary which condenses as it cools and produces precipitation within an elongated band, which is wide and stratiform, meaning falling out of nimbostratus clouds. When moist air tries to dislodge an arctic air mass, overrunning snow can result within the poleward side of the elongated precipitation band. In the Northern Hemisphere, poleward is towards the North Pole, or north. Within the Southern Hemisphere, poleward is towards the South Pole, or south.

Southwest of extratropical cyclones, curved cyclonic flow bringing cold air across the relatively warm water bodies can lead to narrow lake-effect snow bands. Those bands bring strong localized snowfall which can be understood as follows: Large water bodies such as lakes efficiently store heat that results in significant temperature differences between the water surface and the air above. Because of this temperature difference, warmth and moisture are transported upward, condensing into vertically oriented clouds which produce snow showers. The temperature decrease with height and cloud depth are directly affected by both the water temperature and the large-scale environment. The stronger the temperature decrease with height, the deeper the clouds get, and the greater the precipitation rate becomes.

In mountainous areas, heavy snowfall accumulates when air is forced to ascend the mountains and squeeze out precipitation along their windward slopes, which in cold conditions, falls in the form of snow. Because of the ruggedness of terrain, forecasting the location of heavy snowfall remains a significant challenge.

Within the Tropics

The wet, or rainy, season is the time of year, covering one or more months, when most of the average annual rainfall in a region falls. The term green season is also sometimes used as a euphemism by tourist authorities. Areas with wet seasons are dispersed across portions of the tropics and subtropics. Savanna climates and areas with monsoon regimes have wet summers and dry winters. Tropical rainforests technically do not have dry or wet seasons, since their rainfall is equally distributed through the year. Some areas with pronounced rainy seasons will see a break in rainfall mid-season when the intertropical convergence zone or monsoon trough move poleward of their location during the middle of the warm season.

When the wet season occurs during the warm season, or summer, rain falls mainly during the late afternoon and early evening hours. The wet season is a time when air quality improves, freshwater quality improves, and vegetation grows significantly. Soil nutrients diminish and erosion increases. Animals have adaptation and survival strategies for the wetter regime. Tropical cyclones, a source of very heavy rainfall, consist of large air masses several hundred miles across with low pressure at the centre and with winds blowing inward towards the centre in either a clockwise direction or counterclockwise. Although cyclones can take an enormous toll in lives and personal property, they may be important factors in the precipitation regimes of places they impact, as they may bring much-needed precipitation to otherwise dry regions. Areas in their path can receive a year's worth of rainfall from a tropical cyclone passage.

Large-scale Geographical Distribution

On the large scale, the highest precipitation amounts outside topography fall in the tropics, closely tied to the Intertropical Convergence Zone, itself the ascending branch of the Hadley cell. Mountainous locales near the equator in Colombia are amongst the wettest places on Earth. North and south of this are regions of descending air that form subtropical ridges where precipitation is low; the land surface underneath is usually arid, which forms most of the Earth's deserts. An exception to this rule is in Hawaii, where upslope flow due to the trade winds lead to one of the wettest locations on Earth. Otherwise, the flow of the Westerlies into the Rocky Mountains lead to the wettest, and at elevation snowiest, locations within North America. In Asia during the wet season, the flow of moist air into the Himalayas leads to some of the greatest rainfall amounts measured on Earth in northeast India.

Measurement

The standard way of measuring rainfall or snowfall is the standard rain gauge, which can be found in 100 mm plastic and 200 mm metal varieties. The inner cylinder is filled by 25 mm of rain, with overflow flowing into the outer cylinder. Plastic gauges have markings on the inner cylinder down to 0.25 mm resolution, while metal gauges require use of a stick designed with the appropriate 0.25 mm markings. After the inner cylinder is filled, the amount inside it is discarded, then filled with the remaining rainfall in the outer cylinder until all the fluid in the outer cylinder is gone, adding to the overall total until the outer cylinder is empty. These gauges are used in the winter by removing the funnel and inner cylinder and allowing snow and freezing rain to collect inside the outer cylinder. Some add antifreeze to their gauge so they do not have to melt the snow or ice that falls into the gauge. Once the snowfall/ice is finished accumulating, or as 300 mm is approached, one can either bring it inside to melt, or use lukewarm water to fill the inner cylinder with in order to melt the frozen precipitation in the outer cylinder, keeping track of the warm fluid added, which is subsequently subtracted from the overall total once all the ice/snow is melted.

Other types of gauges include the popular wedge gauge, the tipping bucket rain gauge, and the weighing rain gauge. The wedge and tipping bucket gauges will have problems with snow. Attempts to compensate for snow/ice by warming the tipping bucket meet with limited success, since snow may sublimate if the gauge is kept much above freezing. Weighing gauges with antifreeze should do fine with snow, but again, the funnel needs to be removed before the event begins. For those looking to measure rainfall the most inexpensively, a can that is cylindrical with straight sides will act as a rain gauge if left out in the open, but its accuracy will depend on what ruler is used to measure the rain with. Any of the above rain gauges can be made at home, with enough know-how.

When a precipitation measurement is made, various networks exist across the United States and elsewhere where rainfall measurements can be submitted through the Internet, such as CoCoRAHS or GLOBE. If a network is not available in the area where one lives, the nearest local weather office will likely be interested in the measurement.

Return Period

The likelihood or probability of an event with a specified intensity and duration, is called the return period or frequency. The intensity of a storm can be predicted for any return period and storm duration, from charts based on historic data for the location. The term 1 in 10 year storm describes a rainfall event which is rare and is only likely to occur once every 10 years, so it has a 10 per cent likelihood any given year. The rainfall will be greater and the flooding will be worse than the worst storm expected in any single year. The term 1 in 100 year storm describes a rainfall event which is extremely rare and which will occur with a likelihood of only once in a century, so has a 1 per cent likelihood in any given year. The rainfall will be extreme and flooding to be worse than a 1 in 10 year event. As with all probability events, it is possible to have multiple "1 in 100 Year Storms" in a single year.

Role in Köppen Climate Classification

The Köppen classification depends on average monthly values of temperature and precipitation. The most commonly used form of the Köppen classification has five primary types labeled A through E. Specifically, the primary types are A, tropical; B, dry; C, mild mid-latitude; D, cold mid-latitude; and E, polar. The five primary classifications can be further divided into secondary classifications such as rain forest, monsoon, tropical savanna, humid subtropical, humid continental, oceanic climate, Mediterranean climate, steppe, subarctic climate, tundra, polar ice cap, and desert.

Rain forests are characterized by high rainfall, with definitions setting minimum normal annual rainfall between 1,750 millimetres and 2,000 millimetres. A tropical savanna is a grassland biome located in semi-arid to semi-humid climate regions of subtropical and tropical latitudes, with rainfall between 750 millimetres and 1,270 millimetres a year. They are widespread on Africa, and are also found in India, the northern parts of South America, Malaysia, and Australia. The humid subtropical climate zone where winter rainfall is associated with large storms that the westerlies steer from west to east. Most summer rainfall occurs during thunderstorms and from occasional tropical cyclones. Humid subtropical climates lie on the east side continents, roughly between latitudes 20° and 40° degrees away from the equator.

An oceanic climate is typically found along the west coasts at the middle latitudes of all the world's continents, bordering cool oceans, as well as southeastern Australia, and is accompanied by plentiful precipitation year round. The Mediterranean climate regime resembles the climate of the lands in the Mediterranean Basin, parts of western North America, parts of Western and South Australia, in southwestern South Africa and in parts of central Chile. The climate is characterized by hot, dry summers and cool, wet winters. A steppe is a dry grassland. Subarctic climates are cold with continuous permafrost and little precipitation.

Effect on Agriculture

Precipitation, especially rain, has a dramatic effect on agriculture. All plants need at least some water to survive, therefore rain is important to agriculture. While a regular rain pattern is usually vital to healthy plants, too much or too little rainfall can be harmful, even devastating to crops. Drought can kill crops and increase erosion, while overly wet weather can cause harmful fungus growth. Plants need varying amounts of rainfall to survive. For example, certain cacti require small amounts of water, while tropical plants may need up to hundreds of inches of rain per year to survive.

In areas with wet and dry seasons, soil nutrients diminish and erosion increases during the wet season. Animals have adaptation and survival strategies for the wetter regime. The previous dry season leads to food shortages into the wet season, as the crops have yet to mature. Developing countries have noted that their populations show seasonal weight fluctuations due to food shortages seen before the first harvest, which occurs late in the wet season.

Changes Due to Global Warming

Increasing temperatures tend to increase evaporation which leads to more precipitation. Precipitation has generally increased over land north of 30°N from 1900 through 2005 but has declined over the tropics since the 1970s. Globally there has been no statistically significant overall trend in precipitation over the past century, although trends have varied widely by region and over time. Eastern portions of North and South America, northern Europe, and northern and central Asia have become wetter.

The Sahel, the Mediterranean, southern Africa and parts of southern Asia have become drier. There has been an increase in the number of heavy precipitation events over many areas during the past century, as well as an increase since the 1970s in the prevalence of droughts—especially in the tropics and subtropics. Changes in precipitation and evaporation over the oceans are suggested by the decreased salinity of mid-and high-latitude waters along with increased salinity in lower latitudes. Over the contiguous United States, total annual precipitation increased at an average rate of 6.1 per cent per century since 1900, with the greatest increases within the East North Central climate region and the South. Hawaii was the only region to show a decrease.

Changes Due to Urban Heat Island

The urban heat island warms cities 0.6 °C to 5.6 °C above surrounding suburbs and rural areas. This extra heat leads to greater upward motion, which can induce additional shower and thunderstorm activity. Rainfall rates downwind of cities are increased between 48% and 116%. Partly as a result of this warming, monthly rainfall is about 28% greater between 20 miles to 40 miles downwind of cities, compared with upwind. Some cities induce a total precipitation increase of 51%.

Forecasting

The Quantitative Precipitation Forecast is the expected amount of liquid precipitation accumulated over a specified time period over a specified area. A QPF will be specified when a measurable precipitation type reaching a minimum threshold is forecast for any hour during a QPF valid period. Precipitation forecasts tend to be bound by synoptic hours such as 0000, 0600, 1200 and 1800 GMT. Terrain is considered in QPFs by use of topography or based upon climatological precipitation patterns from observations with fine detail. Starting in the mid to late 1990s, QPFs were used within hydrologic forecast models to simulate impact to rivers throughout the United States. Forecast models show significant sensitivity to humidity levels within the planetary boundary layer, or in the lowest levels of the atmosphere, which decreases with height. QPF can be generated on a quantitative, forecasting amounts, or a qualitative, forecasting the probability of a specific amount, basis. Radar imagery forecasting techniques show higher skill than model forecasts within six to seven hours of the time of the radar image. The forecasts can be verified through use of rain gage measurements, weather radar estimates, or a combination of both. Various skill scores can be determined to measure the value of the rainfall forecast.

TRANSPIRATION

Transpiration is a process similar to evaporation. It is a part of the water cycle, and it is the loss of water vapour from parts of plants, especially in leaves but also in stems, flowers and roots. Leaf surfaces are dotted with openings which are collectively called stomata, and in most plants they are more numerous on the undersides of the foliage. The stoma are bordered by guard cells that open and close the pore. Leaf transpiration occurs through stomata, and can be thought of as a necessary "cost" associated with the opening of the stomata to allow the diffusion of carbon dioxide gas from the air for photosynthesis. Transpiration also cools plants and enables mass flow of mineral nutrients and water from roots to shoots.

Mass flow of liquid water from the roots to the leaves is caused by the decrease in hydrostatic pressure in the upper parts of the plants due to the diffusion of water out of stomata into the atmosphere. Water is absorbed at the roots by osmosis, and any dissolved mineral nutrients travel with it through the xylem.

The rate of transpiration is directly related to the evaporation of water molecules from plant surface, especially from the surface openings, or stoma, on leaves. Stomatal transpiration accounts for most of the water loss by a plant, but some direct evaporation also takes place through the cuticle of the leaves and young stems. The amount of water given off depends somewhat upon how much water the roots of the plant have absorbed.

It also depends upon such environmental conditions as light intensity, humidity, winds and temperature. A plant should not be transplanted in full sunshine because it may lose too much water and wilt before the damaged roots can supply enough water. Transpiration occurs as the sun warms the water inside the blade. The warming changes much of the water into water vapour. This gas can then escape through the stomata. Transpiration helps cool the inside of the leaf because the escaping vapour has absorbed heat, the degree of stomatal opening, and the evaporative demand of the atmosphere surrounding the leaf. The amount of water lost by a plant depends on its size, along with surrounding light intensity, temperature, humidity, and wind speed. Soil water supply and soil temperature can influence stomatal opening, and thus transpiration rate. A fully grown tree may lose several hundred gallons of water through its leaves on a hot, dry day. About 90% of the water that enters a plant's roots is used for this process. The transpiration ratio is the ratio of the mass of water transpired to the mass of dry matter produced; the transpiration ratio of crops tends to fall between 200 and 1000. Transpiration rate of plants can be measured by a number of techniques, including potometers, lysimeters, porometers, photosynthesis systems and heat balance sap flow gauges.

Desert plants and conifers have specially adapted structures, such as thick cuticles, reduced leaf areas, sunken stomata and hairs to reduce transpiration and conserve water. Many cacti conduct photosynthesis in succulent stems, rather than leaves, so the surface area of the shoot is very low. Many desert plants have a special type of photosynthesis, termed crassulacean acid metabolism or CAM photosynthesis in which the stomata are closed during the day and open at night when transpiration will be lower.

CONDENSATION

Condensation is the change of the physical state of matter from gaseous phase into liquid phase, and is the reverse of evaporation. When the transition happens from the gaseous phase into the solid phase directly, the change is called deposition.

Upon the slowing-down of the molecules of the material, the overall attraction forces between these prevail and bring them together at distances comparable to their sizes. Since the condensing molecules suffer from reduced degrees of freedom and ranges of motion, their prior kinetic energy must be transferred to an absorbing colder entity—either a center of condensation within the gas volume or some contact surface. Condensation is initiated by the formation of atomic/molecular clusters of that species within its gaseous volume—like rain drop or snow-flake formation within clouds—or at the contact between such gaseous phase and a liquid or solid surface.

A few distinct reversibility scenarios emerge here with respect to the nature of the surface:

- Absorption into the surface of a liquid is reversible as evaporation.
- Adsorption onto solid surface at pressures and temperatures higher than the specie's triple point—also reversible as evaporation.
- Adsorption onto solid surface at pressures and temperatures lower than the specie's triple point—is reversible as sublimation.

Condensation commonly occurs when a vapour is cooled and/or compressed to its saturation limit when the molecular density in the gas phase reaches its maximal threshold. Vapour cooling and compressing equipment that collects condensed liquids is called "condenser".

Psychrometry measures the rates of condensation from and evaporation into the air moisture at various atmospheric pressures and temperatures. Water is the product of its vapour condensation—condensation is the process of such phase conversion.

Applications of Condensation

Condensation is a crucial component of distillation, an important laboratory and industrial chemistry application. Because condensation is a naturally occurring phenomenon, it can often be used to generate water in large quantities for human use. Many structures are made solely for the purpose of collecting water from condensation, such as air wells and fog fences. Such systems can often be used to retain soil moisture in areas where active desertification is occurring—so much so that some organizations educate people living in affected areas about water condensers to help them deal effectively with the situation.

It is also a crucial process in forming particle tracks in a Cloud Chamber. In this case, ions produced by an incident particle act as nucleation centres for the vapour to condense on.

Biological Adaptation

Numerous living organisms use water made accessible by condensation. A few examples are the Australian Thorny Devil, Darkling beetles on the Namibian coast and Coast Redwoods on the west coast of the United States.

ADVECTION

Advection, in chemistry, engineering and earth sciences, is a transport mechanism of a substance, or a conserved property, by a fluid, due to the fluid's bulk motion in a particular direction. An example of advection is the transport of pollutants or silt in a river. The motion of the water carries these impurities downstream. Another commonly advected property is energy or enthalpy, and here the fluid may be water, air, or any other thermal energy-containing fluid material. Any substance, or conserved property can be advected, in a similar way, in any fluid.

The fluid motion in advection is described mathematically as a vector field, and the material transported is typically described as a scalar concentration of substance, which is contained in the fluid. Advection requires currents in the fluid, and so cannot happen in solids.

It does not include transport of substances by simple diffusion. Advection is sometimes confused with the more encompassing process convection, which encompasses both advective transport and diffusive transport in fluids. Convective transport is the sum of advective transport and diffusive transport.

Advection is important for the formation of orographic cloud and the precipitation of water from clouds, as part of the hydrological cycle. In meteorology and physical oceanography, advection often refers to the transport of some property of the atmosphere or ocean, such as heat, humidity or salinity. Meteorological or oceanographic advection follows isobaric surfaces and is therefore predominantly horizontal.

Distinction between Advection and Convection

Occasionally, the term advection is used as synonymous with convection. However, many engineers prefer to use the term convection to describe transport by combined molecular and eddy diffusion, and reserve the usage of the term advection to describe transport with a general flow of the fluid.

An example of convection is flow over a hot plate or below a chilled water surface in a lake. In the ocean and atmospheric sciences, advection is understood as horizontal movement resulting in transport "from place to place", while convection is vertical "mixing".

Meteorology

In meteorology and physical oceanography, advection often refers to the transport of some property of the atmosphere or ocean, such as heat, humidity or salinity. Advection is important for the formation of orographic clouds and the precipitation of water from clouds, as part of the hydrological cycle.

Other Quantities

The advection equation also applies if the quantity being advected is represented by a probability density function at each point, although accounting for diffusion is more difficult.

Mathematics of Advection

The advection equation is the partial differential equation that governs the motion of a conserved scalar as it is advected by a known velocity field. It is derived using the scalar's conservation law, together with Gauss's theorem, and taking the infinitesimal limit.

Perhaps the best image to have in mind is the transport of salt dumped in a river. If the river is originally fresh water and is flowing quickly, the predominant form of transport of the salt in the water will be advective, as the water flow itself would transport the salt. If the river was not flowing the salt would simply disperse outwards from its source in a diffusive manner, which is not advection.

In Cartesian coordinates the advection operator is,

$$u.\nabla = u\frac{\partial}{\partial x} + v\frac{\partial}{\partial z}$$

where the velocity vector u has components u, v and w in the x, y and z directions respectively.

The advection equation for a scalar ψ, such as temperature, is expressed mathematically as:

$$\frac{\partial\psi}{\partial} + \nabla.(\psi u) = 0$$

where ∇ is the divergence operator and u is the velocity vector field. Frequently, it is assumed that the flow is incompressible, that is, the velocity field satisfies $\nabla \cdot u = 0$ If this is so, the above equation reduces to,

$$\frac{\partial\psi}{\partial t} + u \cdot \nabla\psi = 0.$$

For a vector a, such as magnetic field or velocity, in a solenoidal field it is defined as:

$$\frac{\partial a}{\partial t} + (u \cdot \nabla)a = 0.$$

In particular, if the flow is steady, $u \cdot \nabla\psi = 0$ which shows that ψ is constant along a streamline.

The advection equation is not simple to solve numerically: the system is a hyperbolic partial differential equation, and interest typically centers on discontinuous "shock" solutions. Even in one space dimension and constant velocity, the system remains difficult to simulate.

The equation becomes,

$$\frac{\partial\psi}{\partial t} + u\frac{\partial\psi}{\partial x} = 0$$

where ψ = ψ(x,t) is the scalar being advected and u the x component of the vector $u = (u(x),0,0)$.

Numerical simulation can be aided by considering the skew symmetric form for the advection operator.

$$\frac{1}{2}u \cdot \nabla u + \frac{1}{2}\nabla(uu)$$

where ∇(uu) is a vector with components,

$$\left[\nabla\left(uu_y\right),\nabla\left(uu_y\right),\nabla\left(uu_z\right)\right]$$

and the notation $u = \left[u_z, u_y, u_z\right]$ has been used.

Since skew symmetry implies only imaginary eigenvalues, this form reduces the "blow up" and "spectral blocking" often experienced in numerical solutions with sharp discontinuities. Using vector calculus identities, these operators can also be expressed in other ways, available in more software packages for more coordinate systems,

$$u \cdot \nabla u = \nabla\left(\frac{\|u\|^2}{2}\right) + (\nabla \times u) \times u$$

$$\frac{1}{2} u \cdot \nabla u + \frac{1}{2}\nabla(uu) = \nabla\left(\frac{\|u\|^2}{2}\right) + (\nabla \times u) \times u + \frac{1}{2} u (\nabla \cdot u)$$

This form also makes visible that the skew symmetric operator introduces error when the velocity field diverges.

SUBLIMATION

Sublimation refers to the process of transition of a substance from the solid phase to the gas phase without passing through an intermediate liquid phase. Sublimation is an endothermic phase transition that occurs at temperatures and pressures below a substance's triple point in its phase diagram.

At normal pressures, most chemical compounds and elements possess three different states at different temperatures. In these cases the transition from the solid to the gaseous state requires an intermediate liquid state. Note, however, that the pressure referred to here is the partial pressure of the substance, not the total pressure of the entire system. So, all solids which possess an appreciable vapour pressure at a certain temperature usually can sublime in air.

For some substances, such as carbon and arsenic, sublimation is much easier than evaporation from the melt, because the pressure of their triple point is very high, and it is difficult to obtain them as liquids. Sublimation requires additional energy and is an endothermic change. The enthalpy of sublimation can be calculated as the enthalpy of fusion plus the enthalpy of vapourization. The reverse process of sublimation is deposition. The formation of frost is an example of meteorological deposition.

Examples

- *Carbon dioxide*: Solid carbon dioxide sublimes readily at atmospheric pressure at–78.5°C, while liquid CO_2 can be obtained at pressures and temperatures above the triple point.
- *Water*: Snow and ice sublime, although more slowly, below the melting point temperature. This allows wet cloth to be hung outdoors in freezing weather and retrieved later in a dry state. In freeze–drying the material to be dehydrated is frozen and its water is allowed to sublime under reduced pressure or vacuum. The loss of snow from a snowfield during a cold spell is often caused by sunshine acting directly on the upper layers of the snow.

Ablation is a process which includes sublimation and erosive wear of glacier ice.

- *Other compounds*: Iodine produces fumes on gentle heating. It is possible to obtain liquid iodine at atmospheric pressure by controlling the temperature at just above the melting point of iodine. Naphthalene, a common ingredient in mothballs, also sublimes easily. Arsenic can also sublime at high temperatures. Various substances appear to sublime because of undergoing chemical reactions or decomposition; for example, ammonium chloride when heated decomposes into hydrogen chloride and ammonia.

Sublimation Purification

Sublimation is a technique used by chemists to purify compounds. Typically a solid is placed in a sublimation apparatus and heated under vacuum. Under this reduced pressure the solid volatilizes and condenses as a purified compound on a cooled surface, leaving a non-volatile residue of impurities behind. Once heating ceases and the vacuum is removed, the purified compound may be collected from the cooling surface.

Historical Usage

In alchemy, sublimation typically referred to the process by which a substance is heated to a vapour, then immediately collects as sediment on the upper portion and neck of the heating medium. It is one of the 12 core alchemical processes.

EVAPORATION

Evaporation is a type of vapourization of a liquid that occurs only on the surface of a liquid. The other type of vapourization is boiling, which, instead, occurs on the entire mass of the liquid. Evaporation is also part of the water cycle. On average, the molecules in a glass of water do not have enough heat energy to escape from the liquid. With sufficient heat, the liquid would turn into vapour quickly. When the molecules collide, they transfer energy to each other in varying degrees, based on how they collide. Sometimes the transfer is so one-sided for a molecule near the surface that it ends up with enough energy to escape.

Liquids that do not evaporate visibly at a given temperature in a given gas have molecules that do not tend to transfer energy to each other in a pattern sufficient to frequently give a molecule the heat energy necessary to turn into vapour. However, these liquids are evaporating. It is just that the process is much slower and thus significantly less visible.

Evaporation is an essential part of the water cycle. Solar energy drives evaporation of water from oceans, lakes, moisture in the soil, and other sources of water. In hydrology, evaporation and transpiration are collectively termed evapotranspiration. Evaporation is caused when water is exposed to air and the liquid molecules turn into water vapour, which rises up and forms clouds. The temperature that evaporation occurs ranges from 8.33°F-107.33°F

Theory

For molecules of a liquid to evaporate, they must be located near the surface, be moving in the proper direction, and have sufficient kinetic energy to overcome liquid-phase intermolecular forces. Only a small proportion of the molecules meet these criteria, so the rate of evaporation is limited. Since the kinetic energy of a molecule is proportional to its temperature, evaporation proceeds more quickly at higher temperatures. As the faster-moving molecules escape, the remaining molecules have lower average kinetic energy, and the temperature of the liquid, thus, decreases. This phenomenon is also called evaporative cooling. This is why evaporating sweat cools the human body. Evaporation also tends to proceed more quickly with higher flow rates between the gaseous and liquid phase and in liquids with higher vapour pressure. For example, laundry on a clothes line will dry more rapidly on a windy day than on a still day. Three key parts to evaporation are heat, humidity, and air movement.

On a molecular level, there is no strict boundary between the liquid state and the vapour state. Instead, there is a Knudsen layer, where the phase is undetermined. Because this layer is only a few molecules thick, at a macroscopic scale a clear phase transition interface can be seen.

Evaporative Equilibrium

If evaporation takes place in a closed vessel, the escaping molecules accumulate as a vapour above the liquid. Many of the molecules return to the liquid, with returning molecules becoming more frequent as the density and pressure of the vapour increases. When the process of escape and return reaches an equilibrium, the vapour is said to be "saturated," and no further change in either vapour pressure and density or liquid temperature will occur. For a system consisting of vapour and liquid of a pure substance, this equilibrium state is directly related to the vapour pressure of the substance, as given by the Clausius-Clapeyron relation:

$$\text{In}\left(\frac{P_2}{P_1}\right) = \frac{\Delta H_{vap}}{R}\left(\frac{1}{T_2} - \frac{1}{T_1}\right)$$

where P_1, P_2 are the vapour pressures at temperatures T_1, T_2 respectively, ΔHvap is the enthalpy of vapourization, and R is the universal gas constant. The rate of evaporation in an open system is related to the vapour pressure found in a closed system. If a liquid is heated, when the vapour pressure reaches the ambient pressure the liquid will boil. The ability for a molecule of a liquid to evaporate is based largely on the amount of kinetic energy an individual particle may possess. Even at lower temperatures, individual molecules of a liquid can evaporate if they have more than the minimum amount of kinetic energy required for vapourization.

Factors Influencing the Rate of Evaporation

- *Concentration of the substance evaporating in the air*: If the air already has a high concentration of the substance evaporating, then the given substance will evaporate more slowly.
- *Concentration of other substances in the air*: If the air is already saturated with other substances, it can have a lower capacity for the substance evaporating.
- *Concentration of other substances in the liquid*: If the liquid contains other substances, it will have a lower capacity for evaporation.
- *Flow rate of air*: This is in part related to the concentration points above. If fresh air is moving over the substance all the time, then the concentration of the substance in the air is less likely to go up with time, thus encouraging faster evaporation. This is the result of the boundary layer at the evaporation surface decreasing with flow velocity, decreasing the diffusion distance in the stagnant layer.
- *Inter-molecular forces*: The stronger the forces keeping the molecules together in the liquid state, the more energy one must get to escape. This is characterized by the enthalpy of vapourization.
- *Pressure*: Evaporation happens faster if there is less exertion on the surface keeping the molecules from launching themselves.
- *Surface area*: A substance that has a larger surface area will evaporate faster, as there are more surface molecules that are able to escape.
- *Temperature of the substance*: If the substance is hotter, then its molecules have a higher average kinetic energy, and evaporation will be faster.
- *Density*: The higher the density the slower a liquid evaporates.

In the US, the National Weather Service measures the actual rate of evaporation from a standardized "pan" open water surface outdoors, at various locations nationwide. Others do likewise around the world. The US data is collected and compiled into an annual evaporation map. The measurements range from under 30 to over 120 inches per year.

Applications

- When clothes are hung on a laundry line, even though the ambient temperature is below the boiling point of water, water evaporates.

This is accelerated by factors such as low humidity, heat and wind. In a clothes dryer, hot air is blown through the clothes, allowing water to evaporate very rapidly.

- The botijo, a traditional Spanish porous clay container designed to cool the contained water by evaporation.

SUBSURFACE FLOW

Subsurface flow, in hydrology, is the flow of water beneath earth's surface as part of the water cycle.

In the water cycle, when precipitation falls on the earth's land, some of the water flows on the surface forming streams and rivers. The remaining water, through infiltration, penetrates the soil traveling underground, hydrating the vadose zone soil, recharging aquifers, with the excess flowing in subsurface run-off. In hydrogeology it is measured by the Groundwater flow equation.

Run-off

Water flows from areas where the water table is higher to areas where it is lower. This flow can be either surface run-off in rivers and streams, or subsurface run-off infiltrating rocks and soil. The amount of run-off reaching surface and groundwater can vary significantly, depending on rainfall, soil moisture, permeability, groundwater storage, evaporation, upstream use, and whether or not the ground is frozen. The movement of subsurface water is determined largely by the water gradient, type of substrate, and any barriers to flow.

Surface Return

Subsurface water may return to the surface in groundwater flow, such as from a spring, seep, or a water well, or subsurface return to streams, rivers, and oceans. Water returns to the land surface at a lower elevation than where infiltration occurred, under the force of gravity or gravity induced pressures. Groundwater tends to move slowly, and is replenished slowly, so it can remain in aquifers for thousands of years. Mainly, water flows through the ground which leads to the ocean where the cycle begins again.

INFILTRATION

Infiltration is the process by which water on the ground surface enters the soil. Infiltration rate in soil science is a measure of the rate at which soil is able to absorb rainfall or irrigation. It is measured in inches per hour or millimeters per hour. The rate decreases as the soil becomes saturated. If the precipitation rate exceeds the infiltration rate, run-off will usually occur unless there is some physical barrier. It is related to the saturated hydraulic conductivity of the near-surface soil. The rate of infiltration can be measured using an infiltrometer.

Infiltration is governed by two forces: gravity and capillary action. While smaller pores offer greater resistance to gravity, very small pores pull water through capillary action in addition to and even against the force of gravity.

The rate of infiltration is affected by soil characteristics including ease of entry, storage capacity, and transmission rate through the soil. The soil texture and structure, vegetation types and cover, water content of the soil, soil temperature, and rainfall intensity all play a role in controlling infiltration rate and capacity. For example, coarse-grained sandy soils have large spaces between each grain and allow water to infiltrate quickly. Vegetation creates more porous soils by both protecting the soil from pounding rainfall, which can close natural gaps between soil particles, and loosening soil through root action. This is why forested areas have the highest infiltration rates of any vegetative types.

The top layer of leaf litter that is not decomposed protects the soil from the pounding action of rain, without this the soil can become far less permeable. In chapparal vegetated areas, the hydrophobic oils in the succulent leaves can be spread over the soil surface with fire, creating large areas of hydrophobic soil. Other conditions that can lower infiltration rates or block them include dry plant litter that resists re-wetting, or frost. If soil is saturated at the time of an intense freezing period, the soil can become a concrete frost on which almost no infiltration would occur. Over an entire watershed, there are likely to be gaps in the concrete frost or hydrophobic soil where water can infiltrate. Once water has infiltrated the soil it remains in the soil, percolates down to the ground water table, or becomes part of the subsurface run-off process.

Process

The process of infiltration can continue only if there is room available for additional water at the soil surface. The available volume for additional water in the soil depends on the porosity of the soil and the rate at which previously infiltrated water can move away from the surface through the soil. The maximum rate that water can enter a soil in a given condition is the infiltration capacity. If the arrival of the water at the soil surface is less than the infiltration capacity, all of the water will infiltrate. If rainfall intensity at the soil surface occurs at a rate that exceeds the infiltration capacity, ponding begins and is followed by run-off over the ground surface, once depression storage is filled. This run-off is called Horton overland flow. The entire hydrologic system of a watershed is sometimes Analysed using hydrology transport models, mathematical models that consider infiltration, run-off and channel flow to predict river flow rates and stream water quality.

Infiltration in Wastewater Collection

Robert E. Horton suggested that infiltration capacity rapidly declines during the early part of a storm and then tends towards an approximately constant value after a couple of hours for the remainder of the event.

Previously infiltrated water fills the available storage spaces and reduces the capillary forces drawing water into the pores. Clay particles in the soil may swell as they become wet and thereby reduce the size of the pores. In areas where the ground is not protected by a layer of forest litter, raindrops can detach soil particles from the surface and wash fine particles into surface pores where they can impede the infiltration process.

Infiltration in Wastewater Collection

Infiltration is a component of the general mass balance hydrologic budget. There are several ways to estimate the volume and/or the rate of infiltration of water into a soil. Three excellent estimation methods are the Green-Ampt method, SCS method, Horton's method, and Darcy's law.

General Hydrologic Budget

The general hydrologic budget, with all the components, with respect to infiltration F. Given all the other variables and infiltration is the only unknown, simple algebra solves the infiltration question.

$$F = B_I + P - E - T - ET - S - R - I_A - B_O$$

where,

- F is infiltration, which can be measured as a volume or length;
- B_I is the boundary input, which is essentially the output watershed from adjacent, directly connected impervious areas;
- B_O is the boundary output, which is also related to surface run-off, R, depending on where one chooses to define the exit point or points for the boundary output;
- P is precipitation;
- E is evaporation;
- ET is evapotranspiration;
- S is the storage through either retention or detention areas;
- I_A is the initial abstraction, which is the short term surface storage such as puddles or even possibly detention ponds depending on size;
- R is surface run-off.

The only note on this method is one must be wise about which variables to use and which to omit, for doubles can easily be encountered. An easy example of double counting variables is when the evaporation, E, and the transpiration, T, are placed in the equation as well as the evapotranspiration, ET. ET has included in it T as well as a portion of E.

Green-Ampt

Named for two men; Green and Ampt. The Green-Ampt method of infiltration estimation accounts for many variables that other methods, such as Darcy's law, do not. It is a function of the soil suction head, porosity, hydraulic conductivity and time.

$$\int_0^{F(t)} \frac{1-\psi\Delta\theta}{F+\psi\Delta\theta} dF = \int_0^t K dt$$

where,

- ψ is wetting front soil suction head;
- θ is water content;
- K is Hydraulic conductivity;
- F is the total volume already infiltrated.

Once integrated, one can easily choose to solve for either volume of infiltration or instantaneous infiltration rate:

$$F(t) = Kt + \psi\Delta\theta \ln\left[1 + \frac{F(t)}{\psi\Delta\theta}\right].$$

Using this model one can find the volume easily by solving for F(t). However the variable being solved for is in the equation itself so when solving for this one must set the variable in question to converge on zero, or another appropriate constant. A good first guess for F is the larger value of either Kt or $\sqrt{2\psi\Delta\theta Kt}$. The only note on using this formula is that one must assume that h_0, the water head or the depth of ponded water above the surface, is negligible. Using the infiltration volume from this equation one may then substitute F into the corresponding infiltration rate equation below to find the instantaneous infiltration rate at the time, t, F was measured.

$$f(t) = K\left[\frac{\psi\Delta\theta}{F(t)} + 1\right].$$

Horton's Equation

Horton's equation is another viable option when measuring ground infiltration rates or volumes. It is an empirical formula that says that infiltration starts at a constant rate, f_0, and is decreasing exponentially with time, t. After some time when the soil saturation level reaches a certain value, the rate of infiltration will level off to the rate f_c.

$$f_t = f_c + (f_0 - f_c)e^{-kt}$$

Where,

- f_t is the infiltration rate at time t;
- f_0 is the initial infiltration rate or maximum infiltration rate;

- f_c is the constant or equilibrium infiltration rate after the soil has been saturated or minimum infiltration rate;
- k is the decay constant specific to the soil.

The other method of using Horton's equation is as below. It can be used to find the total volume of infiltration, F, after time t.

$$F_t = f_c\, t + \frac{(f_0 - f_c)}{k}\left(1 - e^{-kt}\right)$$

Kostiakov Equation

Named after its founder Kostiakov is an empirical equation which assumes that the intake rate declines over time just as to a power function.

$$f(t) = akt^{a-1}$$

where a and k are empirical parameters.

The major limitation of this expression is its reliance on the zero final intake rate. In most cases the infilration rate instead approaches a finite steady value, which in some cases may occur after short periods of time. The Kostiakov-Lewis variant, also known as the "Modified Kostiakov" equation corrects for this by adding a steady intake term to the original equation.

$$f(t) = akt^{a-1} + f_0$$

in integrated form the cumulative volume is expressed as:

$$F(t) = kt^{a} + f_0\, t$$

where f_0 approximates, but does not necessarily equate to the final infiltration rate of the soil.

Darcy's Law

This method used for infiltration is using a simplified version of Darcy's law. In this model the ponded water is assumed to be equal to h0 and the head of dry soil that exists below the depth of the wetting front soil suction head is assumed to be equal to–ψ–L.

$$f = k\left[\frac{h_0 - (-\psi - L)}{L}\right]$$

where

h_0 is the depth of ponded water above the ground surface;

K is the hydraulic conductivity;

L is the total depth of subsurface ground in question.

In summary all of these equations should provide a relatively accurate assessment of the infiltration characteristics of the soil in question.

SURFACE RUN-OFF

Surface run-off is the water flow that occurs when soil is infiltrated to full capacity and excess water from rain, meltwater, or other sources flows over the land. This is a major component of the water cycle. Run-off that occurs on surfaces before reaching a channel is also called a nonpoint source. If a nonpoint source contains man–made contaminants, the run-off is called nonpoint source pollution. A land area which produces run-off that drains to a common point is called a watershed. When run-off flows along the ground, it can pick up soil contaminants including, but not limited to petroleum, pesticides, or fertilizers that become discharge or nonpoint source pollution.

Generation

Surface run-off can be generated either by rainfall or by the melting of snow, or glaciers. Snow and glacier melt occur only in areas cold enough for these to form permanently. Typically snowmelt will peak in the spring and glacier melt in the summer, leading to pronounced flow maxima in rivers affected by them. The determining factor of the rate of melting of snow or glaciers is both air temperature and the duration of sunlight. In high mountain regions, streams frequently rise on sunny days and fall on cloudy ones for this reason.

In areas where there is no snow, run-off will come from rainfall. However, not all rainfall will produce run-off because storage from soils can absorb light showers. On the extremely ancient soils of Australia and Southern Africa, proteoid roots with their extremely dense networks of root hairs can absorb so much rainwater as to prevent run-off even when substantial amounts of rain fall. In these regions, even on less infertile cracking clay soils, high amounts of rainfall and potential evaporation are needed to generate any surface run-off, leading to specialised adaptations to extremely variable streams.

Infiltration Excess Overland Flow

This occurs when the rate of rainfall on a surface exceeds the rate at which water can infiltrate the ground, and any depression storage has already been filled. This is called infiltration excess overland flow, Hortonian overland flow or unsaturated overland flow. This more commonly occurs in arid and semi-arid regions, where rainfall intensities are high and the soil infiltration capacity is reduced because of surface sealing, or in paved areas. This occurs largely in city areas where pavements prevent water infiltration.

Saturation Excess Overland Flow

When the soil is saturated and the depression storage filled, and rain continues to fall, the rainfall will immediately produce surface run-off. The level of antecedent soil moisture is one factor affecting the time until soil becomes saturated. This run-off is called saturation excess overland flow or saturated overland flow.

Antecedent Soil Moisture

Soil retains a degree of moisture after a rainfall. This residual water moisture affects the soil's infiltration capacity. During the next rainfall event, the infiltration capacity will cause the soil to be saturated at a different rate. The higher the level of antecedent soil moisture, the more quickly the soil becomes saturated. Once the soil is saturated, run-off occurs.

Subsurface Return Flow

After water infiltrates the soil on an up–slope portion of a hill, the water may flow laterally through the soil, and exfiltrate closer to a channel. This is called subsurface return flow or throughflow.

As it flows, the amount of run-off may be reduced in a number of possible ways: a small portion of it may evapotranspire; water may become temporarily stored in microtopographic depressions; and a portion of it may become run-on, which is the infiltration of run-off as it flows overland. Any remaining surface water eventually flows into a receiving water body such as a river, lake, estuary or ocean.

Human Impact on Surface Run-off

Urbanization increases surface run-off, by creating more impervious surfaces such as pavement and buildings, that do not allow percolation of the water down through the soil to the aquifer. It is instead forced directly into streams or storm water run-off drains, where erosion and siltation can be major problems, even when flooding is not. Increased run-off reduces groundwater recharge, thus lowering the water table and making droughts worse, especially for farmers and others who depend on the water wells.

When anthropogenic contaminants are dissolved or suspended in run-off, the human impact is expanded to create water pollution. This pollutant load can reach various receiving waters such as streams, rivers, lakes, estuaries and oceans with resultant water chemistry changes to these water systems and their related ecosystems. A 2008 report by the United States National Research Council identified urban stormwater as a leading source of water quality problems in the U.S.

Effects of Surface Run-off

Erosion and Deposition

Surface run-off causes erosion of the Earth's surface; deposition is the depositing of erosion. There are four principal types of erosion: splash erosion, gully erosion, sheet erosion and stream bed erosion. Splash erosion is the result of mechanical collision of raindrops with the soil surface. Dislodged soil particles becoming suspended in the surface run-off and carried into streams and rivers. Gully erosion occurs when the power of run-off is strong enough that it cuts a well defined channel. These channels can be as small as one centimeter wide or as large as several meters. Sheet erosion is the overland transport of run-off without a well defined channel. In the case of gully erosion, large amounts of material can be transported in a small time period. Stream bed erosion is the attrition of stream banks or bottoms by rapidly flowing rivers or creeks. Reduced crop productivity usually results from erosion, and these effects are studied in the field of soil conservation.

The soil particles carried in run-off vary in size from about.001 millimeter to 1.0 millimeter in diameter. Larger particles settle over short transport distances, whereas small particles can be carried over long distances suspended in the water column. Erosion of silty soils that contain smaller particles generates turbidity and diminishes light transmission, which disrupts aquatic ecosystems.

Entire sections of countries have been rendered unproductive by erosion. On the high central plateau of Madagascar, approximately ten per cent of that country's land area, virtually the entire landscape is devoid of vegetation, with erosive gully furrows typically in excess of 50 meters deep and one kilometer wide. Shifting cultivation is a farming system which sometimes incorporates the slash and burn method in some regions of the world. Erosion cause loss of the fertile top soil and reduces the its fertility and quality of the agricultural produce.

Modern industrial farming is another major cause of erosion. In some areas in the American corn belt, more than 50 per cent of the original topsoil has been carried away within the last 100 years.

Environmental Impacts

The principal environmental issues associated with run-off are the impacts to surface water, groundwater and soil through transport of water pollutants to these systems. Ultimately these consequences translate into human health risk, ecosystem disturbance and aesthetic impact to water resources.

Some of the contaminants that create the greatest impact to surface waters arising from run-off are petroleum substances, herbicides and fertilizers. Quantitative uptake by surface run-off of pesticides and other contaminants has been studied since the 1960s, and early on contact of pesticides with water was known to enhance phytotoxicity.

In the case of surface waters, the impacts translate to water pollution, since the streams and rivers have received run-off carrying various chemicals or sediments. When surface waters are used as potable water supplies, they can be compromised regarding health risks and drinking water aesthetics. Contaminated surface waters risk altering the metabolic processes of the aquatic species that they host; these alterations can lead to death, such as fish kills, or alter the balance of populations present. Other specific impacts are on animal mating, spawning, egg and larvae viability, juvenile survival and plant productivity. Some researches show surface run-off of pesticides, such as DDT, can alter the gender of fish species genetically, which transforms male into female fish.

In the case of groundwater, the main issue is contamination of drinking water, if the aquifer is abstracted for human use. Regarding soil contamination, run-off waters can have two important pathways of concern.

Firstly, run-off water can extract soil contaminants and carry them in the form of water pollution to even more sensitive aquatic habitats. Secondly, run-off can deposit contaminants on pristine soils, creating health or ecological consequences.

Flooding

Flooding occurs when a watercourse is unable to convey the quantity of run-off flowing downstream. The frequency with which this occurs is described by a return period. Flooding is a natural process, which maintains ecosystem composition and processes, but it can also be altered by land use changes such as river engineering. Floods can be both beneficial to societies or cause damage.

Agriculture along the Nile floodplain took advantage of the seasonal flooding that deposited nutrients beneficial for crops. However, as the number and susceptibility of settlements increase, flooding increasingly becomes a natural hazard. Adverse impacts span loss of life, property damage, contamination of water supplies, loss of crops, and social dislocation and temporary homelessness. Floods are among the most devastating of natural disasters.

Agricultural Issues

A common context of run-off deals with agriculture. When farmland is tilled and bare soil is revealed, rainwater carries billions of tons of topsoil into waterways each year, causing loss of valuable topsoil and adding sediment to produce turbidity in surface waters.

The other context of agricultural issues involves the transport of agricultural chemicals via surface run-off. This result occurs when chemical use is excessive or poorly timed with respect to high precipitation.

The resulting contaminated run-off represents not only a waste of agricultural chemicals, but also an environmental threat to downstream ecosystems. The alternative to conventional farming is organic farming which eliminates chemical usage.

Measurement and Mathematical Modeling

Run-off is Analysed by using mathematical models in combination with various water quality sampling methods. Measurements can be made using continuous automated water quality analysis instruments targeted on pollutants such as specific organic or inorganic chemicals, pH, turbidity etc. or targeted on secondary indicators such as dissolved oxygen. Measurements can also be made in batch form by extracting a single water sample and conducting any number of chemical or physical tests on that sample.

In the 1950s or earlier hydrology transport models appeared to calculate quantities of run-off, primarily for flood forecasting. Beginning in the early 1970s computer models were developed to Analyse the transport of run-off carrying water pollutants, which considered dissolution rates of various chemicals, infiltration into soils and ultimate pollutant load delivered to receiving waters.

One of the earliest models addressing chemical dissolution in run-off and resulting transport was developed in the early 1970s under contract to the United States Environmental Protection Agency. This computer model formed the basis of much of the mitigation study that led to strategies for land use and chemical handling controls.

Other computer models have been developed that allow surface run-off to be tracked through a river course as reactive water pollutants. In this case the surface run-off may be considered to be a line source of water pollution to the receiving waters.

Mitigation and Treatment

Mitigation of adverse impacts of run-off can take several forms:

- Land use development controls aimed at minimizing impervious surfaces in urban areas
- Erosion controls for farms and construction sites
- Flood control programmes
- Chemical use and handling controls in agriculture, landscape maintenance, industrial use, etc.

Land use Controls

Many world regulatory agencies have encouraged research on methods of minimizing total surface run-off by avoiding unnecessary hardscape. Many municipalities have produced guidelines and codes for land developers that encourage minimum width sidewalks, use of pavers set in earth for driveways and walkways and other design techniques to allow maximum water infiltration in urban settings. An example land use control programme can be viewed at seen in the city of Santa Monica, California.

Erosion Controls

Erosion controls have appeared since medieval times when farmers realised the importance of contour farming to protect soil resources. Beginning in the 1950s these agricultural methods became increasingly more sophisticated. In the 1960s some state and local governments began to focus their efforts on mitigation of construction run-off by requiring builders to implement erosion and sediment controls. This included such techniques as: use of straw bales and barriers to slow run-off on slopes, installation of silt fences, programming construction for months that have less rainfall and minimizing extent and duration of exposed graded areas. Montgomery County, Maryland implemented the first local government sediment control programme in 1965, and this was followed by a statewide programme in Maryland in 1970.

Flood Control Programmes

Flood control programmes as early as the first half of the twentieth century became quantitative in predicting peak flows of riverine systems. Progressively strategies have been developed to minimize peak flows and also to reduce channel velocities. Some of the techniques commonly applied are: provision of holding ponds to buffer riverine peak flows, use of energy dissipators in channels to reduce stream velocity and land use controls to minimize run-off.

Chemical Use and Handling

Following enactment of the U.S. Resource Conservation and Recovery Act in 1976, and later the Water Quality Act of 1987, states and cities have become more vigilant in controlling the containment and storage of toxic chemicals, thus preventing releases and leakage. Methods commonly applied are: requirements for double containment of underground storage tanks, registration of hazardous materials usage, reduction in numbers of allowed pesticides and more stringent regulation of fertilizers and herbicides in landscape maintenance. In many industrial cases, pretreatment of wastes is required, to minimize escape of pollutants into sanitary or stormwater sewers.

The U.S. Clean Water Act requires that local governments in urbanized areas obtain stormwater discharge permits for their drainage systems. Essentially this means that the locality must operate a stormwater management programme for all surface run-off that enters the municipal separate storm sewer system.

EPA and state regulations and related publications outline six basic components that each local programme must contain:

1. Public education

2. Public involvement
3. Illicit discharge detection and elimination
4. Construction site run-off controls
5. Post–construction storm water management controls
6. Pollution prevention and "good housekeeping" measures.

Other property owners which operate storm drain systems similar to municipalities, such as state highway systems, universities, military bases and prisons, are also subject to the MS4 permit requirements.

SNOWMELT

In hydrology, snowmelt is surface run-off produced from melting snow. It can also be used to describe the period or season during which such run-off is produced. Water produced by snowmelt is an important part of the annual water cycle in many parts of the world, in some cases contributing high fractions of the annual run-off in a watershed. Predicting snowmelt run-off from a drainage basin may be a part of designing water control projects. Rapid snowmelt can cause flooding. If the snowmelt is then frozen, very dangerous conditions and accidents can occur, introducing the need for salt to melt the ice.

Increased water run-off due to snowmelt was a cause of many famous floods. One well-known example is the Red River Flood of 1997, when the Red River of the North in the Red River Valley of the United States and Canada flooded. Flooding in the Red River Valley is augmented by the fact that the river flows north through Winnipeg, Manitoba and into Lake Winnipeg. As snow in Minnesota, North Dakota, and South Dakota begins to melt and flow into the Red River, the presence of downstream ice can act as a dam and force upstream water to rise. Colder temperatures downstream can also potentially lead to freezing of water as it flows north, thus augmenting the ice dam problem. Some areas in British Columbia are also prone to snowmelt flooding as well.

Energy Fluxes Related to Snowmelt

There are several energy fluxes involved in the melting of snow. These fluxes can act in opposing directions, that is either delivering heat to or removing heat from the snowpack. Ground heat flux is the energy delivered to the snowpack from the soil below by conduction. Radiation inputs to the snowpack include net shortwave and longwave radiation.

Net shortwave radiation is the difference in energy received from the sun and that reflected by the snowpack due to the snowpack albedo. Longwave radiation is received by the snowpack from many sources, including ozone, carbon dioxide, and water vapour present in all levels of the atmosphere. Longwave radiation is also emitted by the snowpack in the form near-Black-body radiation, where snow has an emissivity between

0.97 and 1.0. Generally the net longwave radiation term is negative, meaning a net loss of energy from the snowpack. Latent heat flux is the energy removed from or delivered to the snowpack which accompanies the mass transfers of evaporation, sublimation, or condensation. Sensible heat flux is the heat flux due to convection between the air and snowpack.

INTERCEPTION

Interception, or canopy interception, refers to precipitation that does not reach the soil, but is instead intercepted by the leaves and branches of plants. It occurs in the canopy, and in the forest ground litter. Because of evaporation, interception of liquid water generally leads to loss of that precipitation for the drainage basin, except for cases such as fog interception. Intercepted snowfall does not result in any notable amount of evaporation, and most of the snow falls off the tree by wind or melt.

However, intercepted snow can more easily drift with the wind, out of the watershed. Conifers have greater interception than hardwoods. Their needles gives them more surface area for droplets to adhere to, and they have foliage in spring and fall, therefore interception also depends on the type of vegetation in a wooded area.

RESIDENCE TIMES

The residence time of a reservoir within the hydrologic cycle is the average time a water molecule will spend in that reservoir. It is a measure of the average age of the water in that reservoir.

Groundwater can spend over 10,000 years beneath Earth's surface before leaving. Particularly old groundwater is called fossil water. Water stored in the soil remains there very briefly, because it is spread thinly across the Earth, and is readily lost by evaporation, transpiration, stream flow, or groundwater recharge. After evaporating, the residence time in the atmosphere is about 9 days before condensing and falling to the Earth as precipitation. The major ice sheets–Antarctica and Greenland–store ice for very long periods. Ice from Antarctica has been reliably dated to 800,000 years before present, though the average residence time is shorter.

In hydrology, residence times can be estimated in two ways. The more common method relies on the principle of conservation of mass and assumes the amount of water in a given reservoir is roughly constant. With this method, residence times are estimated by dividing the volume of the reservoir by the rate by which water either enters or exits the reservoir. Conceptually, this is equivalent to timing how long it would take the reservoir to become filled from empty if no water were to leave. An alternative method to estimate residence times, which is gaining in popularity for dating groundwater, is the use of isotopic techniques. This is done in the subfield of isotope hydrology.

CHANGES OVER TIME

The water cycle describes the processes that drive the movement of water throughout the hydrosphere. However, much more water is "in storage" for long periods of time than is actually moving through the cycle. The storehouses for the vast majority of all water on Earth are the oceans. It is estimated that of the 332,500,000 mi^3 of the world's water supply, about 321,000,000 mi^3 is stored in oceans, or about 95%. It is also estimated that the oceans supply about 90% of the evaporated water that goes into the water cycle.

During colder climatic periods more ice caps and glaciers form, and enough of the global water supply accumulates as ice to lessen the amounts in other parts of the water cycle. The reverse is true during warm periods. During the last ice age glaciers covered almost one-third of Earth's land mass, with the result being that the oceans were about 400 ft lower than today. During the last global "warm spell," about 125,000 years ago, the seas were about 18 ft higher than they are now. About three million years ago the oceans could have been up to 165 ft higher.

The scientific consensus expressed in the 2007 Intergovernmental Panel on Climate Change Summary for Policymakers is for the water cycle to continue to intensify throughout the 21st century, though this does not mean that precipitation will increase in all regions. In subtropical land areas—places that are already relatively dry—precipitation is projected to decrease during the 21st century, increasing the probability of drought. The drying is projected to be strongest near the poleward margins of the subtropics. Annual precipitation amounts are expected to increase in near-equatorial regions that tend to be wet in the present climate, and also at high latitudes. These large-scale patterns are present in nearly all of the climate model simulations conducted at several international research centers as part of the 4th Assessment of the IPCC.

Glacial retreat is also an example of a changing water cycle, where the supply of water to glaciers from precipitation cannot keep up with the loss of water from melting and sublimation. Glacial retreat since 1850 has been extensive.

Human activities that alter the water cycle include:

- Agriculture
- Industry
- Alteration of the chemical composition of the atmosphere
- Construction of dams
- Deforestation and afforestation
- Removal of groundwater from wells
- Water abstraction from rivers
- Urbanization

EFFECTS ON CLIMATE

The water cycle is powered from solar energy. 86% of the global evaporation occurs from the oceans, reducing their temperature by evaporative cooling. Without the cooling, the effect of evaporation on the greenhouse effect would lead to a much higher surface temperature of 67 °C and a warmer planet.

Aquifer drawdown or overdrafting and the pumping of fossil water increases the total amount of water in the hydrosphere that is subject to transpiration and evaporation thereby causing accretion in water vapour and cloud cover which are the primary absorbers of infrared radiation in the Earth's atmosphere. Adding water to the system has a forcing effect on the whole earth system, an accurate estimate of which hydrogeological fact is yet to be quantified.

EFFECTS ON BIOGEOCHEMICAL CYCLING

While the water cycle is itself a biogeochemical cycle, flow of water over and beneath the Earth is a key component of the cycling of other biogeochemicals. Run-off is responsible for almost all of the transport of eroded sediment and phosphorus from land to waterbodies. The salinity of the oceans is derived from erosion and transport of dissolved salts from the land. Cultural eutrophication of lakes is primarily due to phosphorus, applied in excess to agricultural fields in fertilizers, and then transported overland and down rivers. Both run-off and groundwater flow play significant roles in transporting nitrogen from the land to waterbodies. The dead zone at the outlet of the Mississippi River is a consequence of nitrates from fertilizer being carried off agricultural fields and funnelled down the river system to the Gulf of Mexico. Run-off also plays a part in the carbon cycle, again through the transport of eroded rock and soil.

SLOW LOSS OVER GEOLOGIC TIME

The hydrodynamic wind within the upper portion of a planet's atmosphere allows light chemical elements such as Hydrogen to move up to the exobase, the lower limit of the exosphere, where the gases can then reach escape velocity, entering outer space without impacting other particles of gas. This type of gas loss from a planet into space is known as planetary wind. Planets with hot lower atmospheres could result in humid upper atmospheres that accelerate the loss of hydrogen.

Chapter 4

Water Quality and Standards

INTRODUCTION

Water quality is a term used here to express the suitability of water to sustain various uses or processes. Any particular use will have certain requirements for the physical, chemical or biological characteristics of water; for example limits on the concentrations of toxic substances for drinking water use, or restrictions on temperature and pH ranges for water supporting animals. Consequently, water quality can be defined by a range of variables which limit the water. Although many uses have some common requirements for certain variables, each use will have its own demands and influences on water quality.

Quantity and quality demands of different users will not always be compatible, and the activities of one user may restrict the activities of another, either by demanding water of a specific quality outside the range required by the other user or by lowering quality during use of the water. Efforts to improve or maintain a certain water quality often compromise between the quality and quantity demands of different users. There is an increasing recognition that natural ecosystems have a legitimate place in the consideration of options for water quality management. This is both for their intrinsic value and because they are sensitive indicators of changes or deterioration in overall water quality, providing a useful addition to physical, chemical and other information.

WATER QUALITY

The composition of surface and under groundwater is dependent on natural factors (geological, topographical, meteorological, hydrological and biological) in the drainage basin and varies with seasonal differences in run-off volumes, weather conditions and water levels. Large natural variations in water quality may, therefore, be observed even where only a single watercourse is involved. Human intervention also has significant effects on water quality. Some of these effects are the result of hydrological changes, such as building of dams, draining of diversion of waste waters. More obvious are the polluting activities, such as the discharge of domestic, industrial, urban and other wastewaters into the watercourse (whether intentional or accidental) and the spreading of chemicals on agricultural land in the drainage basin.

Water quality is affected by a wide range of natural and human influences. The most important of the natural influences are geological, hydrological and climatic, since these affect the quantity and the quality of water available.

Their influence is generally higher when available water quantities are low and maximum use is to be made of the limited resource; for example, high salinity is a frequent problem in arid and coastal areas. If the financial and technical resources are available, seawater or saline groundwater can be desalinated but in many circumstances this is not economically feasible. Thus, although water may be available in adequate quantities, its unsuitable quality limits the uses that can be made of it.

Although the natural ecosystem is in harmony with natural water quality, any significant changes to water quality will usually be disruptive to the ecosystem. The effects of human activities on water quality are both widespread and varied in the degree to which they disrupt the ecosystem and/or restrict water use. Pollution of water by human activities, for example, is attributable to only one source, but the reasons for this type of pollution, its impacts on water quality and the necessary remedial or preventive measures are varied. Faecal pollution may occur because there are no community facilities for waste disposal, because collection and treatment facilities are inadequate or improperly operated, or because on-site sanitation facilities (such as latrines) drain directly into aquifers.

The effects of faecal pollution vary appreciably in space and time. In developing countries intestinal disease is the main problem, while organic load and eutrophication may be of greater concern in developed countries (in the rivers into which the sewage or effluent is discharged and in the sea into which the rivers flow or sewage sludge is dumped).

A single influence may, therefore, give rise to a number of water quality problems, just as a problem may have a number of contributing influences. Eutrophication results not only from point sources, such as wastewater discharges with high nutrient loads (principally nitrogen and phosphorus), but also from diffuse sources such as run-off from livestock feedlots or agricultural land fertilized with organic and inorganic fertilizers. Pollution from diffuse sources, such as agricultural run-off, or from numerous small inputs over a wide area, such as faecal pollution from unsewered settlements, is particularly difficult to control.

The quality of water may be described in terms of the concentration and state (dissolved or particulate) of some or all of the organic and inorganic material present in the water, together with certain physical characteristics of the water.

It is determined by in situ measurements and by examination of water samples on site or in the laboratory. The main elements of water quality monitoring are, therefore, on-site measurements, the collection and analysis of water samples, the study and evaluation of the analytical results, and the reporting of the findings.

The results of analyses performed on a single water sample are only valid for the particular location and time at which that sample was taken. One purpose of a monitoring programme is, therefore, to gather sufficient data (by means of regular or intensive sampling and analysis) to assess spatial and/or temporal variations in water quality.

WATER TEMPERATURE

Water temperature is not only important to fisherman and industries, but also to the growth of fish and algae. A lot of water is used for cooling purposes in power plants that generate electricity. They need cool water to start with, and they generally release warmer water back to the environment. The temperature of the released water can affect downstream habitats. Temperature also can affect the ability of water to hold oxygen as well as the ability of organisms to resist certain pollutants.

pH

The pH is a measure of how acidic/basic a water sample is. The range goes from 0–14, with 7 being neutral. The pH of less than 7 indicates acidity, whereas a pH of greater than 7 indicates a base. The pH is really a measure of the relative amount of free hydrogen and hydroxyl ions in the water.

Water that has more free hydrogen ions is acidic, whereas water that has more free hydroxyl ions is basic. Since pH can be affected by chemicals in the water, pH is an important indicator of water that is changing chemically.

The pH is reported in "logarithmic units," like the Richter scale, which measures earthquakes. Each number represents a 10-fold change in the acidity/basicness of the water. Water with a pH of 5 is ten times more acidic than water having a pH of six.

Pollution can change a water's pH, which in turn can harm animals and plants living in the water. For instance, water coming out of an abandoned coal mine can have a pH of 2, which is very acidic and would definitely affect any fish and other micro organisms. By using the logarithm scale, this mine-drainage water would be several hundred times more acidic than neutral water so stay out of abandoned mines.

SPECIFIC CONDUCTANCE

Specific conductance is a measure of the ability of water to conduct an electrical current. It is highly dependent on the amount of dissolved solids (such as salt) in the water. Pure water, such as distilled water, will have a very low specific conductance, and sea water will have a high specific conductance.

Rainwater often dissolves airborne gasses and airborne dust while it is in the air, and thus often has a higher specific conductance than distilled water. Specific conductance is an important water-quality measurement because it gives a good idea of the amount of dissolved material in the water.

TURBIDITY

Turbidity is the amount of particulate matter that is suspended in water. Turbidity measures the scattering effect that suspended solids have on light: the higher the intensity of scattered light, the higher the turbidity.

Material that causes water to be turbid include:

- Clay
- Silt
- Finely divided organic and inorganic matter
- Soluble coloured organic compounds
- Plankton
- Microscopic organisms

Turbidity makes the water cloudy or opaque. Turbidity is measured by shining a light through the water and is reported in Nephelometric Turbidity Units (NTU). During periods of low flow (base flow), many rivers are a clear green colour, and turbidities are low, usually less than 10 NTU. During a rainstorm, particles from the surrounding land are washed into the river making the water a muddy brown colour, indicating water that has higher turbidity values. Also, during high flows, water velocities are faster and water volumes are higher, which can more easily stir up and suspend material from the stream bed, causing higher turbidities.

Turbidity can be measured in the laboratory and also on-site in the river. A handheld turbidity meter measures turbidity of a water sample.

The meter is calibrated using standard samples from the meter manufacturer. The three glass vials shows turbidity standards of 5, 50, and 500 NTUs. Once the meter is calibrated to correctly read these standards, the turbidity of a water sample can be taken.

DISSOLVED OXYGEN

Although water molecules contain oxygen atom, this oxygen is not what is needed by aquatic organisms living in natural waters. A small amount of oxygen, upto about ten molecules of oxygen per million of water, is actually dissolved in water. This dissolved oxygen is breathed by fish and zooplankton and is needed by them to survive. Rapidly moving water, such as in a mountain stream or large river, tends to contain a lot of dissolved oxygen, while stagnant water contains little. Bacteria in water can consume oxygen as organic matter decays. Thus, excess organic material in the tanks, lakes and rivers can cause an oxygen-deficient situation to occur. Aquatic life can have a hard time in stagnant water that has a lot of rotting, organic material in it, especially in summer, when dissolved-oxygen levels are at a seasonal low.

HARDNESS

The amount of dissolved calcium and magnesium in water determines its 'hardness.' If some one live in an area where the water is 'soft,' then you may never have even heard of water hardness. But, if one live, where the water is relatively hard, may notice that it is difficult to get lather up when washing your hands or clothes. And, industries in these area might have to spend money to soften their water, as hard water can damage equipment. Hard water can even shorten the life of fabrics and clothes.

SUSPENDED SEDIMENT

Suspended sediment is the amount of soil moving along in a stream. It is highly dependent on the speed of the water flow, as fast-flowing water can pick up and suspend more soil than calm water. During storms, soil is washed from the stream banks into the stream. The amount that washes into a stream depends on the type of land in the river's watershed and the vegetation surrounding the river. If land is disturbed along a stream and protection measures are not taken, then excess sediment can harm the water quality of a stream. These bunds and fences are supposed to trap sediment during a rainstorm and keep it from washing into a stream, as excess sediment can harm the creeks, rivers, lakes, and reservoirs.

Sediment coming into a reservoir is always a concern; once it enters it cannot get out–most of it will settle to the bottom. Reservoirs can 'silt in' if too much sediment enters them. The volume of the reservoir is reduced, resulting in less area for water storage for agriculture as well as reducing the power-generation capability of the power plant in the dam.

POTABLE WATER QUALITY

Potable water must be free of anything that would degrade human performance. Further, it should not damage the materials used in its transportation and storage. Potable water must be suitable for maintaining human health (personal hygiene and medical treatment). Water quality standards give a basis for selecting or rejecting water intended for human use. These standards provide minimum accepted values for safeguarding human health.

IMPURITIES IN WATER

As water goes through the hydrologic cycle, it gathers many impurities.

Dust, smoke, and gases fill the air and can contaminate rain, snow, hail, and sleet. The rain water picks up silt, chemicals, and disease organisms. As it enters the earth through seepage and infiltration, some of the suspended impurities may be filtered out. However, other minerals and chemicals are dissolved and carried along. As groundwater it may contain disease organisms as well as harmful chemicals.

In addition to the impurities in water resulting from infiltration, many are contributed by an industrialized society. The garbage, sewage, industrial waste, pesticides etc., are all possible contaminants of raw water. Impurities in raw water are either suspended or dissolved. Suspended impurities include diseases organisms, silt, bacteria, and algae.

They must be removed or destroyed before the water is consumed. Dissolved impurities include salts, (calcium, magnesium, and sodium), iron, manganese, and gases (oxygen, carbon dioxide, hydrogen sulphide, pH and nitrogen). These impurities must be reduced to levels acceptable for human consumption.

RAW WATER CLASSIFICATIONS

Water is classified as fresh, brackish, or salt water (seawater) based on the concentration of TDS (Total Dissolved Solids). Fresh water has a TDS concentration of less that 1,500 ppm. Brackish water is high in minerals and has a TDS concentration between 1,500 ppm and 16,000 ppm. Salt water has a TDS concentration greater than 15,000 ppm.

Generally, groundwater (subsurface) has less chemical or biological contaminants than surface water, provided reasonable care is exercised in the selection of the well site. Harmful micro organisms are usually reduced to tolerable levels by passage through the soil.

TREATED WATER CLASSIFICATIONS

- *Palatable water*: Palatable water is water that is pleasing in appearance and taste. It is significantly free from colour, turbidity, taste, and odour. It should also be cool and aerated.
- *Potable water quality standards*: Quality standards for treated water reflect the values of substances allowed in potable water. Standards exist to measure the physical, chemical, microbiologic, and radiologic quality of water and to test for the presence of chemical agents.

Physical Quality

The principal physical characteristics of water are colour, odour and taste, turbidity, and temperature.

These characteristics and their related quality standards are described below:

- *Colour*: Colour in water is derived from coloured substances, such as vegetable matter, dissolved from roots and leaves, from humus, or from inorganic compounds such as iron and manganese salts. The colour standard is designed to make drinking water more palatable.

- *Odour and Taste*: There is no set standards for odour and taste as there are no specific tests for these. Odour and taste found in water are most commonly caused by algae, decomposed organic matter, dissolved gases, or industrial waste.
- *Turbidity*: Turbidity refers to a muddy or unclear condition of water caused by suspended clay, silt, organic and inorganic matter, and plankton and other microorganisms. The turbidity standard was established to improve the efficiency of disinfection by reducing particles to which micro organisms could attach.
- *Temperature:* Warm water tastes flat. Cooling water suppresses odours and tastes and makes it more palatable. Temperature also effects the chlorination and purification of water. Disinfection takes longer when water is colder, and purification capacity is reduced with reverse osmosis treatment equipment. Water having physical characteristics exceeding the limits or making it less palatable should not be used for drinking. Otherwise, reduced consumption and increased risk of dehydration may result. When water of low physical quality has not been used, the appropriate command level will make that decision based on medical recommendations.

CHEMICAL QUALITY

The chemical quality of water depends on the chemical substances it contains. These substances include TDS, chlorides, sulphates and other ions. The chemical quality of water involves its hardness, alkalinity, acidity, and corrosiveness. Chemical substances having an adverse health effect have established standards that will not be exceeded without medical approval.

Potential Hydrogen

The pH is a measure of the acidic or alkaline nature of water. It is technically defined as the negative logarithm of the hydrogen ion concentration. It ranges from 0 to 14 and pH of 7 is neutral. The pH influences the corrosiveness of the water, the amount of chemicals needed for proper disinfection, and the ability of an analyst to detect contaminants. Water with a pH below 7 is regarded as acidic while that with a pH above 7 is regarded as alkaline. The pH standard was established to ensure effective purification and disinfection.

Arsenic

Arsenic can be present in natural water sources in a wide range of concentrations. It can come from either natural or industrial sources. Ingestion of low concentrations of arsenic can cause nausea, vomiting, abdominal pain, or nerve damage.

Chloride

Chloride exists in most natural waters. It is the main anion found in seawater. Chloride comes from natural salt deposits, domestic and industrial waste, and agricultural run-off. Even in low concentrations, chloride can produce an objectionable taste in water. The chloride standard ensures that potable water is also palatable.

Calcium, Magnesium and Hardness

Calcium and magnesium in drinking water, is an important parameter is to be taken into account: water hardness, even if this term is incorrect and obsolete from a strictly chemical point of view. That is to say that both of these elements largely have not been analysed individually in drinking water in the past, but just non-specifically in summary as hardness.

This approach was applied in many studies focused on health effects of this 'water factor'. Since the definition of water hardness is approached either analytically or technologically, it was not and still has not been defined in a unified manner, and as with other parameters, multiple definitions have been available and multiple units have been used to express it (German, French, and English degrees; equivalent $CaCO_3$ or CaO in mg/l). Initially, water hardness was understood to be a measure of the capacity of water to precipitate soap, which is in practice the sum of concentrations of all polyvalent cations present in water (Ca, Mg, Sr, Ba, Fe, Al, Mn, etc.); nevertheless, since the other ions (apart from Ca and Mg) play a minor role in this regard, later it has been generally accepted that hardness is defined as the sum of the Ca and Mg concentrations.

From the technical point of view, multiple different scales of water hardness were suggested (*i.e.*, very soft–soft–medium hard–hard–very hard).

Expectedly, both extreme degrees (*i.e.* very soft and very hard) are considered as undesirable concordantly from the technical and health points of view, but the optimum Ca and Mg water levels are not easy to determine since the health requirements may not coincide with the technical ones.

Calcium and Magnesium Presence in Waters

Water calcium and magnesium result from decomposition of calcium and magnesium aluminosilicates and, at higher concentrations, from dissolution of limestone, magnesium limestone, magnesite, gypsum and other minerals. Anthropogenenic contamination of drinking water sources with calcium and magnesium is not common but drinking water may be intentionally supplemented with these elements while treated, as happens with deacidification of underground waters by means of calcium hydroxide or filtration through different compounds counteracting acidity such as $CaCO_3$, $MgCO_3$ and MgO, and possibly also with stabilization of low-mineralized waters by addition of CaO and CO_2.

In low-and medium-mineralized underground and surface waters (as drinking waters are), calcium and magnesium are mainly present as simple ions Ca^{2+} and Mg^{2+}, the Ca levels varying from tens to hundreds of mg/l and the Mg concentrations varying from units to tens of mg/l.

Magnesium is usually less abundant in waters than calcium, which is easy to understand since magnesium is found in the Earth's crust in much lower amounts as compared with calcium. In common underground and surface waters the weight concentration of Ca is usually several times higher compared to that of Mg, the Ca to Mg ratio reaching up to 10. Nevertheless, a common Ca to Mg ratio is about 4, which corresponds to a substance ratio of 2.4.

Sulphates

Sulphates occur naturally in water as the result of dissolution of sulfurbearing minerals. Significant concentrations also result from industry sources, such as coal mine drainage, pulp paper mills, tanneries, textile mills, and domestic waste water. When ingested, sulphates have a laxative effect. They also can produce a bad taste in water. The sulphate standard was established to prevent chemically induced diarrhea.

Total Dissolved Solids

The TDS of water is composed of mineral salts and small amounts of other inorganic and organic substances. The proportion of each constituent is the result of weathering of rocks found in the drainage basin and of any industrial contributions. Since TDS is composed of chloride, magnesium, sulfate, and other ions, its ingestion in water has the same effects. Therefore, the TDS standard was established to prevent chemically induced diarrhea.

Chemical water quality standards are based on the effect the water will have on the health of living organisms. The effect of a particular chemical substance determines if a limit is established for that substance. Chemical substances having a negative physical effect will have a mandatory limit that should not be exceeded. Some substances, such as iron and manganese, have no significant negative physical effect, but may restrict the use of the water, such as for the laundering of clothes.

MICROBIOLOGICAL QUALITY

The microbiological quality of potable water shows its potential for transmitting waterborne diseases. These diseases may be caused by viruses, bacteria, protozoa, or higher organisms. A microbiological test will reveal the quality of the raw water source and aid in determining any treatment required.

The test is necessary to maintain the quality of the water. The testing for microorganisms in water is extremely difficult. The number of these organisms is usually very low, even in a badly polluted water supply, and the test used to find them is difficult. For these reasons, indicator organisms are used to detect the presence of contamination.

The bacterial organisms used as an indicator of possible contamination are total coliform. These organisms occur in large quantities in the intestines of warm-blooded animals.

The presence of any coliform organism in treated potable water is an indication of either inadequate treatment or the introduction of undesirable materials to the water after treatment.

While the detection of many disease-causing microbes is difficult, the test to detect a surrogate organism, E. coli, is simple and effective for field use.

Because of its relative simplicity and field adaptability, the membrane filter technique has gained wide acceptance as the preferred technique for the presumptive determination of the presence of coliform organisms in potable water. The microbiological standard was established to ensure infectious microorganisms would not cause diseases.

WATER QUALITY STANDARDS

The idea of water quality management is to ensure that water supplied is free from pathogenic organisms, clear, potable and free from undesirable taste and odour, of reasonable temperature, either corrosive nor scale forming and free from minerals which could produce undesirable physiological effect.

The establishment of minimum standards of quality for public water supply is of fundamental importance in achieving this ideal. Standards of quality from the yardstick with which the quality control of any public water supply has to be assessed. In India, certain minimum standards have already been prescribed and given here.

STANDARDS FOR POTABLE AND SAFE WATER

National Drinking Water Mission is an institution which is the in-charge to define the standards for potable water with respects to their locations. It is defined as the water that is free from pathogenic micro-organisms, poisonous substances, excessive amounts of minerals and organic matter which would produce undesirable physiological effects. It should be free from colour, turbidity, taste and odour, of moderate temperature and aerated.

The physical and chemical quality of water should not excess the limits shown in the table below:

Table. Physical and Chemical Standards

Sl. No.	Characteristics	* Acceptable	** Cause for Rejection
1.	Turbidity (Units on JTU scale)	2.5	10
2.	Colour (units on platinum-cobalt scale)	5.0	25
3.	Taste and odour	Unobjectionable	Unobjectionable
4.	pH	7.0 to 8.5	6.5 to 9.2
5.	Total dissolved solids	500	1500
6.	Total hardness (mg/l as $CaCO_3$)	200	600
7.	Chlorides (mg/l as Cl)	200	1000
8.	Sulphates (as SO_4, mg/l)	200	1000
9.	Fluorides (as F, mg/l)	1.0	1.5
10.	Nitrates (as NO_3, mg/l)	45	100
11.	Calcium (as Ca, mg/l)	75	200
12.	Magnesium (as Mg, mg/l)	30	150
13.	Iron (as Fe, mg/l)	0.1	1.0
14.	Manganese (as Mn, mg/l)	0.05	0.5

Notes:

* The figure indicated under the column acceptable are the limits upto which the water is generally acceptable to the customers.

** Figures in excess of those mentioned under acceptable render the water non-acceptable but still may be tolerated in the absence of alternative and better source but upto the limits indicated under column cause for rejection above which the supply will have to be rejected.

Bacteriological Standards

- Coliform count in any sample of 100 ml should be zero. A sample of water entering the distribution system that does not conform to this standard calls an immediate investigation into both efficacy of the purification and the method of sampling.
- *Water in the distribution system shall satisfy all the three criteria indicated below:*
  - E,Coli count in 100 ml of any sample should be zero
  - Coliform organisms not more that 10 per 100 ml shall be present in any sample.
  - Coliform organisms should not be detectable in 100 ml of any

two consecutive samples or more than 50% of the samples collected for the year

The World Health Organization (WHO) is a specialized agency of the United Nations (UN) that acts as a coordinating authority on international public health. Established on 7 April 1948, and headquartered in Geneva, Switzerland, the agency inherited the mandate and resources of its predecessor, the Health Organization, which had been an agency of the League of Nations. Guideline Values for chemicals that are health significance in drinking–water.

Table. WHO Guidelines (Values for Health Related Organic Contaminants)

Chemicals	Guideline Values (mg/litre)
Arsenic	0.01
Fluoride	1.5
Manganese	0.4
Nitrate	50
Nitrite	3
Chlorine	5
Copper	2000
Lead	10
Nickel	20

These Guidelines provide a generally applicable approach to drinking–water safety. Their application to drinking-water supply through piped distribution and through community supplies are described. In applying the Guidelines in specific circumstances, additional factors are also important.

Chapter 5

Oceans and Saltwater

BIOLOGY OF THE OCEANS

All organisms that live in the ocean are subject to the physical factors of the underwater environment. Some of the more important factors that affect marine (ocean) organisms are light levels, nutrients (chemicals required for growth), temperature, salinity (concentration of salt in the water), and pressure. In general, conditions in the ocean are more stable than those on land.

LIGHT

The amount of light in a certain location controls the growth of the single–celled marine algae called phytoplankton. Phytoplankton are the base of the marine food chain, meaning they are the food for other organisms, who then are the food for higher organisms and so forth. These plants convert sunlight and water into the carbohydrates (sugars) they feed on in a process called photosynthesis. Unlike land, where plants generally live on surfaces, in the ocean, light travels through the water allowing phytoplankton to grow over a vertical distance of nearly 500 feet in some locations (about 150 meters). Light intensity decreases with depth in the ocean and some wavelengths (a property of light that determines whether the light is blue, red, ultraviolet, etc.) of light disappear more quickly than others. Blue light extends the deepest into the ocean, while red light is only present near the surface.

Many factors influence how quickly light disappears. Near the coast, there may be high amounts of sediments (particles of sand, gravel, and silt) in the water and light may only extend 50 feet (15 meters). In the open ocean, the water is particularly clear and light may extend down 500 feet (150 meters). Because the phytoplankton must live where there is light, their predators live there as well. Many fish, crustaceans (aquatic animals with no backbone and a hard shell), and mollusks (soft-bodied animals without a backbone enclosed in a shell) are found in surface waters where phytoplankton grow. Many of these species use light to hunt prey and avoid predators. Light is also important in the life cycles of many fish and invertebrates (animals without a backbone) who use the changing length of the day as a calendar to trigger breeding periods.

NUTRIENTS

Just as people need nutrients to grow, phytoplankton in the ocean also need nutrients. Nutrients are substances that are required for an organism to grow. Phytoplankton absorb nutrients that are dissolved in the water around them. Some nutrients are abundant in ocean water: carbon, oxygen, and sulfur. Others are relatively scarce and become even scarcer when phytoplankton are growing rapidly. The two most important of these are nitrate and phosphate. Some phytoplankton, like cocolithophores, develop shells made out of calcium carbonate (the same material as mollusk shells). For these organisms, calcium is a nutrient that is often in short supply. Other phytoplankton, like diatoms, produce shells made out of silica, which is also a nutrient that can be scarce in ocean water. Some other nutrients that are needed in very small quantities are iron, copper, magnesium, and zinc.

TEMPERATURE

With the exception of the hydrothermal vents (natural springs that vent warm or hot water on the seafloor) in the deep ocean, the range of temperatures in the ocean is much narrower than that found on land. Land temperatures vary from above 120°F (49°C) to well below freezing. Ocean water is rarely above 80°F (27°C) in tropical waters and never below 30°F (-1.9°C), as ocean water freezes at that temperature. Common water temperatures in temperate waters (waters that are not exposed to extremely cold or hot climates) are around 60°F (16°C). Most animals-including most fish-that live in the ocean are ectothermic, which means that their body temperature is close to that of the water in which they live.

The root word ecto means "outside" and the root word therm means "temperature." For these animals, their metabolic rate (the rate at which biochemical processes occur in an organism) is linked to the temperature of water in which the animal lives. For example, if two fish of exactly the same species and size are put in two aquariums, but one is three degrees warmer than the other, the fish in the warmer aquarium will eat more, have a faster heart rate, and swim faster.

Some fish and aquatic insects are endotherms (or, like tuna, are ectotherms that act like endotherms). Endothermic animals generate their own body heat via their metabolism (chemical reactions within the body). The root word endo means "internal." Endotherms can survive in environments that have a very large temperature range. This is demonstrated by whales that migrate from tropical regions to the frigid Arctic waters. However, endotherms require large amounts of food to provide the energy they need to keep their bodies warm.

Salinity

Salinity is the amount of salt found in one kilogram of ocean water. The average salinity in the ocean is 35 parts per thousand (ppt). This means that there are 35 grams of salt per kilogram, or 1,000 grams, of ocean water. In places where rivers flow into the ocean or where there is a lot of run-off from rain, salinity can drop to 6 ppt.

In places that receive little fresh water, such as the Red Sea, salinity can be greater than 40 ppt. Most marine invertebrates have salinities within their bodies that are very similar to the salinity of the water around them. If the salinity of the water suddenly changes, then it can harm the animals by interrupting its natural osmosis. Osmosis is the tendency for the concentration of water to always be the same on both sides of a semipermeable barrier. (A semipermeable barrier allows some materials to pass through in both directions.)

The cell barrier of an animal is semipermeable and water can easily flow through it. A change in salinity on the outside of cells will affect the concentration of the water inside cells. For example, if the seawater surrounding a squid suddenly becomes fresher, then osmosis will move water inside squid's cells. If this happens too quickly, the squid's cells can burst. On the other hand, if the squid suddenly moves to an area of high salinity, the water from inside the squid's cells will flow out into the environment. The cells will shrink and the squid could die.

Pressure

At sea level, the pressure is 14.7 pounds per square inch (1 kilogram per square centimeter) or 1 atmosphere. This pressure results from the weight of the atmosphere (mass of air around Earth) pressing down on Earth. Most organisms on land do not notice this pressure since their bodies are built to push upwards with the same force. Water, however, is much heavier than air. For every 33 feet (10 meters) an organism descends in the ocean, an additional atmosphere of pressure is added. Descending to great depths in the ocean is difficult for mammals because of the gasfilled spaces in their bodies. Sinuses and lungs, in particular, are filled with air that has a pressure of 1 atmosphere.

These parts of the body collapse when the external pressure becomes too great. Humans can only descend to about 3 or 4 atmospheres (100-130 feet; 30-40 meters). Some whales, like the sperm whale, are able to descend to depths of 7,380 feet (2,250 meters). This is equivalent to a pressure of more than 223 atmospheres. Most marine animals, like invertebrates and fish, avoid pressure problems by having internal pressures that are the same as those in their surrounding ocean environment. As a result they can move vertically in the ocean without much effect on their bodies.

COASTLINES

Coastlines are boundaries between land and water that surround Earth's continents and islands. Scientists define the coast, or coastal zone, as a broad swath (belt) of land and sea where fresh water mixes with salt water. Land and sea processes work together to shape features along coastlines. Freshwater lakes do not technically have coastal zones, but many of the processes and features found along ocean coastlines also exist in large lakes.

COASTAL ZONE FEATURES

All coastlines include a thin strip of land that is submerged at high tide and exposed at low tide, called the shoreline. The coastal zone, however, extends far inland from the shore, across lowlands called coastal plains, and far seaward to the water depth where ocean waves do not reach the seafloor. The coastal zone includes lagoons, beaches, estuaries, tidal wetlands, tidal inlets, river deltas, barrier bars and islands, sand bars, and other shallow–water ocean features.

- *Lagoons*: Shallow, salt–water bays between barrier islands and the mainland.
- *Beaches*: Sand deposits along shorelines. Intense waves wash fine-grained mud from coastal sediments (particles of sand, gravel, and silt) leaving only sand-sized grains of resistant minerals like quartz and calcium carbonate. Beaches are common on the seaward side of barrier islands where wave energy is intense.
- *Estuaries*: The mouths of rivers and streams that receive a pulse of saltwater with the tides.
- *Tidal wetlands (flats)*: The broad areas of marshy wetlands around lagoons and estuaries that flood with salt water during high tides.
- *Tidal inlets*: Openings through which water and sediment are washed in and out of lagoons by daily tides.
- *Deltas*: Deposits of sediments at the mouths (ends) of rivers that flow into the ocean.
- *Barrier bars and islands*: Long mounds, or bars, parallel to the shore into which near-shore ocean currents carry and deposit sand. Eventually, some barrier bars grow tall enough to stay exposed at high tide and become barrier islands. The outer banks of North Carolina as well as Galveston, Mustang Island, and South Padre Island in Texas are examples of barrier islands.
- *Sand bar*: A ridge of sand in rivers or along the coast built up by water currents.

PROCESSES THAT SHAPE COASTLINES

The coastal zone is constantly changing. Salt water rushes through tidal inlets into bays and estuaries twice daily. Waves, currents (steady flows of water in a prevailing direction), tides, and storms reshape coastal features over days, weeks, and months. Coastlines move landward and seaward as global sealevel rises and falls over hundreds and thousands of years. All coastlines are at least somewhat affected by waves and tides. Waves straighten uneven shorelines by eroding (wearing away) points that extend into the ocean and depositing sediment in bays. They also generate strong, shallow currents that carry and deposit sediment parallel to the shore. Long shoreparallel features like barrier islands, spits (small strips of land that jut out into the sea), and sand bars border coasts where waves are the dominant force. Tides move sediment and water in and out across the shoreline, and tide–dominated coasts have features like tidal inlets, natural jetties (protective rock barriers), and funnel-shaped estuaries that form a 90° right angle to the shore. Most coastlines are shaped by both waves and tides, and have some parallel and perpendicular features.

TYPES OF COASTLINES

All coastlines are affected by waves, tides, storms, and currents, and every coast includes a shoreline. There are, however, many different types of coastlines. Some coastlines receive large amounts of sand and mud from rivers. Others accumulate the skeletal remains of animals like corals and shellfish. In some places, waves are eroding coastlines that are rising from the sea.

Depositional Coastlines

Coasts that receive a steady supply of sediment are called depositional coastlines. Rivers like the Mississippi in the United States and the Nile in Egypt erode sediment from continental interiors and deposit it in huge deltas at their mouths. Waves, currents, and tides spread sediment into thin layers on the submerged continental shelf (the shallow seabed that stretches from the shore to the deeper ocean water). Over time, the weight of the sediment presses down on the edge of the continent, creating space for more sediment. New layers build on to the edge of the continent, and the coast moves seaward. Depositional coastlines typically encompass a broad coastal plain, and a complex shoreline that includes long, wide beaches. These coastlines are almost flat, causing salt water to move far inland across coastal plains and up rivers during high tide.

The Mid–Atlantic and Gulf of Mexico coasts of the United States are depositional coastlines. Waves, tides, and currents sort and distribute incoming sediment into distinctive features along depositional coastlines. Some of these features include: barrier bars and islands, tidal inlets, lagoons, beaches, estuaries, and tidal wetlands. Depositional coastlines also develop where plants and animals called carbonates live in clear, sunlit water away from river deltas. Carbonates like corals and shellfish have skeletons and shells made of the hard mineral calcium carbonate.

The sediment supply on carbonate coastlines comes from the skeletal remains of the animals and plants that live there. Corals build giant ridges of rocks called reefs up from the seafloor. Florida and the Bahamas have carbonate coastlines.

Erosional Coastlines

Erosional coastlines occur where huge sections of the Earth's crust called tectonic plates lift out of the sea. These coastlines are common along far northern coastlines that are bouncing back after being weighed down by thick ice sheets, and along coasts where tectonic plates meet. Erosional coastlines are the norm in Maine and eastern Canada, along the west coast of North America, and in Scandinavia. Along these coasts, waves pound rocky shorelines and cut into the bottoms of cliffs. The shore retreats as blocks of rock and sediment fall into the sea. Isolated remnants of sea cliffs, called stacks, are left standing in the sea. All of these features are caused by movements of plates and wave action over time. The process of shoreline retreat claims much expensive real estate along erosional coastlines.

LIFE IN THE COASTAL ZONE

The plants and animals that live in the coastal zone have adapted to its cycles of change. Residents of tidal wetlands tolerate twice–daily drowning and drying. Fish in estuaries and lagoons adjust to large changes in water salinity (saltiness). Plants grow with their roots in salt water and can survive burial by shifting beach sand and river mud.

Many ocean and land animals spend their early lives in coastal wetlands where there is shelter and plentiful food before moving to dry land or the open ocean as adults. Coastlines are also home to about twothirds of Earth's human population. People continue to work to understand the nature of coastal zones in order to protect coastal populations from their hazards (storms, waves, floods, erosion), but also to protect coastlines from the damaging effects of everyday human activities, such as eroding sand dunes by climbing them, or generating pollution.

CURRENTS AND CIRCULATION PATTERNS IN THE OCEANS

The oceans are in constant motion. Ocean currents are the horizontal and vertical circulation of ocean waters that produce a steady flow of water in a prevailing direction. Currents of ocean water distribute heat around the globe and help regulate Earth's climate, even on land. Currents carry and recycle nutrients that nourish marine (ocean) and coastal plants and animals. Human navigators depend on currents to carry their ships across the oceans. Winds drive currents of surface water. Differences in temperature and salinity (saltiness) cause water to circulate in the deep ocean.

The rotation of the Earth, the shape of the seafloor, and the shapes of coastlines also determine the complex pattern of surface and deep ocean currents. Ocean water is layered. The shallowest water, called surface water, is warmer, fresher, and lighter than deep water, which is colder, saltier, and denser. The boundary between surface and deep water is a thin layer marked by an abrupt change of temperature and salinity. This layer, called the thermocline, exists in most places in the oceans. Surface and deep water only mix in regions where specific conditions allow deep water to rise or surface water to sink.

Many organisms swim freely across the thermocline, and the remains of plants and animals continuously rain down through the deep water to the seafloor. However, most organisms live, or at least feed, close to the ocean surface where microscopic plants called phytoplankton float freely and absorb the sunlight they need to live. Very little light penetrates the surface water.

SURFACE CURRENTS

Earth's atmosphere (mass of air surrounding Earth) and oceans together form a "coupled system." Winds drive circulation of the oceans' thin upper layer of surface water, and temperature differences in the oceans help to generate atmospheric winds. Friction (resistance to the motion of one surface over another) between the moving air and the water surface pushes water in the direction of the blowing winds. Earth's eastward rotation causes currents to deflect (bend) to the right in the northern hemisphere and to the left in the southern hemisphere, a phenomenon called the Coriolis effect. Coriolis deflection causes clockwise circulation of wind-driven surface ocean currents in the northern hemisphere and counterclockwise circulation in the southern hemisphere.

Warm Surface Currents

The subtropical trade winds, or trades, are strong, steady winds that blow warm water from west to east on either side of the equator (an imaginary line around Earth halfway between the North and South Poles), thereby creating west-flowing equatorial currents in the major oceans. The trades and equatorial currents push piles of warm water into the western halves of the Pacific, Indian, and Atlantic Oceans. Water flows down and away from the centers of the mounds, not unlike pancake batter spreading out on a griddle. The trades continue to push the water to the west, and Coriolis deflection guides it northward in the northern hemisphere and southward in the southern hemisphere. Warm, fast currents, called western boundary currents, flow away from the tropical warm pools towards the poles in the western halves of the ocean basins.

The Gulf Stream is the western boundary current in the North Atlantic. It flows north along the southeastern coast of the United States and then crosses the Atlantic on a diagonal path. Warm Gulf Stream waters create unusually mild climates in northern locations. Gulf Stream waters keep Bermuda balmy, Ireland green, and England foggy. The Gulf Stream is the major shipping route from North America to Europe. The western boundary current in the North Pacific, called Kuroshio, likewise warms the islands of Japan and carries ships towards the Pacific Northwest. Western boundary currents in the southern hemisphere, the Brazil Current in the South Atlantic, East Australian Current in the South Pacific, and Aguellas Current in the Indian Ocean flow south from the equator.

Cool Surface Currents

Cold surface currents carry cool water from the poles towards the equator along the west coasts of the continents. The cool eastern boundary currents are generally shallower and weaker than the western boundary currents. Cold water flowing from the Arctic Ocean at the North Pole feeds the eastern boundary currents in the northern hemisphere, namely the California Current in the Pacific and the Canary Current in the Atlantic.

The Antarctic Circumpolar Current (ACC) that encircles the ice-covered Antarctic continent supplies the cool water to the eastern boundary currents of the southern hemisphere-the Peru, Benguela, and West Antarctic Currents. The ACC is an exception to the general rule that surface currents are shallow. It extends from the sea surface to the seafloor in several places.

Gyres

Surface water circulates in oceans in massive circular patterns called gyres. The major surface currents (eastern boundary, western boundary, and equatorial current) in each ocean link to form a circle. Gyres are clockwise in the northern hemisphere and counterclockwise in the southern hemisphere. For example, a rubber duck dropped into the ocean near San Diego might float south on the California Current to the North Equatorial Current, west across the Pacific Ocean to the Kuroshio western boundary current, and then back across the northern Pacific to British Columbia.

In a few years, the California Current might return the duck to San Diego. Ocean researches have actually conducted many such experiments. One important study tracked 29,000 plastic bathtub toys that spilled from a cargo ship in the North Pacific. The infamous "triangle trade" between Europe, Africa, and North America in the eighteenth and nineteenth centuries relied on the North Atlantic Gyre.

European slave ships arrived in Africa via the Canary Current, then carried slaves to the sugar plantations of the Caribbean on the North Equatorial Current. Having left off slaves and picked up sugar, they rode the Gulf Stream north to the rum distilleries of New England and the liquor shops in Europe. Hurricanes that form in the tropical Atlantic ride the North Equatorial Current towards the Caribbean Islands and then follow the Gulf Stream towards the southeast coast of the United States.

DEEP OCEAN CURRENTS

Deep ocean currents are driven by differences in temperature and salinity. They are generally unaffected by surface currents. Deep water is colder, saltier, and denser than surface water. Deep water forms in polar regions where warmer surface water cools and sinks beneath the Arctic ice cap (permanent ice covering) or Antarctic ice shelves (permanent ice large enough to cover most of a land mass). Salinity increases near the ice caps because seawater forms freshwater ice when it freezes. The salt stays behind and the remaining liquid water becomes saltier. A "global conveyor belt" carries deep water south through the Atlantic, around Antarctica, and north into the Pacific, Indian, and Atlantic Oceans. It could take the molecules in a drop of water more than a thousand years to make a complete circuit of this global deep ocean current.

UPWELLINGS AND DOWNWELLINGS

Deep water rises to become surface water at upwellings. Upwellings are most common along coastlines where strong winds blow away from shore, but they also occur in the open ocean where winds blow away from one another. In both cases, winds push the warm surface water away and cold, nutrientrich deep water rises to the sea surface to replace it. Upwellings are common along the west coasts of the continents, particularly in regions beneath the easterly (west-blowing) trade winds. Because they bring important minerals and nutrients from the deep ocean, upwellings typically support abundant marine and coastal life. Upwellings nourish waters rich with life off Peru, California, and southwestern Africa.

A divergence between wind patterns creates a zone of intense upwelling that completely surrounds Antarctica. Downwellings are ocean zones where surface water sinks into the deep ocean. Downwellings can occur at places were winds meet or blow towards shore. However, warm water does not sink, and warm surface currents are more likely to pile water up against obstacles like coastlines and opposing currents than to force it into the deep ocean. Most deep water forms at intense Arctic and Antarctic downwellings where ice cools the seawater and freezing increases its salinity.

EL NIÑO AND LA NIÑA

El Niño and La Niña are changes in the winds and ocean currents of the tropical Pacific Ocean that have far-reaching effects on global weather patterns. Together, El Niño and La Niña are extremes that make up a cycle called the El Niño Southern Oscillation (ENSO). An oscillation is a repeated movement or time period. El Niño and La Niña events do not occur in a regular or seasonal pattern; instead, they repeat about every two to seven years and last for a few months.

HOW EL NIÑO AND LA NIÑA OCCUR

El Niño events occur when the trade winds and equatorial current south of the equator in the Pacific Ocean lessen in intensity. The trade winds, or trades, are strong, steady winds that blow from east to west and drive strong west–flowing ocean currents on either side of the equator. (The trade winds are named for their role in propelling sailing ships carrying cargo to trade around the world.) The equatorial current is a sustained pattern of water flowing westward near the equator. Less dramatic La Niña episodes occur during the opposite conditions, when the tropical winds and currents are unusually strong.

During normal, non–El Niño conditions, the trade winds and equatorial current in the southern Pacific push warm surface water to the west and allow cold water from the deep ocean to rise along the coast of South America. The southeasterly (northwest–blowing) trades south of the equator usually pile a mound of warm water around the islands of Indonesia, and create a zone of cool water that rises called an upwelling off the coasts of Peru and Ecuador. The cold, nutrient-rich waters of the South American upwelling nourish abundant microscopic plants (phytoplankton) and animals (zooplankton) that provide food for larger sea animals.

It is a biologically rich region for fish and land animals, including humans who depend on fish for food. The pool of warm water in the western Pacific creates a warm, rainy climate, and the cold water of the upwelling causes an arid (extremely dry) climate in coastal South America. Occasionally, for reasons not yet fully understood, the trade winds and southern equatorial current in the south Pacific lessen in strength. Warm water sloshes east towards the central coast of South America and shuts down the South American upwelling. The El Niño phase of an ENSO cycle begins with a dramatic warming of the waters off of South America and a decline of marine (ocean) life. La Niña, the opposite phase of an ENSO cycle, occurs when the southeast trades are particularly strong. La Niña events are marked by a strengthening of the South American upwelling and a good fishing season. La Niña events often, but not always, follow El Niño events.

DISCOVERY OF EL NIÑO AND LA NIÑA

Peruvian fishermen who depended on the South American upwelling for their livelihoods recognized and named the El Niño phenomenon in the nineteenth century. The fishermen noticed that every few years, the seawater became much warmer and the pattern of ocean currents would change within about a month of Christmas day. These changes always marked the start of a very poor fishing season. Normally dry areas along the coast would receive abundant rain. As this typically happened close to Christmas, the fishermen dubbed the phenomenon El Niño, Spanish for "the boy child," after the Christ child. The other half of the ENSO cycle was named La Niña, "the girl child," much later. El Niño has been a well-known local occurrence in coastal South America for more than 150 years.

However, scientists only began to realise that the strong El Niño events were part of a disruption that effected the entire Pacific Ocean in the late 1960s. The effects of the southern oscillation were first recognized (and named) in the western Pacific by Sir Gilbert Walker in 1923. Walker was a British scientist who studied the changes in the summer monsoons (rainy seasons) of India. Using meteorological (weather-related) data, he observed that atmospheric pressure (pressure exerted by the air) seesaws back and forth from the Indian Ocean near northern Australia, to the southwestern Pacific near the island of Tahiti. Walker also noticed that the changes in pressure patterns were related to changes in the weather that affected rainfall, fishing, and agricultural harvests in Southeast Asia and India. In the late 1960s, Jacob Bjerknes, a professor at the University of California, first proposed that the Southern Oscillation and the strong El Niño sea warming were related.

EFFECTS OF EL NIÑO AND LA NIÑA

The effects of El Niño on the climate of the tropical Pacific are now well known. As the mound of warm water in the western Pacific collapses and spreads eastward, the area of heavy rain above it shifts to the east. Fewer rain clouds form over the Pacific Islands, Australia, and Southeast Asia. Lush, biologically diverse rain forests dry out and become fuel for forest fires. Usually arid islands in the central Pacific receive heavy rainfall. In the eastern Pacific, the ocean upwelling weakens as the warm surface water flows towards South America. The surface water off Ecuador and Peru runs low on the nutrients that support the ocean food chain.

Many species of fish and birds go elsewhere to find food, and human fishermen face economic hardship. The warmer waters offshore also encourage development of clouds and thunderstorms. Normally dry areas along the west coast of South America experience torrential rains, flooding, and mud slides during the El Niño years. La Niña events are usually less dramatic, but typically cause an opposite effect on the climate (long-term temperature, rainfall, and wind conditions) of the southern Pacific. El Niño and La Niña also seem to cause far-reaching changes in the weather and climate in other parts of the world. The altered pattern of winds and temperatures in the tropical Pacific may change the paths of the jet streams (high-level winds) that steer storms across North and South America, Africa, Asia, and Europe. El Niños have been linked to mild, wet winters along the west coast of North America, strong storms in the Gulf of Mexico, heavy rains in the American Southwest, and droughts (lack of rain) in Central America and northern South America.

In the El Niño years of 1986–87 and 1997–98, California and Chile both experienced torrential rainstorms and heavy snows that led to mudslides. El Niño may also affect the Indian monsoons and bring drought to northern Africa, thereby threatening agricultural harvests in India, Asia, and Africa. During La Niña episodes like 1998–99, the northern part of the United States may experience heavy snows, increased rainfall, and cold temperatures, while tornado activity increases in the southern states.

The far reaching climatic and economic effects of El Niño and La Niña make understanding ENSO a priority for scientists. Improved understanding and forecasting will help populations plan for the effects of El Niños and limit economic suffering and starvation. While El Niño and La Niña do have far-reaching effects, scientists are also careful not to blame all extreme or abnormal weather on the phenomenon, or to draw too many connections between Niño and global climate variations.

FISH (SALTWATER)

There are over thirty thousand different species of fish, and they are the most numerous vertebrates. Vertebrates are animals that have a bony spine that contains a nerve (spinal) chord. Vertebrates usually have an internal skeleton that provides support and protection for internal organs. This spine and skeleton allow vertebrates to move quickly and to have great strength. Fish usually live surrounded entirely by water. They all have gills for breathing and fins for swimming. Most fish are ectotherms, which means that their bodies are nearly the same temperature as the water in which they live. About 60% of all fish live entirely in saltwater, while the rest live in freshwater or both freshwater and saltwater.

One of the most remarkable things about fish is their diversity. Fish can be very large, like the whale sharks that can reach 90,000 pounds (41,000 kilograms) or very small, like gobies that can weigh as little as 0.0004 ounce (0.1 gram). They have a variety of diets, including plants, other fish, invertebrates (animals without a backbone) and microscopic plankton (freefloating plants and animals). Fish find their food in many different ways, including hunting with their eyes, grazing, scraping the sea floor, and digging. Some fish have special organs that bioluminesce (create light) and lure prey towards their mouths, while others use this glow to make themselves blend in with coral or other ocean features, disguising themselves from predators. Fish come in a variety of colours that can serve to scare predators away or blend into their environment. Some fish live their entire lives within one bay or cove, while others may migrate thousands of miles (kilometers) across the ocean. Fish can live alone or they may swim with many other fish, called schools, to protect against predators. Fish are classified into three groups. The jawless fish belong to the class Agnatha. They include hagfish and lampreys. The rays and sharks belong to the class Chondrichthyes. They have skeletons that are made of a tough material called cartilage. The bony fish belong to the class Osteichthyes. This largest group includes many familiar fish like tuna, halibut, anchovy, and cod.

CLASS AGNATHA

There are about fifty species of Agnathans and they are divided into two groups: hagfish and lampreys. These fish have no jaws and their fins are not evenly matched across their bodies, so they are not efficient swimmers. They have mouths that look like suckers with small teeth that are used for grasping on to prey. Organs for smelling and sensing surround their mouths and help them identify prey. These fish have very poor vision. Hagfish live in colonies (groups) on the sea floor. They dig in sediments (sand, gravel, and silt) for worms to eat. If approached by predators, hagfish emit large quantities of foulsmelling slime from glands along the sides of their bodies. This usually discourages or confuses predators.

After the danger has passed, the hagfish will tie itself in a knot, which it slides along the length of its body to scrape off the slime. The mouth of the lamprey is called an oral disc. It is cone-shaped and contains sharp teeth that it uses to bore a hole into the side of another animal. The lamprey then attaches its oral disk to the live animal's wound and feeds off the blood and tissue of its host. Lampreys usually detach after some period of time without killing their host. Lampreys are usually most often found attached to bony fish, but they also have been seen on whales and dolphins.

CLASS CHONDRICHTHYES

The class Chondrichthyes includes about 700 species of sharks and rays. They are an extremely old group, having been in existence for approximately 280 million years. Nearly all members of this class are marine (live in seawater). These fish have skeletons made out of cartilage, which is a tough but flexible tissue. It is the same material that is found in human ears and noses. Sharks and rays do not have gas bladders (internal sacs that fill with gas to help the fish rise or fall in the water so that it does not waste energy by continually swimming). Sharks and rays must continually swim in order to prevent themselves from sinking.

However, the liver of sharks is large and it contains a lot of oily materials. As oil is less dense than water, this special liver helps the shark stay afloat. Cartilaginous fish have several rows of teeth that fall out as they age. They are then replaced with new teeth that grow in from behind. Sharks do not have scales; instead they have rough plates called dentricles embedded in their skin. These dentricles make the skin feel abrasive, like sandpaper. Many sharks have electroreceptors on their heads. These specialized organs allow the shark to sense the electrical currents generated by fish as they swim through the water.

The shark's well–developed nervous system, including a large brain, also helps it locate its prey (animals that are food). Rays have a more flattened shape than sharks. Their fins are attached to their bodies so that they look like triangular or semicircular wings. Large rays, like the manta ray, can measure 22 feet (7 meters) from fin to fin. These huge animals feed on plankton. Other rays, like the stingray, have sharp barbs attached to the base of their tail. These are used as defence against predators. Another family of rays can actually produce an electric current, which they can use to stun prey.

CLASS OSTEICHTHYES

The class Osteichthyes, or the bony fish, make up the majority of fish, with almost 28,000 different species. These fish all have a strong, but lightweight, skeleton that supports their organs. They have gas bladders that help them maintain buoyancy. The teeth of bony fish are fused to their jawbone and do not fall out as do the teeth of the cartilinagous fish. Osteichthyes are found in every type of marine environment from near-shore tidepools and coral reefs to the very bottom of the deep ocean. Nearly 90% of all the bony fish are categorized into one order, Teleostei.

These fish include many common fish, like the cod, tuna, seabass, and perch. Teleostei also includes unusual fish like the mola, which floats near the surface of the ocean in warm currents; the angler fish, which lives on the seafloor and lures its prey using a worm-shaped appendage; and the football fish, which permanently fuses with its mate.

Movement

Many teleost fish have bodies that are shaped to allow them to move easily through water. In particular, fast or constantly swimming fish have body shapes that minimize drag, or resistance to movement. The less surface area comes into contact with water in the forward direction, the less drag the fish will have. A torpedo shape, with a body that tapers towards the rear, is one of the most effective shapes for minimizing drag, and many fast–swimming fish have this sort of shape. An example of a fish with a shape that minimizes drag is the swordfish, which can reach speeds up to 75 miles per hour (120 kilometers per hour) in short bursts. Some fish, like eels, wave their entire bodies back and forth in an "S" shape to move through the water. This is not a very efficient way of swimming as it requires a lot of energy and it increases the surface that confronts the water as the fish swims forward. More advanced swimmers have a stiff body, with a tail that bends back and forth behind the fish. This allows the fish to conserve energy because it only moves its tail, not its entire body. It also minimizes the area that confronts the water to the head of the fish.

Water and Gas Balance

Just like humans, metabolism (the process of cells burning food to produce energy) in fish requires oxygen and produces carbon dioxide as a waste product. Fish breathe through gills, which are found underneath flaps on both sides of the head. Water containing dissolved oxygen is brought in through the mouth and pumped over the gills. The gills are packed with blood vessels that absorb the oxygen from the water and produce the carbon dioxide waste that is generated within the fish. The body fluids of saltwater teleost fish are about one-third as salty as ocean water.

Another way of thinking about this is that the concentration of water inside these fish is greater than the concentration of water in the ocean. Because of osmosis (the passage of a liquid from a weak solution to a more con centrate solution), the water from the inside of the fish is constantly diffusing (moving outward) from the fish. As a result, saltwater fish must regularly drink seawater to replenish the water that is lost by osmosis. The fish have special salt glands in their gills that remove the excess salt that comes inside the fish with the seawater.

LATERAL LINE

Teleost fish have developed an interesting sensory organ that helps to sense vibrations in the water. Along both sides of most teleost fish is a long row of canals (tubes) that are packed with nerve cells. These nerves detect movements of currents, changes in water pressure, and even noises.

GEOLOGY OF THE OCEAN FLOOR

Geology is the study of the solid Earth and its history. Marine geology is the study of the solid rock and basins that contain the oceans. The rocks and sediments (particles of sand, gravel, and silt) that lie beneath the oceans contain a record book of Earth's past. Topographic features (the physical features of the surface of Earth) and geologic processes in the ocean basins hold the keys to plate tectonics, a fundamental theory of geology that explains the movement of the continents and seafloor over time. (A plate is a rigid layer of Earth's crust and tectonics is the large scale movements of the crust.)

Only when scientists began to successfully probe the secrets of the seafloor in the mid-twentieth century did they begin to understand the complex workings of the solid Earth. If marine geology is the study of the ocean soup bowl, then oceanography is the study of the broth, and marine biology is the study of the vegetables and meat in the broth. These three branches of ocean science are closely linked. The mountains and valleys of the seafloor, together with the continental margins (edges), act to guide ocean currents (a steady flow of water in a prevailing direction) that in turn regulate global climate. Moving water shapes the seafloor by eroding (wearing away) and depositing sediments (particles of gravel, sand, and silt).

The seafloor provides shelter and nutrients (food) for marine (ocean) plants and animals, and living organisms play a role in shaping the seafloor. They burrow (dig holes and tunnels), build shelters, and consume nutrients during their lifetimes, and their remains form layers of sediment on the seafloor. Practical knowledge of seafloor topography (called bathymetry) and marine geology is essential for human navigators, coastal and marine engineers, naval tacticians, as well as petroleum (oil and gas) and mineral prospectors.

DEPTH AND SHAPE OF THE SEAFLOOR

Bathymetry beyond shallow coastal waters was a complete mystery until the middle of the 1800s. Until then, navigators used relatively short ropes and chains to make water depth measurements, called soundings, and to construct charts of shallow coastal waters where seafloor topography is a shipping hazard. By the 1860s, however, advances in science had raised a number of intriguing questions about the nature of the deep ocean floor. Royal Society of London naturalists aboard the ship Challenger used newly developed steel cables to take more than 500 soundings and to dredge 133 rock and sediment samples from the deep ocean during their expedition from 1872 to 1876.

The Challenger scientists discovered that the oceans are very deep. They took their deepest sounding in an ocean trench near the Mariana Islands in the western Pacific Ocean. Today, it is known that Earth's lowest point, the Challenger Deep in the Mariana Trench, is 36,201 feet (11,033 meters) below sea level. In comparison, Earth's highest point, the peak of Mt. Everest in the Himalayan Mountains, is a mere 29,035 feet (8,850 meters) above sea level! The average water depth in the main oceans is 12,200 feet (3,729 meters), deeper than the highest points in 38 of the 50 states of the union. Dredge samples from the Challenger expedition showed that ocean rocks and sediments are fundamentally different from those found on land. In spite of the tantalizing clues turned up by nineteenth century British scientists, a full picture of the ocean basins did not come into focus until the late 1950s. In the early twentieth century, the spirit of scientific inquiry during the Challenger era was replaced by more practical reasons to map the seafloor-naval warfare during the first and second world wars. A new technique, called sonar echosounding, replaced expensive, relatively inaccurate wire soundings. An echosounder works by bouncing a sound wave off the seafloor. Sound travels at a constant velocity (speed) in water, so the time it takes for the sound to travel through the water and echo back to the ship gives the distance to the seafloor.

The faster the sound returns, the shallower the water. American ships carrying troops and supplies to Europe and Asia carried echosounders that recorded the water depths along their routes. Civilian scientists were intrigued by what they saw on the wartime bathymetric profiles, and they set out to survey the seafloor using the new, accurate, inexpensive echosounders.

The first complete maps of the Pacific, Atlantic, Indian, and Arctic Ocean basins were compiled by Columbia University marine geologists Bruce Heezen and Marie Tharp and published by the National Geographic Society in the mid–1950s. The features that were clearly visible on these bathymetric maps, globe-encircling chains of underwater volcanoes and deep ocean trenches, led to a revolution in marine geology and the theory of plate tectonics during the 1960s and 1970s.

Sea Floor Features

A bathymetric profile (cross-section) of a major ocean basin like the Pacific Ocean shows the typical features of the seafloor: continental shelf, continental slope and rise, mid–oceanic ridge, ocean trench, and abyssal plain.

- *Continental shelf:* Continental shelves are the relatively shallow, submerged margins of the continents. Some shelves, like the east coasts of North and South America, are very wide. Others, like the west coasts of North and South America, are very narrow. Over geologic time, the shorelines on continental shelves retreat and advance as the ice in the North and South Poles grow and shrink and global sea-level rises and falls.
- *Continental slope and rise*: The continental slope is the steep transition from the continental shelf to the floor of the abyssal (deep) ocean. The slope is cut by huge canyons that carry underwater landslides, called turbidite flows, downslope at speeds of up to 40 miles (64 kilometers) per hour. The continental rise is the deposit of sediments at the base of the continental slope.
- *Mid-oceanic ridge*: The most striking feature of Heezen and Tharp's bathymetric map was the mid–oceanic ridge system, a continuous chain of low, symmetrical volcanoes that extends through all the ocean basins. A mid-oceanic ridge, like the Mid-Atlantic Ridge between South America and Africa, is a broad uplift with a small

valley at its axis (center). Mild volcanic eruptions fill the ridge axis valley with molten lava that cools to become new seafloor. The ocean basins on either side of a mid-oceanic ridge are symmetrical mirror images that are moving away from the ridge axis over time.

- *Ocean trenches:* Trenches are deep, arc–shaped submarine valleys along the edges of the ocean basins. They are the deepest parts of Earth's oceans. Scientists now know that the moving seafloor is recycled into Earth's interior at trenches, a process called subduction. Chains of large volcanoes, called arc volcanoes, form on the outer edges of trenches. The Andes Mountains of South America and the islands of Japan are examples of arc volcanoes. Friction between rocks during subduction also causes very large earthquakes. The geologically active subduction zones that surround the Pacific Ocean are called the "ring of fire."
- *Abyssal plains*: The abyssal plains are vast, flat areas of the deep–ocean floor. In some places, small repeating sets of sharp-peaked ridges, called abyssal hills, interrupt the nearly featureless abyssal seafloor. Cross-sections through the seafloor show that abyssal hills are the tips of tilted blocks of rock beneath a blanket of deep-ocean sediment.

OCEAN ROCKS AND SEDIMENTS

The solid rock, called the basement, that acts as the floor of the deep ocean is different from that of the continents. Earth's rocky outer crust comes in two varieties, continental and oceanic, that have very different properties and compositions. Ocean crust is denser, thinner, darker-coloured, and contains more of the chemical elements iron and magnesium than continental crust. The basement of the ocean basins is mostly made of black, volcanic rock called basalt. Mid–oceanic ridge volcanoes produce basalt. The centers of the continents are composed mainly of coarse–grained, light-coloured rocks like granite.

The blanket of sediment that covers the floors of the abyssal plains is called pelagic ooze. Oozes form by the slow, steady accumulation of silica- and calcium-rich remains of microscopic animals and plants that sink to the deep seafloor. The boundary between deep ocean rock and continental land rock lies beneath massive fans of sediment that form the continental margins.

Continental margins, including the continental shelf and slope, are composed of thick stacks of layered sediment that rivers and glaciers have carried from the continental interior. Some shelves, called carbonate shelves, are composed of the calcium carbonate shells and skeletons of organisms like corals and mollusks. Along many continental margins, continental sediments gradually give way to pelagic oozes at the toe of the continental rise.

ISLANDS

Islands are land areas smaller than a continent that are completely surrounded by water. Islands range in size from islets (small islands) barely exposed at high tide, to vast landmasses almost the size of continents. Islands exist in all the ocean basins (the deep part of the ocean floor), along coastlines, and in freshwater lakes and rivers. Islands come in many sizes and shapes, but they all share the same defining characteristics. There are more similarities than differences between a huge arctic island like Greenland and a small tropical one like Guam. Islands are isolated. The water around them controls their climate and weather.

The British Isles, which include Great Britain and Ireland, have a mild climate for their northern location because they lie in the path of the warm Gulf Stream current (a current is a steady flow of water on a prevailing direction). The Galapagos Islands, located on the equator in the Pacific Ocean, are kept surprisingly cool due to the rising of cold water from the deep ocean. Islands have limited areas for catching rainwater or snow, and fresh water is generally scarce, particularly on small islands in arid (extremely dry) areas.

The majority of islands are remote and difficult for land plants and animals to reach. They often support unique ecosystems (groups of organisms that live in a particular environment) that have evolved (changed over time) with little influence from surrounding areas. Geologic (natural Earth processes) activities form islands through various ways, including by raising a piece of seafloor above the water surface, or by separating an area of land from the edge of a continent. Volcanoes and corals (small crustacean animals that live in shallow parts of the ocean, building coral reefs with their shells) both construct mounds on the seafloor that can become exposed islands. Chunks of land can be separated from continents.

When the global sea level rises or falls because of melting and freezing of ice on the North and South Poles, islands can be exposed or submerged. Islands also sink below sea level under the weight of volcanic lava flows or ice, and rise when the rock is naturally worn away from their surfaces. Many islands form by a combination of these geologic processes.

VOLCANIC ISLANDS

Many islands are the exposed peaks of active and inactive seafloor volcanoes. An active volcano is one that has erupted in the past and is likely to do so again; an inactive volcano is no longer likely to erupt. Chains of volcanic islands, including the Aleutian Islands, Japan, the Philippines, and the Solomon Islands form the "ring of fire" that surrounds the Pacific

Ocean. These island chains, called volcanic arcs, lie along plate tectonic boundaries, where one massive plate of Earth's outer layer is sinking beneath a second plate, in a process called subduction. (Plate tectonics is the theory that Earth's rocky outer layer is broken into pieces, or plates, that move over time.)

The islands of Indonesia and Borneo lie over a subduction zone in the northwestern Indian Ocean. Cuba, Hispanola, Puerto Rico, and the Lesser Antilles are volcanic arcs in the Caribbean Sea. Other types of volcanoes also build islands and seamounts. (A seamount is an underwater mountain whose peak does not extend above the water surface. Most seamounts are volcanoes that never grew tall enough to become exposed islands, but some are former islands that have been submerged by rising sealevel or their own sinking.)

Sicily in the Mediterranean Sea, Catalina Island off southern California, and the remote Kerguelan Islands in the southern Indian Ocean are also volcanic islands. Some islands are created from hot spot volcanoes. A hot spot is a stationary heat source under a moving tectonic plate. As the plate moves over the hot spot, a line of volcanoes forms above it. The Hawaiian Islands, Iceland, and more than six hundred small islands in the southwest Pacific Ocean are hot spot volcanoes.

COASTAL ISLANDS

Erosion (wearing away of material by natural forces) and deposition by nearshore currents and waves shape sediment (sand, gravel, and silt) into islands along coastlines. These processes can form barrier islands, which are long, narrow coastal islands built up parallel to the mainland. Barrier island chains border the edges of continents that receive large amounts of sediment from rivers. In the Gulf of Mexico, a strong current has built a continuous chain of barrier islands along the coasts of Texas and northern Mexico by dragging sediment away from the Mississippi River Delta.

Tides cut narrow strips of water in the land that separate the sandy barrier strip into islands and carry water into the lagoons (a shallow area of water separated from the ocean by a coral reef or sandbar) between the islands and the mainland. Barrier islands like North Carolina's Outer Banks and South Padre Island in Texas are essentially large sandbars (long strips of sand) that are constantly being reshaped by currents, waves and winds. A large hurricane can completely destroy a barrier island–houses and all-and reshape the sediment into a new island within a few days. Not all barrier islands and coastal islands are made of sand. The Florida Keys and barrier islands of northeast Australia are reefs that fringe continental shelves (the edge of land that slopes into the sea) away from major rivers where corals grow in clear waters. Long Island, Nantucket, and the other islands off southern New England are huge piles of rock fragments and boulders left by glaciers (large masses of moving ice) that retreated at the end of the last ice age about twenty thousand years ago. Some coastlines, like those of Maine, northern California, and Norway are rising from the ocean because of shifting tectonic plates on Earth. Islands off these rising coasts are fragments of sea cliffs and rock that resisted erosion towards the mainland.

BIG ISLANDS: CONTINENTAL FRAGMENTS

The movement of tectonic plates breaks the continents into pieces that move across the Earth surface. Many of Earth's largest islands are blocks of continental crust that split from the main continents. Madagascar, for example, is a large island in the Indian Ocean that separated from the east coast of Africa. Greenland, New Guinea, and Tasmania are also broken continental blocks. India was a very large island until it collided with Asia about forty million years ago.

CARBONATE ISLANDS

Some islands were at least partially constructed by animals. Many marine organisms including corals, mollusks, and gastropods have skeletons and shells composed of a mineral called calcium carbonate. Limestones and other rocks that form from the remains of these animals are called carbonates.

The soft, white beaches of some of the world's most beautiful islands—Tahiti, Hawaii, the Bahamas, Bermuda, and Bali to name a few—are composed of the carbonate remains of lagoon and reef species. Corals are carbonate organisms that build elaborate community structures, called reefs, up from the seafloor. A reef is like an apartment building where each resident builds his own unit.

When an individual animals dies, their "apartment" is vacant but still standing, and living corals continue to build the reef upward. Carbonate species generally require clear, shallow water to thrive. Coastal carbonate islands, like the Florida Keys, only develop away from river deltas (where rivers meet the sea) where the water is too muddy. Carbonate islands away from coastlines usually have a volcanic or continental structure that provides an area of shallow water for light-loving species.

The islands of the Bahamas, for example, are the exposed spines of coral reefs that cover a large block of seafloor that was uplifted by plate tectonic forces about 200 million years ago. The Bahamian platform is sinking under the weight of its surfacing limestone layers, but the corals built the reefs up to the sunlight. Volcanic islands like Hawaii and Tahiti are typically surrounded by a ring of reefs and a shallow lagoon that host numerous carbonate species. This explains, among other things, how islands of black volcanic rock can have white beaches.

ISLAND LIFE

Islands are remote. Island plants grew from seeds carried by birds. Some animals walked to islands on exposed land bridges when the sea level was lower during the ice ages, and others rode away from their home continent on a moving chunk of land. However they first arrived, island plants and animals live in closed ecosystems where they interact only with each other. (Some island ecosystems are, of course, more closed than others.

The difficulties of reaching a barrier island across a lagoon are much less than those of reaching a volcanic island at the center of a huge ocean basin.) Throughout most of Earth's history, plants and animals that lived on islands are descendants of organisms that swam, floated, or rode there in the past. Today, humans sail, fly, and take man–made bridges to almost all of Earth's islands, and they bring plants and animals with them. Islands often are homes to extremely rare species that are unique to their island or group of islands. Charles Darwin's (1809–1882) observations of the rare species of the Galapagos Islands inspired his theory of evolution in 1835. He theorized that new species evolve by combinations between individuals with different traits. Organisms with the traits best suited to its environment will flourish and reproduce, causing these traits to be continuously passed to offspring.

Individuals with traits less suited to their environment will not survive and eventually die out. On islands, the number of individuals and species is limited and changes happen more rapidly. Furthermore, because new types of species arrive infrequently, the species that evolve on an island can be totally unique. The Galapagos Islands are home to Earth's only swimming iguanas as well as huge tortoises that can live for two hundred years. Island organisms also evolve to prey on and protect themselves from only the plants and animals on their island. They have no defence mechanisms from non–native predators and little immunity to foreign diseases. When species arrive from afar, island species can suffer.

Humans have been particularly destructive to island species because they are predators themselves, and they import many non-native plants, animals, and diseases. For example, humans brought snakes to Pacific Islands like Guam and Hawaii that decimated the endemic (native) birds and mammals. Humans then imported mongooses to kill the snakes, and the mongooses proceeded to kill not only snakes, but also a large number of the remaining island species. Today, many island governments and conservation groups are attempting to restore endangered island ecosystems.

KELP AND SEAWEED

From the tiniest of bacteria to the massive blue whale, the world's oceans and freshwater support a tremendous variety of life. Often, a beachcomber will find rubbery plants washed up on the shoreline. These exotic–looking plants are seaweed. A dive below the surface of coastal waters in some areas of the world, such as California, reveals a world of towering plants that sway gently in the ocean current. These giants are one form of seaweed called kelp. Kelp make up only about 10% of all the known seaweed species.

The many varieties of seaweed present in the world's fresh-and saltwater provide a habitat and even a food source for creatures. Humans benefit from seaweed as well. For thousands of years in Far East countries like Japan, seaweed has been an important part of the diet, in the form of soup stock, seasoning, and as an integral part of sushi. In addition, seaweed is useful in the laboratory. The artificial growth surfaces used to raise bacteria rely on a seaweed component as a thickening agent, similar to that found in gelatin.

CHARACTERISTICS OF SEAWEED

The leafy–looking seaweed that grows in ocean waters is a type of algae. Other forms of seaweed look grass–like or feathery. Algae are plants; that is, they contain the chemical chlorophyll that converts energy from the Sun into food substances that the plant uses to grow. Algae range in size from microscopic single cells (the fundamental unit of all living things) to huge numbers of cells assembled together to form a much bigger organism. Seaweed is the large collection of algae cells, or macroalgae. The many types of seaweed come in all shapes, sizes, and colours. Depending on the species, shapes range from the mighty tree-like kelp to the smaller and more delicate leafy or ribbonlike seaweed varieties. Most types of seaweed are found in shallow water, from just a foot or so to depths of 100 to 200 feet below the water's surface, as it needs sunlight for growth.

Also, most seaweed is found where there are rocks, as the seaweed clings onto the rock at one end using a structure called a holdfast. Some seaweed can attach to the sandy ocean bottom using a specialized structure that appears similar to the roots of plants that grow on land. Like plants, seaweed can convert the energy from sunlight into the compounds needed for its growth. In order words, seaweed is a photosynthetic organism. Some seaweed contains the light-absorbing compound chorophyll, which gives seaweed its green Colour. Other species of seaweed contain different lightabsorbing chemicals that are coloured red, brown, blue, or gold.

While similar to land-bound plants in its light-absorbing ability, seaweed is distinct from its land cousins in other ways. It can contain a holdfast, a part that anchors it to the seafloor. Many types of seaweed also have hollow, gas–filled structures called floats that help buoy the leaves up nearer to the sunlight.

THE THREE CATEGORIES OF SEAWEED

Seaweed is often grouped into three categories based on its Colour. These groups are the brown, green, and red algae. Brown algae range in size from forms that are a few inches (centimeters) in size to the giant kelps that can reach over 150 feet (46 meters) long. This type of seaweed lives only in salt water, and cannot grow in waters where the temperature varies much. Brown algae is found in waters that stay cold all year, such as the coastal waters of Alaska, or in tropical waters that stay warm all the time. The chemicals that make up brown seaweed are useful in the manufacturing of cosmetics and some medicines. For example, kelp can be used in medicine to treat high blood pressure, thyroid problems, and even arthritis. Seaweed is also a rich source of such minerals as iodine, zinc, copper, sodium, calcium, and magnesium, which are important in a healthy diet.

Additionally, brown seaweed has been an important part of the Japanese diet for centuries; it is used in soups, as an additive to change the taste of other foods, and as a wrap for the raw fish and rice combinations known as sushi. The red algae group dominates in ecosystems such as coastal regions of California, where they can comprise 70% of all the seaweed species present. The group contains about 3,800 different species of seaweed. Because they can absorb even tiny amounts of sunlight, red algae can live deeper in the water than other kinds of seaweed. Many red algae can live at depths of 150 to 200 feet (46 to 61 meters) below the ocean surface, and some species have been found growing even 600 feet (183 meters) below the surface of the ocean. At these depths, the ocean waters are calmer, and the red algae that live there tend to have a more delicate structure. These algae are more easily broken than seaweed that grows in the churning waters nearer to the surface.

A component of red algae is also used to make the solid food (agar) that is used in laboratories to grow many types of bacteria. Green seaweed can be found in both freshwater and the ocean. These types of seaweed feed on water that contains chemicals such as nitrogen and phosphorus that flow into the water from farmer's fields, or that are in sewage. When some types of green algae are present in high concentrations, this may indicate that the water is polluted with too many of these chemicals.

KELP

Kelp are a type of brown seaweed that often appear as big leaves swaying in the underwater current. Some types of kelp can grow to be almost 100 feet (30 meters) long, and can form an underwater forest. Kelp are important to life in the sea. The thick masses of kelp that grow off the coasts of New York, California, Australia, the Arctic, and the Antarctic are home to a variety of creatures including lobsters, snails, octopuses, seahorses, starfish, fish, and seals. These sea creatures use the seaweed forests as a protective haven as well as a source of food. Thus, kelp is important in establishing and sustaining the complex ecosystems that can form. A kelp plant can grow to be dozens of feet (meters) long and grow quickly. It is anchored to the bottom of fairly shallow waters by means of a holdfast and reaches up towards the surface, forming an underwater forest. The leaf-like structures (fronds) that are near the surface have pockets of air built into them, which act as balloons to hold the leaves nearer to the surface where they can capture the Sun's energy. The material that makes up kelp is also part of peoples' everyday lives. Kelp helps thicken ice cream and jelly, and provides the smooth texture present in some frozen drinks.

LAYERS OF THE OCEAN

Oceanographers (scientists who study the oceans) often divide the ocean into horizontal layers. They use the physical characteristics of the water such as temperature, density, and the amount of light at different depths to classify these layers. The most important factor is the density of the water, which is determined by the combination of salinity (the amount of salt in the water) and temperature. All ocean water is salty, but some contains more salt than others. The water that is saltier is heavier and sinks, while the water that is less salty is lighter and floats.

Similarly, warmer water is lighter than colder water, so it floats on top of colder water. Oceanographers generally categorize the ocean into four layers: the epipelagic zone, the mesopelagic zone, the bathypelagic zone, and the abyssopelagic zone. The word "pelagic" refers to the open ocean, away from the coast. The prefix epi means "surface"; the prefix meso means "middle"; the prefix bathy means "deep"; and the prefix abysso means "without bottom." In addition, the transition zone between the epipelagic and the mesopelagic is often called the thermocline.

EPIPELAGIC ZONE

The epipelagic zone refers to the surface of the ocean where light penetrates. This layer is also called the photic zone, referring to the light that is found at these depths. Light is extremely important in the ocean. Just like plants on land, phytoplankton (free–floating plants, generally microscopic) require light to grow. Phytoplankton are the base of the food web (the network of feeding relationships in an ecosystem) in the ocean; they produce food by converting the energy from the Sun into energy they need to live and grow. When phytoplankton are eaten by zooplankton (free–floating animals) and fish, this energy is converted into the materials in their bodies. This transfer of energy continues as each predator (an animal that hunts, kills, and eats other animals) eats its prey (animals that are hunted and eaten by other animals), but it all begins with the energy from the Sun.

Sometimes the photic zone is referred to as the surface mixed layer. This layer is in contact with the wind and air above the ocean. The wind acts as a mixer, moving the water up and down throughout the top layer of the ocean. As a result, all of the water in the surface mixed layer has the same density. Because this water is often in contact with the air, it contains many of the gases required for life, such as oxygen and carbon dioxide. The epipelagic zone extends about 500 feet (150 meters) into the ocean, although this varies depending on location. Only about 2% of the total volume of the ocean falls in the epipelagic zone.

The Thermocline

Just below the surface mixed layer is a layer of water where the temperature and density change very quickly. This layer is called the thermocline. In warm tropical waters, the thermocline is very abrupt, while in cold polar waters the thermocline is often rather gentle. Below the thermocline, the temperature is always about 40°F (5°C). The thermocline acts as a density barrier between the surface, where there is light and phytoplankton growth, and the deeper layers of the ocean, where food is often scarce

MESOPELAGIC ZONE

The mesopelagic zone extends from about 500 feet (150 meters) to about 3,250 feet (1,000 meters) below the surface. It is often referred to as the "twilight zone" because it is between the epipelagic zone where there is light and the deep ocean where light is absent. The majority of light in this part of the ocean comes from bioluminescence, light that is generated by chemical reactions in bacteria, animals, and plants. Because the epipelagic zone is where phytoplankton grow and where zooplankton and fish feed on phytoplankton, some animals that live in the mesopelagic zone migrate upwards at night to feed. Other animals in the mesopelagic zone rely on food that falls through the thermocline into the mesopelagic zone. Still others have developed special adaptations to prey on the animals that live within the mesopelagic zone. Since food is rather scarce in this zone, predators often have sharp teeth and expandable stomachs to take advantage of anything they encounter, even if it is bigger than they are.

BATHYPELAGIC ZONE

The bathypelagic zone is the part of the ocean between about 3,250 feet (1,000 meters) and 13,000 feet (4,000 meters) below the surface of the ocean. Light is almost non–existent in this part of the ocean. What little light there is, is generated from bioluminescence from animals and bacteria. The pressure is also extremely great in this part of the ocean. However, marine life still exists here. Fish, jellyfish, mollusks, and crustaceans (aquatic animals having no backbone with jointed legs and a hard shell) all have representatives that live at these extreme depths. Most of these animals are either black or red. Red appears black at these depths, because the only wavelengths of light (wave patterns of light that are perceived by the eye as colours) available are blue light from bioluminescence.

ABYSSOPELAGIC ZONE

The abyssopelagic zone extends from 13,000 feet (4,000 meters) below the surface to the seafloor. In this zone, the waters are nearly freezing and the pressures are immense. Nonetheless, there are some animals that live in this very deep part of the ocean. Squid and jellyfish can be found swimming through the waters in the deep ocean. Often, they have little Colour but they do have special organs that can produce light. This bioluminescence is used to attract prey and scare away predators. Several kinds of echinoderms (spiny–skinned animals including starfish and sea urchins) are relatively common in the abyssopelagic zone.

The basket star has long arms that it waves above the seafloor in order to catch prey. The sea pig is a kind of sea cucumber that digs holes or tunnels in the mud, digesting dead animals and bacteria. Another sea cucumber called the flying cucumber has flaps like wings to fly through the deep ocean. Crustaceans can also be found in the abyssopelagic zone. Sea spiders and isopods (of the shrimp family) are often found in these deep regions. Very deep trenches (narrow depressions or cracks in the sea floor) can extend down to 35,750 feet (11,000 meters) in some parts of the ocean. Trenches are the deepest parts of the ocean and are classified by some oceanographers as the hadal zone.

MARINE INVERTEBRATES

Invertebrates are animals that do not have a bony internal skeleton, although many do have hard outer coverings that provide structure and protection. More than 90% of all animals are invertebrates and they are classified into at least 33 major groups, or phyla. Nearly every phylum of invertebrates has members that live in the oceans. Six phyla of invertebrates that are commonly found in the oceans are: Porifera (sponges); Cnidaria (corals, jellyfish, and sea anemones); Annelida (segmented worms); Molluska (snails, clams, mussels, scallops, squid, and octopuses); Arthropoda (crabs, shrimp, barnacles, copepods, and euphausids); and Echinodermata (sea stars, sea urchins, and sea cucumbers). Each phyla is divided into smaller groups called classes, which is then split into families and then species.

PHYLUM PORIFERA

The sponges are the least complex multicellular animals. They generally live attached to a surface and have three basic shapes, encrusting, vase–like, and branching. Sponges live in intertidal (between the tides) zones as well as in the deep ocean. They can be a few inches (centimeters) to 10 feet (3 meters) in diameter. There are nearly 10,000 species of sponges and all but two families are only found in ocean environments. The general body of a sponge is a central cavity surrounded by a fleshy body riddled with holes. The cells lining these holes pump water into the central cavity. As the water moves through their bodies, the sponge absorbs nutrients, and filters out particles from the water as their food. The name Porifera means "hole bearer," reflecting the many holes in the animal's bodies.

PHYLUM CNIDARIA

The phylum Cnidaria includes jellyfish, sea anemones, and corals. The word Cnidaria comes from the root word knide, which means "nettle." It refers to the special stinging cells that the animals in this group have for protection and predation (hunting an animal for food). These cells contain coiled threads that are fired at predators and prey. The threads may contain substances that paralyze or sticky substances that entangle their target.

Cnidarians have two body plans: polyp and medusa. Corals and sea anemones are typical polyps. They are jar-shaped animals with a mouth at the opening of the jar. Tentacles rim the mouth and are used to pull food into the stomach, which is located on the inside of the jar. Sea anemones have a basal disc (tooth-like structure), which is located where the bottom of the jar is, and it is used to burrow (dig a tunnel or hole) into the sand or rocks. Corals have a skeleton made of the mineral calcium carbonate, which cements the individual coral polyps together into a colony (group). Jellyfish have a medusa shape, and this can be visualized as a polyp turned upside-down so that it looks like a bell. Jellyfish swim by contracting their bodies and forcing water out of the bell. They have long tentacles (appendages) surrounding their mouths that are used to capture prey.

PHYLUM ANNELIDA

The phylum Annelida includes worms that are segmented, meaning that the body is made up of sections. Each body part may have a specialized purpose, such as reproduction, locomotion, or sensing the environment. These segments are apparent on the outside of the worm's body and make it look ringed. The word Annelida comes from the word annelus, meaning "ringed." The most familiar member of Annelida is the earthworm, which is not a marine species.

The class Polychaeta (meaning "manyfooted") is the largest class of Annelida, with about 10,000 species, most of which are marine (of the ocean). Almost all of them have paired appendages on their segments that can be used for swimming, burrowing, or walking. Polychaetes often have very well developed heads with a variety of sensory organs that detect prey by touch, vision, and smell. They range in appearance from very colourful to very plain-looking. Some swim through the water, others crawl on the sand or rocks, and others live cemented to the seafloor, building and living in tube-like structures.

PHYLUM MOLLUSKA

The mollusks are an extremely large phylum with over one hundred thousand species, most of which are marine. Most mollusks have a head, a foot, and a body that is covered by a shell–like covering called a mantle. The three most common classes of mollusks are the snails; the clams, oysters, scallops, and mussels; and the squid and octopuses. The gastropods are the largest class of mollusks with over eighty thousand species. They include snails, slugs, abalone, and limpets.

Most gastropods crawl along the seafloor among rocks, grazing on algae (tiny rootless plants that grow in sunlit waters). Some, however, hunt for their food among plankton, which are organisms that drift through the ocean. Other gastropods filter water for food particles. The majority of gastropods live in coiled shells, which provides protection from predators and protection from the force of waves. The shells are also used to protect animals that live in the areas between low and high tides from becoming too dry. The bivalves, meaning "two doors," are mollusks that have two shells like clams, scallops, mussels, and oysters. These animals generally live in sediments (sand, gravel, and silt) on the bottom of the ocean and gather food by filtering particles out of the water.

Many clams are able to burrow into sand or clay. Scallops can swim by forcing water through their shells. Oysters tend to cement themselves to hard surfaces. Mussels produce tough strings called byssal threads that attach their shells to surfaces in wave-swept areas. The most complex of the Mollusca phylum are the squid, octopuses, nautiluses, and cuttle fish. Each of these animals has a head that is surrounded by tentacles. They swim through the water using propulsion from their tentacles, much like swimmers kick their legs underwater, and creep along the bottom of the ocean using their tentacles as legs. Cuttle fish and squid have a shell like other mollusks, but it has been internalized. In octopi, the shell is completely absent. The members of this class have excellent eyesight and are considered intelligent.

PHYLUM ARTHROPODA

Arthropods are the most numerous invertebrate phylum with over one million species identified. Some scientists expect that there may be as many as ten million arthropods on Earth. All arthropods have a strong external skeleton that protects them from predation and supports their body structures. They have a type of muscle called a striated muscle that allows them to move quickly. They also have legs, antennae, and other appendages that are jointed. The phylum Arthropoda has three major divisions or subphyla. The subphylum Crustacea includes about thirty thousand different species, most of which live in the ocean. Crustaceans include many different types of marine animals that are divided into several classes. Brine shrimp, which are important fish food and live in very salty water, belong to the class Branchiopoda.

The class Maxillopoda includes barnacles and copepods. Barnacles are specialized crustaceans that spend the adult part of their life cemented head-down on hard surfaces like rocks, piers, the bottoms of ships, and even the undersides of whales. Their legs have developed into featherlike appendages that they use to generate water currents to bring food particles into their mouth. Copepods are small shrimp-like animals that are extremely important to the planktonic food web, the network of plankton that form the base of the food chain in the oceans. They are the most numerous animals in the ocean, sometimes reaching densities of more than a million per cubic yard (meter).

The class Malacostraca includes shrimp, lobsters, crabs, and euphausids. There is an enormous amount of diversity among the members of this class, which includes about twenty–five thousand different species. Some malacostracans spend their lives swimming among plankton, others walk along the ocean floor scavenging for food, while others live in burrows and attack prey that come nearby. Many members of this group live and feed off of fish or even other crustaceans. This class is very important to the economy, both as food for humans and as pets in the aquarium industry.

PHYLUM ECHINODERMATA

All of the six thousand members of the phylum Echinodermata are marine. The root word echino means "spiny" and the root word derma means "skin." The name Echinodermata refers to the bony structures called ossicles found in the skin of these animals. Echinoderms do not have well developed sensory organs or brains. They all have a water vascular system that is used to circulate nutrients and gasses through their bodies. All echinoderms share the same general appearance, which is based on five similar sections that radiate out from a central point. There are five classes of echinoderms: feather stars, sea stars, brittle stars (sea stars), sea urchins, and sea cucumbers. Brittle stars are fairly uncommon.

There are about six hundred species of brittle stars, living mostly in shallow waters. Sea stars (commonly known as starfish) use their water vascular system to operate suction cups located on the bottoms of their legs. These suction cups are called tube feet and they are used both for predation and for gas exchange. Sea stars eat by gripping both shells of a clam or mussel with its tube and pulling the prey open. Then it inverts its stomach inside the shells and digests the victim. Ophiuroids usually look like a small disc surrounded by five long worm-like arms. They are called brittle stars because when they are attacked, they will simply detach the arm that has been the target. Later, the ophiuroid will regenerate its arm. Sea urchins are usually pin cushion–shaped and covered with sharp spines. These spines are used for locomotion as well as for defence. Between the spines, sea urchins have special appendages called pedicellariae that look like tiny claws. These are used for capturing prey and for cleaning. The mouth of the sea urchin is on its underside and is composed of five tooth-like plates.

Sea cucumbers live on the sea floor and look like cucumbers with five soft ridges. Many live on coral reefs and are extremely colourful. Sea cucumbers feed by extruding feathery appendages that can capture prey that swim too close. When attacked, many sea cucumbers will suck in water and then use the water pressure to eject their internal organs, including their digestive and respiratory systems. The predator becomes confused among all the tissues in the water and may even become entrapped in some of the sticky material. After such an attack, a sea cucumber will regenerate its internal organs over a period of several weeks.

MARINE MAMMALS

Mammals are vertebrates (animals with a backbone) that share characteristics of nursing their young with milk, breathing air, having hair at some point in their lives, and being warm–blooded. Marine mammals are the species of mammals that depend on the oceans for all or most of their lives. There are about 115 different species of marine mammals. Marine mammals vary from the small sea otter to the giant blue whale. Some of them live in groups, like dolphins, while others are solitary, like polar bears.

All marine mammals share four characteristics:

1. They have a streamlined body shape that makes them excellent swimmers.
2. They maintain heat in their bodies with layers of fat called blubber.
3. They have respiratory (breathing) systems that allow them to stay underwater for long periods of time.
4. They have excretory (waste) systems that allow them to survive without drinking freshwater. Instead they obtain the water they need from the food they eat. Marine mammals belong to three groups called orders. An order is a classification of a group of organisms, which eventually splits into species. Marine mammals are in the orders Cetecea, Carnivora, and Sirenia.

ORDER CETACEA

There are about seventy–seven species of cetaceans, which include whales, dolphins, and porpoises. All of these animals live their entire lives in the water. Cetaceans probably evolved from the hoofed mammals that were similar to horses and sheep. Their front legs became fins that are used primarily for steering, and their hind legs became extremely small flippers or flukes (tail fins) so as to streamline the animals for swimming. Cetaceans swim by moving their strong fluke up and down. They are grouped into two categories or suborders: Odontoceti, the toothed whales, and Mysteceti, the baleen whales.

Suborder Odontoceti

The toothed whales account for about 90% of all cetaceans, including dolphins and porpoises as well as the orca (killer whale) and the sperm whale. Toothed whales hunt for prey, which they capture with their teeth. They also have one external hole called a blowhole for breathing. The toothed whales have a large brain for their size and are considered to be some of the most intelligent animals. They hunt using sophisticated echolocation, which is the method of detecting objects by listening to the reflected sounds that it calls out.

Suborder Mysteceti

The baleen whales have no teeth. Instead, they have bristly plates called baleen, which hangs like a curtain from their upper jaws. Baleen is made from a protein similar to that which makes up human fingernails. When it is eating, the baleen whale sucks in huge amounts of water and then forces the water out through its baleen, which acts like a sieve. These whales diet primarily on tiny animal plankton, animals that drift through oceans. Microscopic plankton are concentrated behind the baleen and then swallowed. Even though baleen whales feed on some of the smallest animals in the world, baleen whales are some of the largest animals in the world. The blue whale, the fin whale, and the gray whale all weigh more than 2 tons (9 metric tons). Baleen whales are also distinguished from toothed whales because they have two external blowholes instead of one.

ORDER CARNIVORA

The Carnivora include animals that prey on each other (meat-eaters) like dogs, cats, bears, and weasels. There are two suborders of carnivores that have marine representatives: the pinnipeds and the fissipeds.

Suborder Pinnipedia

Pinnipeds include seals, sea lions, fur seals and walruses. They all have flippers that can be used to move around on land as well as on water. Although they swim much more efficiently than they walk on land, they do give birth to their young on land. Seals account for nearly 90% of all pinnipeds. There are nineteen species that live in all of the oceans and even in a few lakes. Most seal species live near Antarctica and in the Arctic Circle. Seals do not have an external ear, although they can hear very well.

They propel themselves with their rear flippers and use their front flippers for steering. Sea lions and fur seals do have small external ears. Their hind legs are more flexible than those of seals, so they can move around better on land.

They propel themselves with their front flippers when swimming. Walruses are the largest of the pinnipeds. They do not have an external ear, but they can use their rear flippers for moving on land. The canine teeth (pointed, in the front) of walruses are enlarged into tusks. Walruses swim along the bottom of the ocean using their tusks like runners as they look for clams to eat.

Suborder Fissipedia

The suborder Fissipedia includes cats, dogs, raccoons, and bears, as well as two marine mammals: sea otters and polar bears. Sea otters are about 4 feet (1.2 meters) long, the smallest of the marine mammals. Their favourite food is sea urchins, which they eat by lying on their back and smashing them on a stone that is balanced on their chests. Polar bears wander long distances across sheets of floating ice in the Arctic hunting for seals and whales. They can swim between patches of ice using their powerful forepaws like oars.

ORDER SIRENIA

The sirenians, also called sea cows, evolved from the hoofed land mammals, like the cetaceans. The Sirenia include the manatees, which are large plant-eating marine mammals, and the dugongs (commonly known as sea cows). They are the only plant-eating marine mammals, eating sea grasses and algae (tiny rootless plants that grow in sunlit waters) in warm waters. They grow to be quite large, up to 15 feet (4.5 meters) and weigh 1,500 pounds (680 kilograms).

ENDANGERED MARINE MAMMALS

The Endangered Species Act, an animal that could become extinct in all or part of its range is endangered. An animal's range is the entire area where it lives. Of the 115 species of marine mammals, 22 were considered endangered as of 2004, including the blue whale, the gray whale, the finback whale, the Hawaiian monk seal, the stellar sea lion, the marine sea otter, and all four species of dugongs and manatees. Much of the threat is due to human activity. People hunt these animals for their pelts, blubber, and meat, and destroy their habitat from overfishing and mining.

PLANKTON

Plankton is a general term that includes plants, animals, and bacteria that drift through lakes and the oceans. Plankton are the foundation for all life in the ocean and produce much of the oxygen that sustains life on Earth. Plankton represents a diverse and large group of organisms.

Often, the only factor that the nearly ten thousand different species of plankton have in common is their poor swimming ability. Rather than control where they are moving, like fish, whales, and turtles, plankton simply float wherever the water currents take them. In fact, the word plankton is derived from the Greek word planktos, meaning "to wander."

STUDYING AND CLASSIFYING PLANKTON

Biologists identify and count plankton found in water samples. Several different methods are used to collect plankton from water samples, the most common of which includes the use of plankton nets. Most plankton nets are made of nylon or synthetic material that is produced so that the size of the holes between the fibres is uniform. The most common shape for a plankton net is cone-shaped, and the large end of the cone is attached to a metal net ring. The diameter of the net ring is usually 39 inches (1 meter).

The small end of the cone is fitted with a plastic bottle, called a net bucket. The net is pulled slowly behind a boat for a specific distance and plankton larger than the holes in the net are trapped in the net bucket. The net is then reeled into the boat and the plankton trapped in the net bucket are removed for study. Once the plankton have been removed from the net bucket, scientists identify and count the different species of animals and plants. Once the number of a certain type of plankton found in the net bucket has been counted, biologists calculate the concentration of that organism. The volume of water sampled can be calculated by multiplying the distance the net was towed by the area spanned, by the net ring. The concentration of different types of plankton gives biologists information about the water quality and ecology (study of the relationships between organisms and their environment). Biologists also use some of the larger and sturdier plankton removed from the net bucket for experiments involving nutrition, reproduction, and different processes in the human body.

Many plankton are extremely small and pass through even the smallest holes in plankton nets. In order to study these plankton, biologists filter a specified amount of sample water through membranes (tissue) that have very small holes. The water will pass through the membranes, but any plankton larger than the holes in the membranes will be concentrated on top. Researchers can then attach the membranes to slides and view them under a microscope. Special dyes are often used to stain the plankton in order to see them more clearly.

In other cases, the membranes with the plankton on them are ground up and are analysed chemically, which helps researchers determine the types of plankton in the water. Because there are so many different types of plankton, counting and identifying them is time-consuming and often difficult. Instead, biologists often classify plankton into broad groups that simplify the process, but still provide important information about the ecology of the water sample. Three criteria often used for plankton classification are size, cell structure, and life history.

Size of Plankton

There are six major size categories of plankton. They range in size from plankton far too tiny to be seen with the naked eye to organisms that are many feet (meters) long.

- *Net plankton*: These plankton include species that are large enough to be caught in plankton nets.
- *Macroplankton*: Plankton larger than 0.79 inches (more than 20 millimeters) are called macroplankton. Macro is the prefix meaning "large." Macroplankton include the larval (immature worm-like stage) forms of many fish, some marine worms, many different types of crustaceans (water animals with an outside skeleton, including shrimp, crabs, lobsters) and jellyfish that can have tentacles stretching 25 feet (8 meters). Some plants are also classified as macroplankton, such as the giant seaweed Sargassum.
- *Mesoplankton*: Plankton between the sizes of 0.79 to 0.0079 inches (20 to 0.2 millimeters) are called mesoplankton. The prefix meso means "medium." Examples of mesoplankton are shrimp-like creatures called euphausids and many types of larval fish.
- *Microplankton*: These plankton range between 0.0079 to 0.000079 of an inch (between 0.2 and 0.02 millimeters. The prefix micro means "small."
- *Nannoplankton*: Plankton between 79 ten-thousandths to 79 millionths of an inch (between 0.02 to 0.002 millimeter) are called nannoplankton. These plankton are so small they must be concentrated on filters in order to be identified. The prefix nanno means "very small." These plankton include many different types of protozoans (a type of onecelled animal), single-celled plants, and the larvae of crabs, sea urchins, and mollusks.
- *Picoplankton*: The smallest group of plankton is the picoplankton, which are less than 79 millionths of an inch (2 thousandths of a millimeter or 0.0002 millimeters) wide. Picoplankton are the smallest and most numerous plankton in the ocean. The prefix pico means "extremely small." Picoplankton include bacteria that ingest organisms for food organisms, as well as a type of bacteria that can gather

energy from the Sun as do plants. Picoplankton also include many different species of single-celled protozoans and single-celled plants.

Cell Structure of Plankton

Plankton are classified into three major groups just as to cell type: phytoplankton, zooplankton, and bacterioplankton.

Phytoplankton

Phytoplankton are plants and oxygen–like bacteria. The prefix phyto means "plant." Most phytoplankton are single–celled organisms, although there are some phytoplankton that form colonies (groups) and others that are multicellular, such as seaweed. In the open oceans, about three–fourths of phytoplankton are nannoplankton. In coastal waters and lakes, phytoplankton tend to be larger, in the microplankton size range.

Diatoms are the most numerous group of phytoplankton. Diatoms are single–celled plants that can be shaped like rods, spools of thread, or pillboxes (round boxes with a top and bottom of equal height). They secrete two shells made of silicon, the same substance that makes up glass.

The plant cell lives inside the silicon shells and produces threads that protrude through perforations in the shells. In some species, the threads join with threads on other diatoms and form long chains. Diatoms are usually found in the surface waters and when conditions are right, they can reach high concentrations that can make the water appear green. After diatoms, dinoflagellates are the next most common group of phytoplankton. Dinoflagellates come in many different shapes, but commonly look like a chocolate candy kiss placed bottom to bottom with another candy kiss that has two peaks instead of one. Most have two flagella (whiplike appendages) that they use like propellers to spin through the water.

Some are covered with protective plates composed of cellulose, the material that makes up the woody part of trees. A few species of dinoflagellates contain chemicals that poison fish and, occasionally, people. Other dinoflagellates contain pigments that make the ocean appear red.

Zooplankton

Zooplankton are animal plankton that wander in the water currents. Because they are poor swimmers, most zooplankton have special feeding structures that allow them to capture food that they bump into as they drift. These structures come in all forms, from sticky hairs to nets made out of mucous, to brush–like appendages that sweep food particles towards the mouth.

Most zooplankton diet on phytoplankton. The most common zooplankton in the ocean are the crustaceans, which include the crabs, shrimp, and lobsters, and account for about 70% of all zooplankton. In particular, a small shrimp–like animal called the copepod is the most numerous type of animal in the plankton family.

In fact, if all the copepods in the world were divided equally among all the people in the world, each person would receive one billion copepods. Another common zooplankton is the shrimp-like crustacean called the euphausid. Euphausids, also called krill, are slightly bigger than copepods, around 2 inches (5 centimeters) long. They are so numerous in the waters around Antarctica that the diet of many whales consists entirely of krill. In fresh water, the tiny crustacean Daphnia is the most numerous zooplankton. Daphnia are particularly interesting zooplankton because they can reproduce without mating in the spring and summer using a process called parthenogenesis, where female Daphnia produce exact copies of female offspring. Other important animal groups found among zooplankton are the jellyfish, the worms, the mollusks (squid and snails), and the echinoderms (sea cucumbers and sea urchins).

Some zooplankton are single-celled organisms called protozoans. For example, the foraminifera are a type of amoeba (a one-celled animal) that has a shell with holes through it. Foraminifera produce sticky spines that extend through the holes, where animals that bump into them are captured and eaten.

Bacterioplankton

The smallest type of plankton, bacterioplankton are microscopic single–celled organisms. Along with other forms, they play an important role in the ecology (living organisms and their environment) of aquatic systems. Bacteria are single–celled microscopic organisms. These organisms are numerous. Bacteria digest dead zooplankton and phytoplankton, producing the nutrients and other materials needed for new life to grow.

Life History of Plankton

Plankton are also classified just as to their life history. Most species of bacterioplankton, phytoplankton, and zooplankton spend their entire life floating and drifting with the currents. These plankton are called holoplankton (holos is the Greek root meaning "entire"). Other plankton live only the early part of their life as plankton; the adult part of their life is spent in a different part of the ocean or lake. These plants and animals are called meroplankton (meros is the Greek root meaning "mixed"). Some examples of meroplankton are sea urchins, sea slugs, lobsters, worms, and some coral reef fish.

Some aquatic plants are also meroplankton. Many meroplankton scatter eggs into the plankton, where they are fertilized. The fertilized eggs develop into larvae, which float in the plankton. Just as a caterpillar looks nothing like a butterfly, in general these larvae look nothing like the adults they will eventually become. When the larvae are in the plankton, they eat other plankton or survive off of the yolk that was with them in the egg.

Depending on the species, the larvae remain in the plankton for varying periods of time from several days to several months. Afterwards, the larvae settle onto the seafloor or swim away from the plankton and change into their adult form.

IMPORTANCE OF PLANKTON

Plankton are vital to the global climate. Phytoplankton perform photosynthesis, which is a process that uses the energy from sunlight to produce food. In the process of photosynthesis, phytoplankton take in carbon dioxide and produce oxygen. About half of the oxygen on the planet comes from phytoplankton photosynthesis. As humans burn oil and gas to keep their cars moving and their houses warm, carbon dioxide is produced. This carbon dioxide holds heat and is one of the leading causes of global warming.

It is estimated that phytoplankton remove three billion tons of carbon dioxide from the atmosphere each year, as much as all the trees on land. Plankton are also key to the ecology of the ocean. Phytoplankton are the base of the marine food chain. In other words, they are the food for zooplankton, corals, and mollusks. Even some sharks, such as basking sharks and nurse sharks, rely on phytoplankton for their diet.

In turn, fish and larger predators eat the zooplankton. In fact, the giant blue whale relies entirely on shrimp-like zooplankton called euphausids for its diet. Humans also eat fish that prey on plankton. After plankton die, they sink to the bottom of the ocean. Over millions of years, the dead plankton are buried by sediments, and then eventually converted into fossil fuels such as oil and gas.

TIDES

Tides are the alternating rise and fall of bodies of water, relative to land. Each 24-hour period, there are two high tides and two low tides. The arrival times and heights of the tides change every day and follow a pattern over days, months, and seasons.

The shape of a coastline, water depth, shape of the seafloor (bathymetry), weather, and other local factors affect the heights and arrival times of tides at specific locations. The daily tides bring ocean nutrients that nourish brackish–water (slightly salty) plants and wildlife that live in tidal wetlands.

EXPLAINING THE TIDES

Humans in maritime (sea-going and coastal) societies have always recognized and measured the daily, monthly, and yearly pattern of water level rise and fall along coastlines. Navigation, construction, and fishing in coastal areas require precise knowledge of the local tides, and tide prediction is an ancient science. The ancient Hawaiian "moon calendar" charts the tides and relates them to fishing and agricultural harvests. John, Abbot of Wallingford, who died in 1213 supposedly authored the oldest European tide chart. One entry predicts the hours of high water at London Bridge ("flod at london brigge") on the Thames River.

The scientific explanations for how the tides work and why they occur are, however, relatively new discoveries. Ancient Chinese and European philosophers theorized that Earth inhaled and exhaled water. Most ancient scientists, including Greek philosopher Aristotle (384–322 B.C.E.), were silent on the subject of tides. (Ancient Egyptians, Greeks, and Romans lived on the Mediterranean Sea, which has relatively insignificant tides.) Isaac Newton's theory of gravity is the basis for understanding tides. Newton (1642–1727), a seventeenth century English mathematician and physicist, theorized that all objects exert an attractive force, called gravity, on other objects.

The strength of the gravitational pull between objects depends on their relative sizes and the distance between them. Earth, a very large object, pulls smaller objects, like people or apples, strongly towards its center. (Newton's theory of gravity was supposedly inspired by his observation of an apple falling from a tree.) Earth's gravitational pull keeps the Moon in orbit around the planet. Newton's ideas were later applied to an explanation of tides by French mathematician Pierre Simon Laplace (1749–1827), and Irish physicist William Thomson (1824–1907), who is also known as Lord Kelvin.

HOW TIDES WORK

The gravitational pull of the Moon and Sun on Earth's oceans, inland seas, and large lakes causes tides. The Moon's pull on the surface of the oceans as Earth spins on its axis causes two high tides and two low tides during each 24-hour day. To visualize the tides, imagine Earth as a ball completely covered with water. Earth's gravity holds the water on the planet's surface. The Moon's gravity pulls a bulge of water towards it. Another force due to the spinning of the Earth and called the centrifugal force also bulges water at the equator in an outward direction, much like a fast-spinning amusement ride pushes your body towards one side of your seat. Centrifugal force causes a second bulge to form on the direct opposite side of Earth to balance the bulge facing the Moon.

As Earth rotates on its axis over 24 hours, the bulges remain stationary with respect to the Moon. Every location on Earth experiences the passing of both bulges in the form of two high tides each day. The low water moments between the bulges cause two daily low tides. The relative positions of the Earth, Moon, and Sun constantly shift. The Moon's monthly circuit around Earth causes the tides to occur slightly later each day. If the Moon were stationary over the spinning Earth, the high tides would be exactly 12 hours apart, and tides would occur exactly every six hours. As it is, the first high tide of a 24-hour day happens about 50 minutes later than the previous day.

The gravitational pull of the Sun also affects the height of the tides. Solar (sun) tides are much weaker than lunar (moon) tides because the Sun, although much larger than the Moon, is much farther away from Earth. The relative positions of the Earth, Moon, and Sun constantly change during Earth's yearlong trip around the Sun. Very high and very low tides, called spring tides, occur when the Sun and Moon are aligned and pulling at the tidal bulges from the same or exact opposite sides of Earth. Spring tides happen twice a month (about every 15 days) during the new and full moons.

The opposite conditions, when high tide is not very high and low tide is not very low, are called neap tides. These happen when the Moon and Sun are at right angles to each other so their gravitational forces cancel one another.

TIDES VARY AROUND THE WORLD

Earth is obviously not a perfect, watercovered sphere. The continents, seafloor, ocean currents and (mass of air surrounding Earth) winds all affect the tidal bulges as they move around Earth each day. Some places, like the Bay of Fundy in Nova Scotia, Canada, and the English Channel between Great Britain and France, experience very large tides. Other places, like the Mediterranean Sea, have barely noticeable tides. Sometimes the shape of an inlet (a narrow body of water between two islands or leading inland), bay, or harbor delays the tides; in the Gulf of Mexico there is only one high and one low tide each day.

A large storm like a hurricane can add to the tidal bulge as it approaches the shore. Along many coastlines, strong tides carry salt water and ocean sediment (particles of sand, gravel, and silt) far inland. Many rivers, bays, and estuaries (coastal wetlands) experience tides many miles from the ocean.

WAVES

Wind creates waves. As an air current (moving stream of air) moves over an undisturbed water surface, friction between air and water creates a series of waves that move across the surface. The size of the waves depends on the wind speed, the duration of the wind, and the distance over which the wind blows. (The distance of open water surface that the wind blows over is called the fetch.) A week–long tropical storm in the Pacific Ocean might produce waves as tall as three–story houses; a tenminute gust blowing across a small lake might make waves that are only a few inches tall.

Waves move away from their point or area of origin in widening circles, like ripples moving away from a pebble dropped into a pool.

In an ocean basin (the deep ocean floor), waves from many different wind events are moving across the sea surface at any given moment. When sets of waves meet they interact to form new patterns. By the time they reach the coastline, waves have been affected by many wind events. Ocean waves may appear that the water is moving forward but in actuality the water is moving in a circle as the water molecules lift and fall. (A molecule is the smallest unit of a substance that has the properties of that substance.) Imagine floating in the ocean in a raft.

When a wave approaches, you rise and fall as it passes, but you don't move towards the beach. The same thing happens to the water molecules below you. As a wave arrives, the water particles rise and fall in small circles as the wave passes, but they are not carried forward. The highest point the wave reaches is called the crest. The lowest point of the wave is called the trough.

The wavelength is the distance from one crest to the next. The water molecules closest to the surface move in the largest circles, and deeper water moves less. Molecules below a depth known as wave base are undisturbed by a passing wave. Wave base is equal to half the horizontal distance between wave crests, or one-half a wavelength.

BREAKING WAVES

Waves change form when they approach a coastline. When the seafloor is shallower than the wave base, it interferes with the circular motion of the water at the bottom of the wave. Waves that were broad, gentle swells in the open ocean grow taller and their crests get closer together. Eventually, the wave grows too tall to support itself and it breaks; the wave crest collapses over the front of the waves. Spilling breakers that gradually become more steep and then crumble typically form along shallowly sloping shores. Plunging breakers that grow tall and curl sharply generally pound steep coastlines. Big waves start to break farther from shore than smaller waves because they have deeper wave bases. Water from breaking waves sloshes forward up the beach.

It returns in an outgoing current along the seafloor called undertow. The forward and back motion of water in the surf zone (area of rough water next to the land, where ocean waves hit the shore) between the breaking waves and the beach is called swash. Like water in a washing machine, water in a surf zone endlessly cycles between the breakers and the beach. Beaches are subjected to relentless swashing that breaks down all but the most resistant sediments (sand, gravel, or silt). The mineral quartz is particularly strong, and beach sand is often composed of identical quartz grains that waves have rounded into perfect spheres and sorted by size.

WAVE REFRACTION

Waves bend when they reach coastlines. It is extremely rare for wind to blow exactly towards a perfectly straight coastline, and waves almost always approach shorelines at an angle. Wave bending or refraction occurs because the end of a wave that reaches shallow water first slows down and breaks before the deeper end. Water moving in the surf zone flows sideways along the beach from the direction of the approaching wave, and gravity pulls the returning water directly downhill. Water and sediment thus move in a zigzag pattern that carries them along the beach. Wave refraction produces longshore currents, which are currents that flow parallel to coastlines in shallow water. If you have ever dropped your towel on the beach and gone for a swim only to discover that you have been carried away from your towel, you have experienced a longshore current. Wave refraction also brings the eroding power of waves onto headlands, the jagged, rocky, narrow strips of land that extend into the ocean. Longshore currents carry the eroded sediment away from headlands and deposit it in bays. Waves thus, straighten irregular coastlines by wearing down the headlands and filling the bays.

A typical arc-shaped bay with headlands at each end has two longshore currents that flow from the headlands towards each other. The shallow, strong, outgoing current that forms at the tip of the bay where they meet is called a rip current. Rip currents also form where large waves pile water between a sand bar (a ridge of sand built up by currents) and a beach. Rip currents, can be dangerous to swimmers because they can form or become strong suddenly. Waves and longshore currents can also mold sand into strings of barrier islands (a long, narrow island parallel to the mainland) formed from sediment deposits. These islands are often called depositional coastlines. In the Gulf of Mexico, waves have washed sand from the Mississippi River Delta. Longshore currents have deposited the sand in a long streamer of barrier islands along the Louisiana and Texas coastlines. Tidal inlets (inlets maintained by the tidal flow) separate barrier islands from each other and shallow bays called lagoons separate them from the mainland. The barrier island of the United States eastern seaboard, included the Outer Banks of the Carolinas, formed in a similar fashion. Wave patterns and coastline conditions are constantly changing, and coastline features are continuously remolded. The waves from a large hurricane, for example, can completely destroy a barrier island, beach houses and all, and reshape a new one in a matter of days.

Chapter 6

Fresh Water

DELTAS

Deltas are deposits of sediments (particles of sand, gravel, and silt) at the mouths of rivers that flow into the ocean. The mouth of a river is the end where the body of water flows into the sea. Deltas are shaped by interactions of the river's fresh water with the ocean water, tides, and waves. Throughout history, deltas have been important places for human settlement. They are also vital habitats for many animals and plants. In the fifth century B.C.E, Greek naturalist Herodotus coined the name delta to describe the triangular shape of the sediments deposited at the end of the Nile River. The capital Greek letter delta resembles a triangle. Most deltas are triangular in shape because rivers deposit larger amounts of sediment where they meet the sea, then fan out into the mouth of the sea to deposit the remaining sediment.

FORMATION OF DELTAS

As rivers flow through their beds (a channel occupied by a river), the water breaks up rocks and pebbles, which the river then carries with it as it flows. These pieces of rock, sediment, include sand, pebbles, and silt. When the fast–flowing waters of the river reach the ocean, they push against the ocean water. This decreases the speed of the river water, which spreads out along the coastline. As the speed of the river water slows, the sediments that it carries settle to the bottom of the ocean. Over time, the sediments accumulate and form a delta. In most deltas, the river water is less dense (packed together) than the sea water because it contains less salt. As it flows out into the ocean, it floats on top of the ocean water. This is called hypopycnal flow. (The prefix hypo means "under" or "less" and the root word pycn means "density.") In places where hypopycnal flow occurs, the salt water that lies under the fresh water is called a salt wedge. The bigger sediments, like gravel and pebbles, are deposited at the tip of the salt wedge (nearest to shore) and the smaller sediments, like sand, are deposited farther out along the salt wedge. The smallest silt grains (fine sediment particles) are transported far offshore with the river water.

THE STRUCTURE OF A DELTA

Paths of flowing water called channels run through the sediments of deltas. These channels are called distributaries and they may be large and relatively permanent or small, transient features. The sides of the channels are made up of piles of sediments called *levées*. The areas between the distributaries are called interdistributary areas.

TYPES OF DELTAS

The structure of the distributaries and interdistributary areas depends on where the river empties into the ocean and the forces that affect the river water as it flows into the ocean. There are three major types of deltas: river–dominated, tide-dominated, and wave-dominated. Many deltas are formed by a combination of these forces. River-dominated deltas extend outward from the coast as the river water jets out into the ocean. The sediments deposited by the river tend to form levées that hold channels of water. The aerial view of a river-dominated delta looks like the foot of a bird with several branching channels. River-dominated deltas often have sand bars (long deposits of sand) that are perpendicular to the river. The Mississippi River Delta in Louisiana is an example of a river–dominated delta. Tide-dominated deltas are deltas where the sediments deposited by a river are redistributed by tides. These deltas have channels cut by the river water as well as channels cut by tidal currents. The result is a shoreline that looks like the fringe on the end of a carpet. Tide-dominated deltas may also have coastal features like sand bars and shoals (a sandbank seen at low tide) oriented parallel to the tidal flow.

The Ganges-Brahmaputra Delta in the Bay of Bengal is an example of a tidedominated delta. Wave-dominated deltas occur in places where the sediments deposited by the river water are redistributed by waves. Because waves tend to move sediments along the shoreline, the shape of wave-dominated deltas is usually a smooth coastline. If the waves that affect the delta break parallel to the shoreline, the sediments of the delta will tend to be symmetrical on either side of the river. If the waves break at an angle to the shore, the sediments of the delta will tend to accumulate on one side of the delta. The Nile River Delta in Egypt is an example of a delta that is wave–dominated.

HUMANS AND DELTAS

Deltas play an important role in human life both historically and in present times. The land on deltas is typically very good for agriculture. On the parts of the delta close to the river, the soil is fertilized each year by nutrient-containing floodwaters when the river floods. In addition, by building aqueducts (canals or pipelines used to transport water) from the wetter lands near the river to the dryer lands farther from the water source, it is easy to expand the amount of land available for farming. This practice is called land reclamation. In particular, the fertile Nile delta has supported much of the agriculture in Egypt for thousands of years. Trade is another reason why humans have lived on deltas. Because of the many distributaries and the access to a major river and the ocean, deltas are regions where goods can be easily exported both inland and overseas. Many of the world's largest ports are in deltas. Also, communication is relatively easy in deltas. Because the land is usually flat, it is easy to build roads. The many waterways make boat travel easy as well.

LIFE ON DELTAS

The interdistributary areas of deltas can support a variety of different habitats, depending on whether they are closer to the freshwater of the river or the salt water of the ocean. If the interdistributary areas are close to the river and affected by annual floods, they are called floodplains. Other interdistributary areas near the river water may be freshwater marshes, freshwater swamps, or lakes. The interdistributary areas that are closer to the ocean are likely influenced by the tides. They may be tidal flats (flat, barren, muddy areas periodically covered by tidal waters), mangrove (a tree that grows in saltwater) swamps, salt marshes, or marine embayments (an indentation in the shoreline of the sea that forms a bay). Because deltas have such a broad range of environments, they host a diversity of species.

Many different species of plants flourish on deltas, from saltwater trees called mangroves, to sea grasses, swamp sedges, and shrubs. Deltas also serve as nursery grounds for many species of fish and invertebrates (animals without a spine) and many land animals such as snakes and birds. The marine (sea) areas are important habitats for burrowing worms and mollusks, crustaceans (sea animals with hard outer shells) that hunt for food along the seafloor, as well as a variety of different species of fish. Deltas also serve important environmental roles. They remove harmful chemicals that are deposited in them by river pollution. These chemicals are absorbed by sediments and trapped as new sediments settle on top. Over time, bacteria break down many of these harmful substances and release chemicals that are not dangerous to the health of humans or other animals. Deltas are also known as nutrient recharge zones. When animals and plants die, they are buried in sediments and bacteria digest them. This converts the chemicals in their bodies into the raw materials needed for plants to grow.

FRESHWATER LIFE

The animals and plants that live in freshwater are called aquatic life. The water that they live in is fresh, which means that it is less salty than the ocean. The terrestrial (land) environment that surrounds the freshwater environment has a large impact on the animals and plants that live there. Some factors that influence the freshwater environment include climate, soil composition, and the terrestrial animals and plants in the area. Just as on land, aquatic plants require carbon dioxide, nutrients (substances such as phosphate and nitrogen needed for growth) and light for photosynthesis, the process where plants make their food from sunlight, water, and carbon dioxide. Aquatic animals need to breathe in oxygen and consume food.

The physical conditions surrounding the body of water or wetland (lands that are covered in water often enough so that it controls the development of the soil) control the availability of these resources. For example, the concentrations of nutrients, oxygen, and carbon dioxide in the water depend on how much air gets into the water and on the chemical composition of the land nearby. The sediments (particles of sand, gravel, and silt) in the water influence how much light reaches the bottom of the lake or river. The temperature of the water affects how quickly animals and plants grow. The characteristics of the bottom of the body of water (sand, mud, rocks) and the speed of the currents (horizontal movement of water) control what kinds of plants and animals can live and reproduce in an area.

In general, freshwater environments are divided into two major categories: lentic waters and lotic waters. Lentic waters are those that are moving, as in rivers and streams. Lotic waters are those that are stationary, as in lakes and ponds. Sometimes, however, rivers and streams flow into lakes and ponds and the two different habitats merge together. Some wetlands may also contain many characteristics of freshwater environments.

LIFE IN RIVERS AND STREAMS

Rivers and streams are characterized by several physical features. They are generally comprised of freshwater that flows in one direction. The flow of water is most often from an area of high altitude (like a mountain range) to an area of low altitude (like an ocean). Usually, the water flows quickly initially and slows as it moves downstream. Streams often join rivers, so there is more water at the end of a river than at the beginning. As rivers flow, they erode (wear away) rocks and pick up sediments, making rivers often murkier at the end. Because rivers and streams change so much from their beginnings to their ends, there are many different types of habitats for animals. As a result, the number of animal species that live in rivers and streams is greater than the number of species that live in lakes and ponds.

Plant Life in Rivers and Streams

A major challenge facing plants that live in rivers and streams is staying in place, especially in swift currents. Plants have several different techniques to overcome the drag (the pull) of the water. Diatoms are a type of algae. Algae are marine organisms that range in size from microscopic phytoplankton to giant kelp and that contain chlorophyll, the same pigment used by land plants to perform photosynthesis. Diatoms avoid currents by using their small size.

They grow in a single layer on the surfaces of rocks. Because of the friction between the rock surface and the water, the water flow slows nearly to a stop within about a tenth of an inch (one–quarter of a centimeter) from the rock's surface. This region is called the boundary layer, and it provides the diatoms with protection from the forces of the current that would otherwise drag them downstream. Typical large river plants include algae, mosses, and liverworts.

These plants overcome the drag of the water by using special adaptations to grip rocks. Large algae often attach themselves to rocks with root-like structures called holdfasts. In addition, plants often anchor themselves in nooks between rocks or where waters pool, to avoid the drag of the river water. River plants that live within the currents have developed techniques to withstand the forces of the water. These forces would quickly snap any plant with rigid stems or leaves. As a result, plants that live in rivers are very flexible so that they can easily bend and move with the currents.

Animal Life in Rivers and Streams

Animals that live in rivers and streams also face the challenge of staying where they are. Many animals have hooks and suckers that they use to attach themselves to rocks. Blackfly larvae that live in streams in the northern United States and Southern Canada have suction cups that they use to stick to rocks in streams. Mayfly larvae have hooks that they use to fasten themselves to the algae growing on rocks. Other animals have streamlined shapes that minimize drag by presenting little resistance to water. Trout, which are extremely common in oxygen-rich fast–flowing waters, are shaped like torpedoes. Limpets are flattened molluscs that cling to the surfaces of rocks.

Their flat shape decreases the currents' drag on them. Animals that live in streams and rivers have developed interesting ways of gathering food in the fast-flowing waters. Snails, limpets, and caddis fly larvae scrape algae from rocks using special mouthparts. Many different insect larvae, as well as freshwater clams, filter the water for small bits of food. They have specialized mouthparts that look like brushes or combs that they use to strain the water and extract the edible plankton (animals and plants that float with currents) that float into their reach. Rivers and streams are homes to a large number of fish. Perch, smallmouth bass, largemouth bass, bullhead, carp, pike, and sunfish prefer the parts of rivers where waters slow. These fish tend to be large, visual predators (animal that hunts another animal for food) that hunt in pools for smaller fish and invertebrates (animals without a backbone). Sculpins and darters prefer the faster moving sections of the river where waters are highly oxygenated. They use the swift current to bring food to them rather than hunting for their prey. Trout are also found in these faster moving parts of the river.

LIFE IN LAKES AND PONDS

Large lakes are often divided into zones. The near-shore area is called the littoral zone. This is the part of the lake that is shallow enough for aquatic plants to grow. The limnetic zone, also called the epilimnion, is the surface water of the lake away from the shore. (The prefix epi means "on the surface" and the root word limn means "lake.") It extends down as deep as sunlight penetrates. The majority of the plant life in this zone is phytoplankton (microscopic plants that float in currents). The deep part of the lake is called the profundal zone or the hypolimnion. (The prefix hypo means "under.") No plant life exists in this zone because of the absence of light. Most of the biological activity is that of bacteria decomposing dead animals and plants.

Seasonal Changes in Lakes

Lakes and ponds are greatly influenced by the temperature changes throughout the seasons. The description below is typical for a lake in a temperate (moderate) climate, which experience seasonal temperature changes. Tropical lakes (those in hot and humid areas) will have less dramatic fluctuations in temperatures. In the summer, the Sun warms the epilimnion. Warmer water is less dense than colder water, so it floats on top of the cooler water in the hypolimnion.

The region between the warm surface waters and the cold deeper waters is a transition zone where the water changes temperature very quickly with depth. This region is called the thermocline. The thermocline acts as a kind of barrier between the surface and the deep waters. In the early summer, the epilimnion is full of life. Phytoplankton can grow quickly because they have plenty of light and nutrients and the water temperature is warm. In turn, zooplankton (animals like crustaceans and small fish that float in the waters) feed on the phytoplankton. These zooplankton are food for larger fish and birds.

As summer progresses, the phytoplankton use up the nutrients in the epilimnion. They begin to die and sink to the bottom of the lake. There, decomposers, like fungi and bacteria, break up the dead phytoplankton and animals and convert them into the nutrients that phytoplankton need to grow. Because the thermocline acts as a barrier between the bottom and the top of the lake, these nutrients are unavailable to the phytoplankton in the epilimnion. Phytoplankton cannot grow in the hypolimnion, where there are nutrients, because there is no light. In the fall, the air temperature cools, which cools the surface of the lake. Eventually the temperature in the epilimnion becomes the same temperature as that of the hypolimnion. The thermocline disappears and the nutrient–rich waters from the hypolimnion mix with the waters in the surface of the lake. This is called the fall turnover. At this time, the nutrients from the bottom of the lake are mixed throughout the lake. However, because the amount of sunlight decreases in the fall and into the winter, the phytoplankton in the surface cannot grow very quickly.

During the winter, the surface of the lake continues to cool. Freshwater is densest at 39°F (4°C). Ice, with a temperature of 32°F (0°C), is less dense than the deeper waters and so it forms on the surface of the lake. This provides fish and other invertebrates room to live under ice-covered lakes. The ice also acts an blanket–like insulation that helps keep the water underneath from freezing. In the springtime, the temperatures warm so the ice melts. Eventually the whole lake becomes 39°F (4°C) and so the waters from the bottom mix with the waters from the surface. This is called the spring turnover. As summer begins, the surface waters warm and the thermocline again separates the epilimnion from the hypolimnion. Because of the fall and spring turnovers, the nutrients from the bottom of the lake are available to the phytoplankton in the surface waters. This sets the lake up for the summer's rapid growth of phytoplankton and all the animals that depend on them.

Plant Life in Lakes and Ponds

Some of the most plants in lakes and ponds are the smallest. These phytoplankton are usually single-celled plants grouped with the algae. Sometimes they connect themselves together into long strings called colonies. Common phytoplankton in lakes and ponds are diatoms, which have beautiful shells made of silica (the same material that comprises sand); dinoflagellates, which move by snapping their flagella (long whip–like cell extensions that can propel an organism; and cyanobacteria, which are bacteria that perform photosynthesis. The larger plants in ponds and lakes include large algae and mosses, cattails, reeds, water lilies, bladderworts, willows, and button bush. These plants often grow in mud where the gases that they need to grow-such as oxygen and carbon dioxide-are scarce.

Many larger plants have stems that are spongy and they pull gases from the air down into their roots. Plants on land use their roots to gather water and nutrients, however aquatic plants are surrounded by water, and nutrients are dissolved in the water. Some aquatic plants have given up their roots. For example, duckweed (or water lentil) and watermeal are small pea-sized plants that float on the surface of lakes and ponds in the spring and summer. They absorb nutrients from the water and produce a lot of starches. By the fall, they are so heavy with nutrients that they sink to the bottom of the lake. They live out the winter in the mud at the bottom of the lake, existing on their stores of starch.

By spring, they have used up so much of the starch, they are light enough to float again. They pop to the surface just in time to use the strong light of spring and summer for photosynthesis and they begin to use their starch stores once again. Other large plants, like milfoil, water soldier, and water hyacinths also float on the surface of lakes and ponds. The edges of lakes are often divided into four zones based on the physical environment and the types of plants found there. Beginning farthest from the water, the swamp plant zone contains plants that have roots in the shallow water.

At times the water can recede from this zone, leaving the plants roots exposed to the air. Typical plants in the swamp plant zone are rushes and sedges (a type of plant that looks like a stiff grass). The next zone is called the floating-leaf and emergent zone.

Here the water never dries up, but the lake is shallow enough that the tops of plants emerge out of the water. A typical plant that lives in this zone is the water lily, which has special gas filled chambers in its leaves that allow it to stay floating on the surface of the water. In the submerged plant zone, plants live entirely underwater. Canadian waterweed and many types of mosses live in this zone. The freefloating plant zone takes up the center of the lake. Here plants without roots, like duckweed and water soldier, float freely on the surface.

Animal Life in Lakes and Ponds

Zooplankton float in the epilimnion of lakes and eat phytoplankton and other zooplankton. Usually, these animals are nearly transparent, in order to avoid being seen by their predators. Typical zooplankton in lakes include the water flea, Daphnia, which can reproduce without mating. Under normal conditions all of its offspring are female. However, when the animals are stressed, by lack of food for example, they will produce males. This mixes up the gene pool of the population and creates individuals that are likely to withstand environmental changes. Another typical freshwater zooplankton is the rotifer, which has bristles on top of its head that it whirls like propellers in order to move through the water and capture prey.

Many insects have juvenile stages that are aquatic. Mayflies, caddis flies, mosquitoes, and dragonflies all live for some period underwater in lakes and ponds. They swim among the rocks and plants in the lake bottom for a season or several years. Then they metamorphose (change in appearance) into their adult form and fly away from the water. The bottom of the lake is also home to many different worms, mussels, and crustaceans. These animals feed on the remains of plants and animals that drop to the bottom of the lake from above. Larger animals live in lakes and ponds.

In particular, fish, birds, and amphibians prey on the invertebrates that live in the lakes. Fish such as bluegills eat juvenile insects that swim in the bottom of the lake, while crappies eat zooplankton near the surface. Birds like flycatchers and warblers fly near the surface of lakes, preying on insects that are hatching from their juvenile stage. Frogs also hunt for insects that live near the pond. Still other birds and fish prey on smaller fish. Bass, salmon, osprey, loons, and heron hunt for fish by using their keen eyesight. Beavers and muskrats are mammals that depend on water for their homes. They build dams and lodges, which provide them with protection from predators.

GROUNDWATER FORMATION

Groundwater is fresh water in the rock and soil layers beneath Earth's land surface. Some of the precipitation (rain, snow, sleet, and hail) that falls on the land soaks into Earth's surface and becomes groundwater. Water-bearing rock layers called aquifers are saturated (soaked) with groundwater that moves, often very slowly, through small openings and spaces.

This groundwater then returns to lakes, streams, and marshes (wet, low-lying land with grassy plants) on the land surface via springs and seeps (small springs or pools where groundwater slowly oozes to the surface). Groundwater makes up more than one-fifth (22%) of Earth's total fresh water supply, and it plays a number of critical hydrological (water-related), geological and biological roles on the continents.

Soil and rock layers in groundwater recharge zones (a entry point where water enters an aquifer) reduce flooding by absorbing excess run-off after heavy rains and spring snowmelts. Aquifers store water through dry seasons and dry weather, and groundwater flow carries water beneath arid (dry) deserts and semi-arid grasslands. Groundwater discharge replenishes streams, lakes, and wetlands on the land surface and is especially important in arid regions that receive limited rainfall. Flowing groundwater interacts with rocks and minerals in aquifers, and carries dissolved rock-building chemicals and biological nutrients. Vibrant communities of plants and animals (ecosystems) live in and around groundwater springs and seeps.

Almost all of the fresh liquid water that is readily available for human use comes from underground. (The bulk of Earth's fresh water is frozen in ice in the North and South Pole regions. Water in streams, rivers, lakes, wetlands, the atmosphere, and within living organisms makes up only a tiny portion of Earth's fresh water.) For thousands of years, humans have used groundwater from springs and shallow wells to fill drinking water reservoirs, and water livestock and crops.

Today, human water needs far exceed surface water supplies in many regions, and Earth's rapidly-growing human population relies heavily upon groundwater to meet its ever larger demand for clean, fresh water.

AQUIFERS: FRESH WATER UNDERGROUND

An aquifer is a body of rock or soil that yields water for human use. Most aquifers are water-saturated layers of rock or loose sediment. With the exception of a few aquifers that have water-filled caves within them, aquifers are not underground lakes or holding tanks, but rather rock "sponges" that hold groundwater in tiny cracks, cavities, and pores (tiny openings in which a liquid can pass) between mineral grains (rocks are made of minerals).

The total amount of empty pore space in the rock material, called its porosity, determines the amount of groundwater the aquifer can hold. Materials like sand and gravel have high porosity, meaning that they can absorb a high amount of water.

Rocks like granite, marble, and limestone have low porosity, and make poor groundwater reservoirs. Aquifers must have high permeability in addition to high porosity. Permeability is the ability of the rock or other material to allow water to pass through it. The pore space in permeable materials is interconnected throughout the rock or sediment, allowing groundwater to move freely through it.

Some high–porosity materials, like mud and clay, have very low permeability. They soak up and hold water, but don't release it easily to wells or other groundwater discharge points, so they are not good aquifer materials. Sandstone, limestone, fractured granite, glacial sediment, loose sand, and gravel are examples of materials that make good aquifers. Water enters aquifers by seeping into the land surface at entry points called recharge zones and leaves at exit points called discharge zones. (Some aquifers discharge into the ocean.)

Influent or "water-losing" streams, ponds, or lakes are bodies of surface water in recharge zones that contribute groundwater from their water supply. Groundwater flows into effluent or "watergaining" streams and ponds in discharge zones. For the water level in an aquifer to remain constant, the amount of water entering at recharge zones must equal the amount leaving at discharge zones. (Imagine a bucket punched with holes under a dripping faucet. If water drips in at the same rate that it drips out, the water level stays the same.)

If water discharges or is pumped from an aquifer more quickly than it recharges, the groundwater level will fall. The time an average water molecule spends within an aquifer is called its residence time. Water in some fast–flowing aquifers spends only a few days underground, while other rock layers can hold water for ten thousand years. Average aquifers have residence times of about two hundred years.

The Water Table and Unconfined Aquifers

Water enters aquifers by moving slowly down through a layer of surface rocks and soil whose pore spaces are partially filled with air (zone of infiltration). The water continues moving downwards until it reaches a level where all the pore spaces are completely filled with water (zone of saturation).

The top of the zone of saturation is called the water table. In some wet, lowland regions, southern Florida for example, the water may be only a few feet (meters) below the surface. In others, like the American Southwest, water–saturated rocks may be hundreds of feet below the land surface. Groundwater reservoirs that have uniform rock or soil properties (porosity and permeability) throughout are called unconfined aquifers. The water table forms the upper surface of an unconfined aquifer. The shape of the water table in an unconfined aquifer mirrors the shape of the land surface, but its slopes are gentler.

In temperate (moderate) climates that receive moderate amounts of groundwater–replenishing rainfall, water infiltrates into unconfined aquifers in hilltop recharge zones and discharges into effluent streams and ponds in low areas where the water table intersects the land surface. Water will only rise to the level of the water table in a well, so a pump or bucket is required to extract water from an unconfined aquifer.

Confined Aquifers and Artesian Flow

Confined aquifers are pressurized groundwater reservoirs that lie beneath layers of non-permeable rock (granite, shale) or sediment (clay). Groundwater enters a confined aquifer in recharge zones beyond the uphill edges of the confining layer and discharges beyond the downhill edges. Groundwater trapped beneath an impermeable barrier cannot rise to the height of the water table, so pressure builds up in confined aquifers. Artesian wells are wells drilled in confined aquifers where the pressure is great enough to make water flow at the surface.

Lakes are large inland bodies of fresh or saline (salty) water. Lakes form in places where water collects in low areas or behind natural or man-made dams (barriers constructed to contain the flow of water). Some lakes are fed by streams (natural bodies of flowing freshwater), and some form where groundwater (water flowing in rock layers beneath the land surface) discharges onto the land surface. Water leaves lakes by flowing into outlet streams, infiltrating (soaking in) into groundwater reservoirs called aquifers, and by evaporating into the atmosphere (mass of air surrounding Earth). Lakes vary in size from large lakes such as the Great Lakes of North America, to small mountain lakes. Lakes are larger than ponds, which are small bodies of fresh water that are shallow enough for rooted plants to grow.

The study of ecology (relationships between living organisms and their environment) in lakes, inland seas, and wetlands is called limnology. Lakes store only a tiny percentage of Earth's fresh water. They are, however, an extremely important water resource for humans.

Freshwater lakes provide water for agricultural irrigation (watering), industrial processes, municipal uses, and residential water supplies. People who live in the continental interiors use lakes for fishing and recreation, and very large lakes have important shipping and transportation routes.

Humans also construct artificial lakes, called reservoirs, by building dams across rivers. In addition to providing the benefits of natural lakes, reservoirs also store water for specific communities, control floods, and generate hydroelectricity (electricity generated from water power).

HOW LAKES FORM AND DISAPPEAR

Earth scientists who study water on the continents (hydrologists and hydrogeologists) see lakes as temporary reservoirs within stream and groundwater systems. All water that falls as precipitation (rain, snow, sleet, hail) on the land surface of continents eventually makes its way to the ocean or evaporates back into the atmosphere. Water collects in lakes because it enters more rapidly than it escapes, but it is never permanently trapped there. As in a tub with a running faucet and an open drain, individual water molecules (smallest unit of water, each containing two hydrogen atoms and one oxygen atom) are constantly entering and escaping.

After arriving in a lake, an average water molecule spends about one hundred years before moving to a new reservoir. (The time that an average water molecule spends in a reservoir is called it residence time. Water resides for about two weeks in rivers, forty years in glaciers (slow moving mass of ice), and between two hundred and ten thousand years in groundwater reservoirs.)

To geologists (Earth scientists), lakes are temporary features. Stream-fed lakes within stream systems are destined for destruction. Every stream seeks to create a constant slope, called a graded profile, between where the stream's waters begin and ends by eroding (wearing away) and depositing sediment (particles of sand, silt, and clay) along its course.

When a natural or man-made obstruction blocks a stream, such as a river, streams deposit sediment in the lake or reservoir behind the obstruction and erode away in front of it. Eventually, the dam will collapse, and the lake will empty. Lakes that fill depressions and have no outlets fill when the regional climate becomes wetter or when warm periods melt mountain snows and glacial ice. They evaporate away during periods of dry weather and dryer climate. It may take thousands, or even tens of thousands of years, but lakes eventually drain, collapse, or dry up.

LAKE LAYERS AND OVERTURNS

Contrary to their common image as evenly mixed pools of unmoving water, lakes are complex, dynamic bodies of moving surface water. Lake water varies within the lake; its temperature, chemical content, light infiltration, and biological habitats vary from top to bottom and side to side.

Furthermore, the vertical layering (stratification), horizontal variations, and circulation patterns within lakes change over time. Waves, currents (a moving mass of water), and even tides affect circulation of water within lakes. Lakes are thermally stratified (layered just as to temperature); they have layers of warm and cool water that are separated by layers where the temperature changes (thermoclines). Like the oceans, many lakes have a thin layer of warm surface water, and a thicker layer of cool deep water that is separated by a thermocline layer. Wind generates currents on lake surfaces and creates some mixing. Unlike the oceans, however, many lakes have seasonal overturns that mix their waters. water is denser than solid water (ice). Water reaches its maximum density at 39°F (4°C). Because of this odd property, the warm less-dense water rises, the cool denser water sinks, ice floats, and lakes overturn.

Lakes that are ice-covered for part of the year undergo overturns that partially or completely mix their waters. Many lakes in temperate (moderate temperatures) regions like the northern United States overturn and mix completely twice a year (dimictic lakes). During the warm summer months, these lakes have a usual temperature profile with warm surface waters, a thermocline, and cool bottom water. In the fall, when the surface water cools down to 39°F (4°C), it becomes denser than the water underneath it and the surface layer sinks to the bottom.

The bottom water rises to the surface, and the lake overturns. Over the winter, the bottom water is the warmest, and the frozen surface water is the coldest. (Plants and animals survive the winter on the lake floor in the chilly, but not frozen, bottom water.) In the spring, when the ice melts, and the water warms to 39°F (4°C), it sinks, and the lake overturns again. Limnologists classify lakes just as to the number of mixing events they undergo each year.

Lake type classifications have the root term mictic, meaning "to mix," and include:

- *Oligomictic*: Warm, ice-free lakes that rarely mix. They are warmest at the top, and coolest at the bottom. Tropical, oligomictic lakes have warm bottom water and very warm surface water. They rarely overturn because their water does not near 39°F (4°C).
- *Meromictic*: Warm, ice–free lakes that mix incompletely. These are deep lakes that are warmest at the top, and coolest at the bottom.
- *Monomictic/dimictic/polymictic*: Lakes with seasonal ice covers that

overturn and mix completely once (monomictic), twice (dimictic), or many times per year (polymictic). Lake overturns are the norm in temperate regions, but local conditions affect the timing and number of overturns in specific lakes.

- *Amictic*: Lakes that never overturn because they are icecovered throughout the year. These lakes exist near the North and South Poles and atop very high mountains. They have cold bottom water that hovers near 39°F (4°C) and frozen surface water.

LAKE CHEMISTRY: SALINE LAKES

Many of Earth's largest and most important lakes contain salt water. All surface water contains some dissolved chemicals, called salts. Groundwater, streams, and freshwater lakes all contain the chemical components of rocks and minerals. Humans can drink fresh water because our bodies can use or at least tolerate the types and concentrations of dissolved chemicals it contains. Salt water, on the other hand, has a very high concentration of dissolved salts, and is undrinkable. The Dead Sea, on the border between Israel and Jordan, is Earth's saltiest body of water. It is truly a dead sea because it is too salty to support life. Saline lakes generally form in arid (dry) regions where surface water evaporates quickly.

When water evaporates, the salts stay behind. Over time, the lake water becomes saltier. Some saline lakes, such as the Great Salt Lake, are all that remains of a much larger fresh water lake that has evaporated over time. Others, like the Caspian Sea in central Asia, began as saltwaterfilled ocean basins that have since closed. Saline lakes are often temporary features that fill during periods of wetter climate and then dry up when stream flow or groundwater discharge slows. Playa lakes are flat desert basins that occasionally fill with water. Desert oases (watering holes) form and disappear with such regularity that thirsty travellers think they imagined them. The Great Salt Lake, Caspian Sea, Aral Sea, and Dead Sea are all presently evaporating. Over time, the dissolved chemicals become so concentrated in drying lakes that they bond together and form solid salt crystals. Thick layers of salt cover dry lake beds.

LAKE BIOLOGY

Lakes support rich communities of plants and animals (ecosystems) that have adapted to live within ever-changing conditions on lake beds, within the water column (water running from the surface to the lake floor, often showing differences in temperature, nutrients, etc.), and along lake shores. Lakes, like islands, are often closed systems that only rarely gain new species or individuals from other lakes.

Many lakes host groups of rare species that have evolved (changed over time) together in their specific lake. These ecosystems are rich and unique, but fragile. They have little defence against foreign predators or diseases. Human alterations and water pollution have threatened many lake species. Environmental groups and government agencies are presently attempting to protect and revive threatened lake species such as cichlids (rare doublejawed fish) that inhabit the lakes of the Great Rift Valley in east Africa. Lake organisms live in zones that are determined by the physical structure of their lakes such as the amount of available light, water depth, and distribution of nutrients. Most lake plants and animals live in shallow, well-lit surface waters called the euphotic zone.

Most plants depend on the Sun's energy to produce food by the chemical process of photosynthesis, and they cannot grow in water that is too deep or too cloudy for light to penetrate. Lake animals such as fish need oxygen that plants give off during photosynthesis, so they live mostly in the euphotic zone as well. Plants with roots grow in shallow water along edges of lakes where light reaches the lake floor (littoral zone) and floating plants perform photosynthesis in the open surface waters (limnetic zone). Oxygen-consuming bacteria inhabit the deepest, darkest parts of lakes (benthic zone) where dead plant and animal materials accumulate. Limnologists also classify lakes by the balance of organisms and nutrients in their waters.

Types include lakes that are described as oligotrophic, eutrophic, and mesotrophic:

- *Oligotrophic*: Nutrient poor lakes that support very few plants and animals. Oligotrophic lakes are typically cool, deep, and have very clear water. Very little organic (relating to or from living organisms) mud accumulates in oligotropic lakes, and they often have sand and gravel beds.
- *Eutrophic*: Lakes rich in plant nutrients that support abundant plant life in their surface waters. Their water is often clouded by microscopic plants, and their beds covered with thick layers of decaying plant material. Bacteria that live on the organic mud use up oxygen, and eutrophic lakes often have oxygen–poor deep water. Plants and bacteria eventually take over eutrophic lakes. They become oxygen-poor bogs and marshes where fish cannot live. Some chemicals that humans use, including fertilizers and detergents, cause a process called eutrophication when they run off into lakes, which causes the population of plants to increase to such an extent that eventually oxygen-starved fish die.
- *Mesotrophic*: Lakes with moderate amounts of nutrients and healthy, balanced communities of plants, animals and bacteria. Mesotrophic lakes receive adequate amounts of fresh water and nutrients, and seasonal overturns allow nutrient–poor and nutrient–rich layers to mix. Mesotrophic lakes are intermediate

between crystal-clear, lifeless oligothrophic lakes and cloudy, muddy eutrophic lakes.

WHERE LAKES FORM: LAKE BASINS

Lakes form where water collects in depressions, or basins. Many lakes fill low areas created by plate tectonic movements (tectonic basins) and volcanic activity. (Plate tectonics is the movement of large, rigid pieces of Earth's outer rock shell called the lithosphere.) Retreating glaciers and ice sheets leave behind large basins and small depressions that fill with meltwater. Though flowing streams and rivers generally act to fill in and drain lake basins, other sedimentary processes can create landscape depressions and natural dams that confine water in lakes.

Lakes in Tectonic Basins

Rift valley lakes fill long, linear valleys within rift zones. (Rifts are areas where the continental lithosphere is stretching and beginning to break into pieces. They are the precursors of ocean basins.) A chain of large lakes including Tanganyika, Naivete, and Malawi follows the Great Rift Valley through eastern Africa. Lake Victoria, the world's second-largest lake, lies between two branches of the rift valley. The Red Sea, Sea of Galilee, Dead Sea, and Gulf of Ababa fill the northern branches of the rift where it crosses the Arabian Peninsula in the Middle East. Russia's Lake Baikal, the world's deepest lake, fills an ancient, inactive rift valley in central Asia. Lakes also form in places where continents are moving towards each other.

The Black Sea, Caspian Sea, and Mediterranean Seas fill a closing ocean basin between Africa and Europe. When continents collide, water fills depressions in the landscape over folded and broken (faulted) rock layers that were caught between the land masses. Slopes that are too steep collapse and block rivers with natural dams. Blocks of uplifted, erosion-resistant rock form bedrock that holds back mountain lakes.

Volcanic Lakes

Volcanoes are mountains that form from eruptions of molten rock (lava) on the land surface. When a volcanic peak collapses into its emptied magma chamber (a pool or room of magna held under tremendous pressure within a volcano prior to a volcanic eruption), it forms a large circular basin called a caldera. (Craters, the small basins near the top of active volcanoes, sometimes also contain small lakes, but most significant volcanic lakes, including inaccurately-named Crater Lake, fill calderas.) Yellowstone Lake in Wyoming and Crater Lake in Oregon are examples of caldera lakes. Volcanic ash, mud, and lava flows also create natural dams in river valleys. A dam of volcanic rock confines Lake Tahoe in a high valley of the Sierra Mountains on the California–Nevada border.

Glacial Lakes

The thick continental ice sheets that covered northern North America, Europe, and Asia during the Pleistocene ice ages (a division of geologic time that lasted from two million to ten thousand years ago) left behind thousands of lake and ponds when they retreated about twenty thousand years ago.

The weight of the ice sheets pushed down on the continents, leaving broad basins that filled with melt water when they retreated. The Great Lakes of North America (Superior, Huron, Michigan, Erie, and Ontario) formed this way. Hundreds of lakes, such as the Winnipeg, Athabasca, Great Slave, and Great Bear cover the central and eastern provinces and territories of Canada that are still rebounding from their heavy ice load. Advancing glaciers also pile tall ridges of sediment, called moraines, at their toes (the end of extensions of glaciers along the ground). When glaciers retreat, moraines hold back meltwater. Small lakes and ponds also form in glacial depressions called kettles that form when blocks of ice buried in glacial sediment melt. Melting mountain glaciers feed many mountain lakes and glacial sediment traps streams and meltwater.

Groundwater Discharge Lakes

Water moving through pore spaces in rock and soil layers discharges on the land surface in places where the water table (level below which pore spaces are saturated with water) intersects the land surface. In regions with wet climates, the water table is near the land surface and ground water discharges in low spots. Groundwater chemically erodes limestone and other rocks and creates caves, cavities, sink holes, and collapse basins called karst features. Florida's many lakes, including Lake Okeechobee, are groundwater filled karst features.

PONDS

A pond is a depression in the ground that is filled with water that remains year round. Ponds range in size from the artificial backyard projects about the size of a bathtub to bodies of water that are about the size of a football field. Ponds support a variety of animal and plant life, and are also used as recreational sites by people. The difference between a pond and a lake involves size and water depth. A lake is big enough to have at least one beach (sand or rock that slopes down to the water) and contains enough water to generate waves from the wind that blows across the surface of the water. In contrast, a pond is usually too small for waves of any size to form. At the center of a lake, the water can reach depths of many hundreds, even thousands of feet (meters). A pond, however, is a shallow and still body of water where sunlight can usually reach down to the bottom.

HOW PONDS FORM

Natural ponds form in shallow depressions where rainwater (including run-off from nearby higher areas) collects. Water from an underground source such as an underground spring can also collect into a pond. Ponds that people enjoy in their backyard are often artificial, created by preparing the hole and adding water and plants to create a backyard oasis. These ponds can provide relaxation and a habitat for attracting insects, birds, and amphibians (such as frogs and salamanders), even in backyards located in a bustling city. Other artificial ponds are workhorses. One example is a sewage treatment pond. This type of pond keeps the sewage in a place where the growth of microorganisms can occur in the shallow and warm water. As microorganisms such as algae grow, they can use some of the materials in the sewage as food. This helps clean the water, and is an example of bioremediation, the process of using natural substances such as bacteria, to clean a contaminated natural resource, such as water.

LIFE IN AND AROUND PONDS

Ponds are havens for plants. Because the sunlight is abundant all through the water, plant can grow from every location in a pond. (Plants need sunlight to live as they convert the Sun's energy into food in a process called photosynthesis.) Often, the surface of a pond will be almost entirely covered with pond-loving plants such as the water lily and other plants that need higher levels of sunlight or that need direct exposure to air. Ponds also support various species of animal life, both in and surrounding its waters. Often, the bottom of a pond will be muddy, rather than rocky, and the mud hosts a variety of living creatures, such as crayfish. The still pond water and muddy bottom are also favourable conditions for the eggs of insects and creatures such as frogs to develop (often attached to the stems or leaves of plants). For microscopic life such as bacteria and algae, a pond offers plenty of food and the sunlit water provides a suitable temperature for the microscopic cells to grow and divide.

Animals such as deer often use natural ponds as a source of drinking water. Birds feed upon fish that live in ponds. Beavers find the still pond waters a good place to build their lodge. Ponds are often a source of relaxation and recreation. In warmer times of the year, a pond's edge can be a place where people picnic or rest outdoors. In the cold winter season of northern climates, ponds can freeze solid and host winter sports such as ice skating.

THE FATE OF PONDS

The flow of water into and out of a pond can be slow. This feature, along with its shallow depth, makes a pond vulnerable to contamination.

If chemicals that upset the natural composition of the pond are introduced, then the water quality necessary to sustain life can be destroyed. Ponds that form in arid (dry) regions where rainfall is briefly heavy then sparse throughout the rest of the year continue a cycle of filling up, then slowly drying. These ponds attract animal life only when water is abundant, which can sometimes cause conflicts with humans. In some regions of Africa, crocodiles return during the rainy season to newly filled ponds that form near populated villages. Hungry after hibernating (being in an inactive state) the rest of the year, the crocodiles pose a threat to livestock that also drink from the pond, and the people who tend the livestock. As time passes, the vast majority of ponds will naturally fill in, as sediment (particles of gravel, sand, and silt) and other debris collect in the shallow water. After about one hundred years, what was once a pond often becomes a field, and the water source of the pond is diverted by the changing landscape or by changes in rainfall amounts.

RIVERS

Rivers are bodies of flowing surface water driven by gravity. Hydrologists, scientists who study the flow of water, refer to all bodies of flowing water as streams. In common language, it is accepted to refer to rivers as larger than streams. Water flowing in rivers is only a very small portion of Earth's fresh water. The oceans contain about 96% of the water on Earth, and most fresh water is bound up in glacial ice near the North and South Poles. Rivers shape the landscape and are integral to the hydrologic cycle (circulation of water on and around Earth) on the continents. Rivers shape the lands as they erode (wear away) and deposit sediment (particles of gravel, sand, and silt) along their courses. Running river water acts to level the continents.

When geologic forces slowly raise (uplift) mountain ranges, rivers wear them away. The streams that form the Ganges River of India (headwater streams), for example, are presently tearing down the Himalayas almost as quickly as they are uplifted by the movements of Earth's crustal plates (plate tectonics). When geologic forces create depressions or low areas on the continents, rivers act to fill them. River sediment replenishes floodplain (Flat land next to rivers that are subject to flooding) soils and coastal sands. Earth's major rivers, including the Nile, Amazon, Yangtze, and Mississippi, drain the waters of vast continental areas and set down (deposit) huge deposits of sediments at the ends of rivers that flow into the ocean (for example, in deltas at the end of many rivers) Rivers host vibrant communities of plants and animals, and refill groundwater reservoirs and wetlands that support biological life far beyond their banks. Rivers are a main focus of human interaction with the natural environment.

Human agriculture, industry, and biology require fresh, accessible water from rivers. Ancient human civilizations first arose in the fertile valleys of the world's great rivers: the Yangtze and Yellow Rivers in China, the Tigris and Euphrates Rivers in the Middle East, and the Nile River in Egypt. The distribution of Earth's rivers and systems of rivers has influenced human population patterns, commerce, and conquest since ancient times. Rivers flow through the great cities of the world, and the imagery of rivers is deeply embedded in our language, culture, and history. Today, billions of people depend directly and indirectly on rivers for food and water, transportation and recreation, and spiritual and religious inspiration. Almost all major rivers are today confined by man–made dams and levees (walls along the banks) that provide people with the means to generate electricity and protection from floods. These alterations to rivers have come at an environmental cost.

When floodwaters are contained by levees or other flood–control dams, they no longer supply nutrients and sediment to floodplain soils that support agriculture. Furthermore, dams and levees that upset a river's natural path and profile (side view) cause changes to the patterns of erosion and deposition (depositing sediments) throughout the entire river system. Dams have contributed to beach erosion on many coastlines because dams trap sediment in reservoirs.

Agricultural and urban development along riverbanks has threatened many species of plants and animals that live in riverside wetlands. Also, the very dams and levees that prevent frequent small floods create an increased risk of infrequent, disastrous flooding. The city of New Orleans, for example, lies at a lower elevation than the bed of the Mississippi River that runs through the center of the city in an artificial channel behind massive levees. If the levees failed, a flash flood would engulf the city and potentially threaten the lives of its residents.

MAJOR RIVERS

Earth's largest river systems define the natural and human environment within their watersheds. A watershed is the land area that drains water into a river or other body of water. A list of the world's major rivers is also a list of the major natural and cultural geographic regions on six continents. (The continent Antarctica is too cold for liquid water. Its fresh water is bound up in large masses of moving ice called glaciers.)

- *Africa*: The Nile is, by most measurements, the world's longest river. (River lengths are difficult to measure because rivers constantly shift their courses and change length. There is also disagreement about which branches of water (tributaries) are considered part of the main river. By some measurements, the Amazon River in South America is actually slightly longer than the Nile.)

The Nile has sustained life in the inhospitable Sahara desert of eastern Africa for thousands of years. Its headwater (uphill end) streams flow from lakes in Ethiopia and Uganda and feed two branches, the White Nile and the Blue Nile, which meet in the Sudanese city of Khartoum. From there, the Nile cuts a green–bordered lifeline through the Egyptian desert. It flows through Cairo, the bustling capital of modern–day Egypt, past the pyramids of Giza and the ancient Egyptian capital of Thebes, to its outlet in the Mediterranean Sea. The Congo River (called the Zaire River from 1971 to 1997) makes a long loop through the equatorial rainforests and war-torn nations of central western Africa.

The Congo is the main trade and travel route into the African interior, and it is the setting for Joseph Conrad's famous novel Heart of Darkness. The Limpopo, Okavango, Ubangi, and Zambezi are other major African rivers.

- *Asia*: Huge rivers drain water from the massive Asian continent into the Pacific, Indian, and Arctic oceans. In China, the Yangtze (Chang Jiang), Yellow (Huang He) and Pearl Rivers carry flowing waters (run-off) from the northern slope of the Himalayan Mountains and western China to the East China Sea. Hundreds of millions of Chinese people depend on these rivers for their electricity, food, and livelihoods. Water moving south from the Himalayas flows into the rivers of India and South Asia, including the Ganges–Bramaputra system and the Mekong River.

  The Ganges River of northern India is sacred in the Hindu religion. Hindus travel to its banks to meditate and wash away their sins. Upon death, cremated remains are placed into the Ganges in hopes of improving the deceased's fortunes in the afterlife. The Ob, Ikysh, Amur and Lena Rivers run across the northern forests and wind-swept tundra (treeless arctic plains) of Siberia (the Asian portion of Russia) into the icy Arctic Ocean. In the Middle East, rivers play an important role in the history and mythology of western civilization.

  The ancient civilizations of Sumeria and Mesopotamia arose in the "fertile crescent" between the Tigris and Euphrates Rivers (Shat-al-Arab) in what is today Iraq. Along with the Jordan River, they play major roles in Jewish, Christian, and Islamic history.
- *Australia*: The island continent of Australia has only a few major rivers, and its central desert, the outback, is extremely dry. The Murray River and its major tributary (major branch), the Darling, make up Australia's largest river system. The Murray drains water from the southeastern states of Victoria, New South Wales and

southern Queensland and its floodplains are Australia's most productive farmlands. ü Europe: Rivers are intertwined in the history, culture, and geography of Europe.

The capital cities of Europe are synonymous with their rivers (London and Thames, Paris and Seine, Vienna, Budapest and Danube). By their very names, the Rhone (France), Rhine (Germany), Volga (Russia), Oder and Elbe (Germany, Poland, Czech Republic), Po and Tiber (Italy), and Ebro (Spain) conjure images of great art and fine wine, desperate battles and bloody conquests, grand castles and ancient hamlets.

- *North America*: The Mississippi and its major tributaries, the Missouri, Ohio, and Arkansas Rivers, collect water from a huge drainage basin that spans the central plains of North America between the Rocky Mountains and the Appalachians.

  Canada's Mackenzie and Churchill Rivers empty into the Arctic Ocean, and the St. Lawrence River empties the Great Lakes into the Atlantic Ocean. The mighty Yukon River of northern Canada and Alaska carried prospectors to mines and mills during the Alaskan gold rush (1898–99). Many of the great ports of the Atlantic seaboard and Gulf of Mexico lie near river mouths (the end of a river where the river empties into a larger body of water): New York (Hudson), Philadelphia and Washington, D.C. (Potomac, Susquahana), Norfolk (Delaware), New Orleans (Mississippi), and Houston (Brazos). Rivers, including the Mississippi, Missouri, Colorado, Rio Grande, and Columbia, played central roles in European exploration and settlement of the American West. Today, the rivers that carried explorers Meriwether Lewis, William Clark, John Wesley Powell, and other legendary frontiersmen across the continent are used for agricultural irrigation, drinking water, recreation, and power generation. Their water is a valuable and heavily-sought resource.
- *South America*: The Amazon is the largest river in the world. It flows from the Andes Mountains of Peru, across the Brazil and empties into Atlantic Ocean on the northeast coast of Brazil. The Amazon has more than 1,100 tributaries, 17 of which are longer than 1,000 miles (1,609 kilometers) long. The main river runs from west to east just a few degrees south of the equator, and its massive watershed lies entirely within the warm, wet tropical zone. The central Amazon contains Earth's lushest, wettest, most biologically diverse rainforest. The Orinoco (Venezuela), Sao Francisco (Brazil), Parana (Argentina, Paraguay) and Uruguay (Uruguay, Brazil) rivers are other major waterways of South America.

STREAM SYSTEMS

Streams are any size body of moving surface fresh water driven towards sea level by gravity (force of attraction between two masses). Water scientists refer to all bodies of flowing sur-face water as streams regardless of size, yet in common language, streams are considered smaller than rivers.

Stream systems are networks that collect fresh water run-off from the land and carry it to the ocean. Together, tree-shaped systems of small branch streams drain vast areas of the continents into large rivers. Stream systems of all sizes erode (wear down) sediment (particles of gravel, sand, and silt) along their courses and carve complex patterns into the landscape.

They wear down slowrising mountains and fill valleys and lowlands (low and level lands) with layers of sediment. Stream systems change character along their courses. Steep mountain streams feed shallow elevated streams that in turn flow into meandering rivers that snake across broad floodplains (flat, low-lying land near a stream that is covered with water when the stream overflows its banks). Deposits of sediment form at river mouths, the area where fresh river water enters the ocean.

If a rubber duck was dropped into a mountain stream on Pike's Peak in Colorado, it might tumble down the mountainside in whitewater rapids to Cripple Creek. From there, the duck would rush over gravel beds where Colorado miners once panned for gold, and then float serenely across Kansas, Oklahoma, and Arkansas on the Arkansas River. It would pause to drift across huge man-made reservoirs, and then plunge through the spillways of dams before entering the swift, muddy waters of the Mississippi River. A few weeks or months later, you might spot the duck heading out to sea amid barges and river boats in New Orleans.

WATERSHEDS AND DRAINAGE PATTERNS

The land area that drains water into a stream is called a watershed or a drainage basin. A basin is a natural depression in the surface of the land. Watersheds can be as small as a hillside that feeds a wet-weather creek, and as large as a drainage system like the Amazon Basin that carries the run-off from most of a continent. Large watersheds are composed of many smaller drainage basins. The boundaries between watersheds, called drainage divides, are ridge lines or high points where water flows down and away in all directions.

A divide can be limited, like a ridge between two mountain gullies (deep ditches or channels cut in the earth by running water, usually after a rainstorm), or extensive, like the North American Continental Divide along the spine of the Rocky Mountains. Water that falls east of the Continental Divide eventually flows into the Atlantic Ocean, and water that falls west of the Rockies ends up in the Pacific Ocean. Streams are arranged within watersheds in networks that feed water into larger and larger streams. Tree-shaped (dendritic) systems composed of small branch tributaries (small streams that flow into larger streams) that join and flow into large trunk streams are the most common type of stream drainage pattern.

Less common drainage patterns develop where rock layers and geologic features affect the paths of streams. Drainage patterns shaped like cross-hatched garden trellises develop in hilly areas where there are ridges and valleys, and streams flow out from round volcanic mountains in radial patterns like spokes on wheels.

VALLEY AND CHANNELS

Streams cut down into the land surface and create valleys. A stream valley includes the entire area between hills on either side of a stream. The water–filled path of the stream at a specific point in time is called a channel. Over time, channels migrate back and forth and fill stream valleys with thick layers of river sediment. Some streams, particularly those in steep, mountainous terrain have narrow, V-shaped valleys and channels that fill most of the valley floor. Others, including most streams in gently–sloping basins and coastal lowlands have narrow channels that snake across wide sediment filled valleys. For example, the Mississippi River has carved a valley more than 100 miles (161 kilometers) wide and filled it with sediment hundreds of feet (meters) thick over thousands of years.

CHANNEL PATTERNS

Stream channels assume different patterns within their valleys: straight, braided and meandering. While many channels have straight segments between meanders or braids, truly straight channels are quite rare. They develop in steep, mountainous areas where geologic forces are slowing lifting up the land surface. Water flowing rapidly downhill from mountains saws straight channels down into solid rock. Braided streams have many intertwined channels and islands of loose gravel that constantly shift across gravel-filled valley floors. They are common in streams that receive large pulses of water and course-grained sediment.

The sediment-choked streams that carry water from the toes of melting glaciers are typically braided. Streams that bend and curve across gently sloping valleys and coastal plains are called meandering streams. (Individual loops and bends are called meanders.) During normal weather conditions, water flows in a narrow channel that snakes across broad plains of soft sediment. During floods, muddy water overflows the banks of the channel and deposits layers of mud and silt on the surrounding floodplains. River floodplains are typically fertile farmlands that have been replenished by floodwaters.

The coarser grained sediment settles out of flood waters closer to the channel builds natural levees (walls along the banks of a stream channel) along its banks. The path of a meandering channel changes over time. Meanders grow from slight bends into nearly-circular loops. At a river bend, fast-flowing water erodes the outer channel bank and sediment accumulates on the inside of the curve in a deposit called a point bar. Eventually, the bends at the neck of the meander grow so close that the water bypasses the loop.

This process strands crescent-shaped segments of the former channel and round point bar deposits called oxbows on the floodplain. Oxbow lakes are abandoned meanders that contain water. Channel patterns change down the course of a stream system between headwater streams and lowland trunk rivers. They also change over time as streams adjust to changing conditions of water flow, land incline, and amounts of sediment. Stream waters continuously erode and deposit sediment over time, and stream channels constantly shift across valley floors.

STREAM WATER FLOW

Water flows downhill due to Earth's gravity (force of attraction between two masses) pulling it. Streams, like rivers, are gravity-driven bodies of moving surface water that drain water from the continents. Water scientists, called hydrologists, refer to all bodies of running water as streams, no matter their size so, in one sense, rivers are large, well-established streams). In everyday communication, it is common to refer to streams as smaller than rivers. Streams transfer water that falls on the land as precipitation (rain, snow, sleet, and hail) to the oceans. Streams, again like rivers, constantly shift their courses and change length.

The stream is carried along a defined path, called a channel. Water flowing in stream channels is a powerful sculptor that carves landscapes and molds sediment (particles of rock, sand, and silt). It wears down mountain ranges and cuts deep canyons through solid rock. Stream waters support vibrant communities of plants and animals, and they have been the lifeblood of human civilization for thousands of years. Streams shape the land and are also integral to the hydrologic cycle (circulation of water on and around Earth).

EROSION AND DEPOSITION

Streams are the main agent of erosion (wearing away) on land. Water in fast-moving streams is usually turbulent. The flowing water is filled with swirls and small localized whirlpools of swirling water called eddies. Turbulent water picks up particles of sediment that have weathered from rock and soil and carries them downstream. (Weathering is the breaking up of rocks by physical and chemical processes, such as being exposed to the actions of water, ice, chemicals, and changing temperature.) Faster-moving water can carry more sediment in the water, and can push larger stones along the bottom of the channel.

Some mountain streams move huge boulders, while sluggish lowland (low country and level) streams carry only fine grains of silt and mud. The sand grains and larger rock fragments that slide and bounce along stream beds wear away solid rock. In a straight stream, the fastest-moving water and area of greatest erosion is generally in the middle of the channel. Where a stream bends, the strongest current (a moving mass of water) is on the outside of the curve. When water slows down, it drops its sediment load, causing sedimentary deposits to form along stream courses in areas of slowmoving water.

The slower the current, the finer the sediment it deposits. In straight channels, stream water lays down sediment along the stream banks. In bending channels, sedimentary deposits called point bars form on the inside of the bends. Individual sediment grains travel downstream like hitchhikers. Sometimes the grains are picked up by a strong current or flood that moves them far downstream, but usually they don't go very far in a single trip.

The grains of sand on a beach each made a long trip with many stops before they arrived at the ocean. Whether an individual grain of sediment moves depends on the speed of water currents that vary as the amount of water moving through a stream changes. As water currents become faster they can move larger grains. Stream waters also erode rocks by dissolving its minerals, which causes them to crumble. Chemical weathering, also called dissolution, occurs when the slightly acidic water chemically alters the minerals in rocks, which causes them to break down. Clear stream water carries the chemical components (parts) of the rock's minerals called ions (electrically charged particles).

When conditions in the water change (the water slows or cools), the ions recombine into solid mineral crystals. This form of sedimentary deposition is called precipitation. Limestone, salt, and gypsum form by precipitating from water. Ocean animals like corals and shellfish take in ions and use them to build their shells. Some types of rocks, including chalk and flint (also known as chert) form from the remains of organisms.

GRADED STREAMS AND BASE LEVEL

All streams strive to reach a constant slope (incline) called a graded profile by eroding and depositing sediment. The profile (side-view) of a graded stream (a stream with a graded profile) is steep near the uphill end and gently sloping near the point at the end where a stream pours its water into a larger body of water. The position of the downstream end of the profile is determined by the water level at the outlet, called base level. Streams cannot erode below base level. Almost all stream systems run to the sea, so sea level is the ultimate base level for most streams. Conditions change constantly in all streams, and the process of readjustment by erosion and deposition is ongoing.

As conditions change along its course, a stream will readjust its profile by eroding sediment in some places and depositing it in others. If base level falls, stream waters cut down into the land surface. If it rises, they deposit more sediment. If the movements of the underlying plates of Earth's crust rise (geologic uplift) to steepen the upper part of a stream, it will erode down to regain its graded profile. Streams also attempt to level out obstructions along their path. They work to tear down dams, both natural and man-made, by erosion and filling the reservoir behind it with sediment. Lakes are, therefore, only temporary features of stream systems, and dams are interrupting the natural flow of a stream's water.

Chapter 7

Estuaries and Wetlands

ESTUARIES

Estuaries are the areas where rivers run into oceans. They often exist where the opening to the sea is somehow obstructed, for example by a sandbar or a lagoon (sandbars are ridges of sand built up by water; lagoons are shallow areas of water separated from the ocean by sandbars or coral). The water in estuaries is dominated by the flow of the tides. When tides are high, the ocean water washes through the estuary bringing with it sediments (particles of sand, silt, and gravel), nutrients, and organisms from the ocean.

When the tide is low, the freshwater of the river floods the area, releasing its load into the estuary. Because estuaries exist where two different types of water come together and where the land meets the water, estuaries provide many different types of habitats for animals and plants. In addition, both the river and the ocean bring estuaries nutrients such as nitrate and phosphate, which plants need to grow. This results in a complex range of plants and animals that thrive there. Estuaries are also important to human settlement and economics. As a result, estuaries are often subject to pollution and other environmental stresses.

GENERAL STRUCTURE OF AN ESTUARY

The part of the estuary farthest from the ocean is often called a salt marsh. (A marsh is a wetland dominated by grasses.) Water usually flows through salt marshes in tidal creeks. Unlike river water, the water in tidal creeks can flow in two directions. When the tide comes in, the water runs into the salt marsh and when the tide goes out, the water runs the opposite direction, away from the salt marsh. The part of an estuary closer to the ocean may contain mudflats (a thick, flat layer of mud or sand that is usually underwater at high tide) and sandbars. These areas are exposed when the tide is out and may be covered with water when the tide is in. They are often covered with a layer of thin algae, which are tiny rootless plants that grow in sunlit water. Many different types of burrowing (digging holes or tunnels) creatures, like clams and worms, live on mudflats and sandbars.

Birds often walk along mudflats and sandbars when the tide is out, hunting for prey (animals hunted for food) buried in the ground. The ocean edge of the estuary is almost always covered with water, although its depth changes with the tides.

In this region, river water and ocean water mix and the resulting water has a salinity (the concentration of salt in water) that is neither fresh nor seawater. This type of water is called brackish. Brackish water includes water of a large range of salinities, from freshwater, which is about 0.5 part salt per thousand parts of water (ppt) to seawater, which is about 35 ppt. The ways that the freshwater and the ocean water mix within the estuary is often very complicated. Sometimes the freshwater sits on top of the ocean water, because it is less dense. When this occurs, a halocline forms between the two types of water. (The root word halo means "salt" and the root word cline means "change.") A halocline is a layer of water where the salinity changes very quickly. The halocline can act as a physical barrier between the freshwater on top and the saline water below, blocking the exchange of nutrients, and even organisms, between them.

LIFE IN ESTUARIES

Brackish waters pose one of the most important challenges for many animals and plants living in estuaries. Because the salinity of the water is constantly changing, their cells must be able to handle osmotic changes. Osmosis is the tendency of water to have the same concentration on both sides of a material that allows liquid to pass (like a cell membrane, the structure surrounding a cell). When exposed to fresher water, cells that have grown accustomed to waters that are more saline will take in water, expand and even burst.

When exposed to more saline conditions, cells that have grown accustomed to fresher water will release water, shrivel, and perhaps die. There are a variety of animals and plants that have special adaptations so that they can live in waters with changing salinities and these organisms thrive in estuaries. A second problem facing organisms that live in estuaries is the ever-changing water level. Because the tide goes in and out, animals and plants must be able to handle waterlogged environments as well as environments that are dry.

Many animals burrow in the sand and mud in estuaries. For example, sea cucumbers and polychaete worms live in holes in the mud. They expose their tentacles to the water where they capture plankton (free-floating organisms) and small prey that float into their reach. When the tide goes out, they burrow into their holes where they can stay moist.

Plant Life in Estuaries

The salt marsh region of the estuary is characterized by plants that are adapted to salty conditions. The high salt marsh cordgrass has special organs on its leaves that remove the salt it takes up from its roots. The eelgrass, Spartina, looks like a grass with very tough leaves and stems that help it retain moisture in saltwater. It can be found in salt marshes throughout the East Coast of the United States. Other common salt marsh plants are sea-lavender, scurvy grass, salt marsh grass and sea-aster. Farther out in the deeper waters of the estuary, microscopic phytoplankton (tiny plants that float in fresh or saltwater) are some of the most important plants.

These single–celled algaetype plants float near the surface of the water where sunlight is available. Because the ocean water and the river water both deposit the nutrients that phytoplankton needs to grow quickly, phytoplankton in estuaries flourish. The large populations of phytoplankton are food for zooplankton (free-floating animals, often microscopic). In turn, the phytoplankton and zooplankton are meals for worms, clams, scallops, oysters and crustaceans (aquatic animals with jointed limbs and a hard shell).

Animal Life Iin Estuaries

Because the types of habitats in estuaries are so diverse, estuaries are home to many different species of animals. Worms, clams, oysters, sea cucumbers, sea anemones and crabs all make their homes in the muddy floor of the estuary. Many of them burrow in the mud and filter the water for plankton and small fish that swim within the grasp of their tentacles and claws. In some places, the clams and oysters become so numerous that their shells provide special habitats for other small animals. Barnacles grow on oyster shells in oyster beds. Small fish, snails, and crabs will hide from larger predators in the crevasses between clamshells. Mosses and algae will grow on the surfaces of some molluscs, providing food for the animals that take refuge there.

A variety of fish live in estuaries. Very small fish called gobies hunt along muddy and rocky surfaces for small crustaceans like shrimp. Long slender fish called pipefish swim among the grasses in the marsh, their shape blending in with the long blades of the plants. Larger fish like halibut and flounder, swim along the muddy floor, their flattened shape allowing them to move into the shallow regions of the estuary. Large predatory fish like redfish, snook, striped bass, mullet, jack, and grouper make their way into estuaries to feed on the rich supply of fish that can be found. Salmon pass through estuaries on their way up rivers to breed. Many fish and invertebrates (animals without a backbone) use the estuary as a nursery ground for their young. For example, in Florida, a variety of species of shrimp spawn in the ocean, and their larvae (immature young) travel to the mouth of the estuary, where they develop into young shrimp.

At a certain stage of their development, they ride the tide into the estuary, where they live among the eelgrass. The eelgrass provides them with protection from predators and the rich nutrients in the estuary produce plenty of food for them to eat. Once the shrimp become adults, they swim back to the ocean, where they spawn, producing young that will move back to the estuaries again. Birds are extremely numerous in estuaries. During low tides, a variety of shorebirds walk along mudflats, pecking their beaks into holes where worms, crabs and clams are buried. Herons scour the shallow waters for shrimp and small fish. Brown pelicans, an endangered species, use estuaries as breeding grounds and nesting areas for their young.

IMPORTANCE OF ESTUARIES

Estuaries are a unique habitat for a large variety of animals and plants. Because of their complexity a broad variety of species live in estuaries, either for part of their lives or for their entire life. The U.S. Department of Fisheries estimates that three–quarters of the fish and shellfish that people eat depend on estuaries at some point during their lives. Oysters, clams, flounder, and striped bass may live their entire lives within estuaries. Estuaries serve as a buffer from flooding and storm surges. The soil and mud in estuaries is absorbent and can absorb large quantities of water. In addition, the roots of the grasses and sedges (grass-like plants) in estuaries are able to hold together sediments and protect against erosion (wearing away of land). Estuaries provide important protection to the real estate in many coastal communities. As water moves through an estuary it is naturally filtered and cleaned. The many plants and bacteria that live in the estuary use pollutants, like agricultural fertilizers, to grow. Sediments that are transported to estuaries by rivers tend to settle into the estuary, where they act as filters, allowing cleaner water to flow into the ocean.

DANGER TO ESTUARIES

Bacteria can break down some, but not all pollutants and many pollutants are not taken in by plants. Pollutants can build up to harmful concentrations within estuaries that threaten the health of the birds, fish, and humans that live nearby. There are four major types of environmental stresses that affect Chesapeake Bay, the largest estuary in the United States. The most damaging type of pollution to the Bay is the input of nutrients like phosphate and nitrate, which are fertilizers used in agriculture. High concentrations of nutrients enter the Bay as rainwater run-off from land and from sewage treatment facilities. Although they are required for plants to grow, high concentrations can cause overgrowth of algae and marsh plants. This overgrowth can result in the plants using up all the oxygen in the water, causing the fish to die.

A second type of pollution is the input of sediments like clay, sand,

and gravel that enter the Bay through river run-off. Although sedimentation is a natural occurrence, increased rates of erosion sometimes cause large amounts of sediments to be deposited in the Bay. Sediments can clog the feeding apparatus of filter-feeding animals and can cloud the water making it more difficult for plants to get light. Air pollution is a third source of stress on the bay. Pollutants released from factories and cars as exhaust eventually make their way to the bay.

Some of these pollutants produce acid rain, which changes the acidity of the bay, while others contribute to the concentration of nitrogen in the bay. Although evidence shows that dangerous pollutants, a fourth stress on the environment of the bay, are currently not as damaging as the other forms of pollution, the release of chemicals into the bay from some of the industries in the region can be deadly to both animals and plants.

WETLANDS

Wetlands are areas of land where water covers the surface for at least part of the year and controls the development of soil. Plants and animals that live in wetlands are adapted to living in conditions where the soil is waterlogged. There are many different types of wetlands, but they fall into five general classifications: freshwater marshes, freshwater swamps, salt marshes, mangrove swamps, and bogs (also called fens). In general, swamps have trees, while marshes have plants that have soft stems like grasses, reeds, and sedges (grass-like plants). Bogs are characterized by thick mats of peat, which is made of mosses growing on decayed plants and animals.

FRESHWATER MARSHES

Freshwater marshes are found on the edges of lakes, ponds, and rivers. They are particularly common near the floodplains of rivers. About one-quarter of all the wetlands in the United States are freshwater marshes. Plants that grow in freshwater marshes have adapted to survive in waterlogged conditions.

They do not need stiff stems to hold them up because the water provides buoyancy (the ability to float). Water lilies are typical plants that grow in freshwater marshes and they have flexible stems that bend and straighten as the water level changes.

The leaves of the water lily float on the surface of the marsh. Because the roots of freshwater marsh plants are often underwater or in mud, they are frequently exposed to low oxygen concentrations. They have developed spongy stems full of air spaces that can transport oxygen to the plant roots. Some marsh plants, like duckweed, don't have roots at all. They float and absorb minerals from the water. Animals that live in freshwater marshes have also adapted to use their wet environment to their advantage. Many of the insects that live in freshwater marshes have aquatic stages that take up a majority of their life. For example, dragonflies spend up to five years as larvae that live underwater.

The adult fly stage only lasts for one season. Many birds make the marsh their home. Often, birds like the American bittern are striped and well-camouflaged by the marsh grasses. Other birds, like the herons, have extremely long legs and wade through the marsh hunting for fish. Ducks are especially common marsh birds. They have special organs that produce oils to waterproof their feathers. Nearly three-quarters of all ducks in North America breed in freshwater marshes in prairies.

Freshwater Swamps

It is estimated that about two-thirds of all wetlands in the United States are swamps. Trees are the dominant plant in this environment and the trees species found in swamps vary depending on location. In the northern part of North America, evergreens such as spruce, fir, and cedar are commonly found in swamps. In New England, red maple dominates. Black willows and cottonwoods dominate swamps along the Mississippi and Connecticut rivers.

In the southern part of the United States, bald maple, tupelo, and sweetgum trees are commonly found in swamps. The bald maple produces knobs that look like knees that are probably used to acquire oxygen for the roots of the trees, which are buried in mud that has no oxygen. Bald maples, tupelos, and sweetgum trees also grow buttresses (support limbs) that help hold up the trees because their roots are close to the surface of the mud where oxygen is a little more accessible. Trees are crucial to the ecology (relationships between living organisms and their environment) of the freshwater swamp. Because many of the trees that live in swamps lose their leaves each fall, swamp water is highly enriched from decomposing plant material.

Many worms, newborn insects, and crustaceans (aquatic animals with no backbone and a hard shell) live by cutting apart fallen tree leaves for food. Bacteria break down the leftovers and produce nutrients (substances necessary for plant and animal growth) for other animals. These animals attract birds, snakes, and mammals to swamps. In addition, trees provide important hiding places for birds. Woodpeckers drill holes in trees and use them for nests. After the woodpeckers abandon their holes, other birds like owls, titmice, and wrens take over. As trees die and fall over into the swamp waters, ducks and snakes use rotted holes for their homes.

Salt Marshes

Salt marshes are marshes that are found along ocean coasts. One of the most important features influencing salt marshes is the tide that goes in and out each day. When the tide is high, marsh grasses, algae, mussels, crabs, snails, and worms are covered in cool, salty water. When the tide is low, these same organisms may bake in the sun. The animals that live in salt marshes have developed different adaptations to overcome these extreme changes in environment. Land snails crawl up and down the stems of marsh grasses keeping away from the water as the tide goes up and down. Land crabs dig burrows (tunnel or hole) in the mud, which they seal up to protect themselves from the waters of the high tide.

Marine (ocean) snails scurry into pools in the sand and close their shells tightly as protection from drying out during low tide. Salt marshes are places of great biodiversity (variety of life) and productivity. Nutrients circulate into salt marshes from the ocean and rivers. In addition, as plants and animals die and are buried in the marsh, bacteria digest and convert them into nutrients. All of these nutrients support an enormous amount of plant growth. One acre of salt marsh can produce 4.8 tons (4.3 metric tons) of plant material each year, almost twice that of a corn field given fertilizers. This plant growth, in turn, supports many species of invertebrates (animal without a backbone), fish, and birds.

MANGROVE SWAMPS

Mangrove swamps are swamps found on the edge of oceans in tropical (hot) regions. Mangrove swamp water is salty and the plants and animals that live in mangrove swamps have unique characteristics that allow them to handle these salty conditions. Many of the plants that live in mangrove swamps have special organs that remove salt from their tissues.

In addition, many of the sediments (particles of sand, gravel and silt) in swamps are low in oxygen, so plants must be able to adapt to low oxygen levels. Mangroves are trees that can tolerate saltwater. Many mangroves have special adaptations to gather oxygen from the air. There are a variety of different species of mangroves and they each grow best in slightly different environments. In North America, red mangroves are found closest to the water. They have tough roots called prop roots that help anchor them to unstable sandy soils in shallow water. These roots trap sediments and also provide habitats for juvenile fish and invertebrates.

In particular, oysters, anemones, snails, and snakes all live attached to red mangrove roots. Red mangroves have special pores called lenticels in their stems that they use to take in oxygen. Black mangroves are found behind the red mangroves, closer to land. They have special roots that grow like spikes out of the swamp water to gather oxygen from the air. White mangroves are found closer to the land behind black mangroves. On the edge of the swamp that is closest to land, a mangrove called a buttonwood is most common. Many birds and mammals make their homes in mangrove swamps. Roseate spoonbills, ibis, and pelicans are all commonly found digging their bills into the swamp sand hunting for fish and invertebrates.

Egrets, herons, and spoonbills also use the mangrove canopy for nesting, in order to avoid predators such as mammals and fish. Rabbits and raccoons are land mammals that can be found in mangrove swamps. The manatee is a marine mammal that makes its home in the swamps of Florida. It is herbivorous (plant-eating), and grows to enormous sizes (3,500 pounds or 1,600 kilograms). Manatees are classified as an endangered species; the greatest threat to their existence is boat propellers.

BOGS

Bogs are unusual environments. They occur near springs, slow-moving streams or small ponds. Water in bogs is usually very acidic, and because bacteria generally do not grow well in acidic environments, bog water has few bacteria. As a result, dead plant material is not decomposed; instead it piles up, pressing down on layers of plant material below it. This produces peat, which is so thick and solid that it can be used like coal for fuel. As bogs are waterlogged, they do not become solid and walking across a bog feels like walking across a waterbed. Some species of moss and sedge (a grass-like plant) can grow on top of the peat. Most of these plants have adapted to take in oxygen from the air because the soil in a bog is usually low in oxygen. One of the most important bog plants is called sphagnum moss. Sphagnum moss is acidic and thus, few bacteria grow on it. Besides its antiseptic qualities, sphagnum moss. is highly absorbent and was used extensively for bandages during World War I (1914–18).

It is soft and spongy; dry sphagnum moss can absorb water until it weighs as much as twenty times its original weight. Native Americans frequently used sphagnum moss for baby diapers. With few bacteria living in acidic bog waters, decomposition occurs slowly and nutrients remain locked inside dead plant material. Plants that do grow in bogs must be creative in order to get nutrients. Many species of carnivorous (meat-eating) plants live in bogs. They receive nutrients from the prey that they capture, mostly small insects, but still use photosynthesis (process of producing food using sunlight) to generate sugars. Some examples of carnivorous bog plants are Venus flytraps, bladderworts, and pitcher plants.

IMPORTANCE OF WETLANDS

Wetlands are places of great biological diversity. In swamps and marshes, the water is shallow enough that light can penetrate to the bottom, meaning that a great deal of photosynthesis can take place. This, in turn, leads to many different types of herbivorous animals, considered the primary consumers, and animals that in turn eat the herbivores, considered the secondary consumers. Wetlands are very important breeding grounds for both fish and invertebrates. Nearly two-thirds of all shellfish and bony fish in the ocean rely on wetlands for some part of their lives.

About one-third of all endangered species spend time in wetlands, just as to the U.S. Fish and Wildlife Service. Wetlands have several important ecological roles. Because of their ability to trap water, they act as sponges during heavy rains, slowing the flow of water into rivers and minimizing floods. This decreases erosion (wearing away of land) while protecting homes and other structures that could otherwise be flooded with water.

In addition, this slowing of rainwater allows some of it to seep into aquifers (underground basins containing fresh water), which replenishes the source of much drinking water. Wetlands also greatly reduce pollution and a process called eutrophication. Eutrophication is the process of water becoming enriched with nutrients, causing a population explosion of aquatic plants and reducing the oxygen in the water. This results in the loss of animal and plant life. Wetlands absorb many of these nutrients, allowing plants—which take in pollutants—to grow. Wetlands are not permanent features. The natural cycle of a wetland is to convert to a land habitat as sediments accumulate.

The rate at which this occurs depends on the sediment accumulation rate, the rate of precipitation (rain and other water), and changes in sea level. Since humans began using wetlands for agriculture in ancient Mesopotamia (a region in modernday southwest Asia), they have influenced how wetlands have changed. Agriculture uses fertilizers, which add enormous concentrations of nutrients to the soil. When it rains, these fertilizers run into rivers and oceans and initiate eutrophication. Building pipelines used to transport water and dams, and draining wetlands have accelerated the destruction of wetland habitats throughout the world. This causes a loss of habitats for many animals and plants, and is a loss to the environment and people.

Chapter 8

Ice

ARCTIC AND SUBARCTIC REGIONS

The region encircling the North Pole is called the Arctic Circle, an invisible circle of latitude (imaginary line around the Earth parallel to the equator) at 66°33' North. The arctic region sits inside the Arctic Circle and the subarctic region lies just below it. Earth's arctic and subarctic regions are extremely cold, icy areas of land and sea that receive almost no sunlight during their long, dark winters. Temperatures rarely rise above freezing. This is true even during summer in the "land of the midnight sun."

The Sun's rays hit the poles at a very shallow angle and the summer sunlight—while long-lasting—is too weak to provide much heat. Arctic and subarctic regions, however, support diverse groups of land and marine (ocean) plants and animals, including humans that have learned how to survive in their harsh climates. Water, both frozen and liquid, plays a vital role in arctic and subarctic environments. Arctic ice cools warm ocean currents and generates cold deep ocean water. Deep, cold currents flowing south from the Arctic Ocean distribute nutrients and control temperatures in the Atlantic and Pacific Oceans. The permanent covering of ice over its large area is called the Arctic ice cap.

The Arctic ice cap helps to regulate Earth's temperature by reflecting sunlight. Global sea level and ocean water chemistry change when the amount of glacial ice (large masses of moving ice) on the northern continents and islands changes. The North Pole lies under a zone of dry, sinking air in the atmosphere (mass of air surrounding Earth), and the Arctic and is a windblown, frozen desert that receives very little snow each year (it almost never rains there because it is too cold).

GEOGRAPHY OF THE ARCTIC

A ring of continental land masses and large islands surrounds the ice-covered Arctic Ocean over the North Pole. The far northern portions of Asia (Russia, Siberia), Europe (Scandinavia) and North America (Canada, Alaska) lie within the Arctic Circle. Large islands extend even farther north: Nova Zemlya (Russia), Spitsbergen (Norway), Iceland, Greenland (Denmark), and the Queen Elizabeth Islands (Canada). Treeless plains called tundra cover the arctic and subarctic zones between the edge of cool, snowy, northern forests (boreal forests) and the coastlines of the Arctic Ocean.

ARCTIC ICE

Sea ice covers the entire Arctic Ocean in the winter. It surrounds many of the Arctic islands and merges with glaciers and ice shelves (thick ice tht extends out from the land over water) along the edges of the continents. Huge, floating slabs of ice called pack ice crack and buckle as they constantly readjust to winds and currents. In the spring, the ice begins to melt back towards the North Pole. It breaks into large chunks of ice called icebergs and fleets of icebergs float in the open ocean. (Some icebergs, including the infamous 1912 sinker of the cruise ship Titanic, float south into "iceberg alley" in the North Atlantic Ocean where they are a hazard to shipping.)

In the summer, the Arctic ice cap melts towards the pole. Fish, whales, and other ocean species travel north to feed and mate in nutrient-rich open waters along the Arctic coastlines. Sea ice makes up most of the Arctic ice cap. However, glacial ice covers many of the Arctic islands, including the almost continent-sized island of Greenland. The thick ice that covers about 80% of Greenland is the last remnant of massive ice sheets that covered much of North America, Europe, and Asia during the Ice Ages of the Pleistocene era (starting 1.6 million years before the present and ending about 10,000 years ago). Ironically, Greenland is very icy and not particularly green, while neighbouring Iceland is green and not particularly icy. (Warmth from Iceland's volcanoes melts ice and supports grasslands).

LIFE ON THE TUNDRA

In spite of its harsh climate, the subarctic tundra supports a wide range of biological species. No trees grow on the tundra because a layer of frozen soil beneath the top layer of soil, called permafrost, prevents them from taking root. The tundra is also relatively dry; it receives only a few inches (centimeters) of precipitation (rain, snow, and any other form of water) each year. Strong winds sweep away snow and dry the soil. The tundra has only a few year-round residents. The birds (ptarmigan) and mammals (lemmings, hares, foxes, wolves, muskoxen, polar bears, and humans) there exhibit a number of traits that allow them to survive the cold, dark winter: thick, fluffy fur or feathers that turn white in winter and brown in summer offer camouflage that insulates them from the cold; hibernating to conserve energy; and strategies for finding shelter in the snow.

Some species, including lemmings, drastically reduce their population by committing mass suicide during lean times. Polar bears and humans venture onto the pack ice in winter and early spring to hunt for seals, sea lions, whales, and fish. The tundra comes to life in late spring when the pack ice begins to retreat and the permafrost thaws. Birds and mammals travel north to join those that stayed for the winter.

They feed on insects, shallow-rooted grasses and shrubs, lichens (plants that grow on bare rock), bird eggs, other animals, and especially marine life. Birds nest on the tundra. Herd animals like caribou travel north from the boreal forests of Asia, Europe, and North America to graze and bear their young. Wolves, foxes, eagles, and humans hunt the plains. In fall, when the sea ice reforms and the ground freezes solid, most of the birds and animals fly, swim, or walk south while the permanent arctic residents stock up for the long, dark, cold winter.

HUMANS IN THE ARCTIC

Arctic humans, like polar bears, survive the winter by insulating themselves. in warm fur and eating fatty seal meat. Eskimo tribes of Siberia, Arctic North America, and Greenland live in snow shelters, hunt the tundra, and use boats called kayaks to fish the icy Arctic waters. Native residents of Arctic Siberia and Alaska today prefer to be called Inuit, which means "the people." The term Eskimo has come to represent an over-simplified image of Eskimos as fur-clad hunters that live in igloos instead of a diverse group of loosely-related arctic cultures that have their far-northern latitude in common.

GLACIER

A glacier is a large persistent body of ice. Originating on land, a glacier flows slowly due to stresses induced by its weight. The crevasses and other distinguishing features of a glacier are due to its flow. Another consequence of glacier flow is the transport of rock and debris abraded from its substrate and resultant landforms like cirques and moraines. A glacier forms in a location where the accumulation of snow and sleet exceeds the amount of snow that melts. Over many years, often decades or centuries, a glacier will eventually form as the snow compacts and turns to ice. A glacier is distinct from sea ice and lake ice that form on the surface of bodies of water.

The word glacier comes from French. It is derived from the Vulgar Latin glacia and ultimately from Latin glacies meaning ice. The processes and features caused by glaciers and related to them are referred to as glacial. The process of glacier establishment, growth and flow is called glaciation. The corresponding area of study is called glaciology. Glaciers are important components of the global cryosphere. On Earth, 99% of glacial ice is contained within vast ice sheets in the polar regions, but glaciers may be found in mountain ranges of every continent except Australia.

In the tropics, glaciers occur only on high mountains. Glacial ice is the largest reservoir of freshwater on Earth. Many glaciers store water during one season and release it later as meltwater, a water source that is especially important for plants, animals and human uses when other sources may be scant. Because glacial mass is affected by long-term climate changes, *i.e.*, precipitation, mean temperature, and cloud cover, glacial mass changes are considered among the most sensitive indicators of climate change and are a major source of variations in sea level.

TYPES OF GLACIERS

Glaciers are categorized in many ways including by their morphology, thermal characteristics or their behaviour. Alpine glaciers form on the crests and slopes of mountains and are also known as "mountain glaciers", "niche glaciers", or "cirque glaciers". An alpine glacier that fills a valley is sometimes called a valley glacier. Larger glaciers that cover an entire mountain, mountain range, or volcano are known as an ice cap or ice field, such as the Juneau Icefield. Ice caps feed outlet glaciers, tongues of ice that extend into valleys below far from the margins of the larger ice masses.

The largest glacial bodies, ice sheets or continental glaciers, cover more than 50,000 km^2 (20,000 $mile^2$). Several kilometers deep, they obscure the underlying topography. Only nunataks protrude from the surface. The only extant ice sheets are the two that cover most of Antarctica and Greenland. These regions contain vast quantities of fresh water. The volume of ice is so large that if the Greenland ice sheet melted, it would cause sea levels to rise six meters (20 ft) all around the world. If the Antarctic ice sheet melted, sea levels would rise up to 65 meters (210 ft). Ice shelves are areas of floating ice, commonly located at the margin of an ice sheet.

As a result they are thinner and have limited slopes and reduced velocities. Ice streams are fast-moving sections of an ice sheet. They can be several hundred kilometers long. Ice streams have narrow margins and on either side ice flow is usually an order of magnitude less. In Antarctica, many ice streams drain into large ice shelves. However, some drain directly into the sea, often with an ice tongue, like Mertz Glacier. In Greenland and Antarctica ice streams ending at the sea are often referred to as tidewater glaciers or outlet glaciers, such as Jakobshavn Isbræ.

Tidewater glaciers are glaciers that terminate in the sea. As the ice reaches the sea pieces break off, or calve, forming icebergs. Most tidewater glaciers calve above sea level, which often results in a tremendous splash as the iceberg strikes the water. If the water is deep, glaciers can calve underwater, causing the iceberg to suddenly leap up out of the water. The Hubbard Glacier is the longest tidewater glacier in Alaska and has a calving face over 10 km (6 mi) long. Yakutat Bay and Glacier Bay are both popular with cruise ship passengers because of the huge glaciers descending hundreds of feet to the water. This glacier type undergoes centuries-long cycles of advance and retreat that are much less affected by the climate changes currently causing the retreat of most other glaciers. Most tidewater glaciers are outlet glaciers of ice caps and ice fields. In terms of thermal characteristics, a temperate glacier is at melting point throughout the year, from its surface to its base. The ice of a polar glacier is always below freezing point from the surface to its base, although the surface snowpack may experience seasonal melting. A sub-polar glacier has both temperate and polar ice, depending on the depth beneath the surface and position along the length of the glacier.

FORMATION

Glaciers form where the accumulation of snow and ice exceeds ablation. As the snow and ice thicken, they reach a point where they begin to move, due to a combination of the surface slope and the pressure of the overlying snow and ice. On steeper slopes this can occur with as little as 15 m (50 ft) of snow–ice. The snow which forms temperate glaciers is subject to repeated freezing and thawing, which changes it into a form of granular ice called firn. Under the pressure of the layers of ice and snow above it, this granular ice fuses into denser and denser firn. Over a period of years, layers of firn undergo further compaction and become glacial ice. Glacier ice has a slightly reduced density from ice formed from the direct freezing of water. The air between snowflakes becomes trapped and creates air bubbles between the ice crystals. The distinctive blue tint of glacial ice is often wrongly attributed to Rayleigh scattering due to bubbles in the ice. The blue colour is actually created for the same reason that water is blue, that is, its slight absorption of red light due to an overtone of the infrared OH stretching mode of the water molecule.

ANATOMY

The location where a glacier originates is referred to as the "glacier head". A glacier terminates at the "glacier foot", or terminus. Glaciers are broken into zones based on surface snowpack and melt conditions. The ablation zone is the region where there is a net loss in glacier mass. The equilibrium line separates the ablation zone and the accumulation zone. At this altitude, the amount of new snow gained by accumulation is equal to the amount of ice lost through ablation. The accumulation zone is the region where snowpack or superimposed ice accumulation persists.

A further zonation of the accumulation zone distinguishes the melt conditions that exist:

- The dry snow zone is a region where no melt occurs, even in the summer, and the snowpack remains dry.
- The percolation zone is an area with some surface melt, causing meltwater to percolate into the snowpack. This zone is often marked by refrozen ice lenses, glands, and layers. The snowpack also never reaches melting point.

- Near the equilibrium line on some glaciers, a superimposed ice zone develops. This zone is where meltwater refreezes as a cold layer in the glacier, forming a continuous mass of ice.
- The wet snow zone is the region where all of the snow deposited since the end of the previous summer has been raised to 0°C.

The upper part of a glacier that receives most of the snowfall is called the accumulation zone. In general, the glacier accumulation zone accounts for 60-70% of the glacier's surface area, more if the glacier calves icebergs. The depth of ice in the accumulation zone exerts a downward force sufficient to cause deep erosion of the rock in this area. After the glacier is gone, its force often leaves a bowl or amphitheater-shaped isostatic depression ranging from large lake basins, such as the Great Lakes or Finger Lakes, to smaller mountain basins, known as cirques.

The "health" of a glacier is usually assessed by determining the glacier mass balance or observing terminus behaviour. Healthy glaciers have large accumulation zones, more than 60% of their area snowcovered at the end of the melt season, and a terminus with vigourous flow.

Following the Little Ice Age, around 1850, the glaciers of the Earth have retreated substantially through the 1940s. A slight cooling led to the advance of many alpine glaciers from 1950-1985. However, since 1985 glacier retreat and mass balance loss has become increasingly ubiquitous and large.

MOTION

Glaciers move, or flow, downhill due to the internal deformation of ice and gravity. Ice behaves like an easily breaking solid until its thickness exceeds about 50 meters (160 ft). The pressure on ice deeper than that depth causes plastic flow. At the molecular level, ice consists of stacked layers of molecules with relatively weak bonds between the layers. When the stress of the layer above exceeds the inter–layer binding strength, it moves faster than the layer below.

Another type of movement is through basal sliding. In this process, the glacier slides over the terrain on which it sits, lubricated by the presence of liquid water. As the pressure increases towards the base of the glacier, the melting point of water decreases, and the ice melts. Friction between ice and rock and geothermal heat from the Earth's interior also contribute to melting. This type of movement is dominant in temperate, or warm-based glaciers. The geothermal heat flux becomes more important the thicker a glacier becomes. The rate of movement is dependent on the underlying slope, amongst many other factors.

Fracture Zone and Cracks

The top 50 meters of the glacier, being under less pressure, are more rigid; this part is known as the fracture zone, and mostly moves as a single unit, over the plastic–like flow of the lower section. When the glacier moves through irregular terrain, cracks up to 50 meters deep form in the fracture zone. The lower layers of glacial ice flow and deform plastically under the pressure, allowing the glacier as a whole to move slowly like a viscous fluid. Glaciers flow downslope, usually this reflects the slope of their base, but it may reflect the surface slope instead.

Thus, a glacier can flow rises in terrain at their base. The upper layers of glaciers are more brittle, and often form deep cracks known as crevasses. The presence of crevasses is a sure sign of a glacier. Moving ice–snow of a glacier is often separated from a mountain side or snow–ice that is stationary and clinging to that mountain side by a bergshrund. This looks like a crevasse but is at the margin of the glacier and is a singular feature.

Crevasses form due to differences in glacier velocity. As the parts move at different speeds and directions, shear forces cause the two sections to break apart, opening the crack of a crevasse all along the disconnecting faces. Hence, the distance between the two separated parts, while touching and rubbing deep down, frequently widens significantly towards the surface layers, many times creating a wide chasm.

Crevasses seldom are more than 150 feet (46 m) deep but in some cases can be 1,000 feet (300 m) or even deeper. Beneath this point, the plastic deformation of the ice under pressure is too great for the differential motion to generate cracks. Transverse crevasses are transverse to flow, as a glacier accelerates where the slope steepens. Longitudinal crevasses form semi-parallel to flow where a glacier expands laterally. Marginal crevasses form from the edge of the glacier, due to the reduction in speed caused by friction of the valley walls. Marginal crevasses are usually largely transverse to flow.

Crevasses make travel over glaciers hazardous. Subsequent heavy snow may form fragile snow bridges, increasing the danger by hiding the presence of crevasses at the surface. Below the equilibrium line, glacier meltwater is concentrated in stream channels. The meltwater can pool in a proglacial lake, a lake on top of the glacier, or can descend into the depths of the glacier via moulins. Within or beneath the glacier, the stream will flow in an englacial or sub-glacial tunnel. Sometimes these tunnels reemerge at the surface of the glacier.

Speed

The speed of glacial displacement is partly determined by friction. Friction makes the ice at the bottom of the glacier move more slowly than the upper portion. In alpine glaciers, friction is also generated at the valley's side walls, which slows the edges relative to the center. This was confirmed by experiments in the 19th century, in which stakes were planted in a line across an alpine glacier, and as time passed, those in the center moved farther. Mean speeds vary greatly. There may be no motion in stagnant areas, where trees can establish themselves on surface sediment deposits such as in Alaska. In other cases they can move as fast as 20-30 meters per day, as in the case of Greenlands's Jakobshavn Isbræ (Kalaallisut: Sermeq Kujalleq), or 2–3 m per day on Byrd Glacier, the largest glacier in the world in Antarctica. Velocity increases with increasing slope, increasing thickness, increasing snowfall, increasing longitudinal confinement, increasing basal temperature, increasing meltwater production and reduced bed hardness. A few glaciers have periods of very rapid advancement called surges. These glaciers exhibit normal movement until suddenly they accelerate, then return to their previous state. During these surges, the glacier may reach velocities far greater than normal speed. These surges may be caused by failure of the underlying bedrock, the ponding of meltwater at the base of the glacier—perhaps delivered from a supraglacial lake—or the simple accumulation of mass beyond a critical "tipping point". In glaciated areas where the glacier moves faster than one kilometer per year, glacial earthquakes occur. These are large scale tremblors that have seismic magnitudes as high as 6.1. The number of glacial earthquakes in Greenland show a peak every year in July, August and September, and the number is increasing over time. In a study using data from January 1993 through October 2005, more events were detected every year since 2002, and twice as many events were recorded in 2005 as there were in any other year. This increase in the numbers of glacial earthquakes in Greenland may be a response to global warming.

Seismic waves are also generated by the Whillans Ice Stream, a large, fast-moving river of ice pouring from the West Antarctic Ice Sheet into the Ross Ice Shelf. Two bursts of seismic waves are released every day, each one equivalent to a magnitude 7 earthquake, and are seemingly related to the tidal action of the Ross Sea. During each event a 96 by 193 kilometer (60 by 120 mile) region of the glacier moves as much as.67 meters (2.2 ft) over about 25 minutes, remains still for 12 hours, then moves another half-meter. The seismic waves are recorded at seismographs around Antarctica, and even as far away as Australia, a distance of more than 6,400 kilometers. Because the motion takes place of such along period of time 10 to 25 minutes, it cannot be felt by scientists standing on the moving glacier. It is not known if these events are related to global warming

Ogives

Ogives are alternating dark and light bands of ice occurring as narrow wave crests and wave valleys on glacier surfaces. They only occur below icefalls, but not all icefalls have ogives below them. Once formed, they bend progressively downglacier due to the increased velocity towards the glacier's centerline. Ogives are linked to seasonal motion of the glacier as the width of one dark and one light band generally equals the annual movement of the glacier. The ridges and valleys are formed because ice from an icefall is severely broken up, thereby increasing ablation surface area during the summertime. This creates a swale and space for snow accumulation in the winter, which in turn creates a ridge. Sometimes ogives are described as either wave ogives or band ogives, in which they are solely undulations or varying Colour bands, respectively.

GEOGRAPHY

Glaciers occur on every continent and approximately 47 countries. Extensive glaciers are found in Antarctica, Chilean Patagonia, Canada, Alaska, Greenland and Iceland. Mountain glaciers are widespread, *i.e.*, in the Andes, the Himalaya, the Rocky Mountains, the Caucasus, and the Alps. On mainland Australia no glaciers exist today, although a small glacier on Mount Kosciuszko was present in the last glacial period, and Tasmania was extensively glaciated. The South Island of New Zealand has many glaciers including Tasman, Fox and Franz Josef Glaciers. In New Guinea, small, rapidly diminishing, glaciers are located on its highest summit massif of Puncak Jaya. Africa has glaciers on Mount Kilimanjaro in Tanzania, on Mount Kenya and in the Ruwenzori Range. Permanent snow cover is affected by factors such as the degree of slope on the land, amount of snowfall and the winds. As temperature decreases with altitude, high mountains—even those near the Equator—have permanent snow cover on their upper portions, above the snow line. Examples include Mount Kilimanjaro and the Tropical Andes in South America; however, the only snow to occur exactly on the Equator is at 4,690 m (15,387 ft) on the southern slope of Volcán Cayambe in Ecuador.

Conversely, areas of the Arctic, such as Banks Island, and the McMurdo Dry Valleys in Antarctica are considered polar deserts, as they receive little snowfall despite the bitter cold. Cold air, unlike warm air, is unable to transport much water vapour. Even during glacial periods of the Quaternary, Manchuria, lowland Siberia, and central and northern Alaska, though extraordinarily cold with winter temperatures believed to reach–100°C (–148°F) in parts, had such light snowfall that glaciers could not form.

In addition to the dry, unglaciated polar regions, some mountains and volcanoes in Bolivia, Chile and Argentina are high (4,500 metres (14,800 ft)–6,900 m (22,600 ft)) and cold, but the relative lack of precipitation prevents snow from accumulating into glaciers. This is because these peaks are located near or in the hyperarid Atacama desert.

GLACIAL GEOLOGY

Rocks and sediments are added to glaciers through various processes. Glaciers erode the terrain principally through two methods: abrasion and plucking.

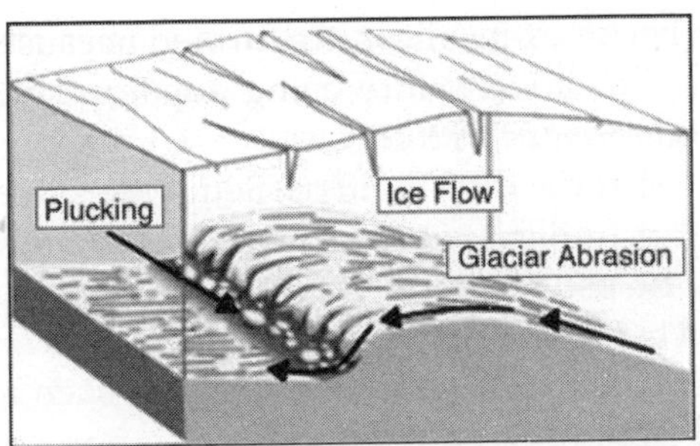

Diagram. Glacial of Plucking and abrasion

As the glacier flows over the bedrock's fractured surface, it softens and lifts blocks of rock that are brought into the ice. This process is known as plucking, and it is produced when subglacial water penetrates the fractures and the subsequent freezing expansion separates them from the bedrock. When the ice expands, it acts as a lever that loosens the rock by lifting it. This way, sediments of all sizes become part of the glacier's load. The rocks frozen into the bottom of the ice then act like grit in sandpaper. Abrasion occurs when the ice and the load of rock fragments slide over the bedrock and function as sandpaper that smooths and polishes the surface situated below.

This pulverized rock is called rock flour. The flour is formed by rock grains of a size between 0.002 and 0.00625 mm. Sometimes the amount of rock flour produced is so high that currents of meltwaters acquire a grayish Colour. These processes of erosion lead to steeper valley walls and mountain slopes in alpine settings, which can cause avalanches and rock slides. These further add material to the glacier.

Visible characteristics of glacial abrasion are glacial striations. These are produced when the bottom's ice contains large chunks of rock that mark scratches in the bedrock. By mapping the direction of the flutes, researchers can determine the direction of the glacier's movement. Chatter marks are seen as lines of roughly crescent-shape depressions in the rock underlying a glacier, caused by the abrasion where a boulder in the ice catches and is then released repetitively as the glacier drags it over the underlying basal rock. The rate of glacier erosion is variable.

The differential erosion undertaken by the ice is controlled by six important factors:

1. Velocity of glacial movement;
2. Thickness of the ice;
3. Shape, abundance and hardness of rock fragments contained in the ice at the bottom of the glacier;
4. Relative ease of erosion of the surface under the glacier;
5. Thermal conditions at the glacier base; and
6. Permeability and water pressure at the glacier base.

Material that becomes incorporated in a glacier are typically carried as far as the zone of ablation before being deposited.

Glacial deposits are of two distinct types:

1. *Glacial till*: Material directly deposited from glacial ice. Till includes a mixture of undifferentiated material ranging from clay size to boulders, the usual composition of a moraine.
2. *Fluvial and outwash*: Sediments deposited by water. These deposits are stratified through various processes, such as boulders' being separated from finer particles.

The larger pieces of rock which are encrusted in till or deposited on the surface are called "glacial erratics". They may range in size from pebbles to boulders, but as they may be moved great distances, they may be of drastically different type than the material upon which they are found. Patterns of glacial erratics provide clues of past glacial motions.

Moraines

Glacial moraines are formed by the deposition of material from a glacier and are exposed after the glacier has retreated. These features usually appear as linear mounds of till, a non-sorted mixture of rock, gravel and boulders within a matrix of a fine powdery material. Terminal or end moraines are formed at the foot or terminal end of a glacier. Lateral moraines are formed on the sides of the glacier.

Medial moraines are formed when two different glaciers, flowing in the same direction, coalesce and the lateral moraines of each combine to form a moraine in the middle of the merged glacier. Less apparent is the ground moraine, also called glacial drift, which often blankets the surface underneath much of the glacier downslope from the equilibrium line. Glacial meltwaters contain rock flour, an extremely fine powder ground from the underlying rock by the glacier's movement. Other features formed by glacial deposition include long snake-like ridges formed by streambeds under glaciers, known as eskers, and distinctive streamlined hills, known as drumlins.

Stoss-and-lee erosional features are formed by glaciers and show the direction of their movement. Long linear rock scratches (that follow the glacier's direction of movement) are called glacial striations, and divots in the rock are called chatter marks. Both of these features are left on the surfaces of stationary rock that were once under a glacier and were formed when loose rocks and boulders in the ice were transported over the rock surface.

Transport of fine-grained material within a glacier can smooth or polish the surface of rocks, leading to glacial polish. Glacial erratics are rounded boulders that were left by a melting glacier and are often seen perched precariously on exposed rock faces after glacial retreat.

The term moraine is of French origin. It was coined by peasants to describe alluvial embankments and rims found near the margins of glaciers in the French Alps. In modern geology, the term is used more broadly, and is applied to a series of formations, all of which are composed of till.

Drumlins

Drumlins are asymmetrical, canoe shaped hills with aerodynamic profiles made mainly of till. Their heights vary from 15 to 50 meters and they can reach a kilometer in length. The tilted side of the hill looks towards the direction from which the ice advanced (stoss), while the longer slope follows the ice's direction of movement (lee). Drumlins are found in groups called drumlin fields or drumlin camps.

An example of these fields is found east of Rochester, New York, and it is estimated that it contains about 10,000 drumlins. Although the process that forms drumlins is not fully understood, it can be inferred from their shape that they are products of the plastic deformation zone of ancient glaciers. It is believed that many drumlins were formed when glaciers advanced over and altered the deposits of earlier glaciers.

Glacial Valleys

Before glaciation, mountain valleys have a characteristic "V" shape, produced by downward erosion by water. However, during glaciation, these valleys widen and deepen, forming a "U"–shaped glacial valley. Besides the deepening and widening of the valley, the glacier also smooths the valley due to erosion. In this way, it eliminates the spurs of earth that extend across the valley. Because of this interaction, triangular cliffs called truncated spurs are formed. Many glaciers deepen their valleys more than their smaller tributaries. Therefore, when the glaciers recede from the region, the valleys of the tributary glaciers remain above the main glacier's depression, and these are called hanging valleys.

In parts of the soil that were affected by abrasion and plucking, the depressions left can be filled by lakes, called paternoster lakes. At the 'start' of a classic valley glacier is the cirque, which has a bowl shape with escarped walls on three sides, but open on the side that descends into the valley. In the cirque, an accumulation of ice is formed. These begin as irregularities on the side of the mountain, which are later augmented in size by the coining of the ice. Once the glacier melts, these corries are usually occupied by small mountain lakes called tarns.

There may be two glacial cirques 'back to back' which erode deep into their backwalls until only a narrow ridge, called an arête is left. This structure may result in a mountain pass. Glaciers are also responsible for the creation of fjords (deep coves or inlets) and escarpments that are found at high latitudes.

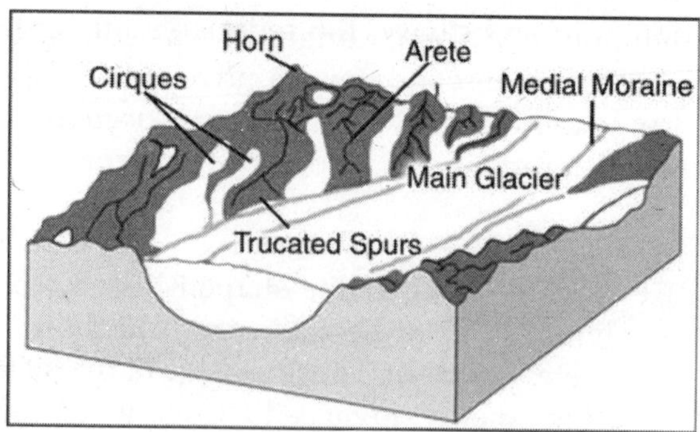

Fig. Features of a Glacial Landscape

Aretes and Horns (Pyramid Peak)

An arete is a narrow crest with a sharp edge. The meeting of three or more aretes creates pointed pyramidal peaks and in extremely steep-sided forms these are called horns. Both features may have the same process behind their formation: the enlargement of cirques from glacial plucking and the action of the ice. Horns are formed by cirques that encircle a single mountain. Aretes emerge in a similar manner; the only difference is that the cirques are not located in a circle, but rather on opposite sides along a divide. Aretes can also be produced by the collision of two parallel glaciers. In this case, the glacial tongues cut the divides down to size through erosion, and polish the adjacent valleys.

Roche Moutonnee

Some rock formations in the path of a glacier are sculpted into small hills with a shape known as roche moutonnee or "sheepback" rock. An elongated, rounded, asymmetrical, bedrock knob can be produced by glacier erosion.

It has a gentle slope on its up-glacier side and a steep to vertical face on the down–glacier side. The glacier abrades the smooth slope that it flows along, while rock is torn loose from the downstream side and carried away in ice, a process known as 'plucking'. Rock on this side is fractured by a combination of various forces, such as water, ice in rock cracks, and structural stresses.

Alluvial Stratification

The water that rises from the ablation zone moves away from the glacier and carries with it fine eroded sediments. As the speed of the water decreases, so does its capacity to carry objects in suspension. The water then gradually deposits the sediment as it runs, creating an alluvial plain. When this phenomenon occurs in a valley, it is called a valley train.

When the deposition is to an estuary, the sediments are known as "bay mud".

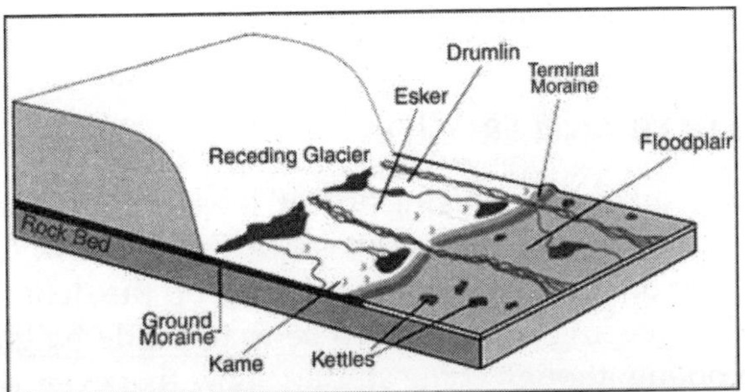

Fig. Landscape Produced by a Receding Glacier

Outwash plains and valley trains are usually accompanied by basins known as "kettles". These are glacial depressions produced when large ice blocks are stuck in the glacial alluvium. After they melt, the sediment is left with holes. The diameter of such depressions ranges from 5 m to 13 km, with depths of up to 45 meters. Most are circular in shape due to the melting blocks of ice becoming rounded. The lakes that often form in these depressions are known as "kettle lakes".

Deposits in Contact with Ice

When a glacier reduces in size to a critical point, its flow stops, and the ice becomes stationary. Meanwhile, meltwater flows over, within, and beneath the ice leave stratified alluvial deposits. Because of this, as the ice melts, it leaves stratified deposits in the form of columns, terraces and clusters.

These types of deposits are known as "deposits in contact with ice". When those deposits take the form of columns of tipped sides or mounds, they are called kames. Some kames form when meltwater deposits sediments through openings in the interior of the ice. In other cases, they are just the result of fans or deltas towards the exterior of the ice produced by meltwater. When the glacial ice occupies a valley, it can form terraces or kame along the sides of the valley. A third type of deposit formed in contact with the ice is characterized by long, narrow sinuous crests, composed fundamentally of sand and gravel deposited by streams of meltwater flowing within, or beneath the glacier. After the ice has melted, these linear ridges or eskers remain as landscape features. Some of these crests have heights exceeding 100 meters and their lengths surpass 100 km.

Loess Deposits

Very fine glacial sediments or rock flour is often picked up by wind blowing over the bare surface and may be deposited great distances from the original fluvial deposition site. These eolian loess deposits may be very deep, even hundreds of meters, as in areas of China and the Midwestern United States of America. Katabatic winds can be important in this process.

TRANSPORTATION AND EROSION

- Entrainment is the picking up of loose material by the glacier from along the bed and valley sides. Entrainment can happen by regelation or by the ice simply picking up the debris.
- Basal ice freezing is thought to be to be made by glaciohydraulic supercooling, though some studies show that even where physical conditions allow it to occur, the process may not be responsible for observed sequences of basal ice.
- Plucking is the process involves the glacier freezing onto the valley sides and subsequent ice movement pulling away masses of rock. As the bedrock is greater in strength than the glacier, only previously loosened material can be removed. It can be loosened by local pressure and temperature, water and pressure release of the rock itself.
- Supraglacial debris is carried on the surface of the glacier as lateral and medial moraines. In summer ablation, surface melt water carries a small load and this often disappears down crevasses.
- Englacial debris is moraine carried within the body of the glacier.
- Subglacial debris is moved along the floor of the valley either by the ice as ground moraine or by meltwater streams formed by pressure melting.

DEPOSITION

- Lodgement till is identical to ground moraine. It is material that is smeared on to the valley floor when its weight becomes too great to be moved by the glacier.
- Ablation till is a combination of englacial and supraglacial moraine. It is released as a stationary glacier begins to melt and material is dropped in situ.
- Dumping is when a glacier moves material to its outermost or lowermost end and dumps it.
- Deformation flow is the change of shape of the rock and land due to the glacier.

ISOSTATIC REBOUND

This rise of a part of the crust is due to an isostatic adjustment. A large mass, such as an ice sheet/glacier, depresses the crust of the Earth and displaces the mantle below. The depression is about a third the thickness of the ice sheet. After the glacier melts the mantle begins to flow back to its original position pushing the crust back to its original position. This post-glacial rebound, which lags melting of the ice sheet/glacier, is currently occurring in measurable amounts in Scandinavia and the Great Lakes region of North America.

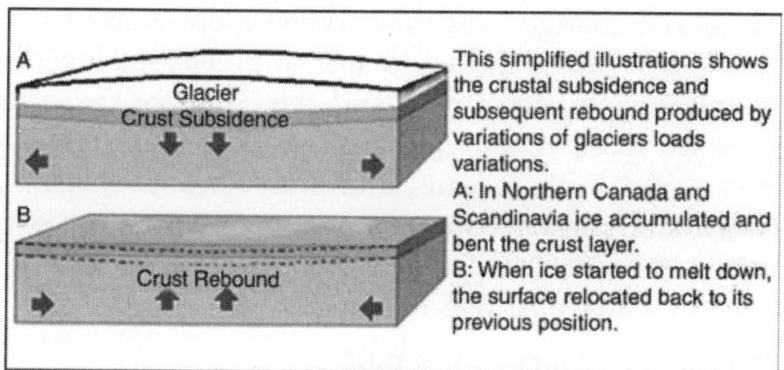

Fig. Isostatic Pressure by a Glacier on the Earth's Crust

An interesting geomorphological feature created by the same process, but on a smaller scale, is known as dilation–faulting. It occurs within rock where previously compressed rock is allowed to return to its original shape, but more rapidly than can be maintained without faulting, leading to an effect similar to that which would be seen if the rock were hit by a large hammer. This can be observed in recently de-glaciated parts of Iceland and Cumbria.

GLACIERS ON MARS

Elsewhere in the solar system, the vast polar ice caps of Mars rival those of the Earth and show glacial features. Especially the south polar cap is compared to glaciers on Earth. Other glacial features on Mars are glacial debris aprons and the lineated valley fills of the fretted terrain in northern Arabia Terra. Topographical features and computer models indicate the existence of more glaciers in Mars' past. Martian glaciers are affected by the thin atmosphere of Mars. Because of the low atmospheric pressure, ablation near the surface is solely due to sublimation, not melting. As on Earth, many glaciers are covered with a layer of rocks which insulates the ice. A radar instrument onboard the Mars Reconnaisance Orbiter found ice under a thin layer of rocks in formations called Lobate Debris Aprons (LDA's).

ICE, SEA–LEVEL, AND GLOBAL CLIMATE

The amount of water frozen in Earth's ice has changed throughout the planet's history. Earth's ice budget (total ice volume) grows when the planet's average temperature falls and shrinks when it rises. During colder periods called ice ages, ice caps (large dome-shaped glaciers) extend far from the

North and South Poles, and mountain glaciers (large masses of moving ice) advance into lowlands.

During warmer periods, ice retreats back towards the poles and up mountain valleys. Earth has even been completely ice-free several times during its long history. Earth's water budget, the total amount of water on the planet, does not change over time. When more water freezes into ice on land, there is less water in the oceans, so global sea level falls during ice ages. Shorelines move seaward, and the edges of the continents are exposed above sea level. When polar ice caps and mountain glaciers melt, more water is stored in the oceans, and global sea level rises. Shorelines move landward and coastal regions are submerged below sea level. (Ice floating in the ocean does not affect global sea level when it melts or forms because no water is added or removed.)

GLOBAL WARMING AND COOLING

Energy from the Sun ultimately determines Earth's average temperature and thus, its ice budget. The planet warms and cools as its position changes relative to the Sun and it receives more and less sunlight over time. The shape of Earth's orbit (the path it revolves around other objects), the tilt of the planet as it spins in its orbits, and the intensity of its seasons all change in repeating patterns over thousands of years.

Glacial advances and retreats in the past have matched the timing of these astronomical cycles with global warming and cooling. (They are called Milankovitch cycles after the Serbian mathematician who developed the theory relating Earth motions and longterm climate change.) Global warming and cooling also occur in response to changes on Earth's surface and in its atmosphere (mass of air surrounding Earth). Gases like carbon dioxide and water vapour keep Earth warm by allowing incoming sunlight to pass and then trapping escaping heat.

The layer of insulating gases in the atmosphere is called the greenhouse layer because it works like glass in a garden greenhouse. Without it, Earth would be a frigid, icy planet that could not support biological life. Processes like forest growth lead to green plants that take in the greenhouse gas carbon dioxide and release oxygen. The removal of greenhouse gases from the atmosphere leads to global cooling, advance of glaciers, and sea level fall. When more greenhouse gases escape into the atmosphere, Earth warms, ice melts, and sea level rises.

ICE AGES AND SEA LEVEL LOWS

Massive ice sheets (flat layers of fresh water ice covering extensive regions of the world advanced and retreated across northern Europe, Asia, and North America many times during the ice ages of the Pleistocene Epoch (a division of geologic time that lasted from 2 million to 10,000 years ago).

Canada and the northern portion of the United States are strewn with evidence of the last Pleistocene ice sheet in North America. The advancing ice polished, etched, and carved the solid rock layers of northern Canada. It tore rock from the ground and carried it south. When the ice melted, it stranded boulders and dropped piles of Canadian sediment (sand, grain, or silt) across the American Midwest and New England. The last North American ice sheet was miles (kilometers) thick and reached as far south as Long Island, New York. Like an overloaded ship, the continent sank under the weight of the ice. The Great Lakes are meltwater-filled depressions created by the ice sheet, and the coastlines of New England and eastern Canada have been rising out of the sea since the ice melted away. When the last Pleistocene glaciers reached their maximum extent about 20,000 years ago, ice covered about 30% of the Earth surface.

Global sea level was about 350 feet (107 meters) lower than its present–day level. Rivers flowed across exposed continental shelves (gently sloping shallow seabed extending into the ocean from a continent) to shorelines far seaward of their present locations. Land animals lived on regions that are submerged below sea level today, and migrated between islands and continents on exposed land bridges. Land bridges are strips of dry land that connects islands or continents. Some of North America's first human inhabitants arrived via a land bridge between Siberia and Alaska.

GLOBAL WARMING AND SEA LEVEL HIGHS

Earth was almost (if not completely) ice-free about 100 million years ago during the Cretaceous period of geologic time (this period lasted from 144 to 65 million years ago). High mountains and continental areas that were over the poles bear no trace of deposits from glaciers. Geologists (Earth scientists) have even found fossils of tropical plants that resemble palm trees in rocks that formed near the poles. Sea level was much higher in the Cretaceous period than it is today. Only central highlands of the continents remained exposed. A shallow seaway covered the interior of North America.

HUMANS AND GLOBAL CLIMATE CHANGE

Today, Earth is in a period that is moving away from the "icehouse" conditions of the Pleistocene. Global temperature and sea level have been rising and ice has been melting since the last glacial maximum. Ice now covers only about 10% of Earth's surface. Most continental ice is bound up in the massive ice sheets covering the continent of Antarctica at the South Pole. Small mountain glaciers and the ice sheet covering Greenland are all that remain of the massive Pleistocene ice sheets of the northern hemisphere.

Much of Earth's continental ice cover is presently retreating, including the Greenland ice sheet, the West Antarctic ice sheet, and many mountain glaciers. The melting is probably adding to a rise in sea level of a few inches per century. Scientists and environmentalists are concerned that humans are contributing to naturally-occurring global warming by releasing greenhouse gases into the atmosphere. Burning fuels like wood, coal, oil, and natural gas releases carbon dioxide, and scientists agree that booming human population growth and industrialization over the last century have significantly raised the level of carbon dioxide in the atmosphere. Most researchers also accept the conclusion that the additional carbon dioxide will have some effect on Earth's climate. Scientists do not, however, agree on what the effects of adding carbon dioxide and other gases to the atmosphere will be in the future. Some argue that enhanced greenhouse warming could lead to rapid sea level rise and dramatic global climate change that would threaten human property, food supplies, and even lives. Others predict that the additional greenhouse gases may have little effect, and that slow, steady warming will continue at a rate that allows humans to gradually adjust.

There are also climate scientists who theorize that upsetting the balance of greenhouse gases, ice cover, and sea levels could trigger the start of a new ice age. The question of human–caused greenhouse warming is one of the most pressing and controversial environmental issues of our time. Scientists are working to understand the complex nature of Earth's past and present climate in hopes of predicting changes in the future. Policy makers, environmentalists, and business leaders hope to use scientists' predictions and recommendations to design workable solutions.

POLAR ICE CAPS

Ice covers Earth's North and South Poles. Weak sunlight and long, dark winters at high latitudes (imaginary lines on Earth that tell how far north or south a place is from the equator) create frigid conditions that support the formation of year–round glaciers (large masses of moving ice) on land and seasonal ice in the oceans. An ice cap is a permanent covering of ice over a large area. The arctic ice cap at the North Pole includes sea ice floating in the Arctic Ocean; glaciers in northern Asia, Europe, North America; and sheets of ice on the island of Greenland. The ice cap at the South Pole is made up of the massive Antarctic ice covering and ice in the Southern Ocean. Polar ice caps play a vital role in regulating global climate, temperature, ocean currents, and sea level. They keep nutrientrich waters of the Arctic and Southern Oceans at a livable temperature for rich communities of biological life. A total of 99.997% of Earth's fresh water is bound up in polar ice. Humans have lived in the North Pole area, the Arctic, for thousands of years and navigators have long been challenged by ice that blocks shipping and travel routes. Humans discovered the South Pole region, Antarctica, in the early 1800s. With the exception of explorers, scientists, and tourists, it remains uninhabited by humans today.

ICE SHEETS

Most of Earth's ice is bound up in immense ice sheets that cover the southern continent of Antarctica and the large arctic island of Greenland. Ice sheets, also called continental glaciers, are dome-shaped masses of ice that completely cover the underlying landscape. Ice sheets are huge; they cover areas larger than 31,069 square miles (50,000 square kilometers), about the size of Switzerland.

They have thick portions called domes where snow accumulates. Ice sheets move when the dome spreads out and eventually compacts into thin, dense layers of ice. Continental glaciers can spread across flat ground because the weight of the thickest ice pushes their edges outward. (Imagine a blob of honey spreading on a slice of flat toast.) All but a tiny portion of the frigid continent of Antarctica. (Antarctica is about 1.5 times the size of the United States.

It is roughly circular in shape and the South Pole lies near its center.) Antarctica lies under a region of very dry air in the atmosphere (air surrounding Earth), and its wind–blown interior plains are Earth's largest deserts (Antarctica receives only a few inches of precipitation per year in the form of snow). Each year, only a few inches (centimeters) of new snow accumulate on the Antarctic ice sheet. It has taken hundreds of thousands of years for thin layers of snow to compact into thousands of feet (meters) of glacial ice. A range of tall mountains, called the Trans–Antarctic mountains, divides Antarctica and its overlying ice sheet into two geographic regions.

The larger piece, East Antarctica, is a single continental block covered by a slow-moving, stable ice sheet with an average depth of more than 1.5 miles (2.4 kilometers)! West Antarctic is composed of numerous small continental blocks, sediment-filled basins (basins filled with sand, grain, or silt), and even several volcanoes. The West Antarctic ice sheet has many faster–moving portions called ice streams that slide over a slippery layer of sediment and water. Ice streams drain ice from central domes towards the coast where they flow out over the ocean to form floating platforms of ice, called ice shelves.

Scientists are concerned that the West Antarctic ice sheet is melting faster than it is growing. Ice cools the climate by reflecting the heat energy from the Sun back towards space. A significant reduction in ice, therefore, would lead to warmer temperatures in the region. The world's second largest ice sheet covers the far northern island of Greenland. The Greenland ice sheet is the last remnant of the ice sheets that covered much of the northern hemisphere during the numerous Pleistocene (2 million–10,000 years ago) ice ages. Like the West Antarctic ice sheet, the thick ice on Greenland appears to be melting more rapidly than it is growing.

ICE SHELVES

Ice shelves form where the protruding edges of ice sheets extend over the ocean. Ice shelves grow when more ice accumulates on the ice sheet, and shrink by melting and breaking off icebergs into the ocean. Icebergs are the splintered chunks of ice shelves. Ice shelves cover the coastal bays and inlets (narrow strips of water running into the land or between islands) of Antarctica. The fastflowing ice streams of the West Antarctic ice sheet produce ice shelves over two very large Antarctic bays called the Ross Sea and the Weddell Sea.

SEA ICE

Sea ice is frozen sea water that makes up a large portion of the polar ice caps. (Ice shelves and icebergs are not considered sea ice because they are composed of fresh water.) Sea ice is seasonal; it forms in winter and melts in summer. Huge slabs of saltwater ice continuously crack and buckle as they move with ocean currents (a steady flow of ocean water in a prevailing direction) in the Arctic and Southern Oceans. The mass of sea ice in the Arctic Ocean extends to the northern edges of North America, Asia, and Europe during the winter months, and melts back towards the North Pole in the summer. Antarctica is surrounded by a halo of sea ice that extends far north into the Southern Ocean in winter and melts almost completely away in summer. Sea ice does not affect global sea level because it is already floating in the ocean. It is, however, extremely important to the climate and biology of polar regions. Many animals in the Arctic (polar bears, seals, and whales) and Antarctic (penguins, whales, seals, sea lions) depend completely on seasonal sea ice for their habitat. The amount of sea ice varies from year to year, affecting ocean temperatures and currents worldwide.

Chapter 9

Water, Weather, and Climates

CLIMATE

Climate encompasses the statistics of temperature, humidity, atmospheric pressure, wind, rainfall, atmospheric particle count and other meteorological elemental measurements in a given region over long periods of time. Climate can be contrasted to weather, which is the present condition of these same elements and their variations over shorter time periods.

The climate of a location is affected by its latitude, terrain, and altitude, as well as nearby water bodies and their currents. Climates can be classified just as to the average and the typical ranges of different variables, most commonly temperature and precipitation.

The most commonly used classification scheme was originally developed by Wladimir Köppen. The Thornthwaite system, in use since 1948, incorporates evapotranspiration in addition to temperature and precipitation information and is used in studying animal species diversity and potential impacts of climate changes. The Bergeron and Spatial Synoptic Classification systems focus on the origin of air masses that define the climate of a region.

Paleoclimatology is the study of ancient climates. Since direct observations of climate are not available before the 19th century, paleoclimates are inferred from proxy variables that include non-biotic evidence such as sediments found in lake beds and ice cores, and biotic evidence such as tree rings and coral. Climate models are mathematical models of past, present and future climates. Climate change may occur over long and short timescales from a variety of factors; recent warming is discussed in global warming.

DEFINITION

Climate is commonly defined as the weather averaged over a long period of time. The standard averaging period is 30 years, but other periods may be used depending on the purpose. Climate also includes statistics other than the average, such as the magnitudes of day-to-day or year-to-year variations.

The Intergovernmental Panel on Climate Change (IPCC) definition is:

- Climate in a narrow sense is usually defined as the "average weather," or more rigorously, as the statistical description in terms of the mean and variability of relevant quantities over a period of time ranging from months to thousands or millions of years. The classical period is 30 years, as defined by the World Meteorological Organization (WMO). These quantities are most often surface variables such as temperature, precipitation, and wind. Climate in a wider sense is the state, including a statistical description, of the climate system.

The difference between climate and weather is usefully summarized by the popular phrase "Climate is what you expect, weather is what you get." Over historical time spans there are a number of nearly constant variables that determine climate, including latitude, altitude, proportion of land to water, and proximity to oceans and mountains. These change only over periods of millions of years due to processes such as plate tectonics. Other climate determinants are more dynamic: for example, the thermohaline circulation of the ocean leads to a 5°C (9°F) warming of the northern Atlantic ocean compared to other ocean basins. Other ocean currents redistribute heat between land and water on a more regional scale.

The density and type of vegetation coverage affects solar heat absorption, water retention, and rainfall on a regional level. Alterations in the quantity of atmospheric greenhouse gases determines the amount of solar energy retained by the planet, leading to global warming or global cooling. The variables which determine climate are numerous and the interactions complex, but there is general agreement that the broad outlines are understood, at least insofar as the determinants of historical climate change are concerned.

CLIMATE CLASSIFICATION

There are several ways to classify climates into similar regimes. Originally, climes were defined in Ancient Greece to describe the weather depending upon a location's latitude. Modern climate classification methods can be broadly divided into genetic methods, which focus on the causes of climate, and empiric methods, which focus on the effects of climate. Examples of genetic classification include methods based on the relative frequency of different air mass types or locations within synoptic weather disturbances.

Examples of empiric classifications include climate zones defined by plant hardiness, evapotranspiration, or more generally the Köppen climate classification which was originally designed to identify the climates associated with certain biomes. A common shortcoming of these classification schemes is that they produce distinct boundaries between the zones they define, rather than the gradual transition of climate properties more common in nature.

Bergeron and Spatial Synoptic

The most simple classification is that involving the concept of air masses. The Bergeron classification is the most widely accepted form of air mass classification. Air mass classification involves three letters. The first letter describes its moisture properties, with c used for continental air masses (dry) and m for maritime air masses (moist). The second letter describes the thermal characteristic of its source region: T for tropical, P for polar, A for Arctic or Antarctic, M for monsoon, E for equatorial, and S for superior air (dry air formed by significant downward motion in the atmosphere). The third letter is used to designate the stability of the atmosphere. If the air mass is colder than the ground below it, it is labeled k. If the air mass is warmer than the ground below it, it is labeled w. While air mass identification was originally used in weather forecasting during the 1950s, climatologists began to establish synoptic climatologies based on this idea in 1973.

Based upon the Bergeron classification scheme is the Spatial Synoptic Classification system (SSC). There are six categories within the SSC scheme: Dry Polar (similar to continental polar), Dry Moderate (similar to maritime superior), Dry Tropical (similar to continental tropical), Moist Polar (similar to maritime polar), Moist Moderate (a hybrid between maritime polar and maritime tropical), and Moist Tropical (similar to maritime tropical, maritime monsoon, or maritime equatorial).

Thornthwaite

Devised by the American climatologist and geographer C. W. Thornthwaite, this climate classification method monitors the soil water budget using the concept of evapotranspiration. It monitors the portion of total precipitation used to nourish vegetation over a certain area. It uses indices such as a humidity index and an aridity index to determine an area's moisture regime based upon its average temperature, average rainfall, and average vegetation type. The lower the value of the index in any given area, the drier the area is. The moisture classification includes climatic classes with descriptors such as hyperhumid, humid, subhumid, subarid, semi-arid (values of–20 to–40), and arid (values below–40).

Humid regions experience more precipitation than evaporation each year, while arid regions experience greater evaporation than precipitation on an annual basis. A total of 33 per cent of the Earth's landmass is considered either arid of semi-arid, including southwest North America, southwest South America, most of northern and a small part of southern Africa, southwest and portions of eastern Asia, as well as much of Australia.

Studies suggest that precipitation effectiveness (PE) within the Thornthwaite moisture index is overestimated in the summer and underestimated in the winter. This index can be effectively used to determine the number of herbivore and mammal species numbers within a given area. The index is also used in studies of climate change.

Thermal classifications within the Thornthwaite scheme include microthermal, mesothermal, and megathermal regimes. A microthermal climate is one of low annual mean temperatures, generally between 0°C (32°F) and 14°C (57°F) which experiences short summers and has a potential evaporation between 14 centimetres (5.5 in) and 43 centimetres (17 in).

A mesothermal climate lacks persistent heat or persistent cold, with potential evaporation between 57 centimetres (22 in) and 114 centimetres (45 in). A megathermal climate is one with persistent high temperatures and abundant rainfall, with potential annual evaporation in excess of 114 centimetres (45 in).

RECORD

Modern

Details of the modern climate record are known through the taking of measurements from such weather instruments as thermometers, barometers, and anemometers during the past few centuries. The instruments used to study weather conditions over the modern time scale, their known error, their immediate environment, and their exposure have changed over the years, which must be considered when studying the climate of centuries past.

Paleoclimatology

Paleoclimatology is the study of past climate over a great period of the Earth's history. It uses evidence from ice sheets, tree rings, sediments, coral, and rocks to determine the past state of the climate. It demonstrates periods of stability and periods of change and can indicate whether changes follow patterns such as regular cycles.

CLIMATE ZONES

Earth's climate zones are defined by their average yearly rainfall (or snowfall) and temperature. In general, they are alternating, east-west oriented, wet and dry zones under the rising and falling Hadley cells. If Earth were a simple, water-covered ball, without complicating factors like continents, high mountain ranges, and ocean currents, there would be five climate zones in each hemisphere: tropical, arid, temperate, cold, and polar.

1. *Tropical (hot, wet)*: Lush, biologicallydiverse rainforests thrive in the tropical zone at the equator. The jungles of central Africa, the

Amazon basin in South America, and south Pacific Islands like Borneo lie in the tropical zone.

2. *Subtropical arid and semi-arid (hot, dry)*: Earth's great deserts lie in arid zones north (Saharan, Arabian) and south (Kalahari, Australian Outback) of the equator. Dry, semi-arid (mostly arid) grasslands form the bordering lands around the subtropical deserts. The African savannah, Asian steppe, and Great Plains of North America are semi-arid grasslands that support large mammals like elephants, horses, and buffalo.
3. *Temperate (mild temperatures, moderate rainfall)*: A large percentage of Earth's population lives in mild and temperate regions of North and South America, Europe, and Asia. These climates often have warm, dry summers and cool, wet winters. Coastal regions are usually wetter and have less extreme temperature variations than inland temperate regions.
4. *Cold (cold, moderate rainfall)*: The cold, snowy, northern forests of North America, Scandinavia, and Asia are called the boreal zone. The treeless plain of the sub-arctic between the boreal forest and the polar ice cap (the thick covering of permanent ice and snow at the North and South Poles) is called tundra.
5. *Polar (very cold, very dry)*: The North and South Poles are cold, dry deserts. The polar ice caps have formed from the accumulation of light snows over thousands of years.

The positions of continental land masses, ocean currents, and high mountain ranges also affect the pattern of climate zones. Land heats up and cools down faster than water. Many coastal areas have climates affected by wet onshore winds that bring rain, dry offshore winds that create coastal deserts, or reversing winds (monsoons) that cause alternating wet and dry seasons. Warm ocean currents keep some regions that are far from the equator warm, and cold water upwelling (rising up) from the deep ocean cools some tropical coastlines and islands. Winds that flow from the oceans onto land generally lose their moisture as they travel inland or uphill. The interiors of large continents like Asia, Australia, and North America are generally dry. When moist air reaches a tall mountain range, it drops rain as it rises and cools. The slopes of mountain ranges exposed to the wind are typically wetter than the sides away from the wind. Arid deserts and semiarid grasslands form in the rainshadows (an area of decreased precipitation on the downwind side of a mountain), behind tall mountains. In the United States, high winds called the jet stream carry moisture-rich air from west to east. It is rainy in the Pacific Northwest, and snowy on the western slopes of the Cascade, Sierra Nevada, and Rocky Mountains. The air is bone dry when it reaches the Mojave, Sonoran, and Chihuahuan deserts of the American Southwest, and northern Mexico.

CLIMATE CHANGE

Climate change is a long–term change in the statistical distribution of weather patterns over periods of time that range from decades to millions of years. It may be a change in the average weather conditions, or in a change of the distribution of events around that average (*i.e.,* more or fewer extreme weather events). Climate change may be limited to a specific region, or may occur across the whole Earth.

Terminology

The most general definition of climate change is a change in the statistical properties of the climate system when considered over long periods of time, regardless of cause, this type of climate change and its effects have been documented in the past. Fluctuations over periods shorter than a few decades, such as El Niño, do not represent climate change. The term sometimes is used to refer specifically to climate change caused by human activity; for example, the United Nations Framework Convention on Climate Change defines climate change as "a change of climate which is attributed directly or indirectly to human activity that alters the composition of the global atmosphere and which is in addition to natural climate variability observed over comparable time periods." In this latter sense, used especially in the context of environmental policy, climate change is synonymous with anthropogenic global warming.

Causes

Factors that can shape climate are climate forcings. These include such processes as variations in solar radiation, deviations in the Earth's orbit, mountain-building and continental drift, and changes in greenhouse gas concentrations. There are a variety of climate change feedbacks that can either amplify or diminish the initial forcing. Some parts of the climate system, such as the oceans and ice caps, respond slowly in reaction to climate forcing because of their large mass. Therefore, the climate system can take centuries or longer to fully respond to new external forcings.

Human Influences

In the context of climate variation, anthropogenic factors are human activities which affect the climate. The scientific consensus on climate change is, "that climate is changing and that these changes are in large part caused by human activities." and that it "is largely irreversible".

- "Science has made enormous inroads in understanding climate change and its causes, and is beginning to help develop a strong understanding of current and potential impacts that will affect people today and in coming decades. This understanding is crucial because it allows decision makers to place climate change in the context of other large challenges facing the nation and the world. There are still some uncertainties, and there always will be in understanding a complex system like Earth's climate. Nevertheless, there is a strong, credible body of evidence, based on multiple lines of research, documenting that climate is changing and that these changes are in large part caused by human activities. While much remains to be learned, the core phenomenon, scientific questions, and hypotheses have been examined thoroughly and have stood firm in the face of serious scientific debate and careful evaluation of alternative explanations."—United States National Research Council, Advancing the Science of Climate Change

Consequently, the debate is shifting onto ways to reduce further human impact and to find ways to adapt to change that has already occurred and is anticipated to occur in the future. Of most concern in these anthropogenic factors is the increase in CO2 levels due to emissions from fossil fuel combustion, followed by aerosols (particulate matter in the atmosphere) and cement manufacture. Other factors, including land use, ozone depletion, animal agriculture and deforestation, are also of concern in the roles they play–both separately and in conjunction with other factors—in affecting climate, microclimate, and measures of climate variables.

Plate Tectonics

Over the course of millions of years, the motion of tectonic plates reconfigures global land and ocean areas and generates topography. This can affect both global and local patterns of climate and atmosphere-ocean circulation. The position of the continents determines the geometry of the oceans and therefore influences patterns of ocean circulation. The locations of the seas are important in controlling the transfer of heat and moisture across the globe, and therefore, in determining global climate. A recent example of tectonic control on ocean circulation is the formation of the Isthmus of Panama about 5 million years ago, which shut off direct mixing between the Atlantic and Pacific Oceans.

This strongly affected the ocean dynamics of what is now the Gulf Stream and may have led to Northern Hemisphere ice cover. During the Carboniferous period, about 300 to 360 million years ago, plate tectonics may have triggered large–scale storage of carbon and increased glaciation. Geologic evidence points to a "megamonsoonal" circulation pattern during the time of the supercontinent Pangaea, and climate modeling suggests that the existence of the supercontinent was conducive to the establishment of monsoons.

The size of continents is also important. Because of the stabilizing effect of the oceans on temperature, yearly temperature variations are generally lower in coastal areas than they are inland. A larger supercontinent will therefore have more area in which climate is strongly seasonal than will several smaller continents or islands.

Solar Output

The sun is the predominant source for energy input to the Earth. Both long-and short-term variations in solar intensity are known to affect global climate. Three to four billion years ago the sun emitted only 70% as much power as it does today. If the atmospheric composition had been the same as today, liquid water should not have existed on Earth. However, there is evidence for the presence of water on the early Earth, in the Hadean and Archean eons, leading to what is known as the faint young sun paradox.

Hypothesized solutions to this paradox include a vastly different atmosphere, with much higher concentrations of greenhouse gases than currently exist Over the following approximately 4 billion years, the energy output of the sun increased and atmospheric composition changed, with the oxygenation of the atmosphere around 2.4 billion years ago being the most notable alteration. These changes in luminosity, and the sun's ultimate death as it becomes a red giant and then a white dwarf, will have large effects on climate, with the red giant phase possibly ending life on Earth.

Solar output also varies on shorter time scales, including the 11-year solar cycle and longer-term modulations. Solar intensity variations are considered to have been influential in triggering the Little Ice Age, and some of the warming observed from 1900 to 1950.

The cyclical nature of the sun's energy output is not yet fully understood; it differs from the very slow change that is happening within the sun as it ages and evolves. While most research indicates solar variability has induced a small cooling effect from 1750 to the present, a few studies point towards solar radiation increases from cyclical sunspot activity affecting global warming. Interestingly, a 2010 study suggests, "that the effects of solar variability on temperature throughout the atmosphere may be contrary to current expectations."

Orbital Variations

Slight variations in Earth's orbit lead to changes in the seasonal distribution of sunlight reaching the Earth's surface and how it is distributed across the globe. There is very little change to the area–averaged annually averaged sunshine; but there can be strong changes in the geographical and seasonal distribution. The three types of orbital variations are variations in Earth's eccentricity, changes in the tilt angle of Earth's axis of rotation, and precession of Earth's axis. Combined together, these produce Milankovitch cycles which have a large impact on climate and are notable for their correlation to glacial and interglacial periods, their correlation with the advance and retreat of the Sahara, and for their appearance in the stratigraphic record.

Volcanism

Volcanism is a process of conveying material from the crust and mantle of the Earth to its surface. Volcanic eruptions, geysers, and hot springs, are examples of volcanic processes which release gases and/or particulates into the atmosphere.

Eruptions large enough to affect climate occur on average several times per century, and cause cooling (by partially blocking the transmission of solar radiation to the Earth's surface) for a period of a few years. The eruption of Mount Pinatubo in 1991, the second largest terrestrial eruption of the 20th century (after the 1912 eruption of Novarupta) affected the climate substantially. Global temperatures decreased by about 0.5°C (0.9°F). The eruption of Mount Tambora in 1815 caused the Year Without a Summer. Much larger eruptions, known as large igneous provinces, occur only a few times every hundred million years, but may cause global warming and mass extinctions. Volcanoes are also part of the extended carbon cycle. Over very long (geological) time periods, they release carbon dioxide from the Earth's crust and mantle, counteracting the uptake by sedimentary rocks and other geological carbon dioxide sinks. The US Geological Survey, however, estimates are that human activities generate 100–300 times the amount of carbon dioxide emitted by volcanoes.

Ocean Variability

The ocean is a fundamental part of the climate system. Short-term fluctuations (years to a few decades) such as the El Niño-Southern Oscillation, the Pacific decadal oscillation, the North Atlantic oscillation, and the Arctic oscillation, represent climate variability rather than climate change.

On longer time scales, alterations to ocean processes such as thermohaline circulation play a key role in redistributing heat by carrying out a very slow and extremely deep movement of water, and the long-term redistribution of heat in the world's oceans.

Physical Evidence for Climatic Change

Evidence for climatic change is taken from a variety of sources that can be used to reconstruct past climates. Reasonably complete global records of surface temperature are available beginning from the mid-late 19th century. For earlier periods, most of the evidence is indirect—climatic changes are inferred from changes in proxies, indicators that reflect climate, such as vegetation, ice cores, dendrochronology, sea level change, and glacial geology.

Historical and Archaeological Evidence

Climate change in the recent past may be detected by corresponding changes in settlement and agricultural patterns. Archaeological evidence, oral history and historical documents can offer insights into past changes in the climate. Climate change effects have been linked to the collapse of various civilizations.

Glaciers

Glaciers are considered among the most sensitive indicators of climate change. Their size is determined by a mass balance between snow input and melt output. As temperatures warm, glaciers retreat unless snow precipitation increases to make up for the additional melt; the converse is also true. Glaciers grow and shrink due both to natural variability and external forcings. Variability in temperature, precipitation, and englacial and subglacial hydrology can strongly determine the evolution of a glacier in a particular season. Therefore, one must average over a decadal or longer time–scale and/or over a many individual glaciers to smooth out the local short–term variability and obtain a glacier history that is related to climate.

A world glacier inventory has been compiled since the 1970s, initially based mainly on aerial photographs and maps but now relying more on satellites. This compilation tracks more than 100,000 glaciers covering a total area of approximately 240,000 km, and preliminary estimates indicate that the remaining ice cover is around 445,000 km. The World Glacier Monitoring Service collects data annually on glacier retreat and glacier mass balance From this data, glaciers worldwide have been found to be shrinking significantly, with strong glacier retreats in the 1940s, stable or growing conditions during the 1920s and 1970s, and again retreating from the mid 1980s to present.

The most significant climate processes since the middle to late Pliocene (approximately 3 million years ago) are the glacial and interglacial cycles. The present interglacial period (the Holocene) has lasted about 11,700 years. Shaped by orbital variations, responses such as the rise and fall of continental ice sheets and significant sea–level changes helped create the climate. Glaciers leave behind moraines that contain a wealth of material-including organic matter, quartz, and potassium that may be dated-recording the periods in which a glacier advanced and retreated. Similarly, by tephrochronological techniques, the lack of glacier cover can be identified by the presence of soil or volcanic tephra horizons whose date of deposit may also be ascertained.

Vegetation

A change in the type, distribution and coverage of vegetation may occur given a change in the climate. Some changes in climate may result in increased precipitation and warmth, resulting in improved plant growth and the subsequent sequestration of airborne CO_2. Larger, faster or more radical changes, however, may result in vegetation stress, rapid plant loss and desertification in certain circumstances. An example of this occurred during the Carboniferous Rainforest Collapse (CRC), an extinction event 300 million years ago. At this time vast rainforests covered the equatorial region of Europe and America. Climate change devastated these tropical rainforests, abruptly fragmenting the habitat into isolated 'islands' and causing the extinction of many plant and animal species.

Ice Cores

Analysis of ice in a core drilled from a ice sheet such as the Antarctic ice sheet, can be used to show a link between temperature and global sea level variations. The air trapped in bubbles in the ice can also reveal the CO_2 variations of the atmosphere from the distant past, well before modern environmental influences. The study of these ice cores has been a significant indicator of the changes in CO_2 over many millennia, and continues to provide valuable information about the differences between ancient and modern atmospheric conditions.

Dendroclimatology

Dendroclimatology is the analysis of tree ring growth patterns to determine past climate variations. Wide and thick rings indicate a fertile, well–watered growing period, whilst thin, narrow rings indicate a time of lower rainfall and less-than-ideal growing conditions.

Pollen Analysis

Palynology is the study of contemporary and fossil palynomorphs, including pollen. Palynology is used to infer the geographical distribution of plant species, which vary under different climate conditions. Different groups of plants have pollen with distinctive shapes and surface textures, and since the outer surface of pollen is composed of a very resilient material, they resist decay. Changes in the type of pollen found in different layers of sediment in lakes, bogs, or river deltas indicate changes in plant communities. These changes are often a sign of a changing climate. As an example, palynological studies have been used to track changing vegetation patterns throughout the Quaternary glaciations and especially since the last glacial maximum.

Insects

Remains of beetles are common in freshwater and land sediments. Different species of beetles tend to be found under different climatic conditions. Given the extensive lineage of beetles whose genetic makeup has not altered significantly over the millennia, knowledge of the present climatic range of the different species, and the age of the sediments in which remains are found, past climatic conditions may be inferred.

Sea Level Change

Global sea level change for much of the last century has generally been estimated using tide gauge measurements collated over long periods of time to give a long-term average. More recently, altimeter measurements—in combination with accurately determined satellite orbits—have provided an improved measurement of global sea level change. To measure sea levels prior to instrumental measurements, scientists have dated coral reefs that grow near the surface of the ocean, coastal sediments, marine terraces, ooids in limestones, and nearshore archaeological remains. The predominant dating methods used are uranium series and radiocarbon, with cosmogenic radionuclides being sometimes used to date terraces that have experienced relative sea level fall.

CLOUDS

Clouds are made of very small drops of water of water, ice crystals, and other small particles in the atmosphere (mass of air surrounding Earth). The water comes from condensation, a process that allows small drops of water to form as the air cools. Cloud shapes and the way clouds form give scientists important clues about local weather and conditions in the atmosphere around the world. Clouds are divided into several types or families of clouds. These families of clouds are named just as to where or how they form, and include high-level clouds, middle-level clouds, and low–level clouds. In addition to belonging to a family, clouds are also named for their shape. Puffy clouds are known as cumuliform clouds, and flat sheet-like clouds are known as stratoform clouds.

HOW CLOUDS FORM

In general, as warm, moist air rises upward through the atmosphere, the air cools. As the air cools, ice crystals or water drops appear and clouds form. Meteorologists (scientists who study weather and climate) name clouds based on how they form, where they form, and the shape of clouds. Cloud classifications are organized into groups or families.

FAMILIES AND TYPES OF CLOUDS

The altitudes (heights above the ground) used to describe cloud families change and become lower as one moves from the equator towards the North or South pole. As one moves north or south from Earth's equator (imaginary circle around Earth between the North and South Pole), high altitude family clouds can be observed at much lower altitudes. High level clouds include cirrus, cirrostratus, and cirrocumulus clouds. These clouds are found at altitudes between 16,000 and 45,000 feet (4,877 and 13,716 meters) above the ground. In comparison, a jumbo passenger jet usually cruises at about 36,000 feet (10,973 meters) above the ground. Middle level clouds include altostratus, altocumulus, and nimbostratus clouds, and are found between 6,500 and 22,000 feet (1,981 and 6,706 meters) above the ground. These clouds include include altostratus, altocumulus, and nimbostratus clouds.

As with many cloud families, the altitudes are not exact, and they can vary depending on the type of terrain (sea or mountains) over which the clouds form or travel. Low–level clouds include stratus and stratocumulus clouds that are found below 6,000 feet (1,829 meters). Clouds that form rapidly in a vertical (up and down) direction are known as vertical development clouds. Vertical development clouds include cumulus and cumulonimbus clouds, which are the clouds that form a thunderstorm. Vertical development clouds form rapidly as air rises from the Earth's surface. They are found anywhere from the surface of the ground to 45,000 feet (13,716 meters). In some very strong thunderstorms, the clouds may reach even higher. One factor that contributes to the development of thunderstorms is unstable warm and humid air that quickly rises through the atmosphere to great heights where the surrounding air is very cold.

SHAPE AND COLOUR OF CLOUDS

The shape of a cloud is determined by the manner in which the water drops condense and the forces of winds that can act to tear away pieces of the cloud as it builds and moves in the atmosphere. Whether a cloud is light or dark depends upon how much light can pass through the cloud. Water droplets bend or block light. Thicker clouds block more light than thinner clouds, and so appear darker than thinner clouds.

NAMES OF CLOUDS

When clouds are widely separated from other members of their family, the term fracto is added to their name. When a cloud produces rain (precipitates) it is also called a nimbus cloud and the term nimbus is added to the cloud name.

High Clouds

Cirrus clouds occur at high levels and are usually wispy and long. If air rises directly upward through the atmosphere, the air cools very quickly. As the air cools, ice crystals or water drops appear and cumulus clouds form. Cumulus clouds are billowy, puffy clouds that resemble cotton balls. Stratus clouds look as if they are blankets or layers of clouds. Because it is very cold at high altitudes, high clouds including cirrus clouds, cirrostratus clouds, and cirrocumulus clouds are composed of ice crystals. Particles of dust or pollution often form the center around which the ice crystals grow. For this reason, dust or particles of pollution are often called centers of crystallization if ice grows around them, or condensation nuclei (nuclei meaning the center) if water drops form around them. Cirrus clouds often produce a shape that looks like a horse tail. These "mare's tails" are wisps of ice crystals. Cirrostratus clouds, because they are thin and because their ice crystals act to both reflect and bend sunlight, sometime appear to form a circle or halo around the Sun or Moon. Cirrocumulus clouds often appear as patch-like thin clouds.

Middle Level Clouds

Middle level clouds include altostratus clouds, altocumulus clouds, and nimbostratus clouds. These clouds are composed of water drops with some ice crystals near the top of the clouds. Sometimes both middle level and low level clouds contain water that is still in liquid water drops even though the air around them is well below the freezing temperature. This super–cooled water (water below freezing that has not yet formed an ice crystal) needs only a seed, usually a particle of dust or pollution around which to form ice. At one time, scientists experimented with making rain by seeding clouds with a chemical called silver iodide.

It was hoped that the silver iodide would provide a center around which large water droplets would form. When the water droplets grew large enough, they would fall as rain. Cloud seeding thus offered hope that it might be possible to produce rain in dry regions. The results of these early experiments were disappointing, however, and produced little rain beyond the amounts that fell without cloud seeding. Because of the way ice crystals reflect and deflect light, altostratus clouds often present a bluish-layered appearance. Depending on thickness, altocumulus clouds often have white or gray layers that appear in washboard or wave-like formations.

Warm moist air that rises can also result in the formation of castlelike altocumulus castellanus clouds, a form of altocumulus that often appear as isolated cumulous clouds with billowing tops. Another form of altocumulus cloud is called a standing cloud (properly termed a lenticular altocumulus cloud), and is formed by condensation in currents of air that cool as they move upwards to cross mountains and ridges. Although constantly forming and disappearing, the standing lenticular altocumulus cloud formations appear not to change and thus seem to stand over the mountain or ridge lifting the air. Nimbostratus clouds often appear as heavy, gray, moistureladen cloud layers.

Low Level Clouds

Low-level stratus clouds are the gray clouds that often produce rain and some types of fog. Stratocumulus clouds present the familiar, cotton ball-like cumulus shapes in an elongate form (a cumulus shape drawn out by shearing winds). Clouds that pass through many levels of the atmosphere, the cumulus and cumulonimbus clouds, often have a widely varyingmixture of ice and water. These clouds often have swirling currents of air that move upwards and downwards. These rapid updrafts and downdrafts of air allow ice crystals to appear at much lower levels than normal. As they cycle through the cloud, the ice crystals can grow large enough to fall to the ground as hail. Although formed from air rising upward from the ground and lower levels of the atmosphere, cumulus clouds often form in fair weather and do not form violent updraft or downdraft currents of air. These cumulus clouds have flat bases and curved tops that look like domes of buildings.

When strong and violent updrafts and downdrafts of air form, however, the air is said to be unstable and the cumulus clouds are said to be more developed. These cumulus clouds have mushroom or cauliflower-like tops, and they often produce rain. When cycles of air moving upwards and downwards become very violent, cumulonimbus clouds form. Cumulonimbus clouds are dark clouds with anvil-like tops (very flat tops with trailing clouds spreading out like a tabletop) that are often cut off (sheared) by strong winds in the upper atmosphere. Cumulonimbus clouds often have heavy turbulence (rough and violent disturbances of air), rains, lightning, and thunder. The most unstable and violent cumulonimbus clouds can occur in cells or groups capable of forming tornadoes.

MONSOON

A monsoon is a regional wind that reverses directions seasonally. In southern Asia, wet, hot monsoon winds blow from the southwest during the summer months and bring heavy rains to a large area that includes India, Bangladesh, Sri Lanka, Pakistan, and Nepal. Northeasterly winds (winds are named by the direction from which they blow) that blow down from the Himalaya Mountains in the winter are cool and dry. Monsoon winds occur in many regions around the world, in Africa, Australia, and in North America, where the Mexican monsoon brings over half of the year's total rain to Northern Mexico, Arizona, and New Mexico each June through August. The Mexican monsoon is a smaller version of the classic and wellknown southern Asian monsoon. The word monsoon comes from the Arabic word mausim which means "season." In southern Asia, the monsoon controls the seasons: hot and wet in the summer, cool and dry in the winter. Plants and animals of southern Asia have adapted ways to survive the annual cycles of flood and drought (prolonged period of dry weather). Humans depend on the rains to fill storage reservoirs and to water crops, especially water-intensive staples like rice and cotton.

HOW MONSOONS WORK

The Asian monsoon works like a large version of reversing land-sea breezes along coastlines. Water heats and cools more slowly than dry land. On a sunny day at the beach, the air over land heats more quickly than air over the water. Warmer air expands and rises. Cool air moves in from the ocean to replace the rising warm air, and this movement creates a nonshore breeze. When the sun goes down in the evening, the land cools more quickly than the sea, and the wind changes direction. The Indian peninsula is a piece of low land surrounded on three sides by the waters of the northern Indian Ocean; it separates the Arabian Sea from the Bay of Bengal.

The massive Himalaya Mountains to the north isolate it from winds and weather in the rest of Asia. India lies just north of the equator, so it receives very intense sunlight and heats up beginning in April. When the hot air over the peninsula rises, wet air from the tropical Indian Ocean flows onshore to take its place. The wind flows from the southeast, across India and into the Himalayan foothills where it is forced upward. Warm air holds more moisture than cool air. The moist ocean air blown in by these winds cools and condenses (changes into liquid from a gas) as it rises, resulting in heavy rains. Monsoon rains are often hard, sustaining rains.

THE RAINY SEASON

The "season of the peacock" begins in mid-May as the monsoon wet phase reaches southern India. Peacocks are the symbol of life-giving rains in India. Male peacocks begin courting females by flashing their brilliant tail feathers a few weeks before clouds form and the arrival of downpours that transform the parched, brown landscape to a lush, green paradise. As the rainy season progresses, two arms of wet weather extend across the region: one reaches up from the tear-drop-shaped island of Sri Lanka (formerly Ceylon), to the tip of India, and towards Pakistan; the other comes in from the Bay of Bengal and Bangladesh, and crosses central and northern India. By early July, the two arms have merged, and the torrential monsoon rains extend throughout south Asia. Animals mate and seeds germinate. Rivers, lakes, reservoirs, and wells fill with water. Humans plant and water their crops. Monsoon rains continue through the summer months and begin to let up by the end of September. During many years, the rains that were so welcome in spring have become "too much of a good thing" by late summer, when flooding threatens crops, buildings, and lives throughout southern Asia.

Soil and rock layers that hold water are saturated (completely full of water), and rainwater runs directly off the land surface. Small streams in the Himalayan foothills flow over their banks and flood crops and towns. Small floods run downhill to join others, and then flow together as a very large pulse of water into the mighty Brahmaputra and Ganges Rivers. The Ganges-Brahmaputra system collects water from most of South Asia; the Indian province of West Bengal and the country of Bangladesh cover its massive delta (the fan-shaped area of land at the river's mouth).

The residents of the Ganges Delta have developed some strategies for surviving the annual deluge, but there are years when particularly strong monsoon rains and heavy snows in the Himalayas create huge, uncontrollable floods that devastate the area. During floods in September 1998, almost 70% of Bangladesh, an area about the size of the state of Tennessee, was underwater. The 1998 floods in Bangladesh killed hundreds of people and caused millions to lose their homes. Poor sanitation and ruined crops led to widespread disease and starvation.

THE DRY SEASON

In the fall, cold air flows down from the high peaks of the Himalayas as the continent begins to cool. A dry northwest wind blows across India and the rain moves offshore into the ocean. Floodwaters recede and leave behind new layers of silt (fine soil particles) and nutrients to fertilize the flooded agricultural lands. By the following July, the land is parched and dry, and south Asians look forward to another drenching rainy season.

ASIAN MONSOON IS A BLESSING AND A THREAT

Plants and animals of South Asia depend on the summer rains and have evolved (changed over time) to survive the floods and droughts of the monsoon. Plentiful rain gives rise to lush forests and grasslands that provide food and shelter for some of Earth's most exotic animal species, including Bengal tigers and Indian elephants. South Asian plants and animals have adapted strategies to reproduce and thrive during the rainy season and then lie dormant (inactive) or survive on stored water during the dry months.

Approximately one quarter of the world's population depends on the monsoon rains and is threatened by related floods and droughts. The monsoon counties are very heavily populated. The land area of India is about the same size as the United States but its population is more than three times as large. Bangladesh, a relatively poor country, is one of the most densely settled places on Earth. (Imagine the entire population of the United States living in Oregon.) Scientists have discovered links between the Asian monsoon and global climate. They worry that natural or human–induced global warming or cooling could affect the monsoon pattern and lead to increased drought or flooding. More than one billion people face starvation, illness, and the loss of their homes during exceptionally rainy or dry monsoon years. Natural ecosystems (interaction of living organisms and their environment in a community) of plants and animals that are suffer as well. The governments of monsoon countries, the United Nations, scientists, and non-profit organizations are working to better understand the monsoon and to develop solutions to economic and environmental problems of South Asia.

STORMS

Storms are disturbances in the atmosphere (air surrounding Earth) that bring severe weather: heavy rain and snow, high winds, lightning and thunder, tornadoes, and hail. There are storms that are mild, such as rainstorms, which are beneficial, bringing needed rainfall for plants, animals, and waterways. Yet storms also have the potential to cause great harm. Hurricanes batter coastlines and islands with high winds, drenching rain, and waves. Thunderstorms and blizzards can cause floods and dangerous traveling conditions.

During thunderstorms, lightning can ignite brush fires, and hail can destroy crops. Tornadoes can cut swaths of destruction across anything in its path. Storms occur in unstable or changing areas of the atmosphere where warm, light air rises rapidly from the land surface. The general conditions that spawn storms are well known; hot summer days in the American Midwest almost always produce thunderstorms. Cold air low pressure areas cause blizzards in the winter, which sweep eastward and warm seas feeding tropical low pressure systems cause hurricanes that spin from the tropical Atlantic Ocean from June through November. Predicting the exact location, severity, and timing of storms, however, is very difficult. Although weather forecasting and storm warning systems have become more accurate in recent years, severe weather still takes humans by surprise. Storms cause billions of dollars of damage and kill thousands of people each year around the world.

THUNDERSTORMS

Thunderstorms form where plumes or masses of warm, moist air rise into cool air above. In temperate climates like central North America, thunderstorms are most common during the spring and summer, but they can also form in the winter. Temperature differences between rising areas of warm air and cool air surrounding them create air currents (moving stream of air) called updrafts and downdrafts. Vertically (upwards and downwards) circulating thunderstorm clouds have central updrafts (areas of rising air) surrounded by a ring of downdrafts (areas of falling air). Tall, billowing, black clouds form, called cumulonimbus clouds or thunderheads. Heavy rain falls. Moving water and ice particles within the clouds create electrical charges, causing lightning bolts to zap between clouds and the ground. Thunder booms and crackles. Thunder is the sound created by the electrical discharge of lightning.

Three ingredients are a recipe for a thunderstorm: warm, moist air near the land surface; cool, dry air above; and something to lift the warm air. Mountain ranges, moving weather fronts (a line between two air masses with differing characteristics, bringing changing weather), converging winds, and uneven heating of land and sea surfaces can all provide an upward push. Sometimes, the rising air is fairly dry, and clouds form that produce lightning but no rain. A line of thunderstorms can form along the moving front of an air mass.

In the summertime, thunderstorms roll across the American Great Plains each afternoon as the land surface heats unevenly. Afternoon lightning and cloudbursts are very common in the Rocky Mountains when warm, moist air rises up the face of the mountain. Most thunderstorms are short-lived, single cell (brief, small) and multi-cell storms (storms with multiple storm–producing clouds) that may produce lightning and heavy rain, but rarely cause severe damage. The most intense thunderstorms, called supercells, will produce battering hail, flash floods, high winds, and tornadoes.

TORNADOES

Tornadoes, or twisters, are narrow columns of violently spinning air that extend, finger–like, from the bases of cumulonimbus clouds during intense supercell thunderstorms. Tornadoes form when instability within the storm causes spiraling air circulation. The base of the storm cloud lowers, and becomes a spinning cloud called a wall cloud. Wall clouds can sometimes develops protruding lumps called mammatus clouds. Tornadoes are whirlpools of upward-moving air that descend from the parent wall cloud to the ground.

The portion of a tornado that actually touches the ground is usually quite small. Numerous accounts describe twisters that completely destroy a structure while leaving an immediate neighbour's property untouched. Small whirlwinds like dust devils (small, circular, brief winds on land) and some waterspouts (a column of rotating air, similar to a tornado, over a body of water) can also develop away from a parent thunderstorm. Meteorologists (weather scientists) classify tornadoes as weak, strong, or violent. Weak and strong tornadoes spin less than about 200 miles per hour (322 kilometers per hour). They can knock over trees, pick up objects and fling them like missiles, demolish mobile homes, and tear roofs from framed houses.

Violent tornadoes can completely destroy a well built home or lift a large object like a car. Thankfully, these are quite rare; only two twisters out of every hundred have winds that exceed 200 miles per hour (322 kilometers per hour). A tornado like the one that tore Dorothy's house from the ground and lifted it into the air during a dream in L. Frank Baum's story The Wizard of Oz is thus unlikely, but not impossible. Dorothy's home, Kansas, is at the center of "Tornado Alley," where severe thunderstorms spawn tornadoes that rake across the plains between the Rockies and Appalachian mountains during the spring and summer.

TROPICAL CYCLONES

Tropical cyclones are huge, spiral–shaped storm systems that form near the equator in the Atlantic, Pacific, and Indian Oceans. Warm, tropical waters fuel their growth from groups of individual thunderstorms, into massive, organized systems of circulating winds and clouds. Tropical cyclones in the Atlantic and eastern Pacific Ocean are called hurricanes. Western Pacific cyclones are called typhoons, and those in southern Pacific and Indian Oceans are simply called cyclones. Earth's rotation causes winds to blow hurricanes and typhoons in the Northern Hemisphere to spin counter–clockwise (east to west).

In the southern hemisphere, winds move west to east, causing hurricanes and cyclones to spin clockwise. Atlantic hurricanes originate from a near–permanent band of thunderstorms near the equator. Warm water and converging trade winds (surface winds blowing westward in the tropics and sub-tropics) create updrafts of moist air that feed huge thunderstorms and dense rain clouds. The first stage of a developing hurricane, called a tropical depression, forms when a group of thunderstorms organizes around a particularly large storm and begins to rotate. Some, but not all, tropical depressions grow into tropical storms and then into hurricanes.

Tropical storms have more organized spiral patterns and stronger winds than tropical depressions. Although tropical storms are not as powerful as full–fledged hurricanes, they bring very heavy rainfall and often cause severe flooding. Tropical storms officially become hurricanes when its winds exceed 74 miles per hour (119 kilometers per hour). A small area of calm, called the eye, forms at the center of the storm. The eye wall, a ring of intense winds and heavy rain, surrounds the eye. Bands of rain and clouds spiral out to the edges of the storm. Meteorologists rate hurricane intensities from category 1 to category 5. Hurricanes stronger than category 3 (wind speeds greater than 111 miles per hour or 179 kilometers per hour) generally cause extensive damage when they make landfall. Atlantic tropical storms and hurricanes ride the warm Gulf Stream current (a warm northbound surface current that carries Atlantic Ocean water into the Norwegian Sea) northwest from the tropics towards the Caribbean Sea, Gulf of Mexico, and Atlantic coast of the United State.

Tropical cyclones depend on warm ocean water to feed warm, moist air into their central updrafts, so they fade when they move over cool water or land. Tropical cyclones take several weeks to develop and move across the ocean before subsiding, and there may be several storms in a particular ocean at one time. To avoid confusion, meteorologists assign names to tropical storms and hurricanes using alphabetical lists of alternating male and female names.

The first storm of the year has a name starting with A, the second with B, and so on. (There are no names beginning with Q, U, or Z.) The 2004 list for the Atlantic Ocean included such early-in-the-alphabet names as Charley, Frances, and Ivan. There are six lists, so these names will be used again in 2010. The names of very large and destructive hurricanes like Camille (1969), Hugo (1989), and Andrew (1992) are retired from the list.

MID-LATITUDE CYCLONES

Mid-latitude (areas midway between the equator and the poles) cyclones cause most of North America's stormy weather. Like tropical cyclones, mid-latitude cyclones are low-pressure systems that rotate counterclockwise in the Northern Hemisphere. Westerly (east–blowing) winds drive air masses across North America from west to east. Easterlies blow cold air to the west in northern Canada. Mid-latitude cyclones develop when a cool, dry air mass follows a warm, moist one. (The leading edge of the cool air mass is called a cold front.)

Some of the warm air flows north (left) towards Canada, and some of the cold Canadian air blows south (left) creating a counter-clockwise spiral with rising air, and low pressure, at its center. Storms form along the cold front and in the low pressure zone where warmer, moist air is forced up into the overlying cold air. Warm air moving north from the Gulf of Mexico provides moisture to fuel winter blizzards and summer thunderstorms in Great Plains. Cyclones also draw moisture from the Great Lakes and drop heavy rain and snow downwind to the east. When a large cyclone reaches the northeast coast of North America, the spiraling winds extend over the North Atlantic and pick up more moisture and then blow back towards the continent. Nor'easters are cold, wet storms that blow into Maine, Nova Scotia, New Brunswick, and Newfoundland from the northeast.

WEATHER

Weather is the state of the atmosphere (mass of air surrounding Earth) at a particular place and point in time. Rain showers, gusty winds, thunderstorms, cloudy skies, droughts (prolonged period of dry weather), snowstorms, and sunshine are all examples of weather conditions. Weather scientists, called meteorologists, use measurable factors like atmospheric pressure (pressure caused by weight of the air), temperature, moisture, clouds, and wind speed to describe the weather. Meteorologists make predictions of future weather based on observations of present regional weather patterns and past trends. Weather prediction, or forecasting, is an important part of meteorology (weather science). Advance warning of such weather phenomena as extreme hot and cold temperatures, heavy rainfall, drought, and severe storms can protect people's property and save lives.

The weather patterns that a region experiences over tens, hundreds, or thousands of years are called climate. For example, the northeastern United States experiences a wide range of weather during an average year. Below-freezing temperatures and heavy snowfall are typical weather conditions in winter, while warm temperatures and afternoon thunderstorms are common in the summer. Communities of plants and animals (ecosystems) adapt over thousands of years to survive the weather extremes of their particular climate.

In New England, plants lie inactive, mammals grow shaggy coats, and birds fly south during the cold dark winter. In the spring, trees pull sap from their roots and grow leaves, animals bear young, and seeds germinate in time to take advantage of mild temperatures and long, sunny days in the summer. Climate change happens over hundreds and thousands of years, but weather varies from day to day, hour to hour, and sometimes from minute to minute.

WEATHER CONDITIONS: PRESSURE, TEMPERATURE, AND MOISTURE

The atmosphere presses down on Earth's surface. (There is no atmosphere in outer space. Without their pressurized space suits, astronaut's bodies would explode.) The weight of the column of air molecules above a surface is called atmospheric pressure. The average weight of the atmosphere on one square inch of ground at sea level is 14.7 pounds. People do not feel this pressure because their senses are adjusted to it and the human body is designed to withstand it.

Meteorologists use an instrument called a barometer to measure pressure, and atmospheric pressure is also called barometric pressure. Evangelista Torricelli (1608–1647), an Italian physicist, invented the barometer in 1643. His instrument, "Torricelli's tube," was a glass tube full of dense, liquid mercury with its end in an open dish of mercury. His barometer works the same way that mercury barometers work in modern day. Air pressing down on the mercury in the dish pushes some of the mercury upwards into the glass tube. As air pressure increases, the mercury in forced into the tube and the column of mercury rises.

When air pressure decreases, the mercury flows back into the dish and the column of falls. Barometric pressure is often measured in inches of mercury. When a weather forecaster says the mercury is falling, it means that air pressure is falling, and bad weather may be approaching. Atmospheric pressure differs from one place on Earth to another due to temperature, moisture, and topography (physical surface features). Pressure decreases with elevation. There are many fewer air molecules above a square foot (kilometer) on the summit of Mt. Everest than above a square foot of Waikiki Beach. Air currents, better known as winds, blow from areas of high pressure to areas of low pressure.

Rapidly changing patterns of winds, precipitation (any form of water falling), clouds, and storms develop around moving high and low pressure centers in Earth's atmosphere. Temperature affects air pressure and moisture in the atmosphere. Warmer air expands and rises, so pressure falls beneath rising columns of warm air. Warm air also holds more moisture, in the form of water vapour, than cool air. Rising warm air in low pressure zones often carry water vapour high into the atmosphere. When the warm air begins to cool, the moisture condenses into droplets or freezes into ice crystals and clouds form. Precipitation and storms are common in low-pressure centers. As air cools it contracts, causing air pressure to rise under the sinking air. Because cool air holds less moisture, and because sinking air masses are usually already dry, high pressure areas usually are low in humidity (air moisture).

HIGH AND LOW PRESSURE SYSTEMS

Major east and west-blowing winds blow high and low-pressure weather systems around Earth. High–pressure systems, also called anticyclones, consist of winds spiraling out from a high-pressure center under sinking, dry air. Low–pressure systems, or cyclones, have low-pressure centers and winds that spiral towards their centers. High-pressure systems are called anticyclones. A cyclone has a column of warm air rising from its center. In anticyclones, the air sinks towards the center and warms as it descends.

In the northern hemisphere (half of the Earth), anticyclones spin clockwise and cyclones spin counterclockwise, and the reverse is true in the southern hemisphere. Because air travels from high to low pressure areas, high–pressure anticyclones often follow low-pressure cyclones. In North America, the jet stream (high-speed winds that race around the planet at about five miles above the Earth) blows cyclones and anticyclones from west to east. In general, cyclones bring intense weather in the form of rain, snow, clouds, and storms. Dry, clear, calm weather usually accompanies the passage of anticyclones. (The parched residents in deserts of the American Southwest might look forward to the clouds and rain storms a cyclone brings.

A southward dip in the Jet Stream causes a near-permanent zone of high pressure over Arizona, New Mexico, and Southern California, and moisturebearing weather systems tend to bypass the region.) Trade winds (persistent tropical winds that blow generally towards the west) blow low-pressure systems that develop in the tropical Atlantic Ocean west towards the Caribbean Sea and east coast of the United States. These tropical cyclones feed on warm ocean waters and can develop into massive storm systems called tropical storms and hurricanes.

AIR MASSES AND FRONTS

An air mass is a large body of air that has similar temperatures and moisture content throughout. Several air masses contribute to weather patterns in North America: cold, dry air over northern Canada; hot, dry air in the American Southwest; cool, moist air moving east over the Pacific Northwest; and warm, moist air traveling north from the Gulf of Mexico. The boundaries between air masses are called fronts. A cold front occurs where a cold air mass is moving in to replace warm air. Clouds, precipitation, and storms are common at cold fronts. The incoming cold, dense mass lifts the warm, moist air and creates unstable conditions where moisture rapidly condenses and winds organize clouds into storms.

Once a cold front has passed, temperatures and humidity drop and a highpressure system moves in. A warm front precedes an incoming warm air mass. Warm fronts bring moisture and higher temperatures. Stationary fronts separate unmoving air masses. A typical cyclone in the American Mid-West is a rotating pinwheel of three air masses and three fronts moving east towards the Atlantic Ocean. Cold, dry air flows south from Canada behind cool, moist air flowing from the Pacific Northwest. Warm air from the Gulf of Mexico moves north and contributes moisture to the system. Thunderstorms and blizzards develop along cold fronts.

Chapter 10

Science and Technology

AQUEDUCTS

Aqueducts are man-made conduits constructed to carry water. The term aqueduct comes from words meaning "to lead water" in Latin, the language of the Romans who were the first builders of large aqueducts. Aqueducts carry water from natural sources, such as springs, into cities and towns for public use.

THE FIRST AQUEDUCTS

Wells, rivers, lakes, and streams are the oldest sources of water. In the ancient world however, rivers and lakes were also sometimes used as places to dispose of sewage and trash. Water from rivers that flowed though several villages often carried disease—causing organisms. Aqueducts provided a way for a plentiful supply of clean water to be piped into cities. The earliest aqueducts were also used to transport water for irrigation (watering crops). Aqueducts were used in ancient India, Persia, Assyria, and Egypt as early as 700 B.C.E. The Romans, however, are regarded as the most famous ancient aqueduct builders. Between 312 B.C.E. and 230 C.E., the most complex and efficient ancient system of aqueducts was built to supply the city of Rome with water. Outside of the capital city of Rome, the Romans built aqueducts throughout their large empire. Ruins of ancient aqueducts can still be seen in Italy, Greece, North Africa, Spain, and France.

HOW ANCIENT AQUEDUCTS FUNCTIONED

Ancient aqueducts used tunnels and channels (passages for water to flow) to transport water. The earliest irrigation aqueducts were simple canals and ditches dug into the ground. In order to keep water for use by people clean, aqueducts that supplied people with water featured covered channels or pipes. The first aqueduct made of stone-covered waterways was built by the Assyrians around 690 B.C.E. Centuries later, Roman aqueduct builders perfected the closed channel design, building thousands of miles (kilometers) of stone aqueducts throughout the Roman Empire. Ancient aqueducts were carefully planned before they were constructed. Water flowed through the channels by the force of gravity alone. The rate of flow (how many gallons could flow through the conduit in a day) was determined by the force of the spring that fed the aqueduct. Aqueduct channels were constructed with a gradual slope (angle) so that water from the source could flow downhill to its destination. There were no pumps that could move water up a hill or slope. Thus, when crossing hilly terrain, aqueducts were built on stone bridges and in tunnels. Pipes made of stone or a type of baked clay called terra cotta carried water through carved out tunnels. Aqueduct bridges (or elevated spans) were required to withstand the heavy weight of water. Spectacular Roman aqueduct bridges featuring several stories (or tiers) of strong arches can still be seen today.

Some are still in use! After the aqueducts entered the city, water flowed into public cisterns (large pools or wells that store water) or flowed from public fountains. In Rome, some citizens had water from the aqueducts piped directly to their homes. Wastewater was carried by sewer systems that emptied into outlying streams that normally did not feed into the aqueduct. Like modern water supply systems, ancient aqueducts required constant maintenance.

Where aqueducts ran underground, shafts (tunnels) were built to provide access to the aqueduct for repairs. Chalk and other minerals built up in the conduits and required regular cleaning. Wars, earthquakes, storms, and floods sometimes damaged whole sections of aqueducts. Fixing aqueducts was an expensive undertaking and required the work of strong laborers and skilled engineers.

INNOVATIONS IN AQUEDUCT TECHNOLOGY

After the fall of the Roman Empire in the fifth century, aqueduct building ceased in Europe. For centuries, the scientific knowledge necessary to build aqueducts, aqueduct bridges, and sewers was lost. Rome and some other cities continued to use their ancient aqueducts. However, during the Middle Ages (500–1500 C.E.), people mostly used wells and rivers as a source of water. During the Renaissance (1300s–1600s), a renewed interest in classical architecture and engineering led scholars of the day to rediscover how ancient water systems worked and how aqueducts were constructed.

In the 1600s, aqueducts were once again included in public water systems. In France, a system of pumps moved water from a river to an aqueduct system that began on the crest (high point) of a nearby hill. An aqueduct spanning 38 miles (61 kilometers) carried water into the city of London, England. The Chadwell River to London aqueduct flowed over 200 small bridges. In the eighteenth and nineteenth centuries, innovations such the steam pump permitted water to be pressurized. Pressurized water is water that is mixed with air or steam that, with the help of a pump, can be moved forcefully through pipes and conduits. This allowed water systems to move water over any terrain. Aqueducts and water pipe systems carried water over greater distances with the aid of pressurized water. Pressurization also created a need for stronger pipes.

Instead of terra-cotta, pipes were made of metals or concrete. Between the 1830s and 1900, the growing city of New York constructed several aqueducts to bring spring and river water into the city from sources over 120 miles (193 kilometers) away. These aqueducts incorporated new and old aqueduct technology. They employed pumps and deep underground pipe systems, but the Old Croton aqueduct, in use until 1955, also featured a Roman-like aqueduct bridge. Today, the three major aqueduct systems that serve New York City deliver nearly 1.8 billion gallons (approximately 6.8 billion liters) of water per day.

AQUEDUCTS TODAY

Aqueducts remain an important and efficient means of delivering clean water to cities. Today's aqueducts are longer and able to carry more water than ancient aqueducts. Pumps and pressurized water flow permit aqueducts to flow up a slope. Improved pipe materials allows today's aqueducts to be completely hidden deep underground. The largest modern aqueduct system in the world has been under construction since the 1960s. When finished, the aqueduct will carry water 600 miles (966 kilometers) through the state of California, from the northern part of the state south to the Mexican border.

DAMS AND RESERVOIRS

Dams are structures that restrict the flow of water in a river or stream. Both streams and rivers are bodies of flowing surface water driven by gravity that drain water from the continents. Once a body of flowing surface water has been slowed or stopped, a reservoir or lake collects behind the dam. Dams and reservoirs exist in nature, and man-made water control structures are patterned after examples in the natural word. Many lakes are held back by rock dams created by geologic events such as volcanic eruptions, landslides or the upward force of Earth that creates mountains. Humans and beavers alike have discovered how to modify their natural environment to suit their needs by constructing dams and creating artificial lakes. Dams are classified into four main types: gravity, embankment, buttress, and arch. Gravity dams: Gravity dams are massive earth, masonry (brick or stone work), rock fill, or concrete structures that hold back river water with their own weight. They are usually triangular with their point in a narrow gorge (deep ravine).

The Grand Dixence dam in the Swiss Alps is the world's tallest gravity dam. Embankment dams: Embankment dams are wide areas of compacted earth or rock fill with a concrete or masonry core that contains a reservoir, while allowing for some saturation and shifting of the earth around the dam, and of the dam within the earth. Buttress dams: Buttress dams have supports that reinforce the walls of the dam and can be curved or straight. Buttresses on large modern dams, such as the Itaipu dam in Brazil, are often constructed as a series of arches and are made of concrete reinforced with steel. Arch dams: Arch dams are curved dams that depend on the strength of the arch design to hold back water. Like gravity dams, they are most suited to narrow, V-shaped river valleys with solid rock to anchor the structure. Arch dams, however, can be much thinner than gravity dams and use less concrete.

DAMS IN HISTORY

Humans have used dams to trap and store fresh water in reservoirs for more than 5,000 years. Although water is ultimately a plentiful, renewable resource on Earth (Earth is after all "the water planet"), fresh water is scarce or only seasonally available in many regions. Left unregulated, the rivers and streams that provide humans' most essential natural resource are often hazardous to human life and too unpredictable to provide a constant source of fresh water. The ancient civilizations of Egypt, Assyria, Mesopotamia, and China grew and prospered in part because construction of dams and reservoirs allowed for irrigation (watering) of arid (extremely dry) lands, control of seasonal floods, and water storage during dry weather.

If Earth's streams and rivers are veins that support human survival, dams are valves that regulate the flow of water through those vessels. Humans today depend on dams to store water for irrigation, drinking water, and flood control just as they did in the ancient Middle East. Mesopotamians and Sumerians used weirs (low dams built across streams or rivers) and channels (passage for water) to irrigate the land between the Tigris and Euphrates Rivers, called the Fertile Crescent, about 6,500 years ago. Earthen dams that hold drinking and irrigation water in reservoirs for small towns and farms around the world today resemble the earliest known remains of dams. Archeologists estimate that a rock weir and series of small dams and reservoirs near the modern–day town of Jawa in Jordan were constructed about 5,000 years ago. Systems of aqueducts (artificial channels for conveying water) and canals (man-made watercourses designed to carry goods or water) like those constructed during the Roman Empire (1500–2000 years ago) carry water from reservoirs to modern farmlands and cities. Dams and reservoirs have a second important use beyond water storage and regulation of river flows.

They can be used to generate hydropower, one of human's oldest, simplest, and cleanest forms of renewable and reusable energy. Water that is held in a reservoir above the elevation of the river downstream has stored energy called hydrologic potential. When water is released through the dam from the reservoir, its motion can be used to turn a wheel that can then power a mill or an electrical generator. The farther the water falls, the more energy it releases. Water scientists and engineers use the height of the reservoir surface, called the hydraulic head, to estimate the amount of potential energy stored behind a dam. The technology to harness the mechanical power of falling water is almost as old as that for water storage and flood control. Ancient Sumerians and Egyptians used waterwheels with buckets on their blades, called norias, to dip water from streams or rivers. By 2,500 years ago, waterwheels drove grain mills and pumped water from wells in the Greek and Roman Empires.

During the late Middle Ages, water mills in the industrial centers of Germany and Italy ground grain, pulped wood for paper, spun silk for textiles, pounded metal, tanned hides, and crushed ore (mineral deposits) from mines. During the Industrial Revolution of the nineteenth century, British civil and mining engineers constructed 200 dams taller than 49 feet (15 meters, which is about the height of a five-story building) to store water for Britain's rapidly growing cities and to provide hydropower for mining and transport of coal, the energy source that powered industrialization.

MODERN DAMS

Today's dams and reservoirs provide many of the same benefits to humans and rely on the same basic technology as they did in ancient times. However, the size and complexity of modern water control and structures and systems would have astounded ancient Greeks and nineteenth century engineers. In developed nations like the United States, all of the major rivers have been dammed and almost every river system has been altered by humans. Worldwide, there were over 45,000 dams taller than 49 feet (15 meters) in 150 countries at the end of the twentieth century. Today, dams hold water for irrigation, control flooding along rivers, provide water for cities, and generate about one-fifth of the world's electricity.

In the countries with the most dams—China, the United States, and the nations of the former Soviet Union—engineering has given humans almost complete control over the rivers. In fact, one of the main reasons humans can no longer depend on hydropower to meet rising electricity needs is that there are very few large rivers left on Earth to be dammed. Dams are, by nature, destined to fail. A river erodes (wears away) and deposits sediment (particles of sand, silt, and clay) along its path from where it originates to the ocean in an attempt to create a constant slope (slanting contour of the land) called a graded profile. When a dam, natural or otherwise, blocks a river, the river adjusts to a new pattern of erosion and deposition in an attempt to return to its graded profile. In essence, the river attempts to remove the obstacle; reservoirs fill with sediment, and downstream erosion cuts under dams. Dams built before the 1930s were constructed with little knowledge of how rivers work or how structures can be designed to resist failure.

One in ten dams built in the United States before 1930 has collapsed. In 1889, more than 2,200 people were killed when the earthen embankment above Johnstown, Pennsylvania failed and the town was flooded. By the 1930s, use of concrete and metal in dams, arched designs, and an understanding of rivers allowed engineers to build safer, stronger dams. The new technology also led to an era of construction of ever-larger dams that has lasted until the present.

ENVIRONMENTAL AND SOCIAL IMPLICATIONS OF SUPERDAMS

The world's largest dams are massive structures over 492 feet or 150 meters tall (more than three times the height of the Statue of Liberty) that hold back reservoirs that cover a total land area about the size of Nebraska and Kansas combined. Construction of more than 300 super dams since the early twentieth century has created both benefits and problems for people living nearby. The economic, social, and environmental costs of major dams like the Grand Coulee dam on the Columbia River in the United States, the High Aswan dam on the Nile in Egypt, the Itaipu dam on the Paraná River between Brazil and Paraguay, the La Grande dams in Canada, and the Three Gorges Dam across the Yangtze River in China are extremely high and could possibly, just as to many geologists, exceed the long–term benefits of the projects. Problems associated with very large dams are now becoming apparent to geologists. The World Commission on Dams (WCD), between 30 and 60 million people, mostly poor farmers and people in India and China, have been displaced by large hydropower projects. Irreplaceable natural, archeological, and historical sites are drowned beneath huge reservoirs.

Drowned vegetation contaminates reservoir water and fish. Dams like Hoover and Glen Canyon dams on the Colorado River in the United States, or the Aswan High Dam on the Nile, disrupt river systems so large that the ecology (living environment) of an entire region has to adjust. Downstream, agricultural lands may lose their fertility, water quality is poor, and natural ecosystems (interactions between living organisms and their environment) are harmed. Coastal erosion results when rivers no longer replenish deltas (land area before river enters larger body of water) with sediment. Many environmental groups, scientists, and even some governments have begun to seek solutions to the problems presented by large dams. Decreasing the size and number of new dams, discovering new energy alternatives, managing river flows to counteract harmful environmental and social effects, and even removing some dams have all been considered. With the new goal of using dams and reservoirs to create a sustainable human and natural environment, modern and ancient water management technology combined could serve well far into the future.

DESALINATION

Approximately 97% of Earth's water is either sea water or brackish water (a mixture of salt and fresh water). Humans and other animals cannot drink salt water and to do so can bring on dehydration (the loss of the body's existing water) that can lead to illness and in extreme cases, death. Desalination is the process of removing salt from seawater to make it drinkable (drinkable water is also called potable water) or to make it useable for irrigation (watering fields and crops).

Natural desalination occurs everyday as a part of the world's hydrologic cycle. As salt water from the oceans evaporates (changes from liquid to gas), the salt is left behind and the water that moves into the atmosphere is fresh water. Thus, the water in clouds that eventually falls as rain is fresh water. Salt can also be removed from water by a series of processes known as manipulated desalinization, desalting, or saline water reclamation (salt water reclamation).

All of these manmade processes are expensive in terms of how much money and energy they each require to produce a gallon of water. Salt is composed (made up) of sodium and chorine atoms (the smallest particles of each element). Seawater contains the same kind of salt (sodium chloride) used everyday on food and in cooking. In addition, seawater also contains many small particles of the chemicals such as calcium and magnesium that also form chemicals called salts. Some of these salts come from chemicals used by industry, others from natural processes. Between three and four pounds out of every 100 pounds of atoms in saltwater (the hydrogen and oxygen atoms that together form water plus the atoms of all chemicals dissolved in the water) are combined into salts. Public health officials who test water use a different scale and label the salt in water as parts (particles) per million (ppm). Using this scale, seawater contains 35,000 ppm of dissolved salts. Brackish water typically contains less than half the amount of salt that is found in seawater, about 5,000–10,000 ppm of salt. Safe drinking water for humans, and water for most types of crops, must contain only 5,000–10,000 ppm of salt.

METHODS TO REMOVE SALT

There are several ways to remove salt from seawater and the method used is determined by the intended use of the water. Salt can also be removed from groundwater contaminated with saltwater. For example, if the water is to be drinkable then more salt needs to be removed than if the water is to be used for crops. Cost is also an important consideration because the more salt that needs to be removed, the greater the cost. Stories from ancient Greece tell of how sailors obtained fresh water by first removing salt from seawater by evaporating the seawater, and then condensing (changing from a gas to a liquid) the air carrying the evaporated water. This process, because it uses the heat of the Sun is now called solar distillation. Solar distillation is similar to the natural process of the heat of the Sun evaporating water from the oceans that later condense into fresh water drops in clouds.

When the water evaporates, only fresh water moves into the surrounding air because the salts are too heavy and are left behind in the ocean. Only fresh water went into the surrounding air (for example, the air over a bucket of seawater). As the air came into contact with cooler sheets or sails spread over the bucket, drops of fresh water would form and could then be collected in a separate bucket. Other, but far less efficient ways to obtain fresh water included the use of filtering seawater. One method of filtering included the use of a wool wick (a length of rope made of wool) to absorb (siphon) the water. The salts were trapped in the wool and fresh water dripped out.

Water was also poured through sand or clay to remove salts. By the fourth century (400 A.D.) onward, people obtained fresh water by boiling salt water and using sponges to absorb the fresh water in the air above the pot. The first scientific document on desalting was published by Arab scientists in the eighth century. The first desalination efforts for industry started in 1869, as land–based steam distillation plants were established in Britain to prepare fresh water for ships going to sea. Methods of distillation and filtering are still the most widely used methods of desalination used in most areas of the world. Other modern techniques use complex machines that change the temperature at which water boils away by lowering the pressure of the atmosphere over a sealed container of water. This methods reduces the formation of crusty white salts, which appear similar to the sticky white powder found at the bottom of a pan from which all water has boiled away. These crusty white salts can clog machinery and make it more difficult to heat water.

In industry, the crusty residue is called scaling, and the method of lowering the temperature at which water boils is called multistage distillation (multiple stages of distillation). The goal of multistage distillation is to reduce the boiling point of water to a temperature where it will still boil (evaporate) into a collection flask, but that it will not form a crusty salt residue. The residue forms at about 160°F (71°C) so the goal is to reduce the boiling point of water to less than 160°F. In some desalinization plants, distilled water is also filtered of other pollutants to make it ready to drink. A process called reverse osmosis can also be used to remove salt from water. Water molecules are forced through a plastic membrane (a barrier) with very small pores (openings) that allow the passage of water, but not of salts.

HYDROPOWER

Hydropower is energy that is generated by moving water. Today, hydropower facilities make electricity by converting kinetic (moving) energy into mechanical (machine) energy as water flows in a river or over a dam. Electricity made at hydropower facilities can be carried away, via power transmission lines, and sold to homes and businesses. Hydropower is a relatively inexpensive, non–polluting form of renewable energy. Canada and the United States are currently the world's top hydroelectric producers. Other countries that use hydropower on a large scale include Brazil, China, Russia, Norway, Japan, India, Iceland, Sweden, and France.

Hydropower produces about 10% of United State's electricity, in contrast to Norway, who generates nearly 99% of its electricity from hydropower. Hydropower is used nationwide, but is primarily used in the western coastal United States where other energy resources such as coal are limited. Hydropower is important to the United States economy because it supplies electricity to a growing population and industry.

HYDROPOWER IN HISTORY

Humans have harnessed water power for thousands of years, using the mechanical energy of moving water to turn wooden wheels to power mills that sawed lumber and processed grains. Water either fell onto the wheel and caused it to turn or the wheel was placed in the river and the river's current (a steady flow of water in a prevailing direction) turned the wheel. The wheel was attached to other levers and gears inside the mill that did the work needed. During the 1700s to 1800s, mechanical hydropower was used extensively in the United States and elsewhere for milling and pumping. It began to be widely used to supply electricity in the late 1800s.

In 1882 the world's first hydropower facility was built on the Fox River in Appleton, Wisconsin. The Fox River facility generated electricity for local industries. By the 1920s, following the development of the electrical motor and the demand for electricity that followed, hydropower accounted for about 40% of the U.S. electricity supply. Since then, other energy technologies have developed that are less expensive than hydropower. In 1933, President Franklin D. Roosevelt (1882–1945) signed the Tennessee Valley Authority (TVA) Act.

The purpose of the TVA was to construct dams on the Tennessee River that would aid in river navigation, control flooding, provide water for irrigation (watering crops) and drinking, and provide hydropower to Tennessee River Valley residents. In the American West, hydropower aided in the production of the dams themselves, moving and lifting construction materials and providing energy for lights to make round-the-clock work possible. Surplus energy was sold to neighbouring farms and homes, which in turn paid for the operation and building costs of the dams. Hydropower facility development was at its peak during the 1930s and 1940s. The Hoover Dam on the Nevada-Arizona border and the Grand Coulee dam located on the Columbia River in Washington state were constructed during this period. The Grand Coulee dam remains the largest concrete structure in the United States.

HYDROELECTRIC TECHNOLOGY

Electricity is one of the most important types of energy because it allows people to perform the work needed to light homes and power appliances and computers. Hydropower generation utilizes the principles of electromagnetic induction (creating an electrical current, the flow of electricity, by moving a magnet through a wire coil), first described by English chemist and physicist Michael Faraday (1791–1867) in 1831 when he made the first generator. A generator is a machine that converts mechanical energy to electrical energy. Energy is the power to do work. Energy cannot be created or destroyed; it merely changes state, as occurs when electrical energy is harnessed from the mechanical energy of moving water. No matter the type or size of the hydropower facility, electricity is generated in much the same way in each one. The dam or the natural elevation drop in a river creates head, or a certain height over which the water flows as it is released from the dam or as it flows downstream. As water is released from a reservoir in an impoundment or pumped storage facility, or diverted from a river through a control gate, it flows by gravity (the attraction between two masses) to a turbine. A turbine is a device that transforms the energy in moving fluid or wind to rotary mechanical energy.

As the water flows past the turbine blades, the turbine blades spin and turn a rotor (the moving part of an electric generator) much like wind spins a pinwheel. Giant magnets inside the rotor move past coils of copper wire and create an alternating current. An alternating current (AC) is produced when electrons wiggle back and forth between atoms. The used water returns to the river through pipes. The alternating current moves through a series of devices called transformers that can increase the voltage (energy required to move a charge from one point to another, similar to pressure).

The increase in voltage allows the electricity to travel faster and more efficiently through power lines (important when the hydropower facility is in a remote location) from the hydropower facility to a town's electricity facility. At the local facility transformers are used again to reduce the voltage to a level that is safe for the electricity to be used in homes and businesses.

TYPES OF HYDROPOWER FACILITIES

There are three types of hydropower facilities: impoundment, pumped storage, and diversion. Impoundment and pumped storage facilities require dams. In the United States, hydropower is a very small percentage of the primary purpose of dams. Usually a dam is built first for other reasons, such as water storage and flood control, and a hydropower facility is incorporated later if there is a demonstrated need for electricity. Impoundment facilities require a large dam that allows river water to be stored in a reservoir. When water is released from the reservoir, the water flows downward through a penstock (pipe) in the dam and spins turbines, thereby creating mechanical energy that is then used to power electric generators. Transmission lines carry electricity away from the impoundment facility to local distributors who sell the electricity to homes and businesses. Pumped storage hydropower facilities also require dams to operate.

In periods of low electricity demand water is pumped from a lower reservoir to a higher reservoir. When electrical demand increases, water is released from the higher reservoir back into the lower reservoir through a penstock, turning turbines that power electric generators. Pumped storage and impoundment hydropower facilities provide a reliable source of electricity because water flow from reservoirs can be controlled so that electricity production meets demand. Diversion hydropower facilities (sometimes called run-ofriver systems) are smaller than impoundment facilities and do not require a dam. Instead, diversion facilities use a river's natural flow to generate electricity. The amount of electricity produced depends upon the river's rate of flow (volume of water flow within a period of time) and the river's elevation change at the diversion facility site.

In diversion hydropower facilities, river water is channeled through a canal (artificial waterway that controls water flow, in this case, to a turbine) or penstock. Diversion facilities are less reliable than impoundment or pumped storage facilities because flow rates in rivers can change drastically depending on the amount of rainfall or spring meltwater that supplies the river.

SIZES OF HYDROPOWER FACILITIES

The size of a hydropower facility depends upon the amount of energy that can be generated at the facility. Some hydropower plants may produce electricity for only one to a few homes. These facilities are called micro-hydropower plants. In microhydropower facilities the total change in elevation of the flowing water is only about 100 feet (30 meters). Energy output can be increased, however, with higher water flow. Larger hydroelectric power facilities such as the Hoover power plant in Nevada, however, can provide electricity to more than one million consumers.

BENEFITS AND DRAWBACKS OF HYDROPOWER

Hydropower is an efficient, clean, and reliable source of energy. Once hydropower plants are constructed and the technology is in place, the cost of hydropower is the lowest of all energy sources. No fuel is used during hydropower production; water returns to the river. No pollutants are released into the air during electrical generation, so the air around hydropower plants remains clean. Hydropower facilities can respond to high demands for electricity through reservoir storage, even in times of water shortage. The negative effects of hydropower generally come from large–scale facilities. Impoundment or pumped storage facilities often alter wildlife habitats along rivers because large dams must be built to provide a reservoir. Dams can flood areas close to a river or lake, obstruct fish migration (periodic movement from one region to another), and affect water quality and flow downstream.

Humans are sometimes displaced from their homes when dams are built. In diversion facilities, seasonal and annual fluctuations (variations) in water supply can negatively affect electrical production if river flow rates become too low. Regulations and licensing permits for dams and hydropower facilities as well as electrical transportation from remote hydropower are often costly.

THE FUTURE OF HYDROPOWER

Recent energy shortages in the United States has spurred the government to study hydropower's future potential. The Department of Energy has identified 5,677 sites in the United States that have potential for large-scale hydropower development, but many of these sites are not possible because of economic drawback, such as their remote location, environmental issues, and other circumstances. Because a change to hydropower as a primary energy source could help lessen air pollution and reduce demands for fossil fuels (oil and gas), hydropower will likely be further explored.

PORTS AND HARBORS

Peoples of ancient civilizations often built their cities on the shores of natural harbors. A harbor is place on the coastline that is protected from the full effects of tide and currents (a steady flow of ocean waters in a prevailing direction). Harbors are often shaped like horseshoes. They are surrounded by land with a narrow opening through which ships can pass. Manmade harbors use structures such as walls or barriers built into the water to protect anchored ships from tide or storm damage. Building cities near harbors permitted the construction of ports for trade.

A port is a place on a shoreline for the loading and unloading of cargo from shipping vessels. Ports can be located on the ocean coast or on the shores of lakes and rivers. Cities with working ports are also called seaports or port cities. Many of the great cities of the ancient world were seaports.

Seaports allowed cities to grow and flourish. Trade made them wealthy. Ports also sheltered ships of war, sometimes a necessity to guard desirable seaports from invasion. The ancient seaport of Alexandria (Egypt), located along a man–made harbor along the Mediterranean Sea and the end of the Nile River, thrived for centuries. Alexandria was invaded several times and ruled by the Egyptians, Greeks, and Romans. The modern city of Alexandria, Egypt, is located a few miles (kilometers) from the underwater ruins of the ancient seaport.

MODERN PORTS AND HARBORS

Improvements in technology has allowed ships to travel faster, carry more cargo, and load with greater ease. Ports have grown in size to accommodate today's bigger, faster ships. However, many parts of modern ports closely resemble their ancient counterparts. In modern day, there are over 185 seaports in the United States. Trade with other nations, called overseas trade, is important to a nation's economy. Two types of overseas trade occur, imports and exports. Imports are goods and materials that are shipped into United States from other countries. Exports are goods and materials from the United States that are shipped to other nations. Ships move most U.S. overseas trade, and shipping on waterways remains one of the least expensive means of transporting goods.

Modern ports have several special structures that help with the movement of cargo between land and ships. Navigation channels or ship channels that ships travel into port are marked "roadways." Docks and piers permit vessels to moor (secure to the dock) and for cargo to be loaded and unloaded from ships. Cranes and ramps aid the movement of cargo. On land, ports have large warehouses to store cargo. The waterside port area is connected to inland transportation systems such as roads, railways, pipelines, and airports. This allows cargo from ships to be loaded onto trucks, trains, or airplanes for transport to inland destinations.

Cargo is often stored in containers that can be loaded from trucks to trains to ships. For example, metal boxes on railroad cars can be detached and loaded directly onto barges or ships. Liquid cargo, such as oil or gasoline, can be piped from large reservoirs onboard ships into tanker trucks or tanker train cars.

BUILDING AND MAINTAINING SUCCESSFUL PORTS

Modern ports must be built to accommodate several types of ships, from oil tankers and tugboats, to barges and passenger ships. Ports not only aid trade; they are also important centers of travel for people. Passenger ferries and cruise ships carry people from one port to another. Ports usually have separate docks, piers, and places to moor for passenger vessels. Even if a port is located in a natural harbor, river, or lake, the waterways must be maintained. Ship channels must be kept sufficiently deep to permit the passage of ships, barges, and tugboats. Dredging makes waterways deeper by removing the silt (tiny particles of rock, soil, and plant material) and mud that clogs channels. Docks, piers, and jetties (protective rock barriers) need continual maintenance. The U.S. government and local port authorities oversee daily operations at United States ports.

PROBLEMS, CONCERNS, AND THE FUTURE OF PORTS

Ports are vital to the economies of the cities in which they are located. However, the warehouses, docks, cranes, and shipyards of a working port are generally not considered attractive. Ports occupy large amounts of land near waterways. Commercial ships, vessels used for trade, are large and difficult to navigate. They can pose a danger to small recreational boats. Laws often prohibit or limit recreational boats in ship channels and other key port waters. Thus, ports can sometimes restrict individuals' use of shorelines, coastal areas, and waterways. City planners work to carefully balance the needs of the port with the interests of citizens and businesses. Ports also pose special challenges for marine (ocean) and coastal environments.

The constant presence of large vessel traffic can disturb plants, animals, and microorganisms that live in port and harbor waters. The construction of jetties and breakwaters changes the effect of tides and currents within a harbor or port area. This alters the water chemistry of water within the jetties. Water chemistry is the balance of temperature, minerals, salts, and even pollutants in water. Even a slight change in the temperature, muddiness, or salinity (saltiness) of water can harm marine life.

As ships require fuel to operate, port areas have to store and transport large amounts of gas and oil. Leaks, spills, and shipwrecks damage underwater and coastal habitats. Ports also increase the amount of other types of pollutants, such as litter.

However, ships produce less air pollution than airplanes, trucks, or trains loaded with the same tonnage of cargo. Transportation systems are always changing and improving. Ships, trains, and airplanes are becoming bigger, faster, and more efficient. Trucks and airplanes, only invented a century ago, are already essential means of moving goods and materials. Yet a port, an idea centuries old, remains an efficient link for land and ocean transport.

TIDE ENERGY

Tides are twice–daily rises and falls of water level relative to land. Ocean tides can produce strong currents (a steady flow of ocean waters in a prevailing direction) along some coastlines. Humans have sought to harness the kinetic (motion–induced) energy of the tides for hundreds of years.

Residents of coastal England and France have used tidal energy to turn water wheels and generate mechanical energy for grain mills since the eleventh century. In modern day, tidal currents are used to generate electricity. Tidal energy is a non–polluting, renewable energy source. Modern day technologies for exploiting tidal energy are, however, relatively expensive and are limited to a few coastlines with extremely high and low tides. Tidal energy may, in the future, become more widely used and economically practical.

THE POWER IN TIDES

Tides result from the gravitational pulls of the Moon and Sun on the surface of the spinning Earth. Gravity is the force of attraction between all masses. The shape of the shore and adjacent seafloor affects the tidal range (difference between high and low tides) along specific coastlines. Some places, like the English Channel between France and England, and the Bay of Fundy in Nova Scotia, Canada, experience very high and low tides. The tides protected Medieval monasteries in the English Channel since the eighth century. Mont-St.-Michel in western France and Lindisfarne (Holy Island) in northern England are churches built on small islands surrounded by miles of tidal flats (a broad, flat area of coastline alternately covered and exposed by the tides).

Today, they are connected to the mainland by roadways but in Medieval times, only devout pilgrims rushed to make the hurried trip across miles (kilometers) of shifting sand between roaring tidal pulses. For tidal energy to be a practical source for electricity generation, the tidal range in a coastal area must be at least 16.5 feet (5 meters). The greater an area's tidal range, the more electricity will be produced. Although tidal energy is reliable and plentiful, only a handful of suitable tidal power station locations have been proposed worldwide. Two large tidal power plants are in operation today at La Rance in Brittany, France, and in the Canadian town of Annapolis Royal, Nova Scotia. In the United States, tidal energy as a power source is realistic only in Alaska and Maine.

EXPLOITING TIDAL ENERGY

Devices used to exploit tidal energy may be shore-based or ocean–based systems. Both systems use a fluctuating column of water to propel turbine blades and generate electricity. This means that the water on one side of the dam is higher than on the other side. As the water falls from the high side of the dam to the other, turbines turn and produce electricity. A barrage is a shore–based, dam–like structure that is built across a narrow-mouthed estuary (the part of a river where it nears the sea and fresh and salt water mix). As the tide ebbs and flows (moves in and out) through tunnels in the barrage, the water turns large fan–like turbines and generates electricity.

Barrages are expensive to build and can harm estuarine life by restricting water flow over the tidal flats. Electricity produced by tidal energy has no harmful wastes or emissions such as greenhouse gases. Once built, the barrage is easy to maintain and inexpensive to run. Ocean–based systems include tidal fences and offshore turbines. Tidal fences are like giant subway turnstiles built across the sea floor between the mainland and an island or between two islands.

When a tidal fence is built across an open body of water, water is forced to pass through the vertical turbine gates and electricity is generated. Tidal fences may restrict tidal flow and the ability of wildlife to pass through. Offshore turbines are like giant propellers placed on large posts that are set in a line across the sea floor. Ocean currents flow past the turbines and cause the blades to spin and generate electricity. Offshore turbines are like giant propellers placed on large posts that are set in a line across the sea floor. Ocean currents flow past the turbines and cause the blades to spin and generate electricity.

Tidal fence turbines have a much lower initial cost when compared to barrages and are much less harmful to the environment because they do not restrict tidal ebb and flow. Tidal turbines also allow wildlife and small boats to pass through the area. Offshore turbine blades are smaller and less protected than those housed in barrages or tidal fences and so they are more prone to damage from strong tidal currents. Oceanbased systems are much less expensive to install than barrages, but are more expensive to maintain due to their remote location.

WASTEWATER MANAGEMENT

The idea of wastewater management is as old as man himself. Simply put, man has struggled through the ages with the problem of what to do with his waste. The painstaking efforts of plumbers past is evidenced by the ancient drains, grandiose palaces, and bath houses of the Minoan civilization some 4,000 years ago. Man knew instinctively, even in his earliest existence, the importance of allowing animal and human waste to go downstream, yielding to the natural flow of things. He may not have known all of the consequences, but he surely found the prospect of harvesting drinking water from the same area of the stream used for waste distasteful. Now, in the dawning of a new century, wastewater management is still an issue in the forefront. As our ancestors sought to answer that eternal question, we, in a more sophisticated manner today, are still trying to figure out the best way to manage our waste.

WHAT IS WASTEWATER MANAGEMENT?

Imagine that you are opening a new business. It is a considerable investment. You have put a lot of time and hard—earned money into it.

Would you open your new store without a long–term plan, having no control over future sales or purchasing? Although this question may seem rudimentary, in many parts of the country the onsite waste water treatment industry has been functioning just this way, without a long-term plan or management programme. The dictionary defines management as "the act, manner, or practice of managing, supervising, or controlling."

Whether you spend millions or thousands of dollars, or whether the system is part of a public works project or an individual septic tank, there should be some entity responsible for the overall consequences and direction. Most communities already manage their onsite systems to some extent through regulation. But the term "management" as it is used today implies a broader definition. In other words, wastewater systems, particularly onsite systems, need to be managed or controlled, not just technologically, but with a broad concept connecting individuals, communities, local officials, and regulatory agencies if failures and malfunctions are to be avoided.

WHY IS MANAGEMENT IMPORTANT?

Trends and numbers speak volumes about the need for onsite wastewater management today. As we enter the new millennium, population growth is moving more and more homeowners into suburban areas, many relying on onsite wastewater treatment and disposal. The majority of homes in rural America already rely solely on onsite systems. Approximately one fourth of the estimated 109 million housing units in the United States are served with septic tanks or cesspools, just as to a 1995 American Housing Survey (AHS). During that year alone, more than 2.5 million septic tanks in America were reported as malfunctioning (or as having a total breakdown of the system). Graham Knowles of the National Small Flows Clearinghouse's (NSFC) National Onsite Demonstration Project (NODP) authored a report titled "Septic Stats, An Overview," based on the AHS data.

In the report, he combines U.S. Department of Commerce Bureau of Census statistics with the AHS data to establish septic tank trends. Knowle's report projects that by the year 2025 there will be 40 million housing units with septic tanks. If the current trend continues, that could mean as many as 4 million septic systems could be malfunctioning by 2025. If this projection becomes a reality, the necessity for greater control through management programmes should be self-evident. As In Olden Times turning the clock back to study how wastewater systems evolved and how management programmes have fared throughout the years can be a useful tool. Close to 4,000 years ago, approximately 1700 B.C., the Minoan Palace of Knossos on the isle of Crete featured four separate drainage systems that emptied into great sewers constructed of stone.

The palace latrine was the world's first flushing toilet with a wooden seat and a small reservoir of water. From 3000 to 1500 B.C., early plumbers laid sewage and drainage systems. Archaeologists have discovered underground channels that remained virtually unchanged for centuries. Ancient gravity sewers were developed in response to the density of populations living in close proximity or in cities, just as to Peter Casey, programme coordinator for the NSFC. These large central systems were actually analagous to sewers developed in the 1800s in London and other large cities.

During these times, there were many outbreaks of various diseases, such as dysentery, cholera, infectious hepatitis, typhoid and paratyphoid, and various other types of diarrhea. "The biggest health benefit of the 20th century was brought about by the purification of drinking water and treatment of wastewater," Casey added. "It increased life expectancies and had a tremendous impact on man's health and survival." Casey said in 1870, the average person could expect to live to be 40 years old. By 1900, that age climbed to 47 with steady increases throughout the decades since. Today, the average person in a developed country can expect to live into his or her 70s or beyond.

In 1997 U.S. Environmental Protection Agency (EPA) document, onsite wastewater systems have been around since the mid-1800s. "In rural areas during the mid–1800s to the early 1900s, sanitation was not a problem because the water supply was hand carried or pumped. No water was required for the privy," Casey said. "If it became full, they would simply cover it and dig a new privy." Many people are familiar with the early 1900 image of a splintered, wooden shed, usually with one door and a hole in the floor, as the rural family's outhouse. Chamber pots were dumped outside or in the privy.

After the 1930s when electricity and gas became available in rural areas, the need for onsite treatment arose because of the increased volume of liquids in the wastewater, Casey said. "Once farmhouses got electricity and indoor plumbing and the conveniences of the large cities, the flows became too great and caused problems," Casey added. "Suddenly, there was running water in the house, making way for baths, showers, and flush toilets. Cesspools were the earliest form of onsite system in response to increased water use.

They were usually just a large, covered hole dug in an inaccessible area. "With the move to the suburbs in the 1940s, we saw dense housing units trying to use all this water on half-acre lots. There was no place for all of the water to go," Casey said. Septic tank systems, specifically, have been used for wastewater treatment since the turn of the century, just as to a report from a 1994 University of Waterloo, Ontario, conference. The report, by Richard J. Otis and Damann L. Anderson, adds that the use of septic tanks did not become widespread until after World War II when the suburban housing boom outgrew the rate of sewer construction.

REGULATION BEGINS

The Otis and Anderson report notes that in the 1950s, states began to adopt regulations to provide a universal basis for the design and installation of septic tank systems. These early codes did not, however, provide much in the way of broad management or prevention of system failure. The programmes regulating the installation and use of onsite systems could not keep up with the increasing demand. The report adds, "Today, it is generally recognized that past approaches to managing onsite wastewater treatment systems use are no longer adequate.... The failure of these systems to gain acceptance as effective and permanent facilities is due primarily to shortcomings in management programmes." The report notes that the biggest assumption at that time was that onsite systems would ultimately be replaced by central sewerage.

Despite this, some early management programmes did arise. In 1954, Fairfax County, Virginia, established an onsite wastewater management programme when the board of supervisors there directed the health department to develop a programme that would prevent future septic system failures. The management plan focused on the planning, design, and construction review of septic tank systems through an extensive permit programme. Under this early management plan, the county was in charge of site evaluation, design review, installation supervision, monitoring, and public education while the homeowners were responsible for the operation, maintenance, and repair of the systems. With the establishment of the wastewater treatment construction grant programme under the Federal Water Pollution Control Act Amendments in 1956, the focus continued on construction of centralized sewers. Throughout the 1960s, the concept of septic tank systems being a temporary solution continued.

ONSITE SYSTEMS ARE RECOGNIZED

In the 1970s, millions of dollars were still being spent on constructing sewers and centralized wastewater treatment facilities, while at the same time, many federal and state agencies started to consider regulating and managing onsite systems as part of environmental pollution control issues. Throughout the 1970s, management programmes sprang up across the country. By 1974, many states had identified the need for better managed individual onsite systems through studies conducted under the Federal Water Pollution Control Act, Section 208. More importantly for the onsite system industry, EPA regulations required the inclusion of a cost-effectiveness analysis of alternatives by all applications initiated after April 30, 1974, under the Federal Construction Grants Programme.

In fact, the 1977 Clean Water Act (CWA) Amendments required communities to examine alternatives to conventional systems. In addition, the NSFC was established by Congress as part of the amendments to provide technical information and assistance to small communities across the country. In the 1978 "Report to the Congress," the Comptroller General of the U.S. stated that septic systems can function as effectively and permanently as central facilities and are a cost–effective alternative to sewage treatment plants, adding that "EPA and other federal agencies should increase the acceptance of septic systems by requiring established public management entities to control their design, installation, and operation."

If the 1970s could be remembered as the decade septic systems became recognized as permanent wastewater treatment options, then the 1980s might be remembered as the decade of onsite exploration. During the 1980s, the field progressed significantly, and many onsite management system models were developed.

In the 1990s, the issue of management has been tweaked further, focusing on the development of adequate monitoring and comprehensive management systems. In the 1997 "Response to Congress on Use of Decentralized Wastewater Treatment Systems," EPA stated that "adequately managed decentralized wastewater systems are a cost-effective and long-term option for meeting public health and water quality goals, particularly in less densely populated areas." Since then, septic tank systems and other alternative systems generally have been recognized not only as environmentally and technologically sound treatment methods, but have been viewed as viable, permanent methods of treatment.

In the 1997 report, EPA noted that one of the barriers to implementing decentralized systems is a lack of management programmes. To overcome this, EPA recommended development of management programmes "on state, regional, or local levels, as appropriate, to ensure that decentralized wastewater systems are sited, designed, installed, operated, and maintained properly and that they continue to meet health and water quality performance standards." As one of the responses to these 1997 recommendations, EPA launched Phase IV of the NODP in 1998.

ENTER PHASE IV

Phase IV of the NODP is a three-year programme, focusing on establishing the necessary processes to help small communities develop a broad concept of management for onsite systems. Knowles, programme coordinator of Phase IV, has been studying management issues. He commented, "Management programmes are imperative today because they will enable communities to control the effectiveness of wastewater treatment and can help ensure public health, improve water quality, and sustain the environment."

Knowles said Phase IV's mission has three components:

1. To gather data, information, knowledge and insights concerning all aspects of onsite management systems nationwide. Under this component, Knowles said objectives will be to establish a repository of information and expertise on the topic of onsite management, forming a national database of management systems complete with case studies addressing issues of management approaches, compliance, improvement, and prevention perspectives.
2. To create a framework, with tools and educational products for national dissemination to communities through a network of partners. Knowles explained that this component seeks to "develop a framework of guiding ideas to assist communities through the process of moving from their current reality towards increasingly effective onsite wastewater management." He added, "The aim is to establish strategies for change, creating "a network of interested publics to partner with, for disseminating the onsite management idea and delivering products, tools, and services."
3. To analyse, evaluate, review, and refine onsite management models, methods, and materials at strategically selected sites.

Knowles said this component of the project is designed to select suitable sites to pilot onsite management systems. It also will "provide NODP expertise, materials, mentors, management insights, tools, and techniques to communities interested in adopting an onsite management systems approach." This component will document and track management strategies, products, tools, and services to meet differing community needs.

Once these components have been met, Knowles said Phase IV ultimately will provide interested communities with practical, hands-on technological and management expertise facilitating community onsite system management programmes tailor-made to meet a particular local community's needs. To help the project succeed, NODP IV has enlisted an expert panel, made up of talented individuals in the wastewater field who have made and are continuing to make significant contributions to the evolution and development of onsite management plans.

CRANBERRY LAKE'S SUCCESS

One panel member, Jane Schautz, vice president and director of the Small Towns Environment Programme at The Rensselaerville Institute in New York, is working with several communities, studying their onsite management programmes. Schautz's role is that of an observer, documenting the progress and noting the plan's assets and possible defects. She defines onsite management as "the systematic monitoring and maintenance of onsite systems to anticipate and/or correct malfunction in order to preserve the life of the system and prevent environmental degradation."

The challenging task, she said, is to make a management programme work in existing communities that have onsite systems and do not automatically see the benefit of adopting a management system with all of its associated costs that the residents have not paid previously. She cited Cranberry Lake, New Jersey, as an excellent case study of this scenario. One important session she has learned is that residents have to be shown there are innumerable benefits to whatever costs might be incurred. The Cranberry Lake Septic Management System was established in 1990. It is a relatively affluent area. Most of the houses around the lake were built in the 1950s and intended for seasonal use. Because of this, some of the lots are small, approximately 50 by 50 feet. "Year–round occupancy was not expected, but with retirement increasing, more and more people are living there year-round," she said. "With retirees you have to be sensitive to their limited incomes. One fear was that they would be thrown out of their houses if a malfunction were discovered. That made the problem more intense."

Prior to the establishment of the management system, Cranberry Lake had a nitrate problem from failing septic systems and was overgrown by weeds. Schautz said that as a bonus, having the management system for wastewater in place helped the township to successfully secure funding for treating unwanted plants. Schautz believes Cranberry Lake's success should be credited largely to Margaret McGarrity, the "spark plug," or local person who took the initiative to get things moving there. She added that Township Manager Ronald Gatti also played a major role in their success. "Trying to establish a management district is going to be controversial, and people have to be willing to deal with controversy without being damaged," said Schautz. "You have to have savvy people who have the guts to stick with it.

A plan is inert until somebody believes in it. You have to have a champion to give any plan a life." McGarrity, a member of the environmental township commission, was that person for Cranberry Lake. Schautz said McGarrity felt that sewers were inappropriate for the area. "Sewers just take wastewater from one area and move it to someplace else," said Schautz. "McGarrity felt they couldn't afford that for the wells or lake. She looked at all of the components and decided it made no sense to spend money installing septic systems and allowing them to malfunction." Under the management plan, residents pay a flat $15 fee that covers a three-year period, that extends from one date of pumping to another.

By paying the fee, they update their permit by showing proof of pumping. Municipal officers oversee the process. The township board of health is responsible for enforcement. "They have astonishing compliance," said Schautz. Schautz added that the township residents' drinking water is provided by privately owned wells. "People understand that this is all related to maintaining the purity of their lake, as well as preserving their drinking wells. "It took awhile to convince people that this was in their best interest, but they now see that this is an improvement of their relationship with the township government," said Schautz.

In fact, Cranberry Lake's management plan has been the model for other communities in the area. She said, "To me one of the most persuasive evidences of the plan's success is that others in the area have seen the results and are taking steps to follow that model. "In the beginning, people were saying, 'why us?' and now after the evidence, not only are they seeing many upgrades to systems made voluntarily, but people are saying, 'why not us?' and have petitioned the township board to include other areas," she added. Schautz cautioned that management plans cannot be established overnight. She quipped, "Starting a wastewater management district is like planting asparagus-the first rule is the ground should have been prepared three years ago. Cranberry Lake did it faster, but very intensively." Cranberry Lake's first step was education, including presentations at local meetings, seminars, articles in the local newspaper, information booths at community meetings, and insert fliers. Schautz said this process took approximately one year. "McGarrity and Gatti said their success depended on persistence-getting the word out and allowing it to take hold, giving people time to come to their own conclusions," Schautz added. "In this case, their commitment and belief eventually became infectious." Another helpful aspect of gaining acceptance was that the ordinance was relatively mild. "That way, there was less opposition," she said. "There's no reason to make this harder than it has to be. In fact, they went out of their way to accommodate people." The management plan gives the township the authority to fine residents $1,000 per day or order them to do 90 days of community service for noncompliance, but Schautz said there has rarely been a need to impose those punishments.

In addition to having a spark plug, educating the public, and persistence, Schautz believes humour is an imperative component to the key to success. "McGarrity and Gatti livened up their material with graphics and energy," she said. "It really worked for the community." In the end, Schautz said a management plan must be based on the local culture and philosophy. "Some of the purists say it isn't a management system unless you have inspectors there all the time, tearing up the soil. I'm not saying that doesn't work; but for an older established community, it seems that moving in areas of environmental sensitivity makes sense when people come to the understanding that they are at risk."

GETTING UTILITIES INTO THE PLAN

Another expert panel member, Bridget Chard, is a Small Communities

Project coordinator and a township supervisor from Pillager, Minnesota. She works as a consultant for many townships in Minnesota, helping them implement a management model, called an "Environmental" Subordinate Service District, that can be tailored to meet differing needs.

In essence, the model allows local township boards, usually lacking the expertise, time, and experience needed, to develop and maintain a management plan by partnering with the local rural utilities. "The rural utilities are already in place and providing electric power to the rural residents.

These residents are already part of the rural electric co-op. Therefore, these utilities are usually more than willing to provide this management service," she explained. "They have the needed assets to oversee the systems. They do the billing, administrative work, and actually manage the wastewater system and therefore relieve the work that the town board would have to do. Essentially this becomes a public-private management system.

It's a good set of checks and balances. The townships have the authority to levy onto the residents property taxes for any unpaid service charges." In this model, enforcement issues are taken care of through a partnership with the township or county. "This is a choice situation," said Chard. "We continually are building new partnerships and better ways to do things.

As homeowners and township board people become introduced to this new model, it's always an education process. We do a great deal of informational work up front before we create the districts." The model allows for different methods of funding, including service charges and/or a property charge.

She added, however, that a township and its residents sometimes find other alternative and equitable methods for financing their projects. Chard, who is an independent contractor and chairman of her township board, said this model works well for old and new systems.

"This model is usually used to retrofit and replace old groups of non–conforming wastewater systems as well as being used for new conservation-based designed subdivisions. It's a fluid, dynamic model that can change and adapt to the local homeowner's needs. You can come up with different ways of handling old systems versus new systems. We want everybody's environment to be protected," she added. "It also stabilizes the local economy and protects the landowners real estate investment." Like Cranberry Lake, the area has water sources to protect.

As a result, Chard said lake associations are very active in Minnesota with education programmes as well as performing lake monitoring. They have been very supportive of projects that protect their lake quality and well supplies. Most of the areas where Chard works as a consultant are served by individual well systems. Also like Cranberry Lake, many of the lots were platted years ago and originally may have been set up for seasonal homes and have very small lot sizes. These lots are now seeing a need to replace a failing system with nowhere to place it on the property.

Chard has helped township projects, ranging from as small as eight to 200 homeowners, set up management districts. "We continually learn from the evolution of these and older districts," she said. "You should always be improving on the models." Chard was involved with the establishment of Cass County's first management district model, which was included in EPA's 1997 "Response to Congress on Use of Decentralized Wastewater Treatment Systems." The statute used in Minnesota for the framework of this management model is Minnesota Statute 365A for townships. This statute is used to provide many services that residents need within a township including road paving, animal control, and many other services.

The statute was used to develop "Environmental" Subordinate Service Districts, which manage a water or wastewater projects or both at the same time.

Under that plan, the Rural Utilities Services, formerly the Rural Electrification Association, was a major player. Cass County sought out the local utility, Crow Wing Power and Light of Brainerd, Minnesota, and asked them to help with the management programme, including monitoring, monthly inspections, pumping, record keeping, and billing administration.

Chard said this type of plan is typical of the model. She added that there are currently four known wastewater management districts operating in Cass County today with many others being implemented around the state by townships and counties.

The county usually partners with the townships to do all of the enforcement, permitting, and siting of treatment sites as well as implementation of a Geographic Information System (GIS) database for the districts. "They don't want to micromanage small groupings of wastewater systems, but would rather partner with townships. This method keeps them informed about the smaller wastewater management systems.

Now they have started doing planning and zoning, road work, and many other ideas have evolved from this original partnership and dialogue," she added. "The county attends yearly audit meetings with the township boards, residents, and rural utility representatives.

They physically review the system and look at the management logs to see how the wastewater system could be improved." The keys to success, in Chard's view, are education and the ability to keep an open mind. "It all goes back to working with your neighbour, building a trust base.

From there, you are challenged to find answers," she said. "You sit down with the property owners in a meeting and say here is the problem, now what can we do. I have yet to come up against a group that can't find their own solutions." Like Schautz, Chard recommends keeping education material and any documents homeowner friendly and humourous. "Try to make it fun. Homeowners always think of the government as being very imposing, but there is a lot of flexibility in this model that can be used to help the homeowners and town board work together and find solutions they need.

When it's done, all feel that they have ownership in their project," she added. Chard believes the definition for onsite management depends on a person's perspective. "Onsite management from the homeowner's perspective is new," she said. "It means taking care of their system, which is something they have never done before. By taking responsibility of your system, you are also protecting your neighbour. Further, we are managing and protecting a considerable investment and not wasting anyone's money to replace it sooner than is necessary." Chard added that onsite systems will always need some type of management tool, from the simplest "tank management" tools to the more sophisticated technologies that homeowners would not understand. "To me, it is doing it right from cradle to grave. It has gone beyond knowing that there are problems, that central piping is no longer a necessary evil because it's so costly," she added. "The NODP IV theory is truly fourth-generation thinking regarding the evolution of wastewater management for the new millennium," Chard said. "Now we know we not only have the technology, but the tools to manage and maintain any onsite and clusterdesigned wastewater system."

SEWAGE TREATMENT

One of the largest sources of wastewater is that which comes from homes and industries. These wastewaters all flow into sanitary sewers, which direct them into sewage treatment plants. Wastewaters from homes contain human waste, food, soaps, and detergents. They also contain pathogens, which are organisms that can cause diseases. Industrial wastewaters contain toxic (poisonous) pollutants, which can endanger human health and harm other organisms. These include pesticides, polychlorinated biphenyls (PCBs,) and heavy metals like lead, mercury, and nickel. These metals are generally toxic to plant and animal life.

The goal of sewage treatment is to remove all of these pollutants from the wastewater so that it can be returned to natural waters. Sewage treatment involves three stages: primary treatment, secondary treatment, and tertiary treatment. Primary treatment physically separates solids and liquids.

The wastewater passes through a grating that strains out large particles. The remaining water is left to stand in a tank, where smaller sediments (particles of sand, clay, and other materials) settle to the bottom. These sediments are called sludge.

At this point, the liquid part of the wastewater still contains many pollutants and is not safe for exposure to humans or the environment. In the second step, called secondary treatment, the liquid part of the wastewater passes through a trickling filter or an aeration tank. A trickling filter is a set of pipes with small holes in it that dribbles water over a bed of stones or corrugated plastic. Bacteria in the stones or plastic absorb pollutants from the water and break them down into substances that are not harmful. An aeration tank is a tank that contains bacteria that break down pollutants.

The liquid part of the wastewater from primary treatment is pumped into the tank and mixed with the bacteria. Air is bubbled through the tank to help the bacteria grow. As bacteria accumulate, they settle to the bottom of the tank and form sludge. The sludge is removed from the bottom of the tank and buried in landfills. After secondary treatment, the water is generally free from the majority of pathogens and heavy metals. It still contains high concentrations of nitrate and phosphate, minerals that can overstimulate the growth of algae and plants in natural waters, which can ultimately cause them and the surrounding organisms to die.

Tertiary treatment removes these nutrients from the wastewater. One method of tertiary treatment involves using biological, chemical, and physical processes to remove these nutrients. Another method is to pass the water through a wetland or lagoon (shallow body of water cut off from a larger body).

STORM SEWERS

In most cities in the United States, the sewers that carry storm waters are routed through sewage treatment plants. Much of the run-off from storms contains fertilizers, oils, and other chemicals that should be removed from the water before it enters lakes, rivers, and oceans. When there are very heavy rainfalls, however, the sewage treatment plants can become overwhelmed by the volume of water entering the facility. At these times, sewage and wastewater from storms may be dumped directly into natural water bodies. Many cities have programmes underway to separate the storm sewers from sanitary sewers, but these projects are very costly and time consuming.

AGRICULTURAL RUN-OFF

Agricultural run-off occurs when rain falls to the ground and then runs through agricultural fields or livestock-raising farms. The rainwater can accumulate fertilizer, oils, and animal wastes before it runs into rivers, lakes, and oceans. These materials pollute natural waters and can cause fish to die, contaminate drinking water, and speed up the rate of sedimentation (particles settling to the bottom of a waterway) in lakes and streams. In the summer of 1995, run-off from hog farms in North Carolina caused the rapid growth of the algae Pfisteria. This algae released toxins that affected the nervous system of fish as well as humans in the area.

In an attempt to manage agricultural run-off, the Office of Wastewater Management (OWM) of the U.S. Environmental Protection Agency has designated farms as Animal Feeding Operations (AFOS). As of 1998, nearly half a million AFOS had been identified. By designating AFOS, the OWM can regulate the disposal of animal waste products. This moves a large portion of agricultural run-off from the non-point source category to the point source category, and allows for better management of agricultural pollutants.

ACID MINE DRAINAGE

In places where coal is mined, the mineral pyrite is a waste material. A series of complex reactions between pyrite, oxygen, and water result in acid mine drainage. Acid mine drainage is wastewater that is extremely acidic and contains high concentrations of heavy metals. Acid mine drainage is one of the major sources of stream pollution in the Appalachian mountain region.

Acid mine drainage has severely damaged more than half the streams in Pennsylvania and West Virginia. There are at least 200,000 abandoned mines throughout the United States that produce acid mine drainage. Acid mine drainage can be treated using chemical treatments that decrease the acidity of the water, and allow the heavy metals to precipitate (separate from the water).

This type of treatment is often very expensive. Another way to treat acid mine drainage is by passing it through a lagoon or wetlands, which removes heavy metals and decreases the acidity of the water. Acid mine drainage is also treated by passing it through a channel of limestone (a rock that is very alkaline), which also neutralizes the acidity of the water.

URBAN RUN-OFF

When rain falls on natural lands such as forests and meadows, some of it soaks into the soil and then slowly makes its way to rivers, lakes, and oceans. In cities, much of the land is paved with cement and asphalt, and water is unable to sink into the ground. Instead, it quickly moves to storm drains and then into natural waterways. This great volume of water causes much erosion (wearing away of the land) and sedimentation.

In addition, as the rainwater runs over paved surfaces, it gathers oil and grease from cars, fertilizers and pesticides from gardening, pathogens form animal wastes, road salts, and heavy metals.

These are dumped directly into natural waters with urban wastewater. Run-off from urban areas is the largest source of pollution in estuaries (the wide part of a river where it nears the sea) and the third largest source of pollution in lakes. Controlling urban run-off is extremely difficult because its sources are hard to identify.

The Environmental Protection Agency works to influence developers to take into account urban run-off when planning new buildings. Some ideas to minimize run-off include adding vegetation and drainage areas to new construction sites.

Some cities have instituted sewer–stenciling programmes that remind people that rainwater flows directly into natural waters. Gas stations have also been targeted as businesses that can help control car oils and grease. Schools have also developed programmes to teach students about urban run-off and non-point source wastewater.

WAVE POWER

Wave power is the transport of energy by ocean surface waves, and the capture of that energy to do useful work—for example, electricity generation, water desalination, or the pumping of water (into reservoirs). Wave power is distinct from the diurnal flux of tidal power and the steady gyre of ocean currents. Wave power generation is not currently a widely employed commercial technology although there have been attempts at using it since at least 1890. In 2008, the first experimental wave farm was opened in Portugal, at the Aguçadoura Wave Park.

PHYSICAL CONCEPTS

Waves are generated by wind passing over the surface of the sea. As long as the waves propagate slower than the wind speed just above the waves, there is an energy transfer from the wind to the waves. Both air pressure differences between the upwind and the lee side of a wave crest, as well as friction on the water surface by the wind, making the water to go into the shear stress causes the growth of the waves.

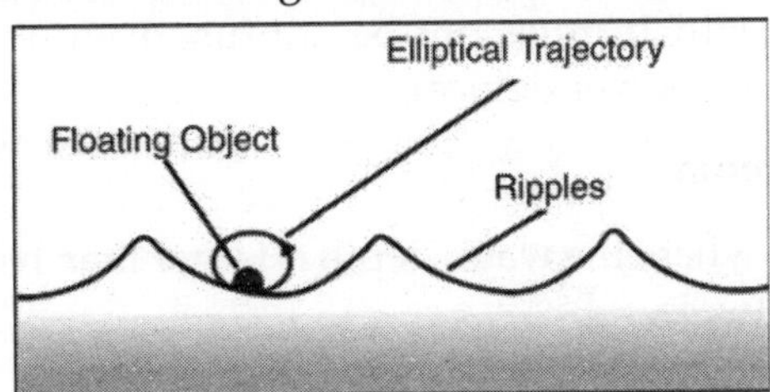

Fig. When an Object Bobs Up and Down on a Ripple in a Pond, it Experiences an Elliptical Trajectory.

Wave height is determined by wind speed, the duration of time the wind has been blowing, fetch (the distance over which the wind excites the waves) and by the depth and topography of the seafloor (which can focus or disperse the energy of the waves). A given wind speed has a matching practical limit over which time or distance will not produce larger waves. When this limit has been reached the sea is said to be "fully developed".

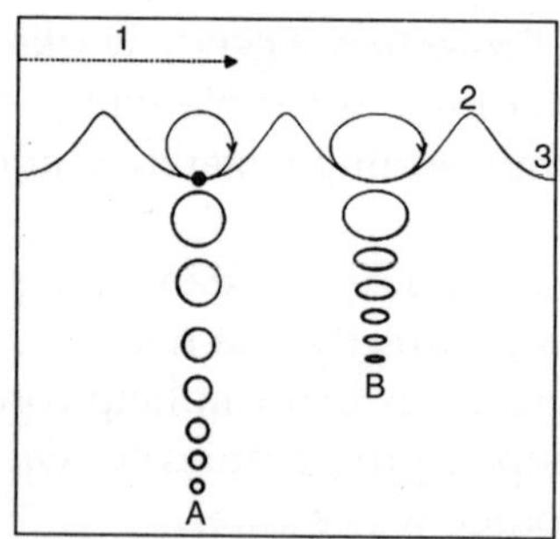

Motion of a Particle in an ocean wave:

A = *At deep water*: The orbital motion of fluid particles decreases rapidly with increasing depth below the surface.

B = *At shallow water (Ocean floor is now at B)*: The elliptical movement of a fluid particle flattens with decreasing depth.

1 = Propagation direction.

2 = Wave crest.

3 = Wave trough.

In general, larger waves are more powerful but wave power is also determined by wave speed, wavelength, and water density. Oscillatory motion is highest at the surface and diminishes exponentially with depth. However, for standing waves (clapotis) near a reflecting coast, wave energy is also present as pressure oscillations at great depth, producing microseisms. These pressure fluctuations at greater depth are too small to be interesting from the point of view of wave power. The waves propagate on the ocean surface, and the wave energy is also transported horizontally with the group velocity. The mean transport rate of the wave energy through a vertical plane of unit width, parallel to a wave crest, is called the wave energy flux (or wave power, which must not be confused with the actual power generated by a wave power device).

Wave Power Formula

In deep water where the water depth is larger than half the wavelength, the wave energy flux is

$$P = \frac{\rho g^2}{64\pi} H^2{}_{m0} T \approx \left(0.5 \frac{kW}{m^3.S}\right) H^2{}_{m0} T,$$

with P the wave energy flux per unit of wave–crest length, H_{m0} the significant wave height, T the wave period, ρ the water density and g the acceleration by gravity. The above formula states that wave power is proportional to the wave period and to the square of the wave height. When the significant wave height is given in meters, and the wave period in seconds, the result is the wave power in kilowatts (kW) per meter of wavefront length.

Example: Consider moderate ocean swells, in deep water, a few kilometers off a coastline, with a wave height of 3 meters and a wave period of 8 seconds. Using the formula to solve for power, we get,

$$P \approx 0.5 \frac{kW}{m^3.S} (3.m)^2 (8.s) \approx 36 \frac{kW}{m},$$

meaning there are 36 kilowatts of power potential per meter of coastline. In major storms, the largest waves offshore are about 15 meters high and have a period of about 15 seconds. Such waves carry about 1.7 MW of power across each meter of wavefront.

An effective wave power device captures as much as possible of the wave energy flux. As a result the waves will be of lower height in the region behind the wave power device.

Wave Energy and Wave Energy Flux

In a sea state, the average energy density per unit area of gravity waves on the water surface is proportional to the wave height squared, just as to linear wave theory:

$$E = \frac{1}{16} \rho g H^2_{m0},$$

where E is the mean wave energy density per unit horizontal area (J/m^2), the sum of kinetic and potential energy density per unit horizontal area. The potential energy density is equal to the kinetic energy, both contributing half to the wave energy density E, as can be expected from the equipartition theorem. In ocean waves, surface tension effects are negligible for wavelengths above a few decimetres.

As the waves propagate, their energy is transported. The energy transport velocity is the group velocity. As a result, the wave energy flux, through a vertical plane of unit width perpendicular to the wave propagation direction, is equal to:

$$P = E c_g,$$

with c_g the group velocity (m/s). Due to the dispersion relation for water waves under the action of gravity, the group velocity depends on the wavelength ë, or equivalently, on the wave period T. Further, the dispersion relation is a function of the water depth h. As a result, the group velocity behaves differently in the limits of deep and shallow water, and at intermediate depths:

Deep Water Characteristics and Opportunities

Deep water corresponds with a water depth larger than half the wavelength, which is the common situation in the sea and ocean. In deep water, longer period waves propagate faster and transport their energy faster. The deep-water group velocity is half the phase velocity. In shallow water, for wavelengths larger than twenty times the water depth, as found quite often near the coast, the group velocity is equal to the phase velocity. The regularity of deep-water ocean swells, where "easy-to-predict long-wavelength oscillations" are typically seen, offers the opportunity for the development of energy harvesting technologies that are potentially less subject to physical damage by near-shore cresting waves.

MODERN TECHNOLOGY

Wave power devices are generally categorized by the method used to capture the energy of the waves. They can also be categorized by location and power take-off system. Method types are point absorber or buoy; surfacing following or attenuator oriented parallel to the direction of wave propagation; terminator, oriented perpendicular to the direction of wave propagation; oscillating water column; and overtopping. Locations are shoreline, nearshore and offshore.

Types of power take-off include: hydraulic ram, elastomeric hose pump, pump-to-shore, hydroelectric turbine, air turbine, and linear electrical generator.

Some of these designs incorporate parabolic reflectors as a means of increasing the wave energy at the point of capture. These capture systems use the rise and fall motion of waves to capture energy. Once the wave energy is captured at a wave source, power must be carried to the point of use or to a connection to the electrical grid by transmission power cables.

These are descriptions of some wave power systems:

- In the United States, the Pacific Northwest Generating Cooperative is funding the building of a commercial wave-power park at Reedsport, Oregon. The project will utilize the PowerBuoy technology Ocean Power Technologies which consists of modular, ocean-going buoys. The rising and falling of the waves moves hydraulic fluid with the buoy; this motion is used to spin a generator, and the electricity is transmitted to shore over a submerged transmission line. A 150 kW buoy has a diameter of 36 feet (11 m) and is 145 feet (44 m) tall, with approximately 30 feet of the unit rising above the ocean surface. Using a three–point mooring system, they are designed to be installed one to five miles (8 km) offshore in water 100 to 200 feet (60 m) deep.
- An example of a surface following device is the Pelamis Wave Energy Converter. The sections of the device articulate with the movement of the waves, each resisting motion between it and the next section, creating pressurized oil to drive a hydraulic ram which drives a hydraulic motor. The machine is long and narrow (snake–like) and points into the waves; it attenuates the waves, gathering more energy than its narrow profile suggests. Its

articulating sections drive internal hydraulic generators (through the use of pumps and accumulators).

- With the Wave Dragon wave energy converter large wing reflectors focus waves up a ramp into an offshore reservoir. The water returns to the ocean by the force of gravity via hydroelectric generators.
- The Anaconda Wave Energy Converter is in the early stages of development by UK company Checkmate SeaEnergy. The concept is a 200 metre long rubber tube which is tethered underwater. Passing waves will instigate a wave inside the tube, which will then propagates down its walls, driving a turbine at the far end.
- The AquaBuOY is a technology developed by Finavera Renewables Inc. In 2009 Finavera Renewables surrendered its wave energy permits from FERC. In July 2010 Finavera announced that it has entered into a definitive agreement to sell all assets and intellectual property related to the AquaBuOY wave energy technology to an undisclosed buyer.
- The FlanSea is a so-called "point absorber" buoy, developed for use in the southern North Sea conditions. It works by means of a cable that due to the bobbing effect of the buoy, generates electricity.
- The SeaRaser, built by Alvin Smith, uses an entirely new technique (pumping) for gathering the wave energy.
- A device called CETO, currently being tested off Fremantle, Western Australia, consists of a single piston pump attached to the sea floor, with a float tethered to the piston. Waves cause the float to rise and fall, generating pressurized water, which is piped to an onshore facility to drive hydraulic generators or run reverse osmosis water desalination.
- Another type of wave buoys, using special polymeres, is being developed by SRI
- Wavebob is an Irish Company who have conducted some ocean trials.
- The Oyster wave energy converter is a hydro-electric wave energy device currently being developed by Aquamarine Power. The wave energy device captures the energy found in nearshore waves and converts it into electricity. The systems consists of a hinged mechanical flap connected to the seabed at around 10m depth. Each passing wave moves the flap which drives hydraulic pistons to deliver high pressure water via a pipeline to an onshore turbine which generates electricity. In November 2009, the first full–scale demonstrator Oyster began producing power when it was launched at the European Marine Energy Centre (EMEC) on Orkney.

- Ocean Energy have developed the OE buoy which has completed (September 2009) a 2–year sea trial in one quarter scale form. The OE buoy has only one moving part.
- The Lysekil Project is based on a concept with a direct driven linear generator placed on the seabed. The generator is connected to a buoy at the surface via a line. The movements of the buoy will drive the translator in the generator. The advantage of this setup is a less complex mechanical system with potentially a smaller need for maintenance. One drawback is a more complicated electrical system.
- An Australian firm, Oceanlinx, is developing a deep–water technology to generate electricity from, ostensibly, easy-to-predict long-wavelength ocean swell oscillations. Oceanlinx recently began installation of a third and final demonstration-scale, grid-connected unit near Port Kembla, near Sydney, Australia, a 2.5 MW_e system that is expected to go online in early 2010, when its power will be connected to the Australian grid. The companies much smaller first-generation prototype unit, in operation since 2006, is now being disassembled.
- An Israeli firm, SDE ENERGY LTD., has developed a breakwater-based wave energy converter. This device is close to the shore and utilizes the vertical motion of buoys for creating an hydraulic pressure, which in turn operates the system's generators. S.D.E. is currently building a new 250 kWh model in the port of Jaffa, Tel Aviv and preparing to construct it's standing orders for a 100mWh power plants in the islands of Zanzibar and Kosrae, Micronesia.
- A Finnish firm, AW–Energy Oy, is developing the WaveRoller device: that is a plate anchored on the sea bottom by its lower part. The back and forth movement of surge moves the plate. The kinetic energy transferred to this plate is collected by a piston pump.
- The United States Air Force Academy is developing a long ocean wave–cancelling electric generator that simulations and small-scale tests show can capture more than 95% of the energy of long ocean waves. Using a cycloidal turbine to convert wave power into lift, the turbine takes the energy of flowing water and converts it through lift to rotational energy. A rotating shaft can generates electricity. The initial concept came from the so-called Voith Schneider propeller, which is used to power tugboats.

POTENTIAL

Deep water wave power resources are significant, between 1 TW and 10 TW, but it is not practical to capture all of this. The useful worldwide resource has been estimated to be greater than 2 TW. Locations with the most potential for wave power include the western seaboard of Europe, the northern coast of the UK, and the Pacific coastlines of North and South America, Southern Africa, Australia, and New Zealand. The north and south temperate zones have the best sites for capturing wave power. The prevailing westerlies in these zones blow strongest in winter. Waves are very predictable; waves that are caused by winds can be predicted five days in advance.

CHALLENGES

- There is a potential impact on the marine environment. Noise pollution, for example, could have negative impact if not monitored, although the noise and visible impact of each design varies greatly.
- In terms of socio–economic challenges, wave farms can result in the displacement of commercial and recreational fishermen from productive fishing grounds, can change the pattern of beach sand nourishment, and may represent hazards to safe navigation.
- Waves generate about 2,700 gigawatts of power. Of those 2,700 gigawatts, only about 500 gigawatts can be captured with the current technology.

WAVE FARMS

The Aguçadoura Wave Farm was the world's first wave farm. It was located 5 km (3 mi) offshore near Póvoa de Varzim north of Oporto in Portugal. The farm was designed to use three Pelamis wave energy converters to convert the motion of the ocean surface waves into electricity, totalling to 2.25MW in total installed capacity. The farm first generated electricity in July 2008 and was officially opened on the 23rd of September 2008, by the Portuguese Minister of Economy. The wave farm was shut down two months after the official opening in November 2008 as a result of the financial collapse of Babcock and Brown due to the global economic crisis. The machines were off-site at this time due to technical problems, and although resolved have not returned to site without financial backing. A second phase of the project planned to increase the installed capacity to 21MW using a further 25 Pelamis machines is in doubt following Babcock's financial collapse. Funding for a 3MW wave farm in Scotland was announced on 20 February 2007 by the Scottish Executive, at a cost of over 4 million pounds, as part of a £13 million funding package for marine power in Scotland. The first of 66 machines was launched in May 2010.

Funding has also been announced for the development of a Wave hub off the north coast of Cornwall, England. The Wave hub will act as giant extension cable, allowing arrays of wave energy generating devices to be connected to the electricity grid. The Wave hub will initially allow 20MW of capacity to be connected, with potential expansion to 40MW. Four device manufacturers have so far expressed interest in connecting to the Wave hub. The scientists have calculated that wave energy gathered at Wave Hub will be enough to power up to 7,500 households. Savings that the Cornwall wave power generator will bring are significant: about 300,000 tons of carbon dioxide in the next 25 years. A CETO wave farm off the coast of Western Australia has been operating to prove commercial viability and, after preliminary environmental approval, is poised for further development.

HARNESSING WAVE ENERGY

There are two types of energy technologies used to capture wave energy and generate electricity: fixed devices and floating devices. Fixed devices are attached to the shore or sea floor. Tapered channels are fixed devices that direct large waves into raised pools on the shore. Water draining from the pools turns turbines that generate electricity. (Turbines are spinning wheels or other devices that convert the energy in falling water to mechanical and electrical energy.) A power plant on the North Coast of Norway uses a Tapchan to harness wave energy. Tapchans are relatively easy and inexpensive to maintain. Water can be stored in the reservoir (body of water) so that power can be generated when needed.

An appropriate Tapchan site however, is difficult to find because it must have consistent waves, small tidal variations, deep water close to the shoreline, and a place to build a reservoir. Oscillating water channels (OWCs) are fixed devices that consist of a small opening, where waves can enter and retreat, attached to a vertical closed pipe. As a wave flows into the OWC, the water forces air up the pipe, turning turbine blades as it passes.

When the wave ebbs (pulls back) the air is sucked back down the pipe and turns the turbine blades again. Adequate locations for oscillating water channels are also difficult to find, in part, because the relatively new technology requires that OWCs be embedded in the shoreline, limiting OWC usage to rocky coasts. Floating devices have several advantages over fixed devices: they have less visual impact on the shoreline, are less likely to change wave patterns and disrupt wildlife, and are quieter than Tapchans. Floating devices use the cyclic motion of waves to generate electricity.

A Salter Duck (developed by University of Edinburgh professor Steven Salter in the 1970s) looks like a row of floating ducks that are anchored to the sea floor. As a wave passes, the duck rotates about a central turning point. This rotational motion causes fluid within the duck to move thereby generating electricity. Japan, Denmark, Norway, England, Spain, and Portugal are all developing new technologies to use wave energy to generate power. As technologies for generating electricity from wave energy are developed, costs for these systems will decrease, and commercial wave power plants will most likely become more common.

Chteapter 11

Science and Research

AQUARIUMS

An Aquariums is any water–filled tank, pool, or pond in which fish, underwater plants, or animals are kept. An aquarium can be as small as a glass bowl for a goldfish and as large as a pool for a whale or a marine museum

HISTORY OF FISH KEEPING

The ancient Sumarians (2500 B.C.E.) were the earliest fish keepers. Fish keeping developed as a way to provide and store food. Fish were caught in rivers and then kept in small ponds until they were used. The ancient Egyptians also kept fish in ponds, but not all of their ponds served a practical purpose. Egyptian hieroglyphs and art depict fish and fishponds as decorative objects. In ancient Iran, China, and Japan, fish keepers bred special types of fish for use in decorative ponds. Fish keepers created koi, a popular decorative fish, by selectively breeding carp (a fish used for food) in pleasing colours and sizes.

The present–day common goldfish, a close relative of the koi, is also a result of these ancient breeding practices. The popularity of fish keeping spread to Europe and the United States in the eighteenth century, when ornamental fish were imported from the East. Fish keepers maintained ponds of ornamental fish, but the fish could only be viewed from above the water's surface. Growing interest in the scientific study of plants and animals sparked curiosity about viewing marine life from below the surface of the water. In the nineteenth century, zoos, circuses, and natural history museums began to add fish and other marine creatures and plants to their exhibits.

DEVELOPMENT OF MODERN AQUARIUMS

The first public aquarium opened in London, England, in 1853. The aquarium was an instant success and several other public aquariums opened in England over the next 10 years. Although popular, these aquariums faced several difficulties. Tanks were limited in size because there were not strong enough materials to construct very large tanks. The tanks also lacked adequate support systems to keep the fish healthy. Without support systems to clean and heat the water, most fish did not survive. Aquariums had to frequently replace the fish in their exhibits.

In the 1870s, the first successful long-term aquariums opened. Improvements in glassmaking and metalwork permitted the construction of larger tanks to house marine exhibits. Scientists began to study aquatic habitats, the water environments in which plants and animals live, to gather information about water chemistry. When caring for fish, water chemistry concerns the temperature of water and its balance of minerals, salt, oxygen, and other particles. Aquarists (people who keep an aquarium) and scientists experimented with the water chemistry of aquarium tanks. New technologies allowed scientists to copy the water chemistry of marine and river environments. Filtration systems cleaned pollutants such as fish waste from the water. Aeration systems added air to the water to aid respiration (breathing) and keep the water moving. Moving water stayed cleaner. Heaters kept water in the tanks at appropriate temperatures. These support systems helped fish and other marine creatures in aquariums remain healthy.

By 1900, support systems permitted aquariums to house exotic tropical fish. Public aquariums competed for the most spectacular creatures and exhibits. Most of the creatures on display came from the ocean. Exotic marine creatures from faraway places were a favourite of aquarium visitors and few exhibits featured local or non-ocean species. The fish and other creatures were the focus of the exhibits. Tanks contained few plants. Several technological advances in the second half of the twentieth century heightened scientific and public interest in the sea. Personal breathing systems for marine divers, nicknamed scuba (self-contained underwater breathing apparatus), permitted more lengthy underwater exploration.

Small submersibles, submarine like vessels usually driven by one to four persons, permitted some of the first glimpses of deep waters inaccessible to scuba divers. Cameras that could film underwater captured underwater images for both scientific study and public broadcast. A television series created by marine biologist Jacques-Yves Cousteau (1910–1997) featured underwater habitats, fish, and animals. The series made aquariums even more popular. Greater scientific and public awareness of underwater life changed aquarium exhibits. Tanks began to copy whole marine habitats. Instead of featuring one type of fish, single tanks would recreate a part of the ocean. Coral reefs (underwater ridges of compacted coral), underwater rocks and coves, and plants were part of ocean landscapes featured in exhibits. Aquarists displayed many species of fish and other marine creatures together in tanks in groupings that appeared in nature. By recreating marine habitats, aquariums sought to show visitors a whole picture of life underwater. Unlike the ornamental fishponds of ancient times, exhibits in aquariums are now designed and managed by scientists, such as marine biologists and ichthyologists (scientists who study fish). Today's aquariums also feature a variety of underwater habitats including oceans, lakes, rivers, bays, and swamps. Some aquariums provide visitors with a look at underwater environments from two angles, above the water and below its surface. Many include exhibits where visitors can touch and handle underwater creatures such as sea stars, clams, and mussels.

Today's aquariums still encourage visitors to have fun, but they also educate visitors about the preservation and protection of underwater habitats and species. Aquariums can show visitors healthy habitats in tanks next to pictures of natural habitats destroyed by pollution. Aquarium exhibits and programmes educate the public about threats to natural underwater habitats and species from litter, chemicals, oil, and over fishing. Captive breeding programmes in aquariums help preserve animals endangered by habitat loss or over fishing. Rescue programmes provide medical care and shelter for animals and fish injured by litter, pollution, and oil spills.

ECOLOGY

Ecology is the study of the relationships between organisms and the relationships between organisms and their environment. Ecology was first recognized as an academic subject in 1869 when German naturalist Ernst Haeckel (1834–1919) first coined the term ecology. The word is derived from the Greek words eco, meaning "house" and logy, meaning "to study," indicating that ecology is the study of organisms in their home. Ecologists often distinguish between two parts of environment as a whole: the living or biotic part and the nonliving or abiotic part.

The biotic part of the environment includes all organisms such as animals, plants, bacteria, and fungi. The abiotic part includes all physical features like temperature, humidity, availability of light, as well as chemical components, such as the concentrations of salts, nutrients, and gases. Ecology, then, is the study of the relationships between and among the biotic and abiotic environments.

ECOLOGY AS PART OF THE BIOLOGICAL SCIENCES

The components of the biological world are often organized along a spectrum. Assuming the spectrum is laid out from left to right with the left being the smallest, atoms are at the far left. When atoms combine together they organize into molecules, which are just to the right of atoms on the spectrum. Moving right along the spectrum, next comes cells, which are the smallest unit of life. Next come tissues, such as muscle tissue and nervous tissue, which are collections of cells that work together to perform a function.

These tissues are then organized into organs, such as the heart and the brain, and these are found to the right of tissues on the spectrum. Organs work together as organ systems, such as the cardiac system, which includes the heart as well as the blood vessels that transport blood throughout the body. Farther to the right along the biological spectrum are organ systems, which come together to form an individual organism, such as a human, fish, or kelp. Towards the center of the biological spectrum, individual organisms are grouped together into populations. These populations are all the members of a species that live together. Even farther to the right of the biological spectrum, populations are grouped into communities, which are all the organisms that are found in a specific location. Communities include members of different species. Communities depend upon the nonliving world in order to survive, so an ecosystem, also called an ecological system, represents all the relationships between a community and the abiotic world.

All the communities of the world together make up the biosphere, which is found near the right side of the biological spectrum. The biosphere interacts with all of the abiotic parts of Earth, including the atmosphere, which are the gasses surrounding the planet; the hydrosphere, which is the water on Earth; and the lithosphere, which is the soil and rock on Earth. The biosphere and its relationships with the atmosphere, hydrosphere and lithosphere make up the ecosphere. The ecosphere is on the extreme right hand side of the biological spectrum. Ecologists generally focus their research on the part of the biological spectrum that is to the right of the individual. For example, a population ecologist may study the ways that populations of sardines off the coast of California differ in their mating habits from populations of sardines off the coast of Chile. On the next level, community ecologist will study the diet of the various populations of sardines. An ecosystem ecologist may study how the populations of sardines are affected by the changes in temperature associated with the warming or cooling of the oceans.

SUBDIVISIONS AND IMPORTANT CONCEPTS OF ECOLOGY

The field of ecology is often subdivided because it incorporates so many different disciplines. Two large groupings within ecology are autecology and synecology. Autecology is the study of the individual organism or an individual species. This part of ecology might focus on the life history of an animal or plant. For example, one could study how the caddis fly grows from an egg into a larva (early stage of insect's life) that builds a house of sand at the bottom of a river and then metamorphoses (changes form) into a fly. Autecology might also investigate how the caddis fly adapts to its environment. For example, a study of how well the caddis fly larvae houses are hidden from predators (animals that hunt others for food) would be an autecological study. Synecology focuses on groups of organisms and how they work together. If a study estimated the amount of energy fish obtained by eating caddis fly larvae, it would be synecological. Synecology tends to ask questions that study the ecosystem on a large scale.

Another way to subdivide the subject of ecology is by the kind of environment. Commonly, environments are grouped into freshwater (lakes, rivers, and streams), marine (oceans), or terrestrial (land-based). Although the fundamental principles of ecology hold in all of these environments, the specific animals and plants vary and it is often convenient to study each type separately.

Finally, ecology can be divided into different types of organisms. This is called a taxonomical grouping. For example, one might study plant ecology, bacterial ecology, or insect ecology. This allows the study to be focused on a specific group and to use similar methods to study the different organisms in the group.

For example, in order to study the environmental factors that influence the growth of marine algae, one could develop several growth environments with different light and different concentrations of the nutrients phosphate and nitrate. These same growth environments could be used to grow several species of algae. The results might be useful in predicting where and when the rapid growth of algae might occur in the ocean.

The Ecosystem

Every living thing has requirements in order to exist: food, water, gases, stable temperatures and a place to live. Living organisms depend the nonliving environment for many of these requirements. The relationships between the biotic and the abiotic are called an ecological system, or an ecosystem.

Inorganic (Non–living) substances such as carbon, nitrogen, phosphorous, carbon dioxide, and oxygen are required for all organisms to produce the molecules in their bodies. Autotrophs are organisms that use inorganic substances to make energy. (The root word auto means "self" and the root word troph means "to eat.") Most often plants are autotrophs, using sunlight, water, and carbon dioxide in a process called photosynthesis to produce energy in the form of carbohydrates that their cells need.

Heterotrophs are organisms that consume autotrophs in order to get their energy and grow. (The root word hetero means "other.") Heterotrophs include animals that eat plants as well as animals that eat other animals. Finally there are the decomposers or saprotrophs, such as bacteria and fungi. (The root word sapro means "to decompose.") These organisms break down dead organisms into inorganic substances, which may then be used by autotrophs. Understanding the ways that substances and energy flow through ecosystems is one of the fundamental principles in ecology.

Homeostasis

Homeostasis is used to describe the tendency for a biological system to resist change. (The root word homeo means "the same" and the root word stasis means "standing.") A principle of ecology is that ecosystems generally remain homeostatic. In other words, if there are no outside influences, the number of organisms that live in any given location will tend to remain the same, and they will have the same food supply and access to shelter over time. Even minor changes in the environment, such as temperature changes or changes in rainfall, will not greatly affect an ecosystem.

One of the ways that an ecosystem maintains homeostasis is through negative feedback mechanisms. For example, kelp forests grow off the coasts of California. Sea urchins eat the kelp, keeping its density relatively constant.

In turn, sea otters eat sea urchins. If the population of sea otters were to suddenly decrease, the population of sea urchins would grow because they would not have any predators. However, the sea urchins would eat a large amount of kelp removing their food supply. Many urchins would starve, decreasing the population of urchins and allowing the kelp to grow back to its former density. Eventually the ecosystem would return to its former state. Of course, large disruptions to ecosystems can be very destructive. For example, hunting sea otters to extinction (as almost occurred during the early part of the twentieth century) would completely disrupt the homeostasis of the California kelp forest.

Energy in the Ecosystem

One of the fundamental principles of physics is that energy cannot be created or destroyed. It can, however, be transformed from one form to another. Light is a form of energy. It can be transformed into heat or chemical energy that is stored in food. In fact, this is the basis for photosynthesis. Some of the energy in light is stored in the chemical bonds of the molecules in plants. Another fundamental principle of physics states that when energy is transformed from one form to another, some of the energy is lost.

In photosynthesis, some of the energy in light is lost as heat. This means that the transformations of energy within ecosystems are never 100% efficient. Energy is always lost as it flows from one organism to the next. In any ecosystem, energy is transferred between organisms as they eat and are eaten by other organisms. Usually the autotrophs or primary producers (such as seaweed and algae) capture light energy from the Sun through photosynthesis. They store this energy in the chemical bonds of the molecules that make up their cells. Herbivores (plant eaters such as urchins and snails) eat the autotrophs.

These grazers convert the energy in the chemical bonds of the primary producers into energy that is stored in the chemical bonds of their own cells. Usually about 80–90% of the energy in the chemical bonds of the primary producer is lost during this process. Next, carnivores (meat eaters such as frogs and fish) eat the herbivores. They convert the energy stored in the chemical bonds of the herbivores into energy stored in the chemical bonds of their own cells. Again, about 80-90% of the energy is lost in this transformation.

The result of the energy loss each time an organism is eaten results in what is called an ecological pyramid. At the base of the pyramid are the primary producers; in the middle are the herbivores and at the top the carnivores. If one measures the weight of each of these groups after the water has been removed (called the dry weight), one gets an idea of how much energy is stored in chemical bonds at each level of the pyramid.

For example, in a lake in Wisconsin, the dry weight of the primary producers is 96 grams per square meter. The dry weight of the herbivores is only 11 grams per square meter. The dry weight of the carnivores is just 4 grams per square meter. The ecological pyramid demonstrates how about 80-90% of the energy stored in chemical bonds is lost every time an organism is eaten. It also shows that there can never be as many predators as there are prey; there is just not enough energy for that to occur.

HYDROLOGY AND HYDROGEOLOGY

Hydrologists and hydrogeologists are water scientists who study the properties of freshwater and its distribution on the continents. (Oceanographers study the physical and chemical properties of salt water in the oceans.) Together, hydrology and hydrogeology provide information on how to manage and protect freshwater, humans most essential natural resource. Hydrology and hydrogeology are distinct fields of study that employ different methods and techniques, but they overlap to provide a complete picture of Earth's freshwater resources.

Hydrology is a branch of engineering that deals with the physical properties of surface freshwater, such as lakes and rivers, and with its chemical interactions with other substances. Hydrogeology is a subfield of geology (study of Earth) that, by definition, specifically addresses groundwater-water moving through tiny openings in rock and soil layers beneath the land surface. In practice, ground and surface water interact as a single system. Surface water seeps into the ground and groundwater emerges to the surface. Hydrogeologists work to explain the geological effects of surface water in rivers, streams and lakes, and hydrologists lend their technical expertise to the mechanics and chemistry of moving groundwater.

HYDROLOGY

Hydrologists use mathematics and experimental techniques to

determine water's general properties, to make specific observations of freshwater environments, and to design water management systems that contain and direct water. Britain's Centre for Hydrology, a government environmental research agency, describes its mission as an effort to answer two questions about the Earth's freshwater: Why is the natural environment as it is? and What is it likely to be in the future?

Today, humans, not natural processes, manage the flow, distribution, and allocation of almost all of Earth's surface waters. Dams, levees, and reservoirs (natural or man-made lakes) direct and contain the water of the world's largest and most heavily used river systems: the Nile, Yangtze, Amazon, Ganges, and Mississippi. U.S. Army Corps of Engineers hydrologists, along with their counterparts at the Bureau of Reclamation (which has jurisdiction over rivers west of the Rockies), control the flow and distribution of all the surface waters of the United States. Hydrological studies generally seek to understand how water moves between and through bodies of freshwater, such as lakes, streams, aquifers (water–bearing soil and rock layers), and reservoirs, and even the atmosphere (mass of air surrounding Earth) over time. Hydrologists usually begin with data from a particular region. They collect numerous measurements of water system conditions such as rainfall totals, lake and reservoir levels, river discharges (the volume of water that flows through a river in a given time), and current speeds, air and water temperatures, and humidity (amount of water vapour in the atmosphere) at specific sites. Then they merge the data into a computerized model that shows how water has moved through the region's various freshwater reservoirs in the past.

Once a hydrologist has constructed a model of a watershed (land area that contributes water to a stream, lake, or aquifer) or reservoir (body of water), the hydrologist tests it by comparing its calculated predictions with actual results in the natural environment. The model is adjusted to better match the real world. The more times the scientist repeats the process of adding new data to the model and retesting it, the more accurate it becomes. Computer models help hydrologists predict the ways that natural and man-made changes like droughts (uncommonly dry weather), heavy snows and rains, and new dams across rivers affect water supplies and flows in the future. Many models also include information about the way water physically and chemically interacts with rocks and minerals on its path through the system.

These models can be used to track contaminants and predict water quality changes over time. One example of how hydrologists' work can affect a regions water supply is in the city of Denver, Colorado when the city had very low reserves of freshwater in its reservoirs. There had been lower than average snowfall in the nearby Rocky Mountains and the city's water supplies ran low without their usual influx of spring melt water.

Hydrologists at the U.S. Geological Survey in Denver used data and computer models to help the city distribute limited water during the drought period, and to plan for future dry periods. Models of Denver's surface water flows and seasonal patterns also help water managers protect and regulate the city's water supply from contamination by such human waste products as agricultural fertilizers and chemicals, industrial wastes, and sewage.

HYDROGEOLOGY

Hydrogeologists are concerned mostly with groundwater and how geologic features affect groundwater storage, flow, and replenishment. Like other geologists, they use observations of rock types and geologic structures on the land surface together with subsurface samples to map folded, faulted (broken), and fractured (cracked) rock layers and bodies beneath the land surface. Hydrogeologic maps show the locations and shapes of aquifers, and the distribution of less permeable (leaky) layers called aquicludes that confine and pressurize groundwater within aquifers.

They identify systems of caves, cavities, and fractures where groundwater may flow more quickly along specific routes. They also show the areas where surface water enters aquifers (recharge zones), and places where groundwater reemerges at springs and seeps (discharge zones).

Once a hydrogeologist has mapped out the dimensions, physical characteristics, and "plumbing" of a groundwater reservoir, he or she sets out to understand how water moves through the system. Measurements of flows at recharge and discharge points show the rates at which water is entering and leaving the aquifer as well as the time an average water molecule (smallest part of water that has its properties) takes to travel through the system (residence time.) Water levels and pressures in wells indicate flow patterns and rates inside the aquifer.

Groundwater can become polluted. Hydrogeologists also use their understanding of groundwater flow patterns to predict how contaminants might enter an aquifer, move through the system, and reemerge in a distant spring or well. Sometimes an aquifer or soil layer acts as a filter that improves water quality as contaminants move through an aquifer. Others transport polluted water quickly to a discharge site. Some rock layers even contribute hazardous dissolved chemicals to the groundwater. Hydrogeologists collect water samples and monitor water quality within aquifers. They also conduct laboratory and computer experiments to better understand groundwater's chemical interactions. Cities and regions that depend on groundwater require detailed hydrogeologic maps, and a good understanding of how water moves through their aquifer, to effectively manage groundwater resources.

In many states and countries, groundwater is common, public property. Most rules and regulations regarding groundwater consumption and contamination were written in an era when groundwater systems were poorly understood and were considered unending sources of clean freshwater. Today, many regions with large human populations and fragile natural ecosystems (communities of plants and animals) depend on limited, shared groundwater resources. As such, the actions of some of a groundwater reservoir's users can negatively affect the water supply and quality for other people, as well as for the aquifer's plants and animals.

LIMNOLOGY

"Limnology is the study of all inland waters and of the external influences that effect the nature of the waters and the processes going on in them.... It is concerned not only with the life in these waters, the fixation and utilization of energy, and the complex interrelationships among the various organisms, but also with the chemistry and physics of the waters, the geology, meteorology, hydrology, and bioecology of their drainage basins, and the progressively greater influence of man on the total complex of life and processes in these waters."

HISTORY OF LIMNOLOGY

Limnology is a relatively new academic subject. Francois–Alphonse Forel (1841–1912), considered the father of limnology, was a Swiss physician who dedicated much of his life to the study of the biology, chemistry, and physics of Lake Geneva. Around 1868, he coined the term limnology to mean the study of lakes. (The root word limn means "lake" and ology means "the study of.") In 1887, American naturalist Stephen Alfred Forbes (1844–1930), a pioneer in the study of lake ecology (the study of the relationships between organisms and their environment), published the document "Lake as a Microcosm," which is still cited as an important study of lake ecosystems. An ecosystem refers to all of the relationships between the living and nonliving parts of an environment. George Evelyn Hutchinson (1903–1991) was a British American biologist and a physicist.

He made great advancements in limnology beginning in the 1950s and summarized much of the field of limnology in a three-volume text. Hutchinson was extremely influential in bringing modern ecological theories to limnology. Today, limnologists focus much of their attention on integrating ideas from geology, physics, chemistry, and biology into understanding lakes and rivers. They also focus much attention on understanding how humans impact these important ecosystems.

GEOLOGICAL LIMNOLOGY

Geological limnology is focused on the formation of lakes and rivers. Many lakes, especially in North America, were formed by the retreat of the glaciers (slow–moving mass of ice) at the end of the Ice Age. As the glaciers melted, they gouged holes in soft parts of the solid rock. When these depressions filled with water, they became lakes. Other lakes form when tectonic plates (mobile pieces of Earth's crust) pull away from each other, leaving rifts called grabens.

When these rifts fill with water, very deep lakes can be formed. The deepest lake in the world, Lake Baikal in Siberia, was formed in a graben. Rivers usually begin as springs in areas of high altitude such as mountains. As they flow downward, rivers gather water from melting snow and other streams, called tributaries, as they flow towards sea level. Geological limnologists are interested in the size and the shape, also called the topography, of the watersheds of lakes and rivers. A watershed is all of the land and water areas that drain into the lake or the river.

PHYSICAL LIMNOLOGY

Physical limnology deals with the physical properties of the water in lakes and rivers. This includes changes in light levels, water temperatures, and water currents. Water absorbs energy from sunlight, which warms the surface waters. Because the intensity of the sun changes throughout the year, the amount of heat absorbed in the summer is much greater than that absorbed in the winter. During the summer, lakes become stratified or layered, with the warmer, lighter water floating on top of the cooler, deeper water. In the winters, the surface of the lake loses its heat and mixes with the cooler waters below. Understanding the cycle of mixing and stratification is extremely important to understanding the biology of the plants, animals, and microorganisms that live in lakes and the movement of chemicals throughout lakes.

CHEMICAL LIMNOLOGY

Chemical limnology focuses on the cycling of various chemical substances in lakes and rivers. Several factors affect the chemistry of lakes and rivers including the chemical composition of the soil in the watershed, the atmosphere (mass of air surrounding Earth) and the composition of the riverbed or lake bottom. In modern day, human activities have had a very important influence on the chemistry of lakes and rivers, and chemical limnologists play an important role in understanding these effects. For instance, building construction near lakes and rivers changes the erosion (wearing away of soil) patterns and influences they type of chemicals that reach the water.

In some areas, rainwater running into lakes and rivers contains large amounts of fertilizers, oils, and heavy metals. The concentration of the hydrogen ion (H^+) in water is one of the most important chemicals to study. An ion has a positive or negative charge and the hydrogen ion indicates the acidity (charge) of the water, which strongly affects which kinds of organisms can live in the water. Other important substances are the sulfate and nitrate ions, which become concentrated in freshwaters as a result of acid rain. Also, the heavy metal mercury (Hg) is a dangerous pollutant that can circulate in the water and affect the health of animals, along with the humans who eat those animals and use the lake or river.

BIOLOGICAL LIMNOLOGY

Biological limnology is directed at understanding the animals, plants, and microorganisms that live in lakes and rivers. The patterns of distribution of these various organisms depend on the geology, physics, and chemistry of the lake or river. For example, plants require light in order to grow. Because water is very effective at absorbing light, plants must either grow near the shore, where the water is shallow or they must float near the surface of the water. Because the intensity of sunlight changes with season, plants usually have a growth season in the spring when the light levels increase and they die off in the fall when light levels decrease.

In the same way, animals require oxygen dissolved in water in order to breath. Warm water holds less dissolved oxygen than cold waters. As a result, trout, which require a lot of dissolved oxygen, are more often found in cold lakes and rivers. Bass, on the other hand, require less dissolved oxygen and can be found in warmer lakes and the surface waters of lakes.

One of the critical challenges facing biological limnologists is the introduction of exotic species into lakes and rivers. Often, humans introduce new species into lakes and rivers. In some cases these species grow faster than the local species and can take over much of the habitat. For example, in 1985 the zebra mussel was released into the Great Lakes of the United States, in the ballast water of a ship coming from the Caspian Sea in Asia. These mussels are able to reproduce extremely fast in the Great Lakes and have become a widespread problem, clogging sewage pipes and overgrowing docks and piers.

MARITIME ARCHAEOLOGY

Maritime archaeology is a discipline within archaeology as a whole that specifically studies human interaction with the sea, lakes and rivers through the study of associated physical remains, be they vessels, shore side facilities, port-related structures, cargoes, human remains and submerged landscapes. A specialty within maritime archaeology is nautical archaeology, which studies vessel construction and use. As with archaeology as a whole, maritime archaeology can be practised within the historical, industrial, or pre-historical periods. An associated discipline, and again one that lies within archaeology itself, is underwater archaeology, which studies the past through any submerged remains be they of maritime interest or not. An example from the pre-historic era would be the examination of remains submerged in ancient wells or cenotes, or of Indigenous sites now lying underwater yet well away from the sea or inland waters. The study of submerged aircraft lost in lakes, rivers or in the sea is an example from the historical or industrial era. Many specialist sub-disciplines within the broader maritime and underwater archaeological categories have emerged in recent years.

Maritime archaeological sites often result from shipwrecks or sometimes seismic catastrophes, and thus represent a moment in time rather than a slow deposition of material accumulated over a period of years, as is the case with port–related structures (such as piers, wharves and jetties) where objects are lost or thrown off structures over extended periods of time. This fact has led to shipwrecks often being described in the media and in popular accounts as 'time capsules'.

Archaeological material in the sea or in other underwater environments is typically subject to different factors than artifacts on land. However, as with land archaeology what survives to be investigated by modern archaeologists can often be a tiny fraction of the material originally deposited. A feature of maritime archaeology is that despite all the material that is lost, there are occasional rare examples of substantial survival, from which a great deal can be learned. This is as a result of the difficulties often experienced in accessing the sites.

There are those in the archaeology community who see maritime archaeology as a separate discipline with its own concerns (such as shipwrecks) and requiring the specialised skills of the underwater archaeologist. Others value an integrated approach, stressing that nautical activity has economic and social links to communities on land and that archaeology is archaeology no matter where the study is conducted. All that is required is the mastering of skills specific to the environment in which the work occurs.

INTEGRATING LAND AND SEA

Prior to the industrial era, travel by water was often easier than over land. As a result, marine channels, navigable rivers and sea crossings formed the trade routes of historic and ancient civilizations. For example, the Mediterranean Sea was known to the Romans as the inner sea because the Roman empire spread around its coasts. The historic record as well as the remains of harbours, ships and cargoes, testify to the volume of trade that crossed it. Later, nations with a strong maritime culture such as the United Kingdom, Holland, Denmark, Portugal and Spain were able to establish colonies on other continents. Wars were fought at sea over the control of important resources. The material cultural remains that are discovered by maritime archaeologists along former trade routes can be combined with historic documents and material cultural remains found on land to understand the economic, social and political environment of the past. Of late maritime archaeologists have been examining the submerged cultural remains of China, India, Korea and other Asian nations.

PRESERVATION OF MATERIAL UNDERWATER

There are significant differences in the survival of archaeological material depending on whether a site is wet or dry, on the nature of the chemical environment, on the presence of biological organisms and on the dynamic forces present. Thus rocky coastlines, especially in shallow water, are typically inimical to the survival of artifacts, which can be dispersed, smashed or ground by the effect of currents and surf, possibly (but not always) leaving an artifact pattern but little if any wreck structure.

Saltwater is particularly inimical to iron artefacts including metal shipwrecks, and sea organisms will readily consume organic material such as wooden shipwrecks. On the other hand, out of all the thousands of potential archaeological sites destroyed or grossly eroded by such natural processes, occasionally sites survive with exceptional preservation of a related collection of artifacts. An example of such a collection is the Mary Rose.Survival in this instance is largely due to the remains being buried in sediment

Of the many examples where the sea bed provides an extremely hostile environment for submerged evidence of history, one of the most notable, the RMS Titanic, though a relatively young wreck and in deep water so calcium-starved that concretion does not occur, appears strong and relatively intact, though indications are that it has already incurred irreversible degradation of her steel and iron hull. As such degradation inevitably continues, data will be forever lost, object's context will be destroyed and the bulk of the wreck will over centuries completely deteriorate on the floor of the Atlantic Ocean. Comparative evidence shows that all iron and steel ships, especially those in a highly oxygenated environment, continue to degrade and will continue to do so until only their engines and other machinery project much above the sea-floor.

Where it remains even after the passage of time, the iron or steel hull is often fragile with no remaining metal within the layer of concretion and corrosion products. The USS Monitor, having been found in the 1970s, was subjected to a programme of attempted in situ preservation, for example, but deterioration of the vessel progressed at such a rate that the rescue of her turret was undertaken lest nothing be saved from the wreck.

Some wrecks, lost to natural obstacles to navigation, are at risk of being smashed by subsequent wrecks sunk by the same hazard, or are deliberately destroyed because they present a hazard to navigation. Even in deep water, commercial activities such as pipe–laying operations and deep sea trawling can place a wreck at risk. Large pipelines can crush sites and render some of their remnants inaccessible as pipe is dropped from the ocean surface to the substrate thousands of feet below. Trawl nets snag and tear superstructures and separate artifacts from their context.

The wrecks, and other archaeological sites that have been preserved have generally survived because the dynamic nature of the sea bed can result in artifacts becoming rapidly buried in sediments. These sediments then provide an anaerobic environment which protects from further degradation. Wet environments, whether on land in the form of peat bogs and wells, or underwater are particularly important for the survival of organic material, such as wood, leather, fabric and horn. Cold and absence of light also aid survival of artifacts, because there is little energy available for either organic activity or chemical reactions. Salt water provides for greater organic activity than freshwater, and in particular, the shipworm, terredo navalis, lives only in salt water, so some of the best preservation in the absence of sediments has been found in the cold, dark waters of the Great Lakes in North America and in the (low salinity) Baltic Sea (where the Vasa was preserved).

While the land surface is continuously reused by man, the sea bed was largely inaccessible until the advent of submarines and scuba equipment in the twentieth century. Salvagers have operated in much earlier times, but much of the material was beyond the reach of anyone. Thus the Mary Rose was subject to salvage from the sixteenth century and later, but a very large amount of material, buried in the sediments, remained to be found by maritime archaeologists of the twentieth century.

While preservation in situ is not assured, material that has survived underwater and is then recovered to land is typically in an unstable state and can only be preserved as a result of highly specialised conservation processes. While the wooden structure of the Mary Rose, and the individual artifacts have been undergoing conservation since their recovery, the Holland 1 provides an example of a relatively recent (metal) wreck for which extensive conservation has been necessary in order to preserve the hull. While the hull remains intact, its machinery remains inoperable. The SS Xantho engine that was recovered in 1985 from a saline environment after over a century underwater is presently considered somewhat anomalous, in that after two decades of treatment it can now be turned over by hand.

A challenge for the modern archaeologist is to consider whether in-situ preservation, or recovery and conservation on land is the preferable option; or to face the fact that preservation in any form, other than as an archaeological record is not feasible. A site that has been discovered has typically been subjected to disturbance of the very factors that caused its survival in the first place, for example, when a covering of sediment has been removed by storms or the action of man. Active monitoring and deliberate protection may mitigate further rapid destruction making in situ preservation an option, but long term survival can never be guaranteed.

For very many sites, the costs are too great for either active measures to ensure in situ preservation or to provide for satisfactory conservation on recovery. Even the cost of proper and complete archaeological investigation may be too great to enable this to occur within a timescale that ensures that an archaeological record is made before data is inevitably lost.

SUBMERGED SITES

Pre-Historic Landscapes

Maritime archaeology studies prehistorical objects and sites that are, because of changes in climate and geology, now underwater. Bodies of water, fresh and saline, have been important sources of food for people for as long as we have existed. It should be no surprise that ancient villages were located at the water's edge. Since the last ice age sea level has risen as much as 400 feet (~120 meters). Therefore, a great deal of the record of human activity throughout the Ice Age is now to be found under water.

The flooding of the area now known as the Black Sea (when a land bridge, where the Bosporus is now, collapsed under the pressure of rising water in the Mediterranean Sea) submerged a great deal of human activity that had been gathered round what had been an enormous, fresh-water lake. Significant cave art sites off the coast of western Europe such as the Grotto Cosquer can be reached only by diving, because the cave entrances are underwater, though the upper portions of the caves themselves are not flooded.

Historic Sites

Throughout history, seismic events have at times caused submergence of human settlements. The remains of such catastrophes exist all over the world, and sites such as Alexandria and Port Royal now form important archaeological sites. As with shipwrecks, archaeological research can follow multiple themes, including evidence of the final catastrophe, the structures and landscape prior to the catastrophe and the culture and economy of which it formed a part. Unlike the wrecking of a ship, the destruction of a town by a seismic event can take place over many years and there may be evidence for several phases of damage, sometimes with rebuilding in between.

COASTAL AND FORESHORE

Not all maritime sites are underwater. There are many structures at the margin of land and water that provide evidence of the human societies of the past. Some are deliberately created for access—such as bridges and walkways. Other structures remain from exploitation of resources, such as dams and fish traps. Nautical remains include early harbours, and places where ships were built or repaired. At the end of their life, ships were often beached. Valuable or easily accessed timber has often been salvaged leaving just a few frames and bottom planking. Archaeological sites can also be found on the foreshore today that would have been on dry land when they were constructed. An example of such a site is Seahenge, a Bronze Age timber circle.

SHIPS AND SHIPWRECKS

The archaeology of shipwrecks can be divided in a three-tier hierarchy, of which the first tier considers the wrecking process itself: how does a ship break up, how does a ship sink to the bottom, and how do the remains of the ship, cargo and the surrounding environment evolve over time? The second tier studies the ship as a machine, both in itself and in a military or economic system. The third tier consists of the archaeology of maritime cultures, in which nautical technology, naval warfare, trade and shipboard societies are studied. Some consider this to be the most important tier. Ships and boats are not necessarily wrecked: some are deliberately abandoned, scuttled or beached. Many such abandoned vessels have been extensively salvaged.

Bronze Age

The earliest boats discovered date from the Bronze Age and are constructed of hollowed out logs or sewn planks. Vessels have been discovered where they have been preserved in sediments underwater or in waterlogged land sites, such as the discovery of a canoe near St Botolphs. Examples of sewn--plank boats include those found at North Ferriby and the Dover Bronze Age Boat which is now displayed at Dover Museum. These may be an evolution from boats made of sewn hides, but it is highly unlikely that hide boats could have survived.

Ships wrecked in the sea have probably not survived, although remains of cargo (particularly bronze material) have been discovered, such as those at the Salcombe B site. A close collection of artefacts on the sea bed may imply that artefacts were from a ship, even if there are no remains of the actual vessel. Late Bronze Age ships, such as the Uluburun Shipwreck have been discovered in the Mediterranean, constructed of edge joined planks. This shipbuilding technology continued through the classical period.

MARITIME ARCHAEOLOGY BY REGION

Mediterranean Area

In the Mediterranean area, maritime archaeologists have investigated several ancient cultures. Notable early Iron Age shipwrecks include two Phoenician ships of c. 750 B.C. that foundered off Gaza with cargoes of wine in amphoras. The crew of the U.S. Navy deep submergence research submarine NR–1 discovered the sites in 1997. In 1999 a team led by Robert Ballard and Harvard University archaeology Professor Lawrence Stager investigated the wrecks. Extensive research has been carried out on the Mediterranean and Aegean coastlines of Turkey. Complete excavations have been performed on several wrecks from the Classical, Hellenistic, Byzantine, and Ottoman periods.

Maritime archaeological studies in Italy illuminate the naval and maritime activities of the Etruscans, Greek colonists, and Romans. After the second century B.C., the Roman fleet ruled the Mediterranean and actively suppressed piracy. During this Pax Romana, seaborne trade increased significantly throughout the region. Though sailing was the safest, fastest, and most efficient method of transportation in the ancient world, some fractional percentage of voyages ended in shipwreck. With the significantly increased sea traffic during the Roman era came a corresponding increase in shipwrecks.

These wrecks and their cargo remains offer glimpses through time of the economy, culture, and politics of the ancient world. Particularly useful to archaeologists are studies of amphoras, the ceramic shipping containers used in the Mediterranean region from the 15th century B.C. through the Medieval period. In addition to many discoveries in the sea, some wrecks have been examined in lakes. Most notable are Caligula's pleasure barges in Lake Nemi, Italy. The Nemi ships and other shipwreck sites occasionally yield objects of unique artistic value. For instance, the Antikythera wreck contained a staggering collection of marble and bronze statues including the Antikythera Youth. Discovered in 1900 by Greek sponge divers, the ship probably sank in the first century B.C. and may have been dispatched by the Roman general, Sulla, to carry booty back to Rome.

The sponge divers also recovered from the wreck the famous Antikythera mechanism, believed to be an astronomical calculator. Further examples of fabulous works of art recovered from the sea floor are the two "bronzi" found in Riace (Calabria), Italy. In the cases of Antikythera and Riace, however, the artifacts were recovered without the direct participation of maritime archaeologists.

Recent studies in the Sarno river (near Pompeii) show other interesting elements of ancient life. The Sarno projects suggests that on the Tyrrhenian shore there were little towns with palafittes, similar to ancient Venice. In the same area, the submerged town of Puteoli (Pozzuoli, close to Naples) contains the "portus Julius" created by Marcus Vipsanius Agrippa in 37 BC, later sunk due to bradyseism.

The sea floor elsewhere in the Mediterranean holds countless archaeological sites. In Israel, Herod the Great's port at Caesarea Maritima has been extensively studied. Other finds are consistent with some passages of the Bible (like the so-called Jesus boat, which appears to be similar to those in use during the first century AD).

MARINE ARCHAEOLOGY AND CONSERVATION

Excavation is sometimes not the preferred method to explore an underwater archaeological site. Mapping, remote sensing, and surface survey are often used without digging or removing artifacts from a site. Excavation is primarily used only when a site is in danger of destruction from modern vessels, severe storms, dredging (removing sediment from the bottom of a waterway, usually to make it wider or deeper), or rapid, natural decay. Leaving a site intact permits future archaeologists to study the same site. Artifacts can be both helped and damaged by the water that surrounds them.

Some artifacts, such as the wooden hull of ancient ships, are better preserved in the salt waters in which they sank. On the other hand, the iron and metal hulls of relatively modern vessels, such as the Titanic that sank in 1912, are destroyed by saltwater. The temperature of water, its salinity (salt content), and depth of a site also affect preservation. Cold water with low salinity does not destroy artifacts as rapidly as salty, warm water destroys artifacts. The hulls of shipwrecks in extremely deep ocean waters can be crushed by intense pressures equal to thousand of pounds per square inch. Artifacts removed from underwater sites require special handling, preservation, and storage, also known as curation.

The goal of proper conservation and curation is to clean and protect artifacts from further damage so that they may be studied by scientists or placed on display in museums. Various pieces of broken pottery jars are grouped together. Sometimes, the pieces are reassembled to restore the object, and sometimes they are left as individual fragments. Some artifacts recovered from the sea are kept moist in special wet or humid cases, other artifacts are carefully dried out once removed. Even a small artifact, such a sailor's wooden pipe, may take months or years of conservation work before ready for a museum display.

MARINE BIOLOGY

Marine biology is the scientific study of organisms in the ocean or other marine or brackish bodies of water. Given that in biology many phyla, families and genera have some species that live in the sea and others that live on land, marine biology classifies species based on the environment rather than on taxonomy. Marine biology differs from marine ecology as marine ecology is focused on how organisms interact with each other and the environment, and biology is the study of the organisms themselves.

Marine life is a vast resource, providing food, medicine, and raw materials, in addition to helping to support recreation and tourism all over the world. At a fundamental level, marine life helps determine the very nature of our planet. Marine organisms contribute significantly to the oxygen cycle, and are involved in the regulation of the Earth's climate. Shorelines are in part shaped and protected by marine life, and some marine organisms even help create new land. Marine biology covers a great deal, from the microscopic, including most zooplankton and phytoplankton to the huge cetaceans (whales) which reach up to a reported 48 meters (125 feet) in length.

The habitats studied by marine biology include everything from the tiny layers of surface water in which organisms and abiotic items may be trapped in surface tension between the ocean and atmosphere, to the depths of the oceanic trenches, sometimes 10,000 meters or more beneath the surface of the ocean. It studies habitats such as coral reefs, kelp forests, tidepools, muddy, sandy and rocky bottoms, and the open ocean (pelagic) zone, where solid objects are rare and the surface of the water is the only visible boundary. A large proportion of all life on Earth exists in the oceans. Exactly how large the proportion is unknown, since many ocean species are still to be discovered. While the oceans comprise about 71% of the Earth's surface, due to their depth they encompass about 300 times the habitable volume of the terrestrial habitats on Earth.

Many species are economically important to humans, including food fish. It is also becoming understood that the well–being of marine organisms and other organisms are linked in very fundamental ways. The human body of knowledge regarding the relationship between life in the sea and important cycles is rapidly growing, with new discoveries being made nearly every day. These cycles include those of matter (such as the carbon cycle) and of air (such as Earth's respiration, and movement of energy through ecosystems including the ocean). Large areas beneath the ocean surface still remain effectively unexplored.

SUBFIELDS

The marine ecosystem is large, and thus there are many sub-fields of marine biology. Most involve studying specializations of particular animal groups, such as phycology, invertebrate zoology and ichthyology. Other subfields study the physical effects of continual immersion in sea water and the ocean in general, adaptation to a salty environment, and the effects of changing various oceanic properties on marine life. A subfield of marine biology studies the relationships between oceans and ocean life, and global warming and environmental issues (such as carbon dioxide displacement). Recent marine biotechnology has focused largely on marine biomolecules, especially proteins, that may have uses in medicine or engineering. Marine environments are the home to many exotic biological materials that may inspire biomimetic materials.

Related Fields

Marine biology is a branch of oceanography and is closely linked to biology. It also encompasses many ideas from ecology. Fisheries science and marine conservation can be considered partial offshoots of marine biology (as well as environmental studies).

LIFEFORMS

Microscopic Life

Microscopic life undersea is incredibly diverse and still poorly understood. For example, the role of viruses in marine ecosystems is barely being explored even in the beginning of the 21st century.

The role of phytoplankton is better understood due to their critical position as the most numerous primary producers on Earth. Phytoplankton are categorized into cyanobacteria (also called blue–green algae/bacteria), various types of algae (red, green, brown, and yellow–green), diatoms, dinoflagellates, euglenoids, coccolithophorids, cryptomonads, chrysophytes, chlorophytes, prasinophytes, and silicoflagellates.

Zooplankton tend to be somewhat larger, and not all are microscopic. Many Protozoa are zooplankton, including dinoflagellates, zooflagellates, foraminiferans, and radiolarians. Some of these (such as dinoflagellates) are also phytoplankton; the distinction between plants and animals often breaks down in very small organisms. Other zooplankton include cnidarians, ctenophores, chaetognaths, molluscs, arthropods, urochordates, and annelids such as polychaetes. Many larger animals begin their life as zooplankton before they become large enough to take their familiar forms. Two examples are fish larvae and sea stars (also called starfish).

Plants and Algae

Plant life is widespread and very diverse under the ocean. Microscopic photosynthetic algae contribute a larger proportion of the worlds photosynthetic output than all the terrestrial forests combined. Most of the niche occupied by sub plants on land is actually occupied by macroscopic algae in the ocean, such as Sargassum and kelp, which are commonly known as seaweeds that creates kelp forests. The non algae plants that survive in the sea are often found in shallow waters, such as the seagrasses (examples of which are eelgrass, Zostera, and turtle grass, Thalassia).

These plants have adapted to the high salinity of the ocean environment. The intertidal zone is also a good place to find plant life in the sea, where mangroves or cordgrass or beach grass might grow. Microscopic algae and plants provide important habitats for life, sometimes acting as hiding and foraging places for larval forms of larger fish and invertebrates.

Marine Invertebrates

As on land, invertebrates make up a huge portion of all life in the sea. Invertebrate sea life includes Cnidaria such as jellyfish and sea anemones; Ctenophora; sea worms including the phyla Platyhelminthes, Nemertea, Annelida, Sipuncula, Echiura, Chaetognatha, and Phoronida; Mollusca including shellfish, squid, octopus; Arthropoda including Chelicerata and Crustacea; Porifera; Bryozoa; Echinodermata including starfish; and Urochordata including sea squirts or tunicates.

Fish

Fish have evolved very different biological functions from other large organisms. Fish anatomy includes a two–chambered heart, operculum, swim bladder, scales, fins, lips, eyes and secretory cells that produce mucous. Fish breathe by extracting oxygen from water through their gills. Fins propel and stabilize the fish in the water. Well known fish include: sardines, anchovy, ling cod, clownfish (also known as anemonefish), and bottom fish which include halibut or ling cod. Predators include sharks and barracuda.

Reptiles

Reptiles which inhabit or frequent the sea include sea turtles, sea snakes, terrapins, the marine iguana, and the saltwater crocodile. Most extant marine reptiles, except for some sea snakes, are oviparous and need to return to land to lay their eggs. Thus most species, excepting sea turtles, spend most of their lives on or near land rather than in the ocean. Despite their marine adaptations, most sea snakes prefer shallow waters nearby land, around islands, especially waters that are somewhat sheltered, as well as near estuaries. Some extinct marine reptiles, such as ichthyosaurs, evolved to be viviparous and had no requirement to return to land.

Seabirds

Seabirds are species of birds adapted to living in the marine environment, examples including albatross, penguins, gannets, and auks. Although they spend most of their lives in the ocean, species such as gulls can often be found thousands of miles inland.

Marine Mammals

There are five main types of marine mammals:

1. Cetaceans include toothed whales (Suborder Odontoceti), such as the Sperm Whale, dolphins, and porpoises such as the Dall's porpoise. Cetaceans also include baleen whales (Suborder Mysticeti), such as the Gray Whale, Humpback Whale, and Blue Whale.

2. Sirenians include manatees, the Dugong, and the extinct Steller's Sea Cow.
3. Seals (Family Phocidae), sea lions (Family Otariidae—which also include the fur seals), and the Walrus (Family Odobenidae) are all considered pinnipeds.
4. The Sea Otter is a member of the Family Mustelidae, which includes weasels and badgers.
5. The Polar Bear (Family Ursidae) is sometimes considered a marine mammal because of its dependence on the sea.

MARINE HABITATS

Marine habitats can be divided into coastal and open ocean habitats. Coastal habitats are found in the area that extends from the shoreline to the edge of the continental shelf. Most marine life is found in coastal habitats, even though the shelf area occupies only seven per cent of the total ocean area. Open ocean habitats are found in the deep ocean beyond the edge of the continental shelf

Alternatively, marine habitats can be divided into pelagic and demersal habitats. Pelagic habitats are found near the surface or in the open water column, away from the bottom of the ocean. Demersal habitats are near or on the bottom of the ocean.

An organism living in a pelagic habitat is said to be a pelagic organism, as in pelagic fish. Similarly, an organism living in a demersal habitat is said to be a demersal organism, as in demersal fish. Pelagic habitats are intrinsically shifting and ephemeral, depending on what ocean currents are doing. Marine habitats can be modified by their inhabitants. Some marine organisms, like corals, kelp and seagrasses, are ecosystem engineers which reshape the marine environment to the point where they create further habitat for other organisms.

Intertidal and Shore

Intertidal zones, those areas close to shore, are constantly being exposed and covered by the ocean's tides. A huge array of life lives within this zone. Shore habitats span from the upper intertidal zones to the area where land vegetation takes prominence. It can be underwater anywhere from daily to very infrequently. Many species here are scavengers, living off of sea life that is washed up on the shore. Many land animals also make much use of the shore and intertidal habitats. A subgroup of organisms in this habitat bores and grinds exposed rock through the process of bioerosion.

REEFS

Reefs comprise some of the densest and most diverse habitats in the world. The best–known types of reefs are tropical coral reefs which exist in most tropical waters; however, reefs can also exist in cold water. Reefs are built up by corals and other calcium–depositing animals, usually on top of a rocky outcrop on the ocean floor. Reefs can also grow on other surfaces, which has made it possible to create artificial reefs. Coral reefs also support a huge community of life, including the corals themselves, their symbiotic zooxanthellae, tropical fish and many other organisms.

Much attention in marine biology is focused on coral reefs and the El Niño weather phenomenon. In 1998, coral reefs experienced the most severe mass bleaching events on record, when vast expanses of reefs across the world died because sea surface temperatures rose well above normal. Some reefs are recovering, but scientists say that between 50% and 70% of the world's coral reefs are now endangered and predict that global warming could exacerbate this trend.

Open Ocean

The open ocean is relatively unproductive because of a lack of nutrients, yet because it is so vast, in total it produces the most primary productivity. Much of the aphotic zone's energy is supplied by the open ocean in the form of detritus. The open ocean consists mostly of jellyfish and its predators such as the mola mola.

Deep Sea and Trenches

The deepest recorded oceanic trenches measure to date is the Mariana Trench, near the Philippines, in the Pacific Ocean at 10,924 m (35,838 ft). At such depths, water pressure is extreme and there is no sunlight, but some life still exists. A white flatfish, a shrimp and a jellyfish were seen by the American crew of the bathyscaphe Trieste when it dove to the bottom in 1960.

Other notable oceanic trenches include Monterey Canyon, in the eastern Pacific, the Tonga Trench in the southwest at 10,882 m (35,702 ft), the Philippine Trench, the Puerto Rico Trench at 8,605 m (28,232 ft), the Romanche Trench at 7,760 m (24,450 ft), Fram Basin in the Arctic Ocean at 4,665 m (15,305 ft), the Java Trench at 7450 m (24,442 ft), and the South Sandwich Trench at 7,235 m (23,737 ft).

In general, the deep sea is considered to start at the aphotic zone, the point where sunlight loses its power of transference through the water. Many life forms that live at these depths have the ability to create their own light a unique evolution known as bio-luminescence.

Marine life also flourishes around seamounts that rise from the depths, where fish and other sea life congregate to spawn and feed. Hydrothermal vents along the mid–ocean ridge spreading centers act as oases, as do their opposites, cold seeps. Such places support unique biomes and many new microbes and other lifeforms have been discovered at these locations.

DISTRIBUTION FACTORS

An active research topic in marine biology is to discover and map the life cycles of various species and where they spend their time. Marine biologists study how the ocean currents, tides and many other oceanic factors affect ocean lifeforms, including their growth, distribution and well–being. This has only recently become technically feasible with advances in GPS and newer underwater visual devices. Most ocean life breeds in specific places, nests or not in others, spends time as juveniles in still others, and in maturity in yet others. Scientists know little about where many species spend different parts of their life cycles. For example, it is still largely unknown where sea turtles and some sharks travel. Tracking devices do not work for some life forms, and the ocean is not friendly to technology. This is important to scientists and fishermen because they are discovering that by restricting commercial fishing in one small area they can have a large impact in maintaining a healthy fish population in a much larger area far away.

MARINE GEOLOGY AND GEOPHYSICS

Marine geology and geophysics are scientific fields that are concerned with solving the mysteries of the seafloor and Earth's interior. Marine geologists, like all geologists, seek to understand the processes and history of the solid Earth, but their techniques differ from geologists who work on land because they study geologic (Earth's) features that are underwater. The oceans cover more than 70% of Earth, and water obscures a wealth of information about the rocks and sediments (particles of rock, sand, and other material) in the ocean basins. Marine geologists rely mainly on physical techniques to uncover the features and processes of the seafloor.

Geophysicists are scientists who study the physical properties of the solid Earth, and often work closely with marine geologists. Geophysicists use experiments and observations to determine how Earth materials such as rock, magma (molten rock), sediments, air, and water affect physical phenomena such as sound, heat, light, magnetic fields (a field of magnetic force), and earthquake tremors (seismic waves). Marine geologists and geophysicists make images and maps of the seafloor, along with maps ofsediment and rock layers below the seafloor. They also use instruments to measure changes in Earth's gravity (the attraction between two masses), magnetic field, and the pattern of heat flow arising from deep in the Earth that help to explain geologic features of the ocean basins. Marine geology and geophysics involve many different fields of science.

Many marine geoscientists (a group including both marine geologists and marine geophysicists) have backgrounds in such diverse academic fields as physics, chemistry, oceanography, engineering, and paleontology (study of biological life in the fossil record). Most marine geologists are familiar with the theories and techniques of geophysics, and most geophysicists understand the geological significance of the processes and features they are working to clarify. Marine geology is also closely linked to the sciences of oceanography and marine biology. Oceanographers study the physical and chemical properties of the water in oceans and marine biologists study the living organisms in oceans. In order to completely understand the cycles, structures and processes of the oceans, scientists from many fields must collaborate.

WHY STUDY THE SEAFLOOR?

The ocean basins hold keys to understanding the two most important theories of geological science: plate tectonics and the sedimentary record of geologic history. Marine geologists and geophysicists were the first to discover the globe–encircling chain of volcanic mountains, called the mid–ocean ridge system, where new ocean floor is created. Using their observations of the seafloor, these scientists developed the theory of plate tectonics, the idea that Earth's outer shell (lithosphere) is made of rigid pieces (plates) that move relative to one another over time. Plate tectonic theory explains the worldwide distribution of mountain ranges, ocean trenches (deep, arc-shaped valleys along the edges of the ocean basins), volcanoes, rock types, and earthquakes. By studying plate tectonics, scientists can better understand and predict geologic actions of today, such as volcanic activity and earthquakes.

Scientists also know from studying plate tectonics that the moving seafloor is recycled into Earth's interior at trenches, a process called subduction. Like the theories of evolution (change over time) in biology and relativity in physics, plate tectonics is a unifying theory that has general significance to all of science. Marine geologists and geophysicists also study layered sedimentary rocks (strata) on the seafloor that hold clues to the chemical, biological, and geographic history of the oceans. The ocean basins hold a vast wealth of economically important minerals, such as manganese and nickel, and hydrocarbons (oil and natural gas). Petroleum (oil and gas) and mining companies hire marine geologists and geophysicists to find offshore sources of petroleum. They rely heavily on marine scientific techniques to locate petroleum reservoirs and mineral deposits.

STUDYING THE SEAFLOOR

Marine geology and geophysics use a number of technologies uniquely adapted for ocean exploration. Many of the methods used are geophysical because they allow a "hands off" approach to seafloor observation. In other words, geophysical technologies allow marine geoscientists to "see" through water, rock, and sediment. (Techniques that involve observing or measuring the properties of land, sea, and seafloor surface from a distance are generally termed remote sensing.) Like all geologists, marine geologists collect rock and sediment samples.

They use dredges, which are metal buckets or claws that are lowered from a ship and dragged along the sea floor, and coring (drilling) devices to bring materials up from the bottom of the sea. Scientists then examine the material's physical, chemical, and biological properties. Seafloor samples are, however, difficult and very expensive to obtain, especially in very deep water. Marine geologists usually collect them from a few critical locations within a study area and then use geophysical images to generate a big picture of the study area.

Sediments and deep rock samples are collected using shipboard drills that bring back cores (metal tubes) that are filled with several meters of sample. By using samples together with seafloor maps and profiles (cross-sections) through the rock and sediment layers below the seafloor, marine geologists construct three–dimensional representations of their study areas.

Although most features that interest marine geologists, such as submarine (underwater) volcanoes, massive sand dunes, and deep trenches are too large to observe from the seafloor, direct observations by divers, submersibles, and remotely-operated vehicles (ROVs) can be useful in some cases. Geologists use waterproof cameras and other instruments carried by divers, lowered on cables from ships, or attached to remotely operated watercraft to capture details of the seafloor environments. Submersibles are small submarines that are capable of carrying passengers to the deep seafloor. ROVs and autonomous underwater vehicles (AUVs) are unmanned robotic submarines equipped with cameras and instruments that operators control from a ship, much like a remote controlled car.

Marine geologists rely on sonar (short for "sound navigation and ranging"), which is the use of underwater sound waves. Sound travels at a constant velocity (speed) in water, so the time it takes for the sound wave to travel through the water and echo back to the ship illustrates variations in the seafloor. Sonar is used to measure bathymetry, the topography or layout of the sea floor. A "chirp" is transmitted from a ship hull and travels until it reaches the sea floor and bounces back to a receiver on the ship where the travel time is recorded.

To determine the distance from sea level to the ocean bottom, scientists multiply the time it takes for the sound wave to travel to the ocean floor and back by the rate (speed) at which the sound wave travels in water. Scientists can also map ocean floor bathymetry using satellite (vehicles in orbit around Earth) instruments. The ocean surface is not completely flat, but mimics the sea floor by bulging upward and downward. Satellite observations reveal detailed patterns of mid-ocean ridges and trenches and underwater volcanoes, thus confirming plate tectonics. Magnetometers, towed behind a ship, measure small changes in Earth's magnetic field.

Sensitive shipboard gravimeters record subtle changes in Earth's field of gravity (the attraction between the Earth and another body).

Marine geoscientists also use seismology (earthquake waves) to make an image of the seafloor. A ship tows several air guns that make an underwater explosion using compressed air. The shock waves from the explosion are the same type as waves made in an earthquake. These waves penetrate layers of rock underlying the surface of the ocean and bounce back to hydrophones (receivers). The waves travel at different speeds depending on the type of rock.

OCEANOGRAPHY

Oceanography has been aptly defined as the study of the world below the surface of the sea: it should include the contact zone between sea and atmosphere. Present–day acceptance it has to do with all the characteristics of the bottom and margins of the sea, of the sea water, and of the inhabitants of the latter. Thus widely combining geophysics, geochemistry, and biology, it is inclusive, as is, of course, characteristic of any 'young' science: and modern oceanography is in its youth. But in this case it is not so much immaturity that is responsible for the fact that these several subsciences are still grouped together, but rather the realization that the physics, chemistry, and biology of the sea water are not only important per se, but that in most of the basic problems of the sea all three of these subdivisions have a part. And with every advance in our knowledge of the sea making this interdependence more and more apparent, it is not likely, that we shall soon see any general abandon ment of this concept of oceanography as a mother science, the branches of which, though necessarily attacked by different disciplines, are intertwined too closely to be torn apart. Every oceanic biologist should, therefore, be grounded in the principles of geophysics and geochemistry; every chemical or physical oceanographer in some of the oceanic aspects of biology.

A feature equally inherent in sea science is that it is no less inclusive from the geographic standpoint, because the subjects with which it is concerned cover so large a part of the surface of our planet.. And the vastness of the areas to be considered, whatever phase of the sea be under consideration, has; determined the paths that the science of oceanography has followed in its advance from, its early beginnings to its present state. Most oceanographers, too, would agree that the geographic factor has likewise been responsible for a failure to progress at a rate commensurate either with the relative importance, of this field of knowledge in the general household of science, or with the amount of energy that has been, devoted thereto during the past quarter-century.

In the nature of things, the oceanographer constantly meets a, twofold obstacle of another sort, when he attempts to extend his investigations out from the shore–line to the high seas, no matter from what headquarters he may work, in the necessity of studying the majority of oceanic phenomena and events within the sea, not merely upon it, or from its borders. Even if his investigation be of a sort that can be carried on in a laboratory on shore, the raw data must be gathered at sea. Therefore he must have a boat, a necessity that places him at a disadvantage as compared with the general biologist who turns to marine animals chiefly for convenience, and so can pick up many things of interest on a stroll along tideline.

If the student is to venture out more than a few miles from land, his craft must be large enough to contain living quarters and to navigate safely in all weathers, for oceanography is impossible unless some one does go out to sea, on short trips or on long. That is to say, for even one investigator, or one party, to gather information ofany kind about the ocean in appreciable amount demands the labours of many, as reflected in the maintenance and operation of a seagoing craft, with crew to man her, with supplies for their subsistence; also with fuel for her propulsion. And as any craft larger than a rowboat is an expensive means of conveyance for a small number of passengers, it follows that exploration into the economy of the high seas is essentially a costly undertaking.

The expense of extended voyages, combined with the necessity for such if large sea areas were to be studied more intensively than could be done from examination of vessel's log books, was no doubt the chief reason that systematic examination, even of the surface of the sea, was not seriously undertaken until the middle of the nineteenth century. But when, at about this period, science awoke to the whole new world for exploration that was offered by the oceans, it was soon learned that no very serious technical difficulties, apart from expense, were involved in extending investigations down into the abyss, whether it was a question of developing the contour of the sea. floor, of gathering samples of the bottom, of sampling the living creatures, or of measuring the physical and chemical characteristics of the water.

It would, indeed, have been quite within the technical abilities of the Romans of Pliny's day to have plumbed the depths of the Mediterranean and to have explored its deep-water biota, though of course examination of the temperature and salinity of the sea must, in any case, have awaited the development of the sciences of physics and chemistry as we now know them. Efficient gear was, in fact, so rapidly developed in the three years 1868–70, as soon as a serious start in that direction was made on the voyages of the 'Lightning' and of the 'Porcupine,' that when the 'Challenger' sailed two years later on the first oceanwide exploration. of the deeps, the scientists on board were already in position efficiently to reap the rich harvest that stimulated many other subsequent expeditions in various parts of the world. From that time forward, with every fresh venture below the surface of the sea, and with constant improvement in technical methods of operating gear of different kinds at great depths, a flood of new facts came pouring in so rapidly that more was learned about the,sea and about the inhabitants of its deeps during the last thirty years of the nineteenth century than had been up till then. This was the heyday of the deep-sea exploring expedition, when one cruise after another was sent out by different maritime nations, when the broad relief of the ocean floors was mapped, when the general nature of the submarine sediments was determined, when the distribution of temperature and salinity was worked 'out in its essential outlines over the high seas, and when the general characteristics of the deep–sea fauna were explored.

While this regional-descriptive era of oceanography will never definitely close so long as the science of the sea is pursued, there came a change, towards the end of the century just past, when persistence in the old discursive methods, determined by established habits of thought, no longer yielded new and wonderful discoveries at the rate that had been the order of the day when no one knew what was to be found at the bottom of the sea.

Thenceforth, with increasing frequency, continued exploration along these preliminary lines yielded results more corroborative than novel. And a period of general oceanographic stagnation might then have succeeded to the preceding peak of activity had there not arisen new schools, centering their attention on the biologic economy of the inhabitants of the ocean as related to their physical–chemical environment, on mathematical analysis of the internal dynamics of the sea water, and on the geologic bearing of submarine topography and sedimentation, rather than on areal surveys of one or another feature of the sea. This conscious alteration of viewpoint, from the descriptive to the analytic, is one of two chief factors that gives to oceanography its present tone: the other is the growth of an economic demand that oceanography afford practical assistance to the sea fisheries.

This demand developed first in northwestern Europe, where, as it chanced, the fisheries were so rapidly expanding, and increasing in intensity through the adoption of more effective methods of fishing, that dread of depletion began to loom in the offing, just when oceanography was approaching the end of its nineteenth–century boom; *i.e.,* just when it needed a fresh stimulus.

The immediate, practical result was a concentration of attention on limited coastwise areas as contrasted with the broad oceans, and the development of an international and official organization—he Conseil International pour l'Exploration de la Mer–with power to coordinate the scientific efforts of the Fisheries Bureaus of the several nations fronting on these areas in northwestern Europe.

It is an interesting speculation whether, without this enforced direction of scientific attention to the North Sea region, we should have come to appreciate, as clearly as we do today, that application to the adjacent oceans of principles established by intensive investigation of such test cases offers the most promising lines of approach to many of the broad, underlying problems of oceanography.

However, this may be, we can hardly doubt that the advance of oceanography on the analytic-synthetic side would have continued slow and halting had not the Conseil and other coordinating institutions of more recent birth, but with similar aims, added their unifying influence to the attempts at synthetic investigation that would in any case have followed the alteration in viewpoint. And it is certain that, today, the most rapid approaches towards an understanding of events in the sea are being made by orderly, intensive, and concerted attacks upon one or another phase, via definitely stated and apparently illustrative problems, rather than by haphazard accumulation of unrelated facts, gathered in the hope that somehow, sometime, these may be fitted together by some one.

This is reflected in the fact that several broad–scale expeditions that have been sent out within the last few years—'Meteor,' 'Dana,' 'Carnegie,' 'Marion'—have devoted their attention chiefly to extending to the high seas special lines of investigation the theoretical basis for which had already been developed from intensive studies falling in the general category just stated.

The foregoing remarks are introductory to the thesis that a discussion of certain of the underlying problems that seem most clearly to illustrate the general fields of research falling within the province of the oceanographer, and that are now most to the fore, is integral in any rational exposition of the scope and present status of this inclusive branch of science.

To list all the problems that await the oceanographer will never be possible so long as science lives, for new ones will constantly unfold, as the boundaries of knowledge are rolled back.

In practice oceanography falls most conveniently into three chief divisions:

1. The geological;
2. The physical-chemical;
3. The biological.

To consider first the problems of the shape and composition of the basins that hold the oceans; next, those associated with the physical character and chemical composition of the waters that fill these basins; and third, those of the nature and activities of the animals and plants that inhabit the waters is therefore a rational order of presentation. Subsequent stages discuss the fundamental unity of these different divisions, and outline certain of the direct economic benefits that may be expected to accrue from the study of oceanography.

A word is perhaps due the reader to explain our omission of any references to authorities. Citations are an essential part of any presentation of the results of investigation, or of any textbook, but this is not necessarily true when the discussion is of things to be studied. In the present case to have mentioned authors would have necessitated a general bibliography of the subject; also roster of oceanographers, for each of the latter has by his published work helped to make this book possible. To cut this Gordian knot personal references are omitted.

HISTORY

Humans first acquired knowledge of the waves and currents of the seas and oceans in pre–historic times. Observations on tides are recorded by Aristotle and Strabo. Early modern exploration of the oceans was primarily for cartography and mainly limited to its surfaces and of the creatures that fishermen brought up in nets, though depth soundings by lead line were taken. Although Juan Ponce de Leon in 1513 first identified the Gulf Stream, and the current was well-known to mariners, Benjamin Franklin made the first scientific study of it and gave it its name. Franklin measured water temperatures during several Atlantic crossings and correctly explained the Gulf Stream's cause. Franklin and Timothy Folger printed the first map of the Gulf Stream in 1769–1770.

When Louis Antoine de Bougainville, who voyaged between 1766 and 1769, and James Cook, who voyaged from 1768 to 1779, carried out their explorations in the South Pacific, information on the oceans themselves formed part of the reports. James Rennell wrote the first scientific textbooks about currents in the Atlantic and Indian oceans during the late 18th and at the beginning of 19th century. Sir James Clark Ross took the first modern sounding in deep sea in 1840, and Charles Darwin published a document on reefs and the formation of atolls as a result of the second voyage of HMS Beagle in 1831-6. Robert FitzRoy published a report in four volumes of the three voyages of the Beagle. In 1841–1842 Edward Forbes undertook dredging in the Aegean Sea that founded marine ecology.

As first superintendent of the United States Naval Observatory (1842–1861) Matthew Fontaine Maury devoted his time to the study of marine meteorology, navigation, and charting prevailing winds and currents. His Physical Geography of the Sea, 1855 was the first textbook of oceanography. Many nations sent oceanographic observations to Maury at the Naval Observatory, where he and his colleagues evaluated the information and gave the results worldwide distribution.

The steep slope beyond the continental shelves was discovered in 1849. The first successful laying of transatlantic telegraph cable in August 1858 confirmed the presence of an underwater "telegraphic plateau" mid–ocean ridge. After the middle of the 19th century, scientific societies were processing a flood of new terrestrial botanical and zoological information.

In 1871, under the recommendations of the Royal Society of London, the British government sponsored an expedition to explore world's oceans and conduct scientific investigations. Under that sponsorship the Scots Charles Wyville Thompson and Sir John Murray launched the Challenger expedition (1872–1876). The results of this were published in 50 volumes covering biological, physical and geological aspects. 4417 new species were discovered. Other European and American nations also sent out scientific expeditions (as did private individuals and institutions). The first purpose built oceanographic ship, the "Albatros" was built in 1882. The four-month 1910 North Atlantic expedition headed by Sir John Murray and Johan Hjort was at that time the most ambitious research oceanographic and marine zoological project ever, and led to the classic 1912 book The Depths of the Ocean.

Oceanographic institutes dedicated to the study of oceanography were founded. In the United States, these included the Scripps Institution of Oceanography in 1892, Woods Hole Oceanographic Institution in 1930, Virginia Institute of Marine Science in 1938, Lamont-Doherty Earth Observatory at Columbia University, and the School of Oceanography at University of Washington. In Britain, there is a major research institution: National Oceanography Centre, Southampton which is the successor to the Institute of Oceanography. In Australia, CSIRO Marine and Atmospheric Research, known as CMAR, is a leading center. In 1921 the International Hydrographic Bureau (IHB) was formed in Monaco.

In 1893, Fridtjof Nansen allowed his ship "Fram" to be frozen in the Arctic ice. As a result he was able to obtain oceanographic data as well as meteorological and astronomical data. The first international organization of oceanography was created in 1902 as the International Council for the Exploration of the Sea.

The first acoustic measurement of sea depth was made in 1914. Between 1925 and 1927 the "Meteor" expedition gathered 70,000 ocean depth measurements using an echo sounder, surveying the Mid atlantic ridge. The Great Global Rift, running along the Mid Atlantic Ridge, was discovered by Maurice Ewing and Bruce Heezen in 1953 while the mountain range under the Arctic was found in 1954 by the Arctic Institute of the USSR. The theory of seafloor spreading was developed in 1960 by Harry Hammond Hess. The Ocean Drilling Project started in 1966. Deep sea vents were discovered in 1977 by John Corlis and Robert Ballard in the submersible "Alvin".

In the 1950s, Auguste Piccard invented the bathyscaphe and used the "Trieste" to investigate the ocean's depths. The nuclear submarine Nautilus made the first journey under the ice to the North Pole in 1958. In 1962 there was the first deployment of FLIP (Floating Instrument Platform), a 355 foot spar buoy. Then, in 1966, the U.S. Congress created a National Council for

Marine Resources and Engineering Development. NOAA was put in charge of exploring and studying all aspects of Oceanography in the USA. It also enabled the National Science Foundation to award Sea Grant College funding to multi–disciplinary researchers in the field of oceanography.

From the 1970s, there has been much emphasis on the application of large scale computers to oceanography to allow numerical predictions of ocean conditions and as a part of overall environmental change prediction. An oceanographic buoy array was established in the Pacific to allow prediction of El Nino events. 1990 saw the start of the World Ocean Circulation Experiment (WOCE) which continued until 2002. Geosat seafloor mapping data became available in 1995. In 1942, Sverdrup and Fleming published "The Ocean" which was a major landmark. "The Sea" edited by M.N. Hill was published in 1962 while the "Encyclopedia of Oceanography" by Rhodes Fairbridge was published in 1966.

CONNECTION TO THE ATMOSPHERE

The study of the oceans is linked to understanding global climate changes, potential global warming and related biosphere concerns. The atmosphere and ocean are linked because of evaporation and precipitation as well as thermal flux (and solar insolation). Wind stress is a major driver of ocean currents while the ocean is a sink for atmospheric carbon dioxide.

- Our planet is invested with two great oceans; one visible, the other invisible; one underfoot, the other overhead; one entirely envelopes it, the other covers about two thirds of its surface.

BRANCHES

The study of oceanography is divided into branches:

- Biological oceanography, or marine biology, is the study of the plants, animals and microbes of the oceans and their ecological interaction with the ocean;
- Chemical oceanography, or marine chemistry, is the study of the chemistry of the ocean and its chemical interaction with the atmosphere;
- Geological oceanography, or marine geology, is the study of the geology of the ocean floor including plate tectonics;
- Physical oceanography, or marine physics, studies the ocean's physical attributes including temperature-salinity structure, mixing, waves, internal waves, surface tides, internal tides, and currents. Of particular interest is the behaviour of sound (acoustical oceanography), light (optical oceanography) and radio waves in the ocean.

These branches reflect the fact that many oceanographers are first trained in the exact sciences or mathematics and then focus on applying their interdisciplinary knowledge, skills and abilities to oceanography. Data derived from the work of Oceanographers is used in marine engineering, in the design and building of oil platforms, ships, harbours, and other structures that allow us to use the ocean safely. Oceanographic data management is the discipline ensuring that oceanographic data both past and present are available to researchers.

BIOLOGICAL OCEANOGRAPHY

Biological oceanographers (or marine biologists) focus on the patterns and distribution of marine organisms. These scientists work to understand why certain animals, plants, and microorganisms are found in different places and how these organisms grow. A variety of factors influence the success of a certain species in any location, including the chemistry and physical properties of the water. In turn, the biological organisms in the ocean affect the oceans on a global and local level. Biological oceanographers study all types of organisms that live in the ocean, from the very small to the very large.

They investigate patterns and distributions of the microscopic organisms including viruses (which are not really organisms, but genetic material such as DNA that do have the ability to reproduce), bacteria, and plankton (free-floating animals and plants). They also study the larger animals and plants, like kelp, seaweed, marine invertebrates (animals without a backbone), fish, and marine mammals. They incorporate information and techniques from a broad range of disciplines including chemistry, physics, remote sensing (the use of specialized instruments, such as satellites, to relay information about one location to another location for analysis), paleontology (study of fossils), and geography (study of Earth's surface) for their research.

CHEMICAL OCEANOGRAPHY

Chemical oceanographers study the chemicals that are dissolved in the ocean waters. Different parts of the ocean contain varying concentrations of gasses, salts, and other chemical components. These variations are due to the impact of the atmosphere, surrounding lands, seafloor, and biological organisms in the ocean water. Chemical oceanographers work to develop theories that explain the various patterns throughout the oceans. One of the more important problems facing chemical oceanographers today is understanding the concentration of and changes in carbon dioxide in the ocean. Carbon dioxide is a major greenhouse gas, meaning it holds a lot of heat when it is found as a gas in the atmosphere. Burning fossil fuels for industry and in cars releases carbon dioxide into the atmosphere, where it contributes to global warming. The ocean, however, can remove a lot of carbon dioxide from the atmosphere. Carbon dioxide readily combines with seawater. It then goes through a series of complex chemical reactions before it becomes a solid material called calcium carbonate. Calcium carbonate can be buried in the sediments (particles of gravel, sand, and clay) at the bottom of the ocean. This means that the ocean has the potential to act as a "sink" for a lot of the carbon dioxide in the atmosphere. Chemical oceanographers are working to determine just how large the sink is and how quickly it can act.

PHYSICAL OCEANOGRAPHY

Physical oceanographers study the physical properties of the ocean. These include temperature, salinity, density, and ability to transmit light and sound. In turn, these fundamental physical characteristics affect the way that ocean currents move, the forces associated with waves, and the amount of energy absorbed by the ocean. The temperature and salinity of the water affect the density of the water. Cooler and saltier water sinks while warmer and fresher water floats. This seemingly simple property of the ocean drives much of the water circulation throughout the globe.

Density also affects the way that sound travels through water and the buoyancy (ability to float) of marine organisms. Some of the projects that physical oceanographers are studying include understanding trends in climate. Satellites measure ocean temperatures over the whole globe to try to discriminate between local changes in ocean temperature, like the El Niño-La Niña, a cycle that brings warm water and storms to the Eastern Pacific every 5 to 7 years, from more large scale changes, like global warming.

MARINE GEOLOGY

Marine geologists study the geological features of the ocean. These scientists try to determine the composition of the inner Earth by looking at special places on the seafloor where the tectonic plates are moving away from each other. In these places, called spreading centers, material from the inner Earth rises to the seafloor. Marine geologists Analyse the chemical and physical makeup of this material to gain an understanding on how the Earth was formed.

The shifting of tectonic plates also can cause earthquakes. Marine geologists also study the movements of the tectonic plates in the ocean to try to predict where and when earthquakes will occur. Another focus for marine geologists is the sediments found on the seafloor. These sediments are made up of particles from the land, dead marine plants and animals, precipitates (solid material) from chemical reactions, and even material from space.

Studying the chemical and physical composition of sediments provides information on how the Earth's climate has changed over time and where valuable resources, like oil and minerals, can be found.

REMOTE SENSING

Remote sensing is the acquisition of information about an object or phenomenon, without making physical contact with the object. In modern usage, the term generally refers to the use of aerial sensor technologies to detect and classify objects on Earth (both on the surface, and in the atmosphere and oceans) by means of propagated signals (*i.e.*, electromagnetic radiation emitted from aircraft or satellites).

There are two main types of remote sensing: passive remote sensing and active remote sensing. Passive sensors detect natural radiation that is emitted or reflected by the object or surrounding area being observed. Reflected sunlight is the most common source of radiation measured by passive sensors. Examples of passive remote sensors include film photography, infrared, charge-coupled devices, and radiometers. Active collection, on the other hand, emits energy in order to scan objects and areas whereupon a sensor then detects and measures the radiation that is reflected or backscattered from the target. RADAR is an example of active remote sensing where the time delay between emission and return is measured, establishing the location, height, speed and direction of an object.

Remote sensing makes it possible to collect data on dangerous or inaccessible areas. Remote sensing applications include monitoring deforestation in areas such as the Amazon Basin, glacial features in Arctic and Antarctic regions, and depth sounding of coastal and ocean depths. Military collection during the cold war made use of stand-off collection of data about dangerous border areas. Remote sensing also replaces costly and slow data collection on the ground, ensuring in the process that areas or objects are not disturbed.

Orbital platforms collect and transmit data from different parts of the electromagnetic spectrum, which in conjunction with larger scale aerial or ground-based sensing and analysis, provides researchers with enough information to monitor trends such as El Niño and other natural long and short term phenomena. Other uses include different areas of the earth sciences such as natural resource management, agricultural fields such as land usage and conservation, and national security and overhead, ground–based and stand–off collection on border areas. By satellite, aircraft, spacecraft, buoy, ship, and helicopter images, data is created to Analyse and compare things like vegetation rates, erosion, pollution, forestry, weather, and land use. These things can be mapped, imaged, tracked and observed. The process of remote sensing is also helpful for city planning, archaeological investigations, military observation and geomorphological surveying.

DATA ACQUISITION TECHNIQUES

The basis for multispectral collection and analysis is that of examined areas or objects that reflect or emit radiation that stand out from surrounding areas.

Applications of Remote Sensing Data

- Conventional radar is mostly associated with aerial traffic control, early warning, and certain large scale meteorological data. Doppler radar is used by local law enforcements' monitoring of speed limits and in enhanced meteorological collection such as wind speed and direction within weather systems. Other types of active collection includes plasmas in the ionosphere. Interferometric synthetic aperture radar is used to produce precise digital elevation models of large scale terrain.
- Laser and radar altimeters on satellites have provided a wide range of data. By measuring the bulges of water caused by gravity, they map features on the seafloor to a resolution of a mile or so. By measuring the height and wave–length of ocean waves, the altimeters measure wind speeds and direction, and surface ocean currents and directions.
- Light detection and ranging (LIDAR) is well known in examples of weapon ranging, laser illuminated homing of projectiles. LIDAR is used to detect and measure the concentration of various chemicals in the atmosphere, while airborne LIDAR can be used to measure heights of objects and features on the ground more accurately than with radar technology. Vegetation remote sensing is a principal application of LIDAR.
- Radiometers and photometers are the most common instrument in use, collecting reflected and emitted radiation in a wide range of frequencies. The most common are visible and infrared sensors, followed by microwave, gamma ray and rarely, ultraviolet. They may also be used to detect the emission spectra of various chemicals, providing data on chemical concentrations in the atmosphere.
- Stereographic pairs of aerial photographs have often been used to make topographic maps by imagery and terrain analysts in trafficability and highway departments for potential routes.
- Simultaneous multi–spectral platforms such as Landsat have been in use since the 70's. These thematic mappers take images in multiple wavelengths of electro–magnetic radiation (multi–spectral) and are usually found on Earth observation satellites, including (for example) the Landsat programme or the IKONOS satellite. Maps of land cover and land use from thematic mapping can be used to prospect for minerals, detect or monitor land usage, deforestation, and examine the health of indigenous plants and crops, including entire farming regions or forests.

- Hyperspectral imaging produces an image where each pixel has full spectral information with imaging narrow spectral bands over a contiguous spectral range. Hyperspectral imagers are used in various applications including mineralogy, biology, defence, and environmental measurements.
- Within the scope of the combat against desertification, remote sensing allows to follow-up and monitor risk areas in the long term, to determine desertification factors, to support decision–makers in defining relevant measures of environmental management, and to assess their impacts.

Geodetic

- Overhead geodetic collection was first used in aerial submarine detection and gravitational data used in military maps. This data revealed minute perturbations in the Earth's gravitational field (geodesy) that may be used to determine changes in the mass distribution of the Earth, which in turn may be used for geological studies.

Acoustic and Near–acoustic

- *Sonar*: Passive sonar, listening for the sound made by another object (a vessel, a whale etc); active sonar, emitting pulses of sounds and listening for echoes, used for detecting, ranging and measurements of underwater objects and terrain.
- Seismograms taken at different locations can locate and measure earthquakes (after they occur) by comparing the relative intensity and precise timing.

To coordinate a series of large–scale observations, most sensing systems depend on the following: platform location, what time it is, and the rotation and orientation of the sensor. High–end instruments now often use positional information from satellite navigation systems. The rotation and orientation is often provided within a degree or two with electronic compasses.

Compasses can measure not just azimuth (*i.e.,* degrees to magnetic north), but also altitude (degrees above the horizon), since the magnetic field curves into the Earth at different angles at different latitudes. More exact orientations require gyroscopic-aided orientation, periodically realigned by different methods including navigation from stars or known benchmarks.

Resolution impacts collection and is best explained with the following relationship: less resolution=less detail and larger coverage, More resolution=more detail, less coverage. The skilled management of collection results in cost–effective collection and avoid situations such as the use of multiple high resolution data which tends to clog transmission and storage infrastructure.

DATA PROCESSING

Generally speaking, remote sensing works on the principle of the inverse problem. While the object or phenomenon of interest (the state) may not be directly measured, there exists some other variable that can be detected and measured (the observation), which may be related to the object of interest through the use of a data–derived computer model. The common analogy given to describe this is trying to determine the type of animal from its footprints. For example, while it is impossible to directly measure temperatures in the upper atmosphere, it is possible to measure the spectral emissions from a known chemical species (such as carbon dioxide) in that region. The frequency of the emission may then be related to the temperature in that region via various thermodynamic relations.

The quality of remote sensing data consists of its spatial, spectral, radiometric and temporal resolutions:

- *Spatial resolution*: The size of a pixel that is recorded in a raster image–typically pixels may correspond to square areas ranging in side length from 1 to 1,000 metres (3.3 to 3,300 ft).
- *Spectral resolution*: The wavelength width of the different frequency bands recorded–usually, this is related to the number of frequency bands recorded by the platform. Current Landsat collection is that of seven bands, including several in the infra-red spectrum, ranging from a spectral resolution of 0.07 to 2.1 μm. The Hyperion sensor on Earth Observing–1 resolves 220 bands from 0.4 to 2.5 μm, with a spectral resolution of 0.10 to 0.11 μm per band.
- *Radiometric resolution*: The number of different intensities of radiation the sensor is able to distinguish. Typically, this ranges from 8 to 14 bits, corresponding to 256 levels of the gray scale and up to 16,384 intensities or "shades" of colour, in each band. It also depends on the instrument noise.
- *Temporal resolution*: The frequency of flyovers by the satellite or plane, and is only relevant in time–series studies or those requiring an averaged or mosaic image as in deforesting monitoring. This was first used by the intelligence community where repeated coverage revealed changes in infrastructure, the deployment of units or the modification/introduction of equipment. Cloud cover over a given area or object makes it necessary to repeat the collection of said location.

In order to create sensor–based maps, most remote sensing systems expect to extrapolate sensor data in relation to a reference point including distances between known points on the ground. This depends on the type of sensor used. For example, in conventional photographs, distances are accurate in the center of the image, with the distortion of measurements increasing the farther you get from the center. Another factor is that of the platen against which the film is pressed can cause severe errors when photographs are used to measure ground distances.

The step in which this problem is resolved is called georeferencing, and involves computer–aided matching up of points in the image (typically 30 or more points per image) which is extrapolated with the use of an established benchmark, "warping" the image to produce accurate spatial data. As of the early 1990s, most satellite images are sold fully georeferenced.

In addition, images may need to be radiometrically and atmospherically corrected:

- *Radiometric correction*: Radiometric correction gives a scale to the pixel values, *i.e.,* the monochromatic scale of 0 to 255 will be converted to actual radiance values.
- *Atmospheric correction*: Atmospheric correction eliminates atmospheric haze by rescaling each frequency band so that its minimum value (usually realised in water bodies) corresponds to a pixel value of 0. The digitizing of data also make possible to manipulate the data by changing gray–scale values.

Interpretation is the critical process of making sense of the data. The first application was that of aerial photographic collection which used the following process; spatial measurement through the use of a light table in both conventional single or stereographic coverage, added skills such as the use of photogrammetry, the use of photomosaics, repeat coverage, Making use of object's known dimensions in order to detect modifications. Image Analysis is the recently developed automated computer-aided application which is in increasing use. Object–Based Image Analysis (OBIA) is a sub-discipline of GIScience devoted to partitioning remote sensing (RS) imagery into meaningful image-objects, and assessing their characteristics through spatial, spectral and temporal scale.

Old data from remote sensing is often valuable because it may provide the only long–term data for a large extent of geography. At the same time, the data is often complex to interpret, and bulky to store. Modern systems tend to store the data digitally, often with lossless compression. The difficulty with this approach is that the data is fragile, the format may be archaic, and the data may be easy to falsify. One of the best systems for archiving data series is as computer–generated machine–readable ultrafiche, usually in typefonts such as OCR–B, or as digitized half–tone images. Ultrafiches survive well in standard libraries, with lifetimes of several centuries. They can be created, copied, filed and retrieved by automated systems. They are about as compact as archival magnetic media, and yet can be read by human beings with minimal, standardized equipment.

Data Processing Levels

To facilitate the discussion of data processing in practice, several processing "levels" were first defined in 1986 by NASA as part of its Earth Observing System and steadily adopted since then, both internally at NASA and elsewhere; these definitions are:

Level	Description
0	Reconstructed, unprocessed instrument and payload data at full resolution, with any and all communications artifacts (*i.e.*, synchronization frames, communications headers, duplicate data) removed.
1a	Reconstructed, unprocessed instrument data at full resolution, time-referenced, and annotated with ancillary information, including radiometric and geometric calibration coefficients and georeferencing parameters (*i.e.*, platform ephemeris) computed and appended but not applied to the Level 0 data (or if applied, in a manner that level 0 is fully recoverable from level 1a data).
1b	Level 1a data that have been processed to sensor units (*i.e.*, radar backscatter cross section, brightness temperature, etc.); not all instruments have Level 1b data; level 0 data is not recoverable from level 1b data.
2	Derived geophysical variables (*i.e.*, ocean wave height, soil moisture, ice concentration) at the same resolution and location as Level 1 source data.
3	Variables mapped on uniform spacetime grid scales, usually with some completeness and consistency (*i.e.*, missing points interpolated, complete regions mosaicked together from multiple orbits, etc).
4	Model output or results from analyses of lower level data (*i.e.*,variables that were not measured by the instruments but instead are derived from these measurements).

A Level 1 data record is the most fundamental (*i.e.*, highest reversible level) data record that has significant scientific utility, and is the foundation upon which all subsequent data sets are produced. Level 2 is the first level that is directly usable for most scientific applications; its value is much greater than the lower levels.

Level 2 data sets tend to be less voluminous than Level 1 data because they have been reduced temporally, spatially, or spectrally. Level 3 data sets are generally smaller than lower level data sets and thus can be dealt with without incurring a great deal of data handling overhead. These data tend to be generally more useful for many applications. The regular spatial and temporal organization of Level 3 datasets makes it feasible to readily combine data from different sources.

HISTORY

The modern discipline of remote sensing arose with the development of flight. The balloonist G. Tournachon (alias Nadar) made photographs of Paris from his balloon in 1858. Messenger pigeons, kites, rockets and unmanned balloons were also used for early images. With the exception of balloons, these first, individual images were not particularly useful for map making or for scientific purposes.

Systematic aerial photography was developed for military surveillance and reconnaissance purposes beginning in World War I and reaching a climax during the Cold War with the use of modified combat aircraft such as the P–51, P–38, RB–66 and the F–4C, or specifically designed collection platforms such as the U2/TR-1, SR-71, A–5 and the OV–1 series both in overhead and stand-off collection. A more recent development is that of increasingly smaller sensor pods such as those used by law enforcement and the military, in both manned and unmanned platforms. The advantage of this approach is that this requires minimal modification to a given airframe. Later imaging technologies would include Infra–red, conventional, doppler and synthetic aperture radar.

The development of artificial satellites in the latter half of the 20th century allowed remote sensing to progress to a global scale as of the end of the Cold War. Instrumentation aboard various Earth observing and weather satellites such as Landsat, the Nimbus and more recent missions such as *RADARSAT* and *UARS* provided global measurements of various data for civil, research, and military purposes. Space probes to other planets have also provided the opportunity to conduct remote sensing studies in extraterrestrial environments, synthetic aperture radar aboard the Magellan spacecraft provided detailed topographic maps of Venus, while instruments aboard SOHO allowed studies to be performed on the Sun and the solar wind, just to name a few examples. Recent developments include, beginning in the 1960s and 1970s with the development of image processing of satellite imagery. Several research groups in Silicon Valley including NASA Ames Research Center, GTE and ESL Inc. developed Fourier transform techniques leading to the first notable enhancement of imagery data.

REMOTE SENSING SOFTWARE

Remote Sensing data is processed and analysed with computer software, known as a remote sensing application. A large number of proprietary and open source applications exist to process remote sensing data. An NOAA Sponsored Research by Global Marketing Insights, Inc. the most used applications among Asian academic groups involved in remote sensing are as follows: ERDAS 36%; ESRI 30%; ITT Visual Information Solutions ENVI 17%; MapInfo 17%. Among Western Academic respondents as follows: ESRI 39%, ERDAS IMAGINE 27%, MapInfo 9%, AutoDesk 7%, ITT Visual Information Solutions ENVI 17%. Other important Remote Sensing Software packages include: TNTmips from MicroImages, PCI Geomatica made by PCI Geomatics, the leading remote sensing software package in Canada, IDRISI from Clark Labs, Image Analyst from Intergraph, RemoteView made by Overwatch Textron Systems, and the original object based image analysis software eCognition from Definiens. Dragon/ips is one of the oldest remote sensing packages still available, and is in some cases free. Open source remote sensing software includes GRASS GIS, ILWIS, QGIS, OSSIM, Opticks (software) and Orfeo toolbox.

ENERGY OF REMOTE SENSING

The first step in remote sensing is to have a source of energy that will be beamed towards the target. The energy comes in the form of light waves of different sizes. Like the waves in an ocean, energy waves can range from waves whose top point (crest) to lowest point (trough) are very tiny to those that are hundreds of feet (meters) long. The distance of one full wave, from crest to crest or trough to trough, is known as the wavelength. The range of waves is known as the electromagnetic spectrum. At one end of the electromagnetic spectrum lie the tiny waves such as gamma rays and X rays. These waves tend to carry large amounts of energy and can penetrate into solid or liquid material more so than other waves. That is why X rays can pass right through skin to reveal images of the bones and teeth underneath.

At the other end of the spectrum lie waves such as the microwaves that can penetrate a short distance to heat up foods, and radio waves that beam music through a radio speaker. Radio waves are not efficient for remote sensing operations. Microwaves are the longest waves with enough energy to be used for remote sensing. The regions of the electromagnetic spectrum that is useful for remote sensing contain the waves known as ultraviolet rays (the same rays that give a suntan or sunburn). The term ultraviolet means that the waves are just beyond the portion of the spectrum that contains the waves that are visible, in particular the region of the spectrum that contain violet–coloured waves.

Indeed, for the visible portion of the electromagnetic spectrum, our eyes are the remote sensors! Shorter, higher energy wavelengths are preferred for remote sensing because the waves have to move through air or water on their way to the target. Passing through air and water causes some of the waves to be absorbed or deflected (bounced) off the target. (The deflection of different wavelengths of light as they pass through Earth's atmosphere, the mass of air surrounding Earth, is the reason why the sky appears blue. Colours with relatively long wavelengths pass straight through the atmosphere. Blue light has a shorter wavelength and the atmosphere scatters it.) A higher energy wave will be better able to blast through any interference to the target, and to bounce back from the target.

The absorption of waves can be useful when trying to figure out the nature of the target. For example, microwaves tend to be absorbed by the gas form of water known as water vapour. The pattern of absorption detected by scientists on their instruments can provide important clues about the amount of water contained in the air above the ground or water.

HOW REMOTE SENSING WORKS

In order to illustrate how remote sensing works, imagine a bathtub full of water. If a bar of soap is dropped into the water, waves will move outward over the surface of the water. As the waves contact the sides of the tub, some the energy will rebound back into the tub. So it is with the energy that is beamed from a satellite, ship or plane. The returning energy is captured by a detector (also known as a sensor). Instruments and computers that are connected to the sensor can analyse the pattern of the returning waves to help scientists understand the distance and shape of the object on the ground or the ocean floor that deflected the waves.

SENDING ENERGY UNDERWATER

To chart the depth of a lake or ocean bottom, a transmitter on a boat will beam energy for a short time (a pulse transmission) straight down into the water. A sensor on the boat detects the returning signal. Using a mathematical formula to account for the presence of water, scientists can then determine the one–way distance of the signal. Other uses of vertical (up and down) sonar include detecting other ships and as an aid in navigating. The energy pulse can also be sent out horizontally through the water, rather than straight down. This is called sidescan sonar, and is useful in determining what lies around a ship. Some systems are so sensitive that they can detect an object in the water that is less than 0.4 inches (1 centimeter) in size. Sidescan sonar is also useful in investigating underwater archaeological sites.

Chapter 12

Economic Uses of Water

AGRICULTURAL WATER USE

The images of seemingly endless crop fields of the American Midwest and the lush San Joaquin Valley of central California are powerful symbols of the agricultural might of the United States. In the past century, the United States has become the greatest producer of food in the world. Water has always been a vital part of agriculture. Just like humans, crops need water to survive and grow. The process where dry land or crops are supplied with water is called irrigation. A century ago, the relatively small fields of a local farmer in many areas of the United States could receive enough moisture from rainwater, along with water that could be diverted from local streams, rivers, and lakes.

The growth of huge corporate farms that are thousands of acres in size has taken the need for water to another scale. For these operations, water needs to be trucked in, pumped up from underground, and obtained from surface water (freshwater located on the surface) sources in large quantities. In modern times, in countries such as the United States and Canada, agriculture is not the largest user of water but is the largest consumer of water. Other activities such as the oil industry use more water than does agriculture. But, in these other industries, much of the water is put back into the ground or surface water after being used. Agriculture consumes water; the water does not go back to the surface or to the groundwater.

USES OF WATER IN AGRICULTURE

There are four main areas of water use in agriculture: growing of crops, supplying drinking water to livestock, cleaning farm buildings and animals, and supplying drinking water for those who work on the farm. The amount of each category varies just as to the type of farm. For example, farms in the eastern part of North America usually receive enough rainfall and water from melting snow to meet most of the water needs. But drier areas, such as the U.S. and Canadian prairies, regions of Mexico, and some mountainous regions of the West do not receive sufficient natural moisture. On these farms, water must be supplied through irrigation.

IRRIGATION

Nearly 60% of the world's freshwater that is used by humans is used for irrigation. Of this water that is applied to crop fields, only about half returns to surface water or groundwater sources. The rest is lost by natural processes such as evaporation (when liquid water changes to water vapour) and transpiration (when water from plant leaves is transformed into water vapour), and accidental occurrences such as leakage from pipes or spillage. There are various methods to irrigate crops. The oldest, "low-tech" way is to flood the field. This flood type of irrigation has been used for centuries and remains popular for crops like rice. Field flooding is very wasteful, since only about half of the water used actually gets to the plant.

The efficiency of flood irrigation can be improved by making the land contoured, such as eliminating small hills and putting steps (terraces) on larger hills to prevent water from flowing over certain portions of the field and gathering in another part of the field. The flooding of a field can also be controlled by releasing water from dams (barriers) alongside the field, adding water to the field only when needed. Water that flows off of a field can be captured in a pond and re-used. A newer and much more efficient technique of water use is called drip irrigation.

In drip irrigation, water runs through pipes that have tiny holes in them. When buried underground, water can ooze out of the pipe into the soil near the roots of the plants. The loss of water is reduced and less water is required to grow the crops. A popular means of irrigation is spraying. Water flows through a tube and is shot out through a system of spray nozzles positioned along the length of the tube. The tube can be fixed in one position or can be moved manually or automatically from place to place.

A visual example of a spray irrigation system is a green circle seen from an airplane passing over farmland. The green circles are crops that are being irrigated by a circular sprayer. Spray irrigation is sometimes wasteful, as water that is sprayed can evaporate or be blown away before hitting the crop. Some farmers now use an irrigation method where water is gently sprayed from pipes that are suspended over the crop. This method allows about 90% of the water to reach the crop. With the knowledge that surface and groundwaters are resources that can be overused, agricultural scientists and modern farmers are paying attention to methods of conserving and re-using water while maintaining the growth of their crops.

AQUACULTURE

Aquaculture, also known as aquafarming, is the farming of aquatic organisms such as fish, crustaceans, molluscs and aquatic plants. Aquaculture involves cultivating freshwater and saltwater populations under controlled conditions, and can be contrasted with commercial fishing, which is the harvesting of wild fish. Mariculture refers to aquaculture practised in marine environments.

The output, as reported, from aquaculture would supply one half of the fish and shellfish that is directly consumed by humans. However, there are issues about the reliability of the reported figures. Further, in current aquaculture practice, products from several pounds of wild fish are used to produce one pound of a piscivorous fish like salmon. Particular kinds of aquaculture include fish farming, shrimp farming, oyster farming, algaculture (such as seaweed farming), and the cultivation of ornamental fish. Particular methods include aquaponics, which integrates fish farming and plant farming.

HISTORY

The indigenous Gunditjmara people in Victoria, Australia may have raised eels as early as 6000 BC. There is evidence that they developed about 100 square kilometres (39 sq mi) of volcanic floodplains in the vicinity of Lake Condah into a complex of channels and dams, that they used woven traps to capture eels, and that capturing and smoking eels supported them year round.

Aquaculture was operating in China circa 2500 BC. When the waters subsided after river floods, some fishes, mainly carp, were trapped in lakes. Early aquaculturists fed their brood using nymphs and silkworm feces, and ate them. A fortunate genetic mutation of carp led to the emergence of goldfish during the Tang Dynasty. Japanese cultivated seaweed by providing bamboo poles and, later, nets and oyster shells to serve as anchoring surfaces for spores.

In central Europe, early Christian monasteries adopted Roman aquacultural practices. Aquaculture spread in Europe during the Middle Ages, since away from the seacoasts and the big rivers, fish were scarce/ expensive. Improvements in transportation during the 19th century made fish easily available and inexpensive, even in inland areas, making aquaculture less popular. Hawaiians constructed oceanic fish ponds. A remarkable example is a fish pond dating from at least 1,000 years ago, at Alekoko. Legend says that it was constructed by the mythical Menehune dwarf people. In 1859 Stephen Ainsworth of West Bloomfield, New York, began experiments with brook trout. By 1864 Seth Green had established a commercial fish hatching operation at Caledonia Springs, near Rochester, New York. By 1866, with the involvement of Dr. W. W. Fletcher of Concord, Massachusetts, artificial fish hatcheries were under way in both Canada and the United States. When the Dildo Island fish hatchery opened in Newfoundland in 1889, it was the largest and most advanced in the world. Californians harvested wild kelp and attempted to manage supply circa 1900, later labeling it a wartime resource.

21ST CENTURY PRACTICE

About 430 (97%) of the species cultured as of 2007 were domesticated during the 20th century, of which an estimated 106 came in the decade to 2007. Given the long-term importance of agriculture, it is interesting to note that to date only 0.08% of known land plant species and 0.0002% of known land animal species have been domesticated, compared with 0.17% of known marine plant species and 0.13% of known marine animal species. Domestication typically involves about a decade of scientific research. Domesticating aquatic species involve fewer risks to humans than land animals, which took a large toll in human lives. Most major human diseases originated in domesticated animals. through diseases such as smallpox and diphtheria, that like most infectious diseases, move to humans from animals. No human pathogens of comparable virulence have yet emerged from marine species.

Harvest stagnation in wild fisheries and overexploitation of popular marine species, combined with a growing demand for high quality protein encourages aquaculturists to domesticate other marine species.

OVER REPORTING

China overwhelmingly dominates the world in reported aquaculture output. They report a total output which is double that of the rest of the world put together. However, there are issues with the accuracy of China's returns. In 2001, the fisheries scientists Reg Watson and Daniel Pauly expressed concerns in a letter to Nature, that China was over reporting its catch from wild fisheries in the 1990s. They said that made it appear that the global catch since 1988 was increasing annually by 300,000 tonnes, whereas it was really shrinking annually by 350,000 tonnes. Watson and Pauly suggested this may be related to China policies where state entities that monitor the economy are also tasked with increasing output. Also, until recently, the promotion of Chinese officials was based on production increases from their own areas.

China disputes this claim. The official Xinhua News Agency quoted Yang Jian, director general of the Agriculture Ministry's Bureau of Fisheries, as saying that China's figures were "basically correct". However, the FAO accepts there are issues with the reliability of China's statistical returns, and currently treats data from China, including the aquaculture data, apart from the rest of the world.

METHODS

Mariculture

Mariculture is the term used for the cultivation of marine organisms in seawater, usually in sheltered coastal waters. In particular, the farming of marine fish is an example of mariculture, and so also is the farming of marine crustaceans, molluscs (such as oysters) and seaweed.

Integrated

Integrated Multi–Trophic Aquaculture (IMTA) is a practice in which the by-products (wastes) from one species are recycled to become inputs (fertilizers, food) for another. Fed aquaculture (for example, fish, shrimp) is combined with inorganic extractive (for example, seaweed) and organic extractive (for example, shellfish) aquaculture to create balanced systems for environmental sustainability (biomitigation), economic stability (product diversification and risk reduction) and social acceptability (better management practices). "Multi–Trophic" refers to the incorporation of species from different trophic or nutritional levels in the same system. This is one potential distinction from the age–old practice of aquatic polyculture, which could simply be the co-culture of different fish species from the same trophic level.

In this case, these organisms may all share the same biological and chemical processes, with few synergistic benefits, which could potentially lead to significant shifts in the ecosystem. Some traditional polyculture systems may, in fact, incorporate a greater diversity of species, occupying several niches, as extensive cultures (low intensity, low management) within the same pond. The "Integrated" in IMTA refers to the more intensive cultivation of the different species in proximity of each other, connected by nutrient and energy transfer through water. Ideally, the biological and chemical processes in an IMTA system should balance. This is achieved through the appropriate selection and proportions of different species providing different ecosystem functions.

The co-cultured species are typically more than just biofilters; they are harvestable crops of commercial value. A working IMTA system can result in greater total production based on mutual benefits to the co-cultured species and improved ecosystem health, even if the production of individual species is lower than in a monoculture over a short term period. Sometimes the term "Integrated Aquaculture" is used to describe the integration of monocultures through water transfer. For all intents and purposes however, the terms "IMTA" and "integrated aquaculture" differ only in their degree of descriptiveness. Aquaponics, fractionated aquaculture, IAAS (integrated agriculture-aquaculture systems), IPUAS (integrated peri-urban-aquaculture systems), and IFAS (integrated fisheries-aquaculture systems) are other variations of the IMTA concept.

SPECIES GROUPS

Fish

The farming of fish is the most common form of aquaculture. It involves raising fish commercially in tanks, ponds, or ocean enclosures, usually for food. A facility that releases juvenile fish into the wild for recreational fishing or to supplement a specie's natural numbers is generally referred to as a fish hatchery. Fish species raised by fish farms include salmon, bigeye tuna, carp, tilapia, catfish and cod. In the Mediterranean, young bluefin tuna are netted at sea and towed slowly towards the shore. They are then interned in offshore pens where they are further grown for the market. In 2009, researchers in Australia managed for the first time to coax tuna (Southern bluefin) to breed in landlocked tanks.

Crustaceans

Commercial shrimp farming began in the 1970s, and production grew steeply thereafter. Global production reached more than 1,600,000 tonnes (1,570,000 LT; 1,760,000 ST) in 2003, representing a value of nearly 9,000 million U.S. dollars. About 75% of farmed shrimp is produced in Asia, in particular in China and Thailand. The other 25% is produced mainly in Latin America, where Brazil is the largest producer. Thailand is the largest exporter. Shrimp farming has changed from its traditional, small-scale form in Southeast Asia into a global industry. Technological advances have led to ever higher densities per unit area, and broodstock is shipped worldwide. Virtually all farmed shrimp are penaeids (*i.e.*, shrimp of the family Penaeidae), and just two species of shrimp, the Pacific white shrimp and the giant tiger prawn, account for about 80% of all farmed shrimp.

These industrial monocultures are very susceptible to disease, which has decimated shrimp populations across entire regions. Increasing ecological problems, repeated disease outbreaks, and pressure and criticism from both NGOs and consumer countries led to changes in the industry in the late 1990s and generally stronger regulations. In 1999, governments, industry representatives, and environmental organizations initiated a programme aimed at developing and promoting more sustainable farming practices. Freshwater prawn farming shares many characteristics with, including many problems with marine shrimp farming. Unique problems are introduced by the developmental life cycle of the main species, the giant river prawn. The global annual production of freshwater prawns (excluding crayfish and crabs) in 2003 was about 280,000 tonnes of which China produced 180,000 tonnes followed by India and Thailand with 35,000 tonnes each. Additionally, China produced about 370,000 tonnes of Chinese river crab.

Molluscs

Abalone farming began in the late 1950s and early 1960s in Japan and China. Since the mid-1990s, this industry has become increasingly successful. Over-fishing and poaching have reduced wild populations to the extent that farmed abalone now supplies most abalone meat.

Echinoderms

Commercially harvested echinoderms include sea cucumbers and sea urchins. In China, sea cucumbers are farmed in artificial ponds as large as 1,000 acres (400 ha).

Algae

Microalgae, also referred to as phytoplankton, microphytes, or planktonic algae constitute the majority of cultivated algae. Macroalgae, commonly known as seaweed, also have many commercial and industrial uses, but due to their size and specific requirements, they are not easily cultivated on a large scale and are most often taken in the wild.

ISSUES

Aquaculture can be more environmentally damaging than exploiting wild fisheries on a local area basis but has considerably less impact on the global environment on a per kg of production basis. Local concerns include waste handling, side-effects of antibiotics, competition between farmed and wild animals, and using other fish to feed more marketable carnivorous fish. However, research and commercial feed improvements during the 1990s and 2000s have lessened many of these.

Fish waste is organic and composed of nutrients necessary in all components of aquatic food webs. In-ocean aquaculture often produces much higher than normal fish waste concentrations. The waste collects on the ocean bottom, damaging or eliminating bottom-dwelling life. Waste can also decrease dissolved oxygen levels in the water column, putting further pressure on wild animals.

Fish Oils

The nutritional value of farm–raised tilapia may be compromised due to the amount of corn included in the feed. Corn contains short chain omega-6 fatty acids that contribute to the buildup of these materials in the fish. "Ratios of long–chain omega-6 to long-chain omega-3, AA to EPA respectively, in tilapia averaged about 11:1, compared to much less than 1:1 (indicating more EPA than AA) in both salmon and trout." The US produced 1.5 million tons of tilapia in 2005, with 2.5 million projected by 2010. Widespread publicity encouraging fish consumption has led to increases in tilapia consumption by those with lower incomes who are trying to eat a balanced diet.

The lower amounts of omega–3 and the higher ratios of omega–6 compounds in farmed tilapia raise questions of the health benefits of consuming this fish. Adequate diets for salmon and other carnivorous fish can be formulated from protein sources such as soy, although soy-based diets may also change in the balance between omega-6 and omega-3 fatty acids.

Impacts on Wild Fish

Salmon farming currently leads to a high demand for wild forage fish. Fish do not actually produce omega–3 fatty acids, but instead accumulate them from either consuming microalgae that produce these fatty acids, as is the case with forage fish like herring and sardines, or, as is the case with fatty predatory fish, like salmon, by eating prey fish that have accumulated omega–3 fatty acids from microalgae. To satisfy this requirement, more than 50 per cent of the world fish oil production is fed to farmed salmon.

In addition, as carnivores, salmon require large nutritional intakes of protein, protein which is often supplied to them in the form of forage fish. Consequently, farmed salmon consume more wild fish than they generate as a final product. To produce one pound of farmed salmon, products from several pounds of wild fish are fed to them. As the salmon farming industry expands, it requires more wild forage fish for feed, at a time when seventy five per cent of the worlds monitored fisheries are already near to or have exceeded their maximum sustainable yield. The industrial scale extraction of wild forage fish for salmon farming then impacts the survivability of the wild predator fish who rely on them for food. Fish can escape from coastal pens, where they can interbreed with their wild counterparts, diluting wild genetic stocks. Escaped fish can become invasive, out competing native species.

Coastal Ecosystems

Aquaculture is becoming a significant threat to coastal ecosystems. About 20 per cent of mangrove forests have been destroyed since 1980, partly due to shrimp farming. An extended cost–benefit analysis of the total economic value of shrimp aquaculture built on mangrove ecosystems found that the external costs were much higher than the external benefits. Over four decades, 269,000 hectares (660,000 acres) of Indonesian mangroves have been converted to shrimp farms. Most of these farms are abandoned within a decade because of the toxin build-up and nutrient loss.

Salmon farms are typically sited in pristine coastal ecosystems which they then pollute. A farm with 200,000 salmon discharges more fecal waste than a city of 60,000 people. This waste is discharged directly into the surrounding aquatic environment, untreated, often containing antibiotics and pesticides." There is also an accumulation of heavy metals on the benthos (seafloor) near the salmon farms, particularly copper and zinc.

Genetic Modification

Salmon have been genetically modified for faster growth, although they are not approved for commercial use, in the face of opposition. One study, in a laboratory setting, found that modified salmon mixed with their wild relatives were aggressive in competing, but ultimately failed.

PROSPECTS

Global wild fisheries are in decline, with valuable habitat such as estuaries in critical condition. The aquaculture or farming of piscivorous fish, like salmon, does not help the problem because they need to eat products from other fish, such as fish meal and fish oil. Studies have shown that salmon farming has major negative impacts on wild salmon, as well as the forage fish that need to be caught to feed them. Fish that are higher on the food chain are less efficient sources of food energy.

Apart from fish and shrimp, some aquaculture undertakings, such as seaweed and filter–feeding bivalve mollusks like oysters, clams, mussels and scallops, are relatively benign and even environmentally restorative. Filter–feeders filter pollutants as well as nutrients from the water, improving water quality. Seaweeds extract nutrients such as inorganic nitrogen and phosphorus directly from the water, and filter-feeding mollusks can extract nutrients as they feed on particulates, such as phytoplankton and detritus. Some profitable aquaculture cooperatives promote sustainable practices. New methods lessen the risk of biological and chemical pollution through minimizing fish stress, fallowing netpens, and applying Integrated Pest Management. Vaccines are being used more and more to reduce antibiotic use for disease control. Onshore recirculating aquaculture systems, facilities using polyculture techniques, and properly sited facilities (for example, offshore areas with strong currents) are examples of ways to manage negative environmental effects. Recirculating aquaculture systems (RAS) recycle water by circulating it through filters to remove fish waste and food and then recirculating it back into the tanks. This saves water and the waste gathered can be used in compost or, in some cases, could even be treated and used on land.

While RAS was developed with freshwater fish in mind, scientist associated with the Agricultural Research Service have found a way to rear saltwater fish using RAS in low-salinity waters. Although saltwater fish are raised in off-shore cages or caught with nets in water that typically has a salinity of 35 parts per thousand (ppt), scientists were able to produce healthy pompano, a saltwater fish, in tanks with a salinity of only 5 ppt. Commercializing low–salinity RAS are predicted to have positive environmental and economical effects. Unwanted nutrients from the fish food would not be added to the ocean and the risk of transmitting diseases between wild and farm-raised fish would greatly be reduced. The price of expensive saltwater fish, such as the pompano and combia used in the experiments, would be reduced. However, before any of this can be done researchers must study every aspect of the fish's lifecycle, including the amount of ammonia and nitrate the fish will tolerate in the water, what to feed the fish during each stage of its lifecycle, the stocking rate that will produce the healthiest fish, etc.

THE AQUACULTURE AND MARICULTURE INDUSTRY

The combined industry of aquaculture and mariculture represents one of the fastest growing economic areas in the world. The United Nations Food and Agricultural Organization (FAO), aquaculture and mariculture have increased by nearly 10% per year since 1970. China has become a world leader in both aquaculture and mariculture. Between 1970 and 2000, China had an annual growth rate of 11.5% in aquaculture and 14% in mariculture. In China, farms produce three times more fish and shellfish for human consumption than fishermen catch. FAO estimates that aquaculture and mariculture revenues were $56.5 billion in 2000, half of which was generated by China. The crops that generated the largest amounts of revenue were the finfish (catfish, salmon, and talapia), which accounted for about half the world's aquaculture and mariculture production. The other two large crops are mollusks (mostly oysters; mollusks are soft bodied aquatic animals generally having a shell) and plants (mostly kelp). Excluding China, FAO estimates that about one-fifth of the world's fish and shellfish supply comes from aquaculture and mariculture.

MAJOR AQUACULTURE AND MARICULTURE CROPS

A large variety of animals and plants are grown by aquaculture and mariculture. Animals are grown for human consumption, for consumption by other animals, for use in aquaria, for stocking of natural waters and as research animals. Catfish are the most important aquaculture crop in the United States with an estimated 750 million fish grown per year. More than half of these are produced in Mississippi. The next most important fish grown as a crop are the salmon, which are usually raised in pens in bays in the ocean.

In 1999, the world mariculture industry grew by more than 1 million tons of salmon. Norway leads the world in salmon farming, followed by Chile. Tilapia is a finfish with mild, tender meat that is becoming an increasingly important mariculture crop. Shellfish are also grown on farms.

The most important crops are oysters, which are grown both for human consumption and for the pearls that they generate. Shrimp, clams, mussels, and abalone are also farmed in marine waters. In freshwaters, the largest shellfish crop is crawfish, followed by shrimp. Many species of aquatic plants are raised on farms.

The major saltwater food crop is kelp, also called wakame in Japan, which is a type of brown algae. This brown algae is also harvested to make agar, a thickening agent used in salad dressings, paint and ink. A red algae, called purple laver or nori, is used in many types of sushi. The most commonly grown freshwater plants for human consumption are watercress and Chinese water chestnuts. Other algae are raised as animal feeds and as mulches and fertilizers (products used by gardeners). Water hyacinth, which efficiently removes excess pollutants from water, is grown for use in wastewater treatment plants.

DRAWBACKS TO AQUACULTURE

Although aquaculture and mariculture have the potential to make great contributions to the world's food supply, there are some drawbacks to the growth of these industries. In some developing countries, natural habitats are destroyed in order to build pens for crops. For example, shrimp farmers often cut down large areas of trees called mangroves. These trees have the ability to live in salt water. The roots of these trees serve important purposes in the tropical marine ecosystem (community of organisms and their environment).

They provide habitats for a variety of juvenile fish and invertebrates (animals without a backbone) that hide from predators in their crevasses. They also prevent erosion (wearing away of soil) during floods and storms, by holding soil in place. Finally they use some pollutants, like nitrogen and phosphorus, that are generated by aquatic organisms as they grow. Pollution is a second problem that aquacultural and maricultural farmers have to confront. Having a large number of animals concentrated in a small area produces much waste. These wastes can stimulate the growth of microorganisms such as phytoplankton and bacteria, which harm animals that live nearby.

Some newer technologies involve growing animals in enclosed tanks where water is cleaned and recycled rather than simply released into the environment.

Although not yet financially practical, these techniques may represent a cleaner way to farm fish and shellfish in the future. Finally, the economics of mariculture and aquaculture play a large role in the expansion of these industries. Building and running a facility that grows freshwater or marine organisms is not always profitable. Just as in farming on land, animals and plants that are grown on farms in the water must have traits that allow for domestication. For example, animals that exhibit territoriality or aggressive behaviours are not good candidates for aquaculture or mariculture. Disease can ruin crops, and expensive antibiotics may need to be used to keep animals healthy. Controlling the reproductive rate of farmed animals is extremely important. If animals reproduce too fast, some can become stunted and unable to be sold. If animals reproduce to slowly, costs can overcome profits.

COMMERCIAL AND INDUSTRIAL USES OF WATER

Besides being vital for human survival, water is also necessary in commerce and in industry. Commercial operations are those that generally do not manufacture a product, but provide a service, such as hospitals, restaurants, and schools. Industry usually involves manufacturing a product. In industry, water helps keep machinery needed for the making of products running smoothly and efficiently. Water can also be a vital part of the product, such as in sports drinks or soft drinks. In the United States, the total amount of fresh and salt water used every day by industry is nearly 410 billion gallons. To illustrate such a huge number, think of that amount of water in terms of weight. A gallon of water weighs a little over 8 pounds (3.6 kilograms). The daily water usage in the United States totals almost 3.5 trillion pounds (1.6 trillion kilograms), about the same as 200 million 200-pound (90.7 kilogram) people!

COMMERCIAL WATER USE

In modern day, water is essential to people's daily lives. Without water, restaurants could not supply meals or even clean up after the meals, cars would go unwashed, and fires could be disastrous, with no means of dousing the blaze. Green parks, recreational fields, and golf courses rely on water to keep the grass and soil moist and healthy. Roadways would become dirty and grimy in the absence of any water-based cleaning programme. Offices would grind to a halt with no water available for drinking and bathrooms, and office buildings, stores, and public and private centers would also be dark places without the water necessary to generate electricity for lighting. The water for these and other commercial uses comes from the surface and from underground (groundwater) sources. The extent to which a community uses a surface or a groundwater source depends on which source is more abundant in the particular area. For example, the drier central portions of the United States and Canada do not have as much surface water as the eastern and western coasts. In the prairies, wells that reach down to tap underground water sources are more common than in coastal regions such as California.

Some of the water that is used for commercial purposes can be reused. The water used in a car wash is one example. Another example is the water that is applied to golf courses. Surface water that is obtained from a lagoon (shallow body of water cut off from a larger body) can be suitable for keeping a golf course lush and green. Other commercial water uses, such as drinking water, demand water that is free of chemicals and harmful microorganisms. Fresh and salt water is home to many living creatures that are harvested by humans. Whether for sport or as a business, fisheries are completely dependent on water.

COMMERCIAL FISHING

Both fresh and salt waters have long supported commercial fisheries in North America. Rivers on the eastern and western portions of the United States and Canada once were the basis of a productive commercial salmon fishery. However, in the past few decades, the number of salmon that return from the ocean to their river homes has been steadily declining. One reason is over–fishing; the catching of more fish than is produced. But other factors may be playing a role. The decline in water quality is one suspected factor.

A century ago, the Grand Banks off the coast of Newfoundland, Canada was the world's premier cod fishing ground. Nets would strain under the weight of untold numbers of cod, often the source of fish used in preparing fish sticks and the traditional 'fish and chips'. However, over–fishing by local fishers and by large factory trawlers have greatly reduced the cod stocks. In the 1990s, the government of Canada ended fishing for cod off the east coast of Canada so that the numbers of cod could again increase in their natural habitat. A decade later, the numbers of cod fish had not recovered, and the cod fishery industry in the area was, at least temporarily, lost.

INDUSTRIAL WATER USE

Industries require large supplies of water. Machinery relies on water to cool it to a temperature that allows the manufacturing process to keep going. The mining industry needs water to wash off the material that has been brought up from underground in order to sort out the genuine product from other particles. Water is also used to clean machinery, buildings, and even, in the case of the meat processing industry, the carcasses of the cattle, pigs, and other animals that will be trimmed into the items found in the meat part of the local supermarket.

In oil producing regions, vast amounts of water are used. As oil wells get older and the underlying oil reserve is tapped, it becomes more difficult to pump out oil that is hiding in cracks in the rock deep underground. One way of getting this oil is to pump water down into the oil formation. The water can make its way into cracks and crevasses and push the oil out in front of it. The oil is then pumped up using another well. Without this industrial use of water, oil and gasoline would be more scarce and more expensive.

The generation of electrical power also makes use of water, to cool equipment and to push the turbines that are the heart of the process that produces electricity. Turbines are turning wheels with buckets, paddles or blades that turn as water moves by converting the energy of moving water to mechanical power. The U.S. Geological Survey, in the year 2000 about 20 billion gallons (76 billion liters) of water were used each day to make electricity. This represents about 53% of all water use in the country. The vast amount of this water comes from surface water sources. Much of this water is eventually returned to the environment for reuse. In contrast, water that is used to irrigate (water) crops usually cannot be recovered after being applied to the crops.

Another big user of water is the pulp and paper industry; millions of gallons of water is used in the various processes that turn a log into a piece of paper. Clean water is also required for papermaking. If the water contains too many solid particles, the paper will not be smooth and the paper-making machinery could be damaged.

For other industries, water may not be a key part of the actual making of the product, but it is nevertheless, required. In the steel making industry, water is needed for cooling equipment. Like in the oil industry, this use of water does not require water of the same quality as drinking water. Care must be taken in disposing of the water, however. For example, water cannot be disposed of immediately after it is used to cool equipment, as the high temperature of the water would damage fish and other life in the natural environment. This water is usually cooled in a holding pond or container before being released.

ECONOMIC USES OF GROUNDWATER

Groundwater is one of human's most valuable natural resources. Groundwater is the water contained in the rock and soil layers beneath Earth's surface, and it makes up most of Earth's supply of fresh, liquid water. (The oceans and ice in the North and South Poles contain 99% of Earth's total water supply. Groundwater accounts for almost all of the remaining 1%.) Throughout history, humans have settled in areas with plentiful and pure groundwater, and have fought to own and protect wells and springs. Today, human water needs in many arid (dry) or heavily populated regions far exceed surface water supplies. Earth's rapidly–growing human population is becoming increasingly reliant on groundwater. Groundwater fills wells and city water supplies. Groundwater irrigates (waters) crops, feeds livestock, and produces farm-raised fish. Groundwater is used to cool nuclear reactors that generate electricity, mix concrete, and manufacture millions of consumer products. In short, groundwater plays a vital role in almost every facet of people's lives, from drinking water, foods, and products people buy to roads and the buildings in which people live and work.

GROUNDWATER RESERVOIRS: AQUIFERS

Water enters underground reservoirs by soaking in through soils, stream beds, and ponds in areas termed recharge zones. Water flows, often very slowly, through interconnected pore (tiny opening) spaces and then remerges onto the land surface at natural discharge points called springs and seeps. When discharge from natural springs and/or human wells exceeds the rate of recharge, the groundwater level falls, shallow wells and springs dry, and eventually, the reservoir empties. Many groundwater reservoirs, particularly those beneath arid deserts and semi-arid grasslands, filled with water many centuries ago when regional climate was wetter.

Groundwater reservoirs that yield water for human use are called aquifers. In part, human economics determine which water-bearing units are exploited as aquifers. In regions where clean surface water is plentiful and inexpensive, groundwater may go unused. In arid regions with scarce or polluted surface water, and in places where human water needs exceed the water supply in streams and lakes, groundwater extraction and purification become economically worthwhile. When conditions change, as during periods of drought (prolonged dry weather) or increased population growth, new groundwater supplies are tapped, thereby elevating them to aquifer status.

WELLS

In addition to collecting groundwater from springs, humans extract water from aquifers by digging or drilling wells that extend from the ground surface to the water table, the level below which all the empty space in the rocks and soil are completely full of water (saturated). When a well reaches the water table, groundwater fills the hole like water filling a hole dug in beach sand.

In wet regions, the water table may lie only a few feet (meters) below the surface. In arid regions, groundwater wells are often hundreds of feet (meters) deep. Most wells require a bucket system or pump to raise the water to the land surface. Some aquifers, however, contain pressurized ground-water that flows to the land surface on its own. Such free-flowing groundwater discharges are called artesian wells and springs. There are a number of ways to construct wells.

Some common types of wells are hand-dug, driven, and drilled wells:

- *Hand-dug wells*: Historically, wells were dug into soil and even rock by hand. Well diggers with shovels or picks would dig a hole below the water table by bailing water faster than it flowed into the well. Once a well was complete, its builders reinforced its walls and fitted it with a bucket system or pump to bring water to the surface. Hand-dug wells are still regularly constructed in many parts of the world, but they are uncommon in developed nations like the United States.
- *Driven wells:* Driven wells are constructed by forcing or hammering a narrow pipe into soft ground. These wells are inexpensive and can reach very deep aquifers, but can only be

used in areas that have loose soil or sediment (particles of sand, gravel, and silt).

- *Drilled wells*: Today, most water wells are drilled using rotary (turning) or percussion (hammering) machines that are mounted on large trucks. Drilled wells that penetrate loose material are lined with plastic or metal pipe called casing, which keeps the sides of the hole from collapsing. An electric pump is placed at the bottom of the well to bring the water to the surface.

HISTORICAL GROUNDWATER USE

Humans in arid regions such as northern Africa, the Middle East, and central Asia have relied on groundwater to provide drinking water and irrigate crops for thousands of years.

DOWSING

Groundwater can be hard to find. Today, hydrogeologists use scientific methods to locate aquifers and productive water wells. Aquifers can be extremely complex and groundwater flow patterns difficult to predict, and it is not uncommon for hydrogeologists to drill dry wells. In the past, water–seekers consulted with spiritually-guided water prospectors called dowsers or water witches.

Dowsers profess special powers that allow them to sense or divine water beneath the ground. While a hydrogeologist searches for groundwater by taking measurements, making observations, and drawing maps, a dowser strolled across the client's land holding a metal or wooden Y-or L–shaped divining rod or a pendulum. When water was present, the rod or pendulum was said to be attracted to the water beneath. Some dowsers even claimed that their divining rods would locate groundwater on maps of the land surface.

The practice of dowsing has its roots in ancient Egypt and China, and its first published reference appeared in 1430. Early dowsers and water witches probably relied on a combination of spiritual guidance and astute scientific observations of groundwater discharge features such as springs, seeps, and vegetation patterns to locate underground water. Like witch doctors in ancient cultures, dowsers used all their available tools, including scientific knowledge, to help their clients solve problems. As such, modern hydrogeologists are perhaps their closest professional descendants.

Modern-day dowsers claim to find water entirely with their spiritually enhanced extrasensory powers. They assert that ground-water has a magnetic field that pulls on their dowsing rods, a theory that has never been scientifically proven. Dowsers do successfully locate groundwater, but without clues to the local groundwater system, their results are statistically no better than random well drilling.

Archeologists have discovered the remains of hand-dug wells, oasis (areas in the desert with a source of water) settlements, and groundwater distribution systems throughout the ancient world. Humans have drunk from groundwater springs at the Oasis of Bahariya in the Sahara desert of western Egypt since the early stone age (Paleolithic Age) more than one million years ago.

Knowledge of groundwater supplies and extraction technologies was critical information for ancient desert empires such as Mesopotamia, Sumeria, and Egypt. Nomads (wandering tribes) in the Saharan and Arabian deserts relied upon fiercely guarded knowledge of groundwater springs and seeps to survive. Egyptians, Mesopotamians, and Chinese who first practiced agriculture dug wells to provide irrigation for water–intensive crops such as rice and cotton, and drinking water for permanent settlements. Groundwater availability affected patterns of conquest and settlement in Greek and Roman Empires. European explorers sought groundwater and white settlers excavated wells that supported settlement and farming throughout North and South America.

MODERN GROUNDWATER USE

Today, people use groundwater for agricultural irrigation, industrial processes, municipal (city) and residential (home) water supplies. In the United States, groundwater accounted for about one quarter (26%) of total water use in the year 2000. (Surface freshwater made up the other 74%.) Groundwater use, however, varies by location, and many U.S. residents and industries depend almost completely upon water drawn from regional aquifers. More than one-third of U.S.' 100 largest cities, including Miami Beach, San Antonio, Memphis, Honolulu, and Tucson get all their water from aquifers. Almost all rural households (98%) draw their water from private wells.

Farmers and ranchers in Midwestern and Western states make heavy use of groundwater for irrigation of crops. In the eastern and southern U.S., most drinking and agricultural water comes from lakes and streams, but industries use vast quantities of groundwater for such activities as refining petroleum, aluminum, and other ores; manufacturing steel and chemicals; producing plastics; and mining. Aquaculture (fish farming) is big business and a significant groundwater consumer in Southeastern states like Mississippi, Alabama, and Louisiana. In the United States, groundwater is particularly important in arid and semi-arid agricultural states in the western half of the nation. Heavily agricultural states such as California, Oregon, and Texas use large quantities of groundwater for irrigation of food crops. The livestock industry also draws heavily upon groundwater supplies in states such as Texas, Nebraska, Kansas, and Colorado. Water drawn from wells not only fills watering troughs, but also irrigates vast tracts of midwestern cropland that produce material for cattle, poultry, pig, and fish feed. Meat processing plants also require water. (It takes about 13 gallons [49 liters] of water to produce 1 pound (0.45 kilogram) of beef, and about 4 gallons [15 liters] of water go into producing 1 gallon [3.8 liters] of milk!)

MINERALS AND MINING

Minerals are defined as naturally occurring solids found in the earth that are composed of matter other than plants or animals. Ore is a naturally occurring source of minerals, such as a rock. A mineral can be composed from one element, such as diamond, which contains only carbon, or several elements, such as quartz, which contains silicon and oxygen. An element is a substance that cannot be divided by ordinary chemical means. Even ice is considered a mineral. Minerals are found everywhere on Earth, from the bottom of the ocean to the highest mountains. Mineral deposits are frequently located underground, and thus they must be mined.

South Africa and Russia hold the largest amount of minerals in the world. Minerals are vital to people's lives, and many of these minerals are critical to countries' industries and economies. The United States is relatively poor in critical minerals, including platinum, cobalt, and gold, but there are sand deposits of titanium ore in Florida and the Pacific Northwest. In the central United States, minerals that contain lead and gallium (used in computer chips) are abundant, and iron ore is found in the states near the Great Lakes. Most of the diamonds are mined in Africa, as is gold, although gold is found in many other locations as well.

Importance of Minerals

Minerals are essential in every aspect of life for humans. Humans need to ingest minerals in order for our bodies to function normally. Most of these required minerals come from the foods people eat. Gold and silver have been valued by civilization since ancient times. Metals became useful for purposes other than money during the Bronze Age, when weapons and tools could be made from metal as people became more educated with how to process the minerals and extract (remove) metals. As metals were not evenly dispersed around the globe, the more powerful nations became that way through military might from weapons made from metal.

Industry also depends upon minerals. Without aluminum (mined mostly in Jamaica), people could not manufacture airplanes, much less soda-pop cans. Titanium is used in the aerospace industry for constructing spacecraft and in medicine for the construction of artificial limbs and joints. Copper is required to make wire that carries electricity to homes and factories. Mineral reserves are of great importance in the marketplace and it is not uncommon to stockpile (save) certain metals extracted from mineral ores. Because large mineral deposits are located in regions of the world that are, at times, politically and economically unstable, the supply of critical minerals is not guaranteed.

The United States stockpiles metals such as platinum, palladium, cobalt, chromium, manganese, and vanadium. These metals are used in the high-technology industries and the military. Chromium, for instance, is used to produce stainless steel. Vanadium is used, along with aluminum, to make forms of titanium that are resistant to fracture (breaking), enabling the manufacture of jet planes that can withstand extreme conditions. Platinum is used in removing the impurities from oil. Palladium is used in the exhaust systems of automobiles to reduce the amount of pollutants. It is advantageous for a highly-industrialized country such as the United States to have these resources at hand, and to purchase reserves when prices are low.

WATER–LAID ORES AND MINERALS

The formation of mineral deposits always involves water. Water is part of the chemical processes of mineral formation and also changes the mineral content of rocks by dissolving certain elements in the ore and transporting them elsewhere. Heat is another ingredient in the formation of many mineral deposits.

Hydrothermal Deposits

Many metalbearing ores are found in veins (cracks in rock filled with minerals) that cut through surrounding rock. In these cases, very hot water reacted with elements and other minerals in the rock, and burst through the layers of rock where there was a weakness. These mineral deposits are called hydrothermal deposits. Hydrothermal deposits form gold, silver, and the platinum–group metals, which are commonly found in veins. The metals themselves are hosted in a vein that is often quartz. Miners follow the vein, extract the ore, and remove the host rock to extract the metals contained within. Mining minerals from veins is an expensive process that is seldom used today.

When hot water flows through porous rock (rock with many small holes), the rock can become a host to a kind of deposit known as porphyry. The host rock containing a porphyry deposit is filled with small veins of (usually) quartz that contain the minerals. Although the mineral content is low, porphyry deposits are large, and most of the copper that is mined comes from these unique deposits. Fool's gold (iron sulfide) is often found in porphyry deposits as well.

Volcanogenic Deposits

Volcanogenic deposits form when magma (molten or melted rock beneath the Earth's surface) from miles (kilometers) down in the earth is transported to the surface in volcanoes. There are two kinds of volcanic eruptions that most concern scientists. One of these brings iron–rich magma to the surface (such as in Hawaii) and one brings explosive plumes of ash and magma to the surface (such as in Mt. Saint Helens). Elements in the water that is in contact with the magma, along with the rock through which the water travels on its way to the surface, determine the kinds of minerals found in volcanogenic deposits.

For the most part, lead and iron ore are found in volcanogenic deposits, along with smaller amounts of cadmium, antimony, and copper. On the floor of the ocean, the same kinds of deposits can form where magma seeps through a crack in the seabed. These features are called black smokers, because the iron-rich magma makes the plume appear like black smoke. The mineral deposits collect near the smoker until the hole becomes plugged or the magma is diverted elsewhere.

MINING FOR MINERALS

The process of mining for minerals begins after a mineral deposit has been identified.

The common types of mines used to excavate minerals are open-pit mines, strip mines, and stope and adit mines:

- *Open-pit mines*: These mines are large craters dug into the earth to extract ore that is near the surface. Open pit mines are usually associated with porphyry deposits, and minerals such as galena (which contains lead), chalcopyrite (which contains copper), and sphalerite (which contains zinc) are commonly mined at open pits. The open pit is excavated using very powerful and large earth-moving equipment, and processing of the ore (crushing, grinding, partial refinement) is often done near the open pit. Open pits are less environmentally friendly than conventional mines because any native vegetation in the area is lost, and abandoned open pit mines eventually pool waters that are frequently contaminated. Open-pit mines are used for copper in Arizona.
- *Strip mines:* These mines are large swaths dug through ore-rich zones (common for coal). Most mines are located below the surface of the Earth, and require drilling shafts to enable workers to reach the ore below and transport it to the surface. Strip mines are used for coal in many states and other areas where the ore is buried deeply, as in Montana, where platinum and palladium ore is extracted (removed).
- *Stope and adit mines*: These mines are bored into the ground. Shafts are bored vertically, and horizontal offshoots (adits) from the shafts that lead into ore-containing portions of the subsurface are dug. Large mining vehicles that crush the rock move along veins

of ore, and this byproduct is transported to the surface by rail or cable lifts. Temperatures in conventional mines are high, and the conditions are dangerous—cave-ins occur often—so safety measures are very strict. The shafts and other structures are reinforced with concrete or metal supports.

OTHER ORE AND MINERAL DEPOSITS

Placer deposits are concentrated metals that have been transported to streambeds (the channel through which the stream runs) or beaches. The most famous placer deposits are gold nuggets, although silver is sometimes found as well. Placer metals must be resistant to water, or they would dissolve again.

The usual way to extract the placer deposits is to scoop sediment (particles of rock, clay, or silt) from the stream and sift it, leaving the larger rock behind and making the gold easier to spot. California, Alaska, Oregon, and Idaho have all had significant placer deposits of gold. Evaporite deposits form by evaporation. As waters that contain dissolved mineral species evaporate, the minerals remain in solid form. Minerals found in evaporites include potassium chloride, sodium chloride (halite or table salt), calcium sulfate (gypsum), barium sulfate (barite), and potassium nitrate (saltpeter).

Most of these deposits are near the surface and are scooped from the ground with large earth-moving equipment. Gypsum is used to make sheetrock, which is used to construct the walls of homes and buildings. It is fire-retardant and easily cast into shapes. Barium sulfate is used as drilling mud in oil-producing wells because it is very dense, and prevents oil gushers from erupting as the drill is lowered. Barium itself has use in medicine. Saltpeter is used as an ingredient in gunpowder and fertilizers. In the oceans, concentrations of dissolved mineral ingredients are very high. Evaporites of the chloride type are most common, and they occur in areas where seawater collects in shallow areas that are confined.

Thus, a pool of salt-rich water forms, evaporation speeds up the process, and salt deposits result. Another widespread mineral formed in marine environments is limestone, or calcium carbonate. This mineral is formed in the same way as the chloride salts, but also includes another source, organisms whose skeletons are made from calcium carbonate. These organisms die and collect on the sea floor, where they add to the content of calcium carbonate. Limestone is used in constructing buildings and as an ingredient in concrete.

MUNICIPAL WATER USE

Many people live in municipalities (cities, towns, and villages with services such as water treatment, police, and fire departments). One benefit of living in a municipality is that potable water (water safe to drink) is usually available at any time by turning on the tap. Part of the responsibility of citizens and municipal officials however, is to manage and protect the local water supply.

If municipal water becomes contaminated, the result can be far-reaching and rapid. Bacteria and viruses in water can spread throughout the underground reservoir of water (the aquifer) or throughout the miles of pipelines that carries water to houses in towns and cities. As well, non-living pollutants such as oil, gasoline and sediment can spread contaminate water. The results of such contamination can be disastrous. In the summer of 2000, the municipal water supply of Walkerton, a town in the Canadian province of Ontario, became contaminated with a certain type of bacteria called Escherichia coli (or E. coli for short).

This type of E. coli caused a serious illness in over a thousand people who drank the town water, and killed seven people. In addition to protecting water for human use, water management also benefits the environment. Polluted water is bad for the many creatures that live in the water and depend on the watercourse in their lives.

PROTECTING MUNICIPAL DRINKING WATER

People who live in a municipality usually have to pay money to the local government for their water. Municipal drinking water may come from wells, which pump water that is located underneath the ground (groundwater) into an underground reservoir. Groundwater is often free of contaminating chemicals and microorganisms because the contaminants are filtered out of the water as it moves downward into the ground, yet the water still must be tested to ensure the absence of contaminants. Once tested, the water is pumped through pipes that run underneath the streets of the municipality. The pipes lead to houses, fire stations, other offices, swimming pools, and the many other places where water is used.

Some municipal drinking water is obtained from streams, rivers, and lakes. This water is called surface water. Surface water must be treated before it can be used for drinking, because there is a greater chance that harmful chemicals or microorganisms could have washed into surface water. Municipalities that rely on surface water will pump the water from the river or lake to a water treatment plant. The water will be cleaned in a series of steps and tested to ensure that it is safe to drink. The treated water can then be pumped to storage tanks until it is used.

In many municipalities, one of the treatment steps is the addition of a chemical called chlorine. This chemical kills bacteria such as E. coli, and so is an effective and inexpensive way to keep the water free from bacteria. The amount of chlorine that is added to water needs to be monitored, since too much chlorine can create taste and odour problems. Furthermore, excess chlorine can combine with organic material in the water (like rotting leaves) to form a compound called trihalomethane that has been linked to the development of cancer in humans.

Some municipalities have installed other means of killing or removing microorganisms. These include the use of ultraviolet (UV) light, which kills microorganisms by breaking apart their genetic material. Another technique is to pass the water through a series of filters (a material that has very tiny holes in it). While the water molecules can pass through these holes, the holes are too small to allow most microorganisms to pass through. After water is used, the chemicals, sewage, and other contaminants must be removed before the water can be reused or returned to a reservoir. In order to accomplish this, wastewater leaves buildings through sewage pipes that lead to the treatment facility, and the treatment cycle begins again.

OTHER USES OF MUNICIPAL WATER

Many municipalities provide golf courses, swimming pools, sports fields, gardens and parks for their residents. All of these places require water. Fire fighters need easy access to water, which is provided by a system of pipes that lead to fire hydrants positioned throughout the municipality. The fire fighters hook their hoses up to the high–pressure hydrants to fight fires with water. Many municipalities have cleaning programmes, where roads and other surfaces are cleared of dirt and other material that piles up during the winter or a dusty, dry summer. Water is sprayed from vehicles that move slowly along the road, to wash away the accumulated grime.

SAFEGUARDING MUNICIPAL WATER

Many municipalities have laws that restrict people from throwing garbage into streams, rivers, and lakes, and to stop the dumping of liquids such as oil and gasoline into the water. Preserving undeveloped areas of riverbanks or lakes also encourages growth of natural vegetation that benefits the water supply. By leaving grass, trees, and other vegetation alongside a stream or river, it makes it more difficult for toxic (poisonous) material to wash into the water. Along with this benefit, the natural stream or river bank often becomes an attractive spot to walk, bike ride, and picnic.

PETROLEUM EXPLORATION AND RECOVERY

Petroleum, also called crude oil, is a thick, yellowish black substance that contains a mixture of solid, liquid, and gaseous chemicals called hydrocarbons. Since its discovery as an energy source in the mid–1800s, petroleum has become one of human's most valuable natural resources. Petroleum is arguably the single–most important product in the modern global economy. Hydrocarbons separated (refined) from crude oil provide fuels and products that affect every facet of life in industrialized nations like the United States. Natural gas and propane are gaseous hydrocarbons that are used to heat homes and fuel stoves. Natural gas actually exists as a gas in underground reservoirs (underground rock formations containing oil or natural gas) and is not refined from crude oil, but it is still considered a petroleum product.

The liquid portion of petroleum becomes such essential products as home heating oil, automobile gasoline, lubricating oil for engines and machinery, and fuel for electrical power plants. Asphalt road surfaces, lubricating oils for machinery, and furniture wax are all composed of semi–solid hydrocarbons. Petroleum products are the building blocks of plastics. The hydrocarbon gas ethylene is even used to help ripen fruits and vegetables! Oil and water don't mix, but these two essential natural resources do have a lot in common.

Naturally occurring petroleum forms from the chemical remains of organisms that lived and died in ancient seas. Petroleum collects in deeply buried rock layers called sedimentary rock that are, more often than not, the geologic remains of water–laid deposits like beds of sand or coral reefs on the sea floor. (Sediment is particles of rock sand or silt.) Petroleum reservoirs are similar to groundwater reservoirs, the water below Earth's surface. Petroleum scientists use many of the same skills and methods to find and extract oil that groundwater scientists use to prospect for underground water. Finally, many untapped petroleum deposits are buried beneath the seafloor, and much of our present and future petroleum supply lies offshore.

FORMATION OF PETROLEUM DEPOSITS

Hydrocarbons are organic (part of or from living organisms) chemicals; they form by the breakdown of microscopic organisms that were once living. (Biological organisms combine the chemical elements hydrogen and carbon during their lives, thus the term hydrocarbon.) Microscopic plants and animals that collect on the seafloor provide the organic material that eventually becomes petroleum. Dark, smelly, organic–rich mud collects where a heavy rain of dead plants and animals accumulates in an oxygen–poor seafloor environment. (Oxygen–rich waters support animals and bacteria that eat or decompose the organic material.)

Unfortunately for petroleum users, oil doesn't simply collect in underground puddles. Organic material must undergo a series of complex changes over many thousands of years before it becomes petroleum that can be extracted for human use. First, organic–rich mud layers become source rocks like shale and mudstone when they are buried beneath thick stacks of newer sediment. Heat, pressure, and bacteria within source rocks chemically transform plant and animal parts into hydrocarbons. Next, pressure squeezes the petroleum out of the source rocks and it migrates (moves) to reservoir rocks where it fills tiny openings, fractures, and cavities.

Productive petroleum reservoir rocks, such as sandstone and some types of limestone, are like swiss cheese. They have lots of empty space between mineral grains (high porosity) and the space is interconnected so petroleum can flow easily through the rock (high permeability). Finally, exploitable petroleum reservoirs are typically contained beneath layers of relatively impermeable rock called cap rock that keeps oil from escaping onto the land surface. Geologic structures like faults (fracture or break along which rocks slip) and folds (bends in rock layers due to the stress imposed by the movement of Earth's tectonic plates) trap petroleum from the sides. Petroleum geologists use maps and rock samples from the land surface as well as images of the subsurface to search for deeply–buried oil and natural gas deposits.

HISTORY OF THE MODERN PETROLEUM INDUSTRY

Petroleum has been known to mankind for thousands of years. Ancient Mesopotamians, Egyptians, Greeks, and Romans collected the sticky black substance called bitumen from tar pits and seeps (an area where groundwater or oil slowly rises to the surface) and used it to pave roads, heal wounds, waterproof buildings and, to a limited extent, for lighting. The modern quest for petroleum began in the mid-1800s when rapid industrialization and population growth prompted a search for a new type fuel that could replace coal in furnaces and whale oil in lamps. (Coal, like petroleum, is an organic fossil fuel that must be mined from underground. Coal beds are the fossilized remains of land plants that grew in ancient swamps.)

North American prospectors seeking inexpensive lamp oil first struck oil in Ontario, Canada in 1858. They made the first major petroleum discovery one year later in Titusville, Pennsylvania in 1859. John D. Rockefeller (1839–1937), a businessman who saw economic potential in Pennsylvania oil, founded the Standard Oil Company in 1865, the same year the American Civil War (1861–65) ended. (Rockefeller went on to become the world's first billionaire. Most major U.S. energy companies, including Exxon-Mobil, Chevron–Texaco, Conoco—Phillips, and the American portion of British Petroleum—AMOCO were originally part of Standard Oil.)

By 1901, when a gusher (fountain of pressurized petroleum) shot up into the air above the famous Spindletop well near Beaumont Texas, the American oil industry was positioned to capitalize on an invention that has changed the face of modern civilization, the internal combustion engine. An internal combustion engine takes the energy in fuel and combusts (burns) it inside the engine to produce motion. Petroleum releases heat energy and gaseous carbon dioxide when combusted. Like wood, coal and other organic fuels, petroleum can be used to heat homes, cook food, and power steam engines in trains, ships, and factories. However, petroleum fuels are more efficient than coal and wood, meaning that they produce more energy and less pollution per unit volume. Smoke-belching nineteenth century steam trains required a carload of coal and a full–time laborer to feed the coal into the just to leave the station. Today, automobiles powered by internal combustion engines drive hundreds of miles (kilometers) using only a few gallons of gasoline. Petroleum–fueled engines and furnaces generate electricity, heat homes, propel ships, and run industrial machinery.

PROBLEMS OF PETROLEUM USE

Petroleum is presently industrial nations' most affordable, efficient, and accessible source of energy. Its use, however, presents a number of grave environmental, economic, and social problems. Petroleum that spills and leaks from oil and natural gas wells, tankers, pipelines, refineries, and storage tanks into ocean, surface, and ground water causes serious water pollution that threatens the health of plants and animals, including humans. (Hydrocarbons are carcinogenic; they cause cancer.) Explosions and fires threaten petroleum workers and people who live near petroleum facilities. Smokestacks and automobile exhaust pipes emit poisonous gases and ash particles that block sunlight, cause acid rain, and negatively affect biological health. Strict regulations and new technologies have made petroleum extraction, processing, and use cleaner and safer in recent years.

However, two more-difficult problems remain as the reliance on petroleum continues to grow. First, only a few regions, including many politically unstable countries in the Middle East, South America, and Africa produce significant amounts of petroleum. Counties that use more oil than they produce, like the United States, are at the mercy of oil producers like Saudi Arabia and Venezuela. Economic and social conflicts often arise over oil, and sometimes these oil-related disagreements escalate to armed conflict. Second, the carbon dioxide gas emitted during petroleum combustion is a greenhouse gas. Scientists have observed rising levels of carbon dioxide in Earth's atmosphere (mass of air surrounding Earth), and worry that it may lead to global climate change. Scientists, energy companies, governments, environmentalists and other groups share a common concern for meeting the needs of Earth's ever more energy-dependent human population while reducing the negative effects of petroleum use.

RESIDENTIAL WATER USE

In the United States, approximately 408 billion gallons (1,544 billion liters) of water are used every day! While power production and irrigation (watering crops) consume the majority of water usage, public and self-supply water systems produce 47 billion gallons (178 billion liters) a day for residential users and businesses. Residential water use includes both indoor and outdoor household water usage. Water is used indoors for showering, flushing toilets, washing clothes, washing dishes, drinking, and cooking. Outdoor water usage includes washing the car, and watering the lawn, pools, and plants.

PUBLIC AND PRIVATE WATER

Nearly 85% of residential water users in the United States receive their water from public supply water systems. A public supply water system is a government facility or private company that collects water from a natural source such as a lake, river, or the ground. Through a process called purification, pollutants, mud, and salt are removed from the water, and then the clean water is delivered to residents for a fee. Public water systems also remove wastewater, all water that goes down a drain, away from homes. Sewer systems carry wastewater to treatment plants where the water is cleaned and then released. County and city water utilities are examples of public supply water systems. The remaining 15% of residential water users in the United States obtain water from a self-supplied water system. A self–supplied water system typically uses a well to obtain clean water from the ground and a septic system to purify wastewater. Since the 1950s, the number of Americans on public supply systems has more than doubled, while the number on self supplied systems has decreased slightly. This pattern of water usage reflects the trend of Americans moving from rural areas, which often must rely on wells, to the cities.

CONSERVING WATER

In many parts of the United States, particularly in the West and Florida, the increased reliance on public water systems has put a strain on the water supply. This has led many local and state governments to ban or limit certain forms of residential water usage. State and local laws may restrict how often residents may water their lawns or wash their cars. Some cities encourage residents to use lawn and garden plants that require less water. Water conservation is important because the average American uses 60 to 70 gallons (227 to 265 liters) of water per day. The high rate of residential water usage led Congress to promote the manufacture of low-flush toilets. The Energy Policy Act of 1992 stated that toilets must operate on 1.6 gallons (6 liters) or less per flush.

While this may appear to be an unusual law, older toilets, which used 3.5–5 gallons (13 to 19 liters) per flush, accounted for almost half of indoor residential water usage. Before the Energy Policy Act, toilets used 4.8 billion gallons (18 billion liters) of water per day or 9,000 gallons (34,069 liters) per person every year. The introduction of other water saving devices may further lessen residential water usage while also saving money on water bills. Showers account for about 20% of indoor water use.

A low-flow showerhead uses 2.5 gallons (9.5 liters) per minute instead of the standard 4.5 gallons (17 liters) per minute. Studies have shown that a low-flow showerhead can save thousands of gallons (liters) of water per person every year. Even a simple and inexpensive water aerator, which mixes air with the water coming out of a kitchen or bathroom sink, can reduce faucet water usage by up to 60% a year.

SALT

Common table salt is a compound. A compound is a chemical substance in which two or more elements are joined together. An element is a substance that cannot be broken down into a simpler substance. Elements, either alone or joined together as compounds, make up every object. The elements sodium and chlorine join together to make table salt. Sodium is represented by the symbol "Na," and chlorine is represented by the symbol "Cl." Because one atom (smallest unit that has all the chemical and physical characteristics of an element) of sodium joins with one atom of chlorine, table salt is represented by the symbol "NaCl."

THE NEED FOR SALT

All animals, including humans, require salt. Salt is needed to regulate many bodily functions including maintaining a regular heart rhythm, blood pressure, and fluid balance in the body. Additionally, salt is required for nerve cells to communicate efficiently, and for regulating the electrical charges moving into and out of cells during processes such as muscle contraction. An adult human has about 9 ounces (about 250 grams) of salt in the body. As the body cannot produce salt, animals must get salt from food and water. If too much salt is consumed, the kidneys remove the salt and flush it out of the body. Salt is also economically important. In ancient societies, salt was often traded for other valuable goods. Early cultures used salt for food preservation and Roman soldiers were paid partially in salt, probably giving rise to the word soldier from the Latin sal dare for "giving salt." Today salt is used in food, on food, to de-ice highways, and in the production of industrial chemicals. Nearly 250 million tons (220 metric tons) of salt are produced worldwide every year.

GETTING SALT

Salt comes from a variety of sources. Salt is known as rock salt, or halite, when it is found in the ground. Rock salt can be mined from beneath the surface through drilling, blasting, and hauling it to the surface. Most mined rock salt is used to de-ice roads in the winter. Salt may also be extracted by solution mining, which involves pumping water underground. The water dissolves the salt, creating brine, which is then pumped back to the surface.

The water is evaporated out of the brine, leaving behind salt deposits.

Solution mining produces purer salt than rock salt mining. Solution mining is often used to produce edible (able to be eaten) salt. Salt can also be removed from seawater through a process called solar salt production. Solar salt production involves removing seawater and allowing the water to evaporate. Salt deposits are left behind, forming sea salt. Sea salt is pure and highly sought after for cooking due to its clean taste. A single cubic foot of ocean water produces 2.2 pounds (1 kilogram) of salt.

The oceans hold more salt than humans could ever use. Salt accounts for about 3.5% of the weight of the oceans. The oceans contain an estimated 39 quadrillion tons (39 million, billion tons, or 35 million, billion metric tons) of salt! The oceans are getting even saltier. Flowing rivers pick up dissolved salts and minerals such as chloride, sodium, sulfate, and magnesium from the rocks and soils. Once rivers flow into the ocean, these salts and minerals are deposited in the ocean. Salts and minerals do not evaporate out of the ocean. Once salts are deposited, they will remain there forever, unless humans remove them. Gradually, the ocean gets saltier as more dissolved salts are carried into it.

SHIPPING ON FRESHWATER WATERWAYS

For thousands of years humans have used freshwater waterways to ship food, building materials, and goods between regions. A freshwater waterway is any low-salt body of water, such as a river, lake, or man–made canal on which ships may travel. The need for freshwater for drinking and irrigation (watering crops) led most early civilizations to develop along rivers. Shipping on freshwater waterways continues to be a reliable and important way to transport goods. Shipping goods over waterways is slower than other forms of shipping, yet it is less expensive and allows larger loads of cargo. Therefore many heavy raw materials such as coal, oil, timber, food products, and metal are often shipped over water. Many modern cities are still located along rivers and lakes.

SHIPPING IN ANCIENT EGYPT

The ancient Egyptians (3000 B.C.E.–30 C.E.) depended upon the Nile River for their survival. The Nile River was the only source of drinking water for most Egyptians. Its yearly floods deposited silt (fine particles smaller than sand) that fertilized Egyptians crops. The Egyptians also used the Nile as their main highway, connecting Upper Egypt in the south with Lower Egypt in the north. Egyptian boats relied on wind or oars to travel on the Nile.

Generally, boats traveled south by wind, as the wind usually blew from the north. Since the Nile flows from south to north, most ships would follow the flow of the river and drift with the current or row north. The Egyptians relied on barges to transport large amounts of goods. A barge is a large, usually flat ship that can carry heavy cargo. Egypt depended on barges to move building materials such as stone from places where it was mined and cut in the south to the major cities such as Cairo and Alexandria in the north. Without transporting goods on the Nile, the Egyptians would not have been able to construct the pyramids or construct large cities.

PROPULSION SYSTEMS

The ancient Greeks, Romans, Phoenicians, and numerous European civilizations also relied on freshwater shipping to move goods. Like the Egyptians, all of these civilizations relied on sails or oars for propulsion. Propulsion is the means by which a ship moves through the water. Sometimes animals, such as horses or mules, walked along the shore and pulled a ship along a slow-moving river or canal using a rope. Over the last few centuries however, humans have developed new forms of propulsion. In the early nineteenth century, the first steamships were developed. Steam propulsion ships, whose power came from boilers that provided steam under pressure to turn turbines or paddlewheels, proved useful on freshwater waterways such as rivers and lakes, which were poorly suited for sailboats. Inland waterways (water bodies away from the coast, such as rivers) often lacked enough wind, or wind blowing in the proper direction, to propel large cargo ships against the current of the river.

The invention of diesel engines in the early twentieth century led to diesel-powered ships that replaced steamboats because diesel engines allowed cheaper shipping without the dangers of pressurized steam. With these new forms of propulsion goods can be shipped on freshwater waterways faster, cheaper, and in greater quantity. For example, in the United States, over 700 million tons (635 million metric tons) of cargo are shipped on freshwater waterways every year. Most of this cargo is carried on freshwater waterways by barges. Modern barges are large, flat boats that are often joined together like railroad cars. A typical string of about fifteen barges is pulled or pushed by small, powerful tugboats, or tugs. A single barge can carry as much cargo as sixty large truck containers or fifteen railroad cars.

TYPES OF FRESHWATER WATERWAYS

Major river systems are the most common form of freshwater waterways used for shipping. A river system is made up of a major river and all of its tributaries, the smaller rivers or creeks that feed into the main river. Lake systems are also often used to ship goods. In the United States, for example, there are two main freshwater systems that are used for shipping, the Mississippi-Missouri river system and the Great Lakes–St. Lawrence River system. The Mississippi–Missouri river system allows shipping in the Midwest and Southeast. The Great Lakes-St. Lawrence River system serves the Midwest and northeastern United States and part of eastern Canada. Over 75% of all materials shipped over freshwater waterways in the United States are shipped on either the Mississippi–Missouri river system or the Great Lakes-St. Lawrence River system. Other smaller river systems are also used for shipping in the United States. The Ohio River system in the Midwest, the Tennessee River system in the Southeast, the Colorado River system in the West, and the Columbia River in the Northwest are important for shipping. There are also major river systems in other parts of the world that are used for shipping: the Danube, Rhine, and Volga river systems in Europe; the Nile in Africa; the Amazon in South America; and the Yangtze in China.

CANALS

What happens when river systems do not quite reach important places? Other forms of shipping, such as trains, airplanes, or trucks may be used, but the best solution may be the construction of a canal. A canal is a man-made deep and wide waterway through which ships may travel. A canal can connect one river or lake with another to allow ships to travel farther inland to reach major cities. The bodies of water that a canal links may be at different elevations (heights) above sea level and therefore, not navigable for ships. Therefore the water level in a canal is usually not the same from one end of the canal to the other. To solve this problem and allow the passage of ships through the canal, a series of gates and locks must be constructed along the course of the canal. A lock is a large area with gates at each end that raises or lowers a ship to areas of the canal that have different water levels. Similar to an elevator with door on opposite sides, locks are an essential part of any canal. When a ship is going from an area of lower water level to an area of higher water level the ship enters the lock and the gate behind it closes. Water is pumped into the lock to raise the ship.

When the water level inside the lock is at the same level as the higher water level in the canal the front gate opens and the ship continues its journey on the canal. When a ship is moving from an area of higher water level in the canal to an area of lower water level, then the process works in reverse. Once the ship enters the lock, water is pumped out of the lock to lower the water level until it matches the lower level on the other side. Once the water level is the same the gate opens and the ship continues its voyage.

PROBLEMS WITH SHIPPING ON FRESHWATER WATERWAYS

One disadvantage of shipping on freshwater is that it is much slower than other forms of shipping such as trucks, railroads, or airplanes. Also, freshwater waterways do not reach everywhere that goods are needed. The construction of canals can extend freshwater waterways to some, but not all areas. These locations must ship their goods to and from the nearest port (seaside) city by another means of shipping. A third disadvantage is that like roads, waterways must be maintained or they fall into disrepair. Tree limbs, trash, and soil clog waterways so that a ship cannot pass. Freshwater waterways must be dredged occasionally to remove buildup on the bottom of the waterway and around bridges and shores.

Dredging is a process where a ship drags a hook or grate along the bottom of a waterway in order to remove the accumulated silt and mud. Dredging makes freshwater waterways more navigable (able for ships and barges to move through the waterway) by deepening the waterway and sometimes helping to smooth areas with strong currents.

SHIPPING ON THE OCEANS

Throughout recorded history, humans have relied on the oceans to ship goods quickly and efficiently. Historically, shipping on the oceans had several advantages over shipping over land. Shipping over land required moving bulky and heavy goods over mountains, across deserts, or through forests. The location of roads often dictated where goods could be shipped. Before vehicles, land travellers also had to carry enough food and water to keep their pack animals alive, adding to the weight of their loads. Two thousand years ago, the power of the Roman Empire was founded on the economic benefit that Rome gained from its control of trade on the Mediterranean Sea.

Most of Rome's empire lay on the shores of the Mediterranean Sea, which served as a highway for the trade of wine, food, timber, spices, and other valuable materials. Rome's power stretched from Gaul (modern France) around the Mediterranean Sea to the Middle East and to North Africa. Rome also had territory in modern Britain. Rome typically imported raw materials from its faraway territory and exported finished goods back to its territory. The expansion of European nations into lands fueled trade with their colonies over vast expanses of oceans. British trade with India and Southeast Asia under the British East India Company in the sixteenth through eighteenth centuries delivered spices and teas to Britain via ships.

Britain's colonies in North America also shipped raw materials such as timber, furs, and cotton across the Atlantic Ocean to Britain, who shipped finished goods back to the United States. In the nineteenth century, ships made of iron and steel used steam power to transport goods across the oceans faster than ever before. The rise of diesel powered vessels in the twentieth century made shipping cheaper and faster. Goods could be shipped to ports on the other side of the world in days instead of weeks and months.

SHIPPING TODAY

Today, merchant ships transport more than 90% of the world's cargo.

There are several reasons that ships move more cargo than any other form of shipping. First, ships are the cheapest form of transportation. Second, in a world in which many countries have poor roads, boats are often the most efficient and reliable means of shipping. Third, boats can move greater amounts of cargo than any other form of shipping. Despite the rise of shipping cargo by aircraft, the ocean shipping industry continued to grow throughout the twentieth century. From the early 1920s through the end of the century, the worldwide number of ships in the merchant shipping fleet increased from under 30,000 to nearly 90,000. Total tonnage increased at an even greater rate. The total tonnage of merchant ships increased from 59 million gross tons (a unit of measurement to describe the size of a ship) to over 500 million gross tons during the same period.

TYPES OF MERCHANT SHIPS

Modern merchant ships serve a variety of purposes. Therefore, shipping vessels come in many different shapes. Some ships, called tankers, are designed to carry liquids. The most common type of tanker is the oil tanker. Over 3,500 oil tankers carry petroleum products to ports around the world. Oil tankers are among the largest ships in the world. Some oil tankers are over 1,300 feet long, or the length of about 4.5 football fields! Similar to oil tankers, chemical tankers carry various liquids such as vegetable oil, acids, and liquid fertilizers. Many of the chemicals carried by chemical tankers are hazardous. Chemical tankers carry smaller loads than oil tankers due to the increased danger of the cargo, and because consumers require greater amounts of oil. Most merchant ships carry dry cargo. Over the last fifty years, container ships have become one of the most important ship designs. Container ships carry sealed cargo containers that can be unloaded directly onto trains or trucks, thus becoming a railway car or a truck trailer.

This allows the container to be loaded only once upon departure and unloaded once upon delivery to its final destination. New designs for cargo ships will soon carry up to 15,000 containers that are each 20 feet in length. Bulk carriers, another type of dry cargo ship, carry large quantities of raw material such as iron ore, steel, coal, or wheat. Bulk carriers transport their goods in large cargo holds without the use of containers. A cargo hold is a part of a ship that is divided from the rest of the ship for the transport of a single type of cargo. Shipping in bulk decreases transportation cost by reducing loading and unloading costs. For example, with modern loading methods, more than 15,000 tons of iron ore can be loaded onto a bulk carrier in one hour. Bulk carriers can carry more than 250,000 tons of goods and may have over 10 individual cargo holds.

PROBLEMS WITH SHIPPING

Shipping on the oceans poses a variety of possible problems, including harm to the environment, loss of cargo, and loss of lives. The major causes of shipping accidents are human mistakes, poor equipment maintenance, and natural disasters. Most accidents are avoidable, and the last 100 years have seen a dramatic decrease in the number of shipping accidents. Increased training and safety regulations have lessened the number of accidents caused by human error and poor maintenance. Weather forecasting has improved greatly in the last century, leading to fewer accidents from natural disasters such as hurricanes. Oil and chemical tankers pose a serious threat to the environment if they lose their cargo. Oil and chemical spills can poison fish and marine mammals.

A well–known oil spill, although not the largest, was the Exxon Valdez accident in 1989, when the tanker Exxon Valdez ran aground on rocks in Prince William Sound, Alaska. Nearly 11 million gallons of oil spilled out into the natural environment, killing fish, birds, and marine mammals. The resulting cleanup of the waters and shore cost about $2 billion. Bulk container ships are involved in more accidents than any other form of cargo ship. Bulk container ships have large hatches that stretch across most of the width of the ship. This decreases the overall strength of the ship, especially in rough seas. About thirteen bulk container ships sink each year. On average, about 70 people lose their lives every year in accidents involving bulk container ships. Although shipping by ocean is far less expensive than shipping by aircraft, it is also slower. Because the large amount of cargo that modern merchant ships can carry means that less than one per cent of the purchase price of a product goes towards ocean shipping, many merchants and consumers choose to wait the extra time for the goods.

SURFACE AND GROUNDWATER USE

Surface water is the water that lies on the surface naturally as streams, rivers, marshes, lagoons, ponds, and lakes. Surface water can also be collected and stored in containers that have been built especially for that purpose. These containers are called reservoirs. Fresh water also collects in areas of soil and rock underground. This is groundwater. Rain falling from the sky and snow melting in the springtime can flow downhill to gather in stream or riverbeds. From there, the water flows to a lake or ocean. In other locations, the rain or melted snow is soaked up by the soil and makes its way further down into the ground because of gravity (the force of attraction between all masses in the universe).

USES OF SURFACE AND GROUNDWATER

Surface water tends to be used by humans more often than groundwater. This is because it is much easier to obtain surface water. Inserting a pipe or tube into the water and then pumping out the water is all that is needed. Sometimes, if the surface water source is located on a hillside, the water flows through the pipeline because of gravity. Surface water makes up almost 80% of the 410 billion gallons of water that is used in the United States every day. Groundwater makes up the remaining approximate 20% This huge amount of water is enough to fill 400,000 Olympic swimming pools, every day of the year!

Drinking Water

The main use of surface and groundwater is for drinking water. Without freshwater to drink, animals such as humans die within days. Much of our drinking water is surface water, which must be treated before drinking. Soil and plant material can wash into surface water in a rainstorm or as the snow melts into the stream, river, pond, or lake. Microorganisms that live in the feces of animals can also be washed into the water.

If the water is not treated to remove the material and the microorganisms, the contaminated water can make humans and animals ill. This is why campers and hikers filter their drinking water or add chemicals that kill the harmful organisms in the water. This is also why the water that comes out of the tap in towns and cities has usually come from a water treatment plant; a place where the water is put through a series of steps to make it potable (drinkable). Groundwater may not require treatment before drinking. This is because the ground itself is a filter. As the water moves down into the soil and rocks, big objects like leaves are left on the surface, and smaller objects including bacteria (a million bacteria could fit on the period at the end of this sentence) either stick to the soil or cannot pass through the even tinier holes in the rock. By the time the water collects in the ground, the harmful microorganisms and chemicals have been removed by the filtering action of the soil and rock layers. This can often mean that potable water can be pumped out of the ground from wells.

However, it is wise for those who have a private well to have their water tested at regular intervals. Community wells are checked every month to ensure that no contamination of the groundwater has occurred that could be harmful to the community that the well supplies.

Recreation

Diving into the swimming pool, water–skiing, and fishing in a lake are all fun (recreational) uses of water that make use of surface water. Groundwater aquifers are sometimes the source of warm or cool springs that come to the surface and are also popular for recreational use. The need to take care of recreational water has been recognized for a long time. In the United States, laws made in the 1960s were designed to help keep surface waters healthy. These laws are known as the Federal Water Project Recreation Act and are still important in maintaining surface water for recreation.

Agriculture

Both surface and groundwater help keep crops growing. Depending on the type of crop being grown, water can be pumped or sprayed onto the field. Additionally, farm owners and their livestock such as cattle, pigs, and poultry all require drinking water to stay healthy, and water is needed to keep the farm clean.

Industry

Industry uses large amounts of water to keep machinery cool, to pump into oil fields to help force the oil up to the surface, to generate electricity, and for other purposes. Much of this water is used and then put back into the ground or onto the surface. Surface water is used to generate electricity by building a wall (dam) across a river. The dam causes water to collect on one side. When gates in the dam are opened, water rushes through. The rushing water turns turbines, a device that converts the fluid into mechanical motion that in turn generates electricity. While dams are necessary to supply the electricity that big cities need, they can sometimes change the river in ways that are not healthy for the animals, plants, and microorganisms that live further in the river.

TOURISM ON THE OCEANS

Human interest in the sea fuels a multi-billion dollar a year ocean tourism industry. Ocean tourism refers to pleasure travel in which the sea is the primary focus of activities. Ocean tourism comes in many forms including cruises, ecotourism, and fishing expeditions.

CRUISING THE OCEANS

Cruises are one of the most popular forms of ocean tourism. In the late nineteenth and early twentieth century, cruise liners were needed to carry passengers across the oceans. Many of these cruise ships—including the ill–fated Titanic, which sank in 1912 killing over 1,500 people—provided passengers a luxurious way to travel. Originally powered by steam-driven engines, most modern cruise ships use diesel fuel to power their engines. While cruise ships were needed for Atlantic Ocean crossings, by the mid–twentieth century, air travel made ocean crossings cheaper and faster. An airplane can cross the Atlantic in several hours instead of the one week required by most cruise ships. Cruise lines could no longer promote their services as providing a means of travel to and from vacation. (A cruise line is a company that owns one or more cruise ships.) With little need for cruise ships for ocean crossings, cruise line operators had to take a different approach to their business.

They began to change the concept of the cruise itself to a vacation. Ships started traveling to exotic locations and offering more services and activities. Today's cruise ships are large ships that serve as floating hotels for vacationers. Cruise ships include restaurants, shops, swimming pools, theaters, and cinemas. Some cruise ships even offer college–level courses onboard. Cruise ships cost hundreds of millions of dollars to construct and may be over 1,000 feet (305 meters) long, over 150,000 gross tons (a term use to describe the size of a boat, ship, or barge), and stand taller than a 20-story building. The length of the largest cruise ship in 2004, the Queen Mary 2, is only 117 feet (36 meters) shorter than the height of the Empire State Building.

The largest cruise ships can carry nearly 4,000 people, including the crew. Tourism on the oceans provides a major boost to the economies of countries that are popular cruise destinations. In the United States, nearly 8 million people take a cruise every year. Cruises contribute an estimated $18 billion per year to the American economy. Cruise lines directly employee over 25,000 Americans. An estimated 250,000 American jobs are supported by the cruise industry. The ocean tourism industry is highly regulated. Every commercial ship, including cruise ships, must be registered with a country in order to sail in international waters.

A country may register ships only if it is a member of the International Maritime Organization (IMO). The IMO is an agency of the United Nations. The United Nations is an organization consisting of most of the independent states of the world and is designed to promote peace and security. Any country that registers ships under the IMO must have adopted the IMO's Resolutions and Conventions on maritime safety. The cruise industry has taken a major role in promoting safety on the seas. The International Council of Cruise Lines (ICCL) is a nongovernmental group that works with the IMO to promote maritime safety and environmental preservation.

In additional to ship registration, the nation where a ship docks, called the port state, may also impose restrictions on cruise ships. The United States has a reputation for strictly enforcing safety rules. The U.S. Coast Guard inspects every ocean–going ship in its ports four times per year. The United States imposes additional restrictions on ships registered in the United States, including that construction and ownership of the ship must be American. This leads many cruise ships to register in other countries, including Norway, Liberia, Panama, and the Bahamas. Over 90 cruise ships are registered in Liberia and Panama.

ECOTOURISM AND FISHING ON THE OCEAN

Cruise ships are not the only form of ocean tourism. Ecotourism of the oceans has become increasingly popular. Ecotourism involves tourism that focuses on the natural environment without harming it. One popular form of ecotourism is scuba diving. Scuba diving involves the use of a selfcontained breathing system that allows a person to remain underwater for long periods. Scuba stands for "self–contained underwater breathing apparatus." Scuba divers enjoy the beauty of fish, coral reefs, and other marine features.

Another form of ecotourism involves cruises to view wildlife such as humpback whales or dolphins, while impacting their environment as little as poissible. Deep–sea fishing expeditions are another popular form of tourism on the oceans. Deep–sea fishing involves taking a boat several miles from shore in order to catch large fish, including tuna, marlin, and dolphin fish. Some species of deep-sea fish can weigh from several hundred to over 1,000 pounds (454 kilograms).

PROTECTING THE ENVIRONMENT

In many areas of the world, such as the Sea of Cortez off the coast of Mexico, numbers of large game fish are reduced, presumably from over-fishing. Many countries, including the United States, have laws stipulating the number, types, and sizes of game fish that may be caught and kept in order to reduce harm to the fish population. Cruise lines have placed an increased emphasis on protecting the environment over the last two decades. Cruise ships must follow the environmental laws of a country when in that country's territorial waters. Ships must follow the Clear Air Act, the Clean Water Act, and the Oil Pollution Control Act when in American waters. These are all laws passed by Congress to control pollution in the United States.

The IMO and the ICCL also set environmental regulations for all registered ships. In 1973, IMO adopted the International Convention for the Prevention of Pollution from Ships at Sea (MARPOL), which it revised in 1978. MARPOL sets environmental standards that all ocean–going ships must meet. Cruise lines have also sought better methods to prevent pollution from the waste that cruise ships generate, including sewage and garbage.

TRANSPORTATION ON THE OCEANS

For thousands of years, oceans provided one of the fastest and most valuable forms of transportation. By 3200 B.C.E., Egyptian ships made of reeds (tall, woody grass) used sails to travel along the coast of northern Africa. Over the centuries, ocean–going ships became larger and faster. Around 1000 B.C.E. the Vikings explored the coast of Canada in sailboats. Spanish ships explored the Americas in the fifteenth and sixteenth centuries. British tall ships carried settlers to the Americas, Asia, Australia, and Africa in the sixteenth through nineteenth centuries. Until the mid–twentieth century, ships were the only mode of transportation for ocean crossings. The rise of air transportation after 1930 reduced the role of ocean-going vessels in transportation. Airplanes provided a quicker and often cheaper way to move people great distances, which caused the types of vessels and purposes of ocean transportation to change.

IMMIGRATION TO THE NEW WORLD

For the first 450 years after the discovery of the New World, ships provided the only form of transportation between Europe and the Americas. Nearly every citizen of the United States is descended from ancestors who traveled to the New World by ship, and immigration to the New World was a major factor in ocean transportation during this time. Immigration patterns to the United States reflect that immigrants came from various countries in waves. The earliest settlers came from the British Isles and Africa. Before 1790, about 500,000 immigrants came to the United States from the British Isles, and 300,000 immigrants came from Africa.

The middle half of the nineteenth century saw a flood of immigrants from Europe with 3 million from the German Empire, 2.8 million from Ireland, and 2 million from England. The United States experienced its greatest influx of immigration between 1880 and 1930. During this period, nearly 20 million immigrants crossed the Atlantic Ocean on ships. These immigrants came primarily from Italy, Russia, Germany, Britain, and the Austro-Hungarian Empire. Twelve million of these immigrants entered the United States through Ellis Island, near New York City. Between 1897 and 1938, Ellis Island served as the main processing point for immigrants.

Today over 100 million Americans can trace their ancestry to an immigrant who landed on Ellis Island. Ocean transportation in America has a dark side. Slave ships transported tens of thousands of Africans to the New World every year. Between the sixteenth and nineteenth centuries, between 15 million and 20 million Africans were involuntarily brought to the Americas as slaves. About 400,000 slaves were transported to the British colonies and the United States. Scholars estimate that as many as 1 million African slaves died during ocean transit to the Americas.

TRANSATLANTIC JOURNEYS

Not all ocean crossing ships were only filled with immigrants. Travellers also used ships to cross the Atlantic Ocean to go between Europe and the Americas. In 1818, New York's Black Ball Line became the first company to offer regular travel across the Atlantic Ocean. The rise of steam ships in the mid-1800s made ocean crossings faster. While these ships focused on luxury travel for wealthy passengers, they also fueled immigration. Cruise liners offered low–cost, no frills transportation for many immigrants. The immigrants stayed in steerage class, the least expensive accommodations, and were often responsible for bringing their meals.

By the early twentieth century, cruise liner companies began to build larger and more luxurious ships, including Olympic, Lusitania, Britannic, and Titanic. These ships emphasized comfort and extravagance over speed. Many of these cruise liners contained swimming pools, dance halls, and tennis courts. Unfortunately, the superliners of the early nineteenth century did not stress safety. Thousands of lives were lost in the sinkings of the Titanic in 1912 and Lusitania in 1915.

THE RISE OF THE CRUISE SHIP

By 1950, airplanes replaced cruise liners as the main mode of transportation across the oceans. Many travellers did not choose to spend days crossing the ocean when it could be done in hours by plane. Cruise liner companies had to change their approach to fit the new reality of air travel. They could no longer market cruise liners as a form of transportation to take while on vacation. Instead, cruise companies began advertising cruise liners as a vacation by themselves. By focusing on exotic locales, such as the Caribbean and Mediterranean Seas, cruise companies found a willing audience.

In modern day cruise ships have swimming pools, cinemas, dance clubs, theatres, and classrooms. Modern cruise ships are subject to many safety regulations. Today nearly 8 million Americans go on cruises every year. Cruises generate about $18 billion every year for the United States' economy. A modern cruise ship carries about 2000 guests and 900 crew members. The largest cruise ship in the world as of 2004, Queen Mary 2, was 1,132 feet (345 meters) long and 151,400 gross tons (term describing the size of a boat, ship, or barge). Queen Mary 2 can carry 2,620 guests and 1,253 crew members. In 2004 Queen Mary 2 was the only passenger ship that made regular transatlantic journeys.

FERRIES

Ferries are one of the most important forms of modern ocean transportation. Ferries are ships that carry people and, occasionally, cars over relative short distances. Some ferries are simple ships that transport only people. Ferries that transport people and cars are called "roll-on, roll-off" ships. Cars can quickly roll on these ferries upon departure and easily roll off upon arrival. While some ferries are simple boats, many ferries are technologically advanced ships, including hovercrafts or hydrofoils. A hovercraft is a ship that floats above the surface of the water on a cushion of air.

A rubber skirt is located between the main ship and the water. Air is pushed into the rubber skirt, creating a cushion of air. Hovercrafts offer smooth rides over rough seas. A hydrofoil is a ship that has wing-like foils (wing-like structures that raises part or all of a powerboat's hull out of the water) underneath the hull of the ship. As the boat increases speed, the foils lift the hull of the ship out of the water. Only the foils skim the top of the water. Like a hovercraft, the main body of a hydrofoil rides above the surface of the water. This reduces drag and increases speed.

Unlike most cruise ships, not all ferries are subject to strict safety regulations. Many passengers die in ferry accidents every year, mostly in the developing world. In 2002, the ferry Joola sank off the coast of Africa near Senegal. Joola was carrying over three times its capacity. Over 1,800 people died in the accident, which is more than the number of people who died on the Titanic.

WHALING

Whaling is the hunting of whales mainly for meat and oil. Its earliest forms date to at least 3000 BC. Various coastal communities have long histories of sustenance whaling and harvesting beached whales. Industrial whaling emerged with organized fleets in the 17th century; competitive national whaling industries in the 18th and 19th centuries; and the introduction of factory ships along with the concept of whale harvesting in the first half of the 20th century.

As technology increased and demand for the resources remained high, catches far exceeded the sustainable limit for whale stocks. In the late 1930s more than 50,000 whales were killed annually and by the middle of the century whale stocks were not being replenished. In 1986 the International Whaling Commission (IWC) banned commercial whaling so that stocks might recover.

While the moratorium has been successful in averting the extinction of whale species due to overhunting, contemporary whaling is subject to intense debate. Pro-whaling countries, notably Japan, wish to lift the ban on stocks that they claim have recovered sufficiently to sustain limited hunting. Anti–whaling countries and environmental groups contend that those stocks remain vulnerable and that whaling is immoral, unsustainable, and should remain banned permanently.

HISTORY OF WHALING

The history of whaling is very extensive, stretching back for millennia. This object discusses the history of whaling up to the commencement of the International Whaling Commission (IWC) moratorium on commercial whaling in 1986.

Prehistoric to Medieval Times

Humans have engaged in whaling since prehistoric times. The oldest known method of catching cetaceans is simply to drive them ashore by placing a number of small boats between the animal or animals and the open sea and to frighten them with noise and activity, herding them towards shore in an attempt to beach them. Typically, this was used for small species, such as Pilot Whales, Belugas, Porpoises and Narwhals. This is described in A Pattern of Islands (1952) by British administrator Arthur Grimble, who lived in the Gilbert and Ellice Islands for several decades.

The next step was to employ a drogue (a semi-floating object) such as a wooden drum or an inflated sealskin which was tied to an arrow or a harpoon in the hope that after a time the whale would tire enough to be approached and killed. Several cultures around the world practiced whaling with drogues, including the Ainu, Inuit, Native Americans, and the Basque people of the Bay of Biscay. Bangudae Petroglyphs, an archaeological evidence from Ulsan in South Korea suggests that drogues, harpoons and lines were being used to kill small whales as early as 6000 BC Petroglyphs (rock carvings) unearthed by researchers at the Museum of Kyungpook National University show Sperm Whales, Humpback Whales and North Pacific Right Whales surrounded by boats. Similarly-aged cetacean bones were also found in the area, reflecting the importance of whales in the prehistoric diet of coastal people.

A description of the assistance a little European technology could bring to skilled indigenous whale hunters is given in the memoir of John R. Jewitt, an Englishman blacksmith who spent three years as a captive of the Mowachaht (Nuu-chah-nulth/Nootka) people in 1802–1805. Jewitt also mentions the importance of whale meat and oil to the diet. Whaling was integral to the cultures and economies of other indigenous peoples of the Pacific Northwest as well, notably the Makah and Klallam. For other groups, most famously the Haida, whales appear prominently as totems.

BASQUE WHALING

The first mention of Basque whaling was made in 1059, when it was said to have been practiced at the Basque town of Bayonne. The fishery spread to the actual Spanish Basque region in 1150, when King Sancho the Wise of Navarre granted petitions for the warehousing of such commodities as whalebone (baleen). At first, they only hunted the whale they called sarda, or the North Atlantic Right Whale, using watchtowers (known as vigias) to look for their distinctive twin vapour spouts.

By the 14th century they were making "seasonal trips" to the English Channel and southern Ireland. The fishery spread to Terranova (Labrador and Newfoundland) in the second quarter of the 16th century, and to Iceland at least by the early 17th century. They established whaling stations at the former, mainly in Red Bay, and probably established some in the latter as well. In Terranova they hunted bowheads and right whales, while in Iceland they appear to have only hunted the latter.

The fishery in Terranova declined for a variety of reasons. Principal among them the conflicts between Spain and other European powers during the late 16th and early 17th centuries, attacks by hostile Inuit, declining whale populations, and perhaps the opening up of the Spitsbergen fishery in 1611. The first voyages to Spitsbergen by the English, Dutch, and Danish relied on Basque specialists, with the Basque provinces sending out their own whaler in 1612. The following season San Sebastian and Saint-Jean-de-Luz sent out a combined eleven or twelve whalers to the Spitsbergen fishery, but most were driven off by the Dutch and English. Two more ships were sent by a merchant in San Sebastian in 1615, but both were driven away by the Dutch.

They continued whale fishing in Iceland and Spitsbergen at least into the 18th century, but Basque whaling in those regions appears to have ended with the commencement of the Seven Years' War (1756–63).

GREENLAND WHALING

Encouraged by reports of whales off the coast of Spitsbergen in 1610, the English Muscovy Company sent a whaling expedition there the following year. The expedition was a disaster, with both ships sent being lost. The crews returned to England in a ship from Hull. The following year two more ships were sent. Other countries followed suit, with Amsterdam and San Sebastian each sending a ship north. The latter ship returned to Spain with a full cargo of oil. Such a fabulous return resulted in a fleet of whaleships being sent to Spitsbergen in 1613. The Muscovy Company sent seven, backed by a monopoly charter granted by King James I. They met with twenty other whaleships (eleven–twelve Basque, five French, and three Dutch), as well as a London interloper, which were either ordered away or forced to pay a fine of some sort. The United Provinces, France, and Spain all protested against this treatment, but James I held fast to his claim of sovereignty over Spitsbergen.

The following three and a half decades witnessed numerous clashes between the various nations (as well as infighting among the English), often merely posturing, but sometimes resulting in bloodshed. This jealousy stemmed as much from the mechanics of early whaling as from straightforward international animosities. In the first years of the fishery England, France, the United Provinces and later Denmark-Norway shipped expert Basque whalemen for their expeditions. At the time Basque whaling relied on the utilization of stations ashore where blubber could be processed into oil. In order to allow a rapid transference of this technique to Spitsbergen, suitable anchorages had to be selected, of which there were only a limited number, in particular on the west coast of the island.

Early in 1614 the Dutch formed the Noordsche Compagnie (Northern Company), a cartel composed of several independent chambers (each representing a particular port). The company sent fourteen ships supported by three or four men-of-war this year, while the English sent a fleet of thirteen ships and pinnaces. Equally matched, they agreed to split the coast between themselves, to the exclusion of third parties. The English received the four principal harbors in the middle of the west coast, while the Dutch could settle anywhere to the south or north. The agreement explicitly stated that it was only meant to last for this season.

In 1617 a ship from Vlissingen whaling in Horn Sound had its cargo seized by the English vice-admiral. Angry, the following season the Dutch sent nearly two dozen ships to Spitsbergen. Five of the fleet attacked two English ships, killing three men in the process, and also burned down the English station in Horn Sound. Negotiations between the two nations followed in 1619, with James I, while still claiming sovereignty, would not enforce it for the following three seasons. When this concession expired, the English twice (in 1623 and 1624) tried to expel the Dutch from Spitsbergen, failing both times.

In 1619 the Dutch and Danes, who had sent their first whaling expedition to Spitsbergen in 1617, firmly settled themselves on Amsterdam Island, a small island on the northwestern tip of Spitsbergen; while the English did the same in the fjords to the south. The Danish–Dutch settlement came to be called Smeerenburg, which would become the centre of operations for the latter in the first decades of the fishery. Numerous place names attest to the various nations' presence, including Copenhagen Bay (Kobbefjorden) and Danes Island, where the Danes established a station from 1631–58; Port Louis or Refuge Français (Hamburgbukta), where the French had a station from 1633–38, until they were driven away by the Danes; and finally English Bay (Engelskbukta), as well as the number of features named by English whalemen and explorers—for example, Isfjorden, Bellsund, and Hornsund, to name a few.

Hostilities continued after 1619. In 1626 nine ships from Hull and York destroyed the Muscovy Company's station in Bell Sound, and sailed to their own in Midterhukhamna. In 1630 both the ships of Hull and Yarmouth, who had recently joined the trade, were driven away clean (empty) by the ships from London. From 1631-33 the Danes, French, and Dutch quarreled with each other, resulting in the expulsion of the Danes from Smeerenburg and the French from Copenhagen Bay.

In 1634 the Dutch burned down one of the Danes' huts. There were also two battles this season, one between the English and French (the latter won) and the other between London and Yarmouth (the latter won, as well). In 1637 and again in 1638 the Danes drove the French out of Port Louis and seized their cargoes.

The latter year they also held two Dutch ships captive for over a month, which led to protests from the Dutch. Following the events of 1638 hostilities for the most part ceased, with the exception of a few minor incidents in the 1640s between the French and Danes, as well as between Copenhagen and Hamburg and London and Yarmouth, respectively.

The species hunted was the Bowhead Whale, a baleen whale that yielded large quantities of oil and baleen. The whales entered the fjords in the spring following the break up of the ice. They were spotted by the whalemen from suitable vantage points, and pursued by shallops, chaloupes or chalupas, which were manned by six men. (These terms derive from the Basque word "txalupa", used to name the whaling boats that were widely utilized during the golden era of Basque whaling in Labrador in the 16th century.) The whale was harpooned and lanced to death and either towed to the stern of the ship or to the shore at low tide, where men with long knives would flense (cut up) the blubber. The blubber was boiled in large copper kettles and cooled in large wooden vessels, after which it was funneled into casks.

The stations at first only consisted of tents of sail and crude furnaces, but were soon replaced by more permanent structures of wood and brick, such as Smeerenburg for the Dutch, Lægerneset for the English, and Copenhagen Bay for the Danes. Beginning in the 1630s, for the Dutch at least, whaling expanded into the open sea. Gradually whaling in the open sea and along the ice floes to the west of Spitsbergen replaced bay whaling. At first the blubber was tried out at the end of the season at Smeerenburg or elsewhere along the coast, but after mid-century the stations were abandoned entirely in favour of processing the blubber upon the return of the ship to port. The English meanwhile stuck resolutely to bay whaling, and didn't make the transfer to pelagic (offshore) whaling until long after.

In 1719, the Dutch began "regular and intensive whaling" in the Davis Strait. Nevertheless, encouraged by import duty exemptions, the South Sea Company financed 172 unprofitable whaling voyages from London's Howland Dock between 1725–32. In 1733 the Government introduced a 'bounty' of £1.00 per ship ton, increasing to £2.00 per ton in 1749. These subsidies along with high oil and whalebone prices encouraged expansion. London sent out six whalers in 1749; 45 in 1777 and 91 in 1788. However, reductions in the bounty, and wars with America and France saw London's Greenland fleet fall to 19 in 1796. The British would continue to send out whalers to the Arctic fishery into the 20th century, sending her last on the eve of the First World War.

Japanese Open-boat Whaling

Because of some evidence of whaling found such as hand harpoons and porpoise skulls in burial mounds, hunting of cetaceans possibly began in the Jômon period (10,000–300 BC) just as to The Institute of Cetacean Research.

The oldest written mention of whaling in Japanese records is from Kojiki, the oldest Japanese historical book written in the 7th century AD. In this book whale meat was eaten by Emperor Jimmu. In Man'yoshu, the oldest anthology of poems in the 8th century, the word "Whaling" was frequently used in depicting the ocean or beaches.

One of the first records of whaling by the use of harpoons are from the 1570s at Morosaki, a bay attached to Ise Bay. This method of whaling, known as the harpoon method (tsukitori-ho) spread to Kii (before 1606), Shikoku (1624), northern Kyushu (1630s), and Nagato (around 1672).

Kakuemon Wada, later known as Kakuemon Taiji, was said to have invented net whaling, or the net method (amitori-ho) sometime between 1675 and 1677. This method soon spread to Shikoku (1681) and northern Kyushu (1684) Using the techniques developed by Taiji, the Japanese mainly hunted four species of whale, the North Pacific right (Semi–Kujira), the humpback (Zato-Kujira), the fin (Nagasu–Kujira), and the gray whale (Ko-Kujira or Koku–Kujira). They also caught the occasional blue (Shiro Nagasu-Kujira), sperm (Makko-Kujira), or sei/Bryde's whale (Iwashi-Kujira).

Whaling has been frequently mentioned in Japanese historical texts:

- Whaling history, Seijun Ohtsuki, 1808.
- Whaling Picture Scroll, Jinemon Ikushima, 1665.
- Whale Hunt Picture Scroll, Eikin Hangaya, 1666.
- Ogawajima Whaling Wars, Unknown, 1667.

In 1853, the US naval officer Matthew Perry forced open Japan's doors to the world. One of the purposes of this was to gain access to ports for the American whaling fleet in the north–west Pacific Ocean. The traditional whaling was eventually replaced in the late 19th century and early 20th century with modern methods.

Yankee Open–boat Whaling

Beginning in the late colonial period, the United States, with a strong seafaring tradition in New England, an advanced shipbuilding industry, and access to the oceans grew to become the pre-eminent whaling nation in the world by the 1830s.

American whaling's origins were in New England, especially Cape Cod, Massachusetts and nearby cities. The oil was in demand chiefly for lamps. Hunters in small watercraft pursued right whales from shore. By the 18th century, whaling in Nantucket had become a highly lucrative deep–sea industry, with voyages extending for years at a time and with vessels traveling as far as South Pacific waters. During the American Revolution, the British navy targeted American whaling ships as legitimate prizes, while in turn many whalers fitted out as privateers against the British. Whaling recovered after the war ended in 1783 and the industry began to prosper, using bases at Nantucket and then New Bedford. Whalers took greater economic risks to turn major profits: expanding their hunting grounds and securing foreign and domestic workforces for the Pacific. Investment decisions and financing arrangements were set up so that managers of whaling ventures shared their risks by selling some equity claims but retained a substantial portion due to moral hazard considerations. As a result, they had little incentive to consider the correlation between their own returns and those of others in planning their voyages. This stifled diversity in whaling voyages and increased industry-wide risk.

Ten thousand seamen manned the ships. More than three thousand African American seamen shipped out on whaling boats from New Bedford between 1800 and 1860, about 20% of the entire whaling force.

In port the most successful of the whaling merchants was Jonathan Bourne, who opened offices in New Bedford in 1848. Chandlery shops and storage rooms for whaling outfits occupied the first floor. Lofts and rigging lofts occupied the upper stories; the counting-rooms were on the second floor, with counters and iron railings fencing off the tall mahogany desks at which the bookkeepers stood up, or sat on high stools; about the walls were models of whaleships and whaling prints.

Early whaling efforts were concentrated on right whales and humpbacks, which were found near the American coast. As these populations declined and the market for whale products (especially whale oil) grew, American whalers began hunting the Sperm Whale. The Sperm Whale was particularly prized for the reservoir of spermaceti (a dense waxy substance that burns with an exceedingly bright flame) housed in the spermaceti organ, located forward and above the skull. Hunting for the Sperm Whale forced whalers to sail farther from home in search of their quarry, eventually covering the globe.

Whale oil was vital in illuminating homes and businesses throughout the world in the 19th century, and served as a dependable lubricant for the machines powering the Industrial Revolution. Baleen (the long keratin strips that hang from the top of whales' mouths) was used by manufacturers in the United States and Europe to make consumer goods such as buggy whips, fishing poles, corset stays and dress hoops. New England ships began to explore and hunt in the southern oceans after being driven out of the North Atlantic by British competition and import duties. Ultimately, American entrepreneurs created a mid-19th-century version of a global economic enterprise. This was the golden age of American whaling.

An early winter in the north Pacific in September 1871 forced the captains of an American whaling fleet in the Arctic to abandon their ships. With 32 vessels trapped in the ice and provisions insufficient to weather the nine-month winter, the captains ordered the abandonment of the ships and the three million dollar's worth of property carried on board but in the process saved the lives of over 1,200 men. From the Civil War, when Confederate raiders targeted American whalers, through the early 20th century, the American whaling industry was overwhelmed by new, crippling economic competition, especially from kerosene, which was a superior fuel for lighting. New Bedford, once the fourth busiest port in the United States, gave up whaling.

Localities

Whaling became important for a number of New England towns, particularly Nantucket and New Bedford, Massachusetts. Vast fortunes were made, and culture of these communities was greatly affected; the results can be seen today in the buildings surviving from the era. Larger cultural influence is evidenced by former whaler Herman Melville's novel Moby–Dick, which is often cited as the Great American Novel.

Nantucket joined in on the trade in 1690 when they sent for one Ichabod Padduck to instruct them in the methods of whaling. The south side of the island was divided into three and a half mile sections, each one with a mast erected to look for the spouts of right whales. Each part had a temporary hut for the five men assigned to that area, with a sixth man standing watch at the mast. Once a whale was sighted, rowing boats were sent from the shore, and if the whale was successfully harpooned and lanced to death, it was towed ashore, flensed (that is, its blubber was cut off), and the blubber boiled in cauldrons known as "trypots." Even when Nantucket sent out vessels to fish for whales offshore, they would still come to the shore to boil the blubber, doing this well into the 18th century.

The South Sea Fishery

Britain

Samuel Enderby, along with Alexander Champion and John St. Barbe, using American vessels and crews, fitted out twelve whaleships for the southern fishery in 1776. More were sent in 1777 and 1778 before political and economic troubles hampered the trade for some time. In 1786, Alexander Champion, with his brother Benjamin, sent the first British whaler east of the Cape of Good Hope. She was the Triumph, Daniel Coffin, master.

On 1 September 1788, the 270 ton whaleship Emilia, owned by Samuel Enderby and Sons and commanded by Captain James Shields, departed London. The ship went west around Cape Horn into the Pacific Ocean to become the first ship of any nation to conduct whaling operations in the Southern Ocean. A crewman, Archelus Hammond of Nantucket, killed the first sperm whale there off the coast of Chile on 3 March 1789. Emilia returned to London on 12 March 1790 with a cargo of 139 tons of sperm oil.

In 1784 the British had fifteen whaleships in the southern fishery, all from London. By 1790 this port alone had sixty vessels employed in the trade. Between 1793 and 1799 there was an average of sixty vessels in the trade. The average increased to seventy-two in the years between 1800 and 1809.

In 1819 the first British whaleship, the Syren (510 tons), under Frederick Coffin of Nantucket, was sent to the Japan grounds, where she began whaling on 5 April 1820. She returned to London on 21 April 1822 with 346 tons of sperm oil. The following year at least nine British whalers were cruising on this ground, and by 1825 the British had twenty-four vessels there. Despite this discovery, the number of vessels being fitted out annually for the southern fishery declined from sixty-eight in 1820 to thirty-one in 1824. In 1825 there were ninety ships in the southern fishery, but by 1835 it had dwindled to sixty-one. Fewer and fewer vessels were being fitted out, so that by 1843 only nine vessels were clearing for the southern fishery. In 1859 the last cargoes of sperm oil from British vessels were landed in London.

France

Having failed in an attempt to establish a colony of Nantucket whalemen in England, William Rotch, Sr. went to France in 1786 and was able to establish his colony in Dunkirk. The first two vessels to be fitted out were the Canton and the Mary. By 1789 Dunkirk had fourteen vessels in the trade sailing to Brazil, Walvis Bay, and other areas of the South Atlantic to hunt sperm and right whales. Just a year later Rotch sent the first French whalers into the Pacific. There were twenty–four vessels sailing out of France for the southern fishery by 1791, but the majority of these ships were lost during the Anglo–French War that broke out two years later. Rotch fled France, keeping subordinates there should war tensions ease and allow them to fit out ships for the southern fishery again.

The trade began to revive after hostilities, but when Napoleon came to power Rotch's holdings in Dunkirk were seized. After the Napoleonic Wars the government issued subsidies in an attempt to revive the trade once more, but it wasn't until 1832, with a further increase in bounties, that several whalers were sent by C. A. Gaudin on sperm whaling voyages.

In 1835 the first French whaleship, the Gange (573 tons), Narcisse Chaudiere, master, reached the Gulf of Alaska and discovered an abundance of right whales. Within a decade a large number of American and French vessels would be cruising on this ground. The following year, 1836, the first French whaler had reached New Zealand, but by the 1840s, with the decline of bay whaling, very few French vessels would make their way here.

In 1851 a law was passed to encourage the trade, at which point the French had seventeen vessels employed in it. It wasn't successful. The last whalers returned in 1868.

Rorqual Whaling

By the 1850s, the Euro–American whalemen made a serious attempt at catching such rorquals as the blue and fin whale. This era was inaugurated by one Thomas Welcome Roys. Roys, while cruising south of Iceland in the 441–ton Hannibal, was able to kill a sulfurbottom (blue whale) with a Brown's bomb gun in 1855. He realised that if he had a better way to dispatch such large rorquals as the sulfurbottom that he could easily fill his ship's hold with whale oil. Due to his ship having taken a beating in a heavy gale in these waters, he was forced to put into Lorient, France. While there, he ordered for "two rifles in pairs for killing [rorqual] whales," staying long enough in France to see them nearly completed, then leaving for home in a steamer, and, when finished, having the guns sent by way of England to the US.

The following spring, he went out in the 175–ton brig William F. Safford to test his experimental whaling guns. The guns Roys had ordered from France were lost on the voyage out, so he had to persuade C. C. Brand of Norwich, Conn., to let him use his bomb lance, but to increase his bomb missiles to three pounds in order to ensure greater success. Roys sailed to Bjornøya, where he encountered vast numbers of blue, fin, and humpbacks. He fired at around sixty, with only a single blue whale being saved. He then sailed to Novaya Zemlya, capturing two humpbacks there. After cruising off Russia and Norway, he came to anchor at Queenstown, Ireland, and thence went to England to reconstruct his lost French–made guns.

He had Sir Joseph Whitworth manufacture him some rifled whaling guns and shells. Roys returned to his ship, sailing from Queenstown on 26 November for the Bay of Biscay. Here, when testing one of the guns, he blew off his left hand, having to amputate it "as well as we could with razors." They sailed to Oporto, Portugal, where Roys's lower arm had to be amputated.

Having failed in securing whales on another cruise in 1857, Roys redesigned his gun. This time, the rocket–powered harpoons proved too weak to penetrate the whales correctly. Undaunted, he made another cruise, this time to South Georgia, but he wasn't able to take any whales. He cruised north to put into Lisbon, sailed to Africa, then west to the West Indies in early 1859, where he was able to capture several humpbacks.

In 1861 Roys joined forces with the wealthy New York pyrotechnic manufacturer Gustavus Adolphus Lilliendahl in order to perfect his "whaling rocket". In mid-May 1862 Lilliendahl purchased the 158–ton bark Reindeer, appointing Roys as her master. Unfortunately, she was seized on suspicion of being a slaver, and when everything was finally cleared up, she sailed to Iceland, but arrived too late for the summer whaling season, and had to return home and wait until next year.

In 1863 Roys refitted the Reindeer and once again sailed to Iceland, but he damaged his rudder while off the coast of the island, and was only able to save one of the many whales he shot that season. Roys was much more successful the following season of 1864, saving eleven of the twenty whales that were shot, in part because he was using stronger harpoons and better lines. In November 1864 Roys obtained the rights to establish a shore station on the coast of Iceland from the Danish government.

He acquired the twelve-ton, sixty-two-foot iron steamer Visionary in Scotland, and returned to Iceland in the spring of 1865. He arrived at Seydisfjordur on 14 May, finding his bark Reindeer had already arrived there in April, loaded with whaling equipment, boilers, steam engines, timber, bricks, and everything necessary for the construction of his shore station. Lilliendahl supplied them with defective rockets, and before the station was built, they were forced to tow the dead whales to the Reindeer, where they were flensed and processed the old fashioned way.

After his rockets were rebuilt, Roys and his crew set out in the Visionary, with whaleboats in tow astern, to search for rorquals. Once a whale was sighted, the crews went to their respective boats, and if a whale was successfully captured, they'd heave the carcass to the surface with a steam winch, fasten it to the side of the ship, and tow it back to Seydisfjordur. For the 1865 season they took twenty or more whales, but also lost another twenty. The next season, 1866, he used the Sileno and the iron steamers Staperaider and Vigilant–identical ship, bark-rigged, 116-feet long, each carrying two whaleboats and equipped with steam tryworks and powerful winches to bring aboard large strips of blubber when flensing whales. They killed ninety whales this season, with forty-three or forty–four being saved to produce 3,000 barrels of oil. Roys and Lilliendahl parted company at the end of the season, with Lilliendahl continuing on in Iceland for another year. Using the Vigilant and Staperaider, he only caught thirty-six whales. After this season, he departed as well.

Roys and Lilliendahl found imitators in Iceland, in the form of the Danish naval officer Cap. Otto C. Hammer and the Dutchman Cap. C. J. Bottemanne. The former formed the Danish Fishing Company in 1865, and wound up operations in 1871; while the latter formed the Netherlands Whaling Company in 1869, closing down operations a year after Hammer. In 1866 James Dawson, a Victorian emigrant from Clackmannanshire, Scotland, and a man named Warren tried catching whales in Saanich Inlet, British Columbia, but lost all three whales they struck to bad weather. In 1868 Dawson joined in a partnership with a 27-year-old from San Francisco, Abel Douglass, along with two other Californians, Bruce and Woodward. They were joined by Roys, who chartered the 83-three-foot, 25-ton steamer Emma. His first cruise was a disaster, while the second cruise from early September to October he reportedly struck four whales, killing three, but lost all three in dense fogs. Dawson began whaling on 26 August with the 47-ton Kate, cruising in Saanich Inlet, where they managed to catch eight whales using bomb lances, despite thick fog.

Persistent as ever, Roys formed the Victoria Whaling Adventurers

Company on 22 October, and in January 1869 he sent the Emma to erect a shore station in Barkley Sound, Vancouver Island. Again, Roys was met with by failure, having made fast to only one whale. The harpoon broke free, and the whale escaped. He was defeated once more by the Dawson and Douglass Whaling Company, who took fourteen whales by mid-September 1869 to produce 20,000 gallons of oil.

Dawson and Douglass then joined forces with a man named Lipsett, forming the Union Whaling Company. They only took four whales during two cruises in the winter of 1869–70, forcing the company to suspend operations as of 3 February 1870. Lipsett reorganized and formed the Howe Sound Company, while Dawson found new partners had formed the new Dawson and Douglass Whaling Company on 27 June 1870. Another unidentified group of whalemen using "the Roys Rocket" arrived in June, charting the schooner Surprise and hunting whales in Barkley Sound. Only one of the companies used a vessel equipped with a whaleboat, while the others apparently sent rowing boats out from their shore camps. The three firms only took thirty-two whales, for a yield of 75,800 gallons of oil.

The next season, seemingly undeterred, Roys returned to British Columbia in the 179-ton brig Byzantium on 10 May 1871. He constructed a station at Cumshewa Inlet in the Queen Charlotte Islands, and fitted out the Byzantium with proper onboard tryworks. Douglass split from Dawson and paired with the Victorian vintner and publican James Strachan, while Dawson rejoined Lipsett and formed the British Columbia Whaling Company. Dawson and Lipsett's company produced 20,000 gallons of oil in 1871, with Douglass and Strachan producing about 15,000. Both companies lost money on their ventures, with the former soon being liquidated.

The Kate and other possessions of the company went on the auction block in March 1872. The schooner and equipment went to former company partners Robert Wallace and James Hutcheson, who unsuccessfully attempted to continue whaling operations. We last hear of them in July 1873, when the Kate was said to have been cruising near Lasqueti Island, in the Strait of Georgia, with little success. By the end of the year the schooner had been sold. As usual, Roys fared the worst. The Byzantium struck the rocks in Weynton Passage, Johnstone Strait, forcing the men to abandon her and row ashore, to spend a frigid night huddled on the beach. Roys never operated a whaling company again.

In 1877, John Nelson Fletcher, a pyrotechnist, and the former Confederate soldier from North Carolina, Robert L. Suits, modified Roys's rocket, marketing it as the "California Whaling Rocket". They used the small five in a half ton steam launch Rocket of San Francisco in 1878, killing 35 humpback, fin, and blue whales with their rocket outside the harbour and north to Point Reyes.

In 1880, Thomas P. H. Whitelaw fitted out the 44-ton steamer Daisy Whitelaw of San Francisco. With the California Whaling Rocket she "very successfully" hunted fin whales though the Farallon Islands to Drakes Bay. That same year, some of the rockets were purchased by the Northwest Whaling Company, or Northwest Trading Company, of Killisnoo Island, on the west coast of Admiralty Island, Southeast Alaska. They hunted fins and humpbacks, firing rockets from the deck of the company's small steamer Favourite, as well as from whaleboats. They established a whaling and trading station on Killisnoo Island, giving a few jobs at the whale processing plant to both Killisnoo and Angoon residents. After a few years of whaling, the station was turned into a herring processing plant, going out of business in 1885.

In the late 1870s schooners began hunting humpbacks in the Gulf of Maine. In 1880, with the decline of the menhaden fishery, steamers began to switch to hunting fin and humpback whales using bomb lances in what has been called a "shoot-and-salvage" fishery because of the high–rate of loss due to whales sinking, lines breaking, etc. The first was the steamer Mabel Bird, which towed whale carcasses to an oil processing plant at the head of Linekin Bay in Boothbay Harbor. Soon there were five such factories in Boothbay Harbour processing whales. At its height in 1885 four or five steamers were engaged in the Menhaden whale fishery, but it dwindled to one by the end of the decade. Fin whales accounted for about half the catch, with over 100 whales being killed in some years. The fishery ended in the late 1890s.

Before Svend Foyn launched the industry into the modern era, there were the Norwegians Jacob Nicolai Walsøe and Arent Christian Dahl. The former was probably the first person to suggest mounting a harpoon gun in the bows of a steamship, while the latter experimented with an explosive harpoon in Varanger Fjord (1857–1860). While they were the first in their class, it was Foyn who successfully adopted these ideas and put them into practice. In 1864, his methods, through trial and error, would lead to the development of the modern whaling trade.

During the 1930s, as German whaling in the Antarctic was coming about, the Nazis maintained that a gunsmith from Bremerhaven, H. G. Cordes, was responsible for Foyn's invention, and should thus receive credit for having brought whaling into the modern era. Foyn had indeed ordered material from Cordes, but he had found it unserviceable, and only experimented with his gun for a season. Cordes, working with John P. Rechten of Bremen, had developed an improved version of the Greener gun in 1856. They made a second version of this swivel gun with two barrels, side by side, with the left barrel shooting a harpoon and the right a bomb lance. Their invention was successfully experimented with in the North Sea in 1867. With this success, Rechten attempted to introduce this idea on the

American market two years later, but it isn't known as to whether he succeeded or not.

Modern Whaling

As early as 1611 the 50,000+ bowhead whales resident between the east coast of Greenland and the island of Spitsbergen were the subject of intensive commercial hunting effort by English whalers. By 1911 the bowheads were nearly extinct. The discovery in 1848 of the rich stock of bowhead whales in the Bering Strait region sparked an oil rush that resulted in more than 2,500 annual whaling cruises to the area between 1848 and 1899.

At first slow whales were caught by men hurling harpoons from small open boats. Early harpoon guns were unsuccessful until Norwegian Svend Foyn invented a new, improved version in 1863 that used a harpoon with a flexible joint between the head and shaft. Norway invented many new techniques and disseminated them worldwide. Cannon–fired harpoons, strong cables, and steam winches were mounted on maneuverable, steam-powered catcher boats. They made possible the targeting of large and fast-swimming whale species that were taken to shore-based stations for processing. Breach–loaded cannons were introduced in 1925; pistons were introduced in 1947 to reduce recoil.

These highly efficient devices were too successful, for they reduced whale populations to the point where large–scale commercial whaling became unsustainable. The shore stations on the island of South Georgia were at the center of the Antarctic whaling industry, from its beginnings in 1904 until the late 1920s when pelagic whaling increased. The activity on the island remained substantial until around 1960, when Norwegian-British Antarctic whaling came to an end.

Finnmark

In February 1864, the Norwegian Svend Foyn set sail from Tønsberg, south of Oslo, in the schooner–rigged, steam–driven whale catcher Spes et Fides (Hope and Faith) on a voyage north to Finnmark to hunt rorquals such as the Blue and Fin Whale. He had her fitted out like a minor man-of-war, with seven guns on her forecastle, each firing a harpoon and grenade separately. Several whales were seen, but only four were captured.

He tried again in 1866 and 1867, but he could not catch a single whale in the former season and only caught one whale the latter, while two others were killed but lost. Experimenting with a harpoon gun that fired a grenade and harpoon at the same time, Foyn was able to catch thirty whales in 1868. He patented his grenade-tipped harpoon gun two years later. Foyn was given a virtual monopoly on the trade in Finnmark in 1873, which lasted until 1882. Despite this, local citizens established a

Foyn was given a virtual monopoly on the trade in Finnmark in 1873, which lasted until 1882. Despite this, local citizens established a whaling company in 1876, and soon others defied his monopoly and formed companies. With the commencement of unrestricted catching in 1883, the number of whaling stations increased from eight to sixteen, and the number of whale catchers from twelve to twenty-three. Catching material peaked in 1886–88 with an average of about thirty-one catchers operating each season, while peak catching was not reached until 1892–93 and 1896–98, when between 1,000 and 1,200 whales were caught each year.

Only half the number of whales were taken in 1899, and catching continued to decline until 1902, when it improved somewhat. By this time most of the catching was done far from the coast. The last station closed down in 1904.

Iceland

In 1883 the first whaling station was established in Alptafjordur, Iceland. In the first season, using an 84 gross ton whale catcher, only eight whales were caught, but in the following season (1884) twenty–five were caught, all of which were Blue Whales, with the exception of two.

In 1889 another station was established. Between 1890 and 1894 three more companies, all Norwegian, established themselves in Iceland. Seeing the success of these companies, another five established whaling stations on the island between 1896 and 1903. Catching peaked in 1902, when 1,305 whales were caught to produce 40,000 barrels of oil. By 1907, only 268 whales were caught, and by 1910 the score stood at a mere 170.

A ban on whaling was imposed by the Alting in 1915. It was not until 1935 that an Icelandic company established another whaling station. It shut down after only five seasons. In 1948, another Icelandic company, Hvalur H/F, purchased a naval base at the head of Hvalfjordur and converted it into a whaling station. Between 1948 and 1975, an average of 250 Fin, 65 Sei, and 78 Sperm Whales were taken annually, as well as a few Blue and Humpback Whales. Unlike the majority of commercial whaling at the time, this operation was based on the sale of frozen meat and meat meal, rather than on oil. Most of the meat was exported to England, while the meal was sold locally as cattle feed.

Faroe Islands

The Norwegian Hans Albert Grøn eestablished the first whaling station in the Faroe Islands in 1894 at Strømæs, situated in the sound between the islands of Strømø and Osterø. He caught forty–six whales his first season, intercepting the whales as they migrated north. He operated alone the first four seasons, until Christian Salvesen and Co. formed a company in Oslo for whaling from the islands.

Grøn established another station in 1901, as did Peter O. Bogen, who set up one on the island of Suderø. Three more companies arrived between 1902 and 1905. One was Norwegian, another Danish, and the last a joint Danish–Norwegian concern.

Peak catching was reached in 1909, when 773 whales were caught to produce 13,850 barrels of oil. By 1913 the production of oil had dropped to 3,515 barrels. In 1917, with the war and poor catches, whaling was suspended from the islands. Four companies resumed catching 1920. The results were disappointing; with only one Norwegian company staying at the islands as late as 1930. Further attempts were made to revive catching in the Faroes during the 1930s and after the Second World War, with the last attempt being made in 1962–64.

Spitsbergen

In 1903, the Norwegian Christen Christensen sent the first factory ship, the wooden steamship Telegraf (737 gross tons), to Spitsbergen. She returned to Sandefjord in September with 1,960 barrels of oil produced from a catch of fifty-seven whales—of which forty-two were Blue Whales.

He sent a larger ship, the 1,517 gross ton Admiralen, to Spitsbergen the following season (1904). She returned with a cargo.of 5,100 barrels from 154 whales. By 1905 there were eight companies operating around Spitsbergen and Bear Island, while seven (using fifteen whale catchers) were there in 1906–07. The peak had been reached in 1905, when 559 whales (337 Blue) were caught to produce 18,660 barrels. Only a quarter of this was produced in 1908. Two companies left in 1907, and another two the following year.

As the three companies remaining produced a dismal amount of oil in 1912, they decided to suspend operations. Two unsuccessful attempts were made in 1920 and 1926–27 to revive catching in Spitsbergen waters—since that time only Northern Bottlenose and Minke Whales have been hunted there by converted Norwegian fishing boats.

CONTROVERSY

Key elements of the debate over whaling include sustainability, ownership, national sovereignty, cetacean intelligence, suffering during hunting, the value of lethal sampling to establish catch quotas, the value of controlling whale's impact on fish stocks and the rapidly approaching extinction of a few whale species.

2010 IWC MEETING

At the 2010 meeting of the International Whaling Commission in Morocco, representatives of the 88 member nations discussed whether or not to lift the 24 year ban on commercial whaling. Japan, Norway and Iceland have urged the organization to lift the ban. A coalition of anti–

whaling nations has offered a compromise plan that would allow these countries to continue whaling, but with smaller catches and under close supervision. Their plan would also completely ban whaling in the Southern Ocean. More than 200 scientists and experts have opposed the compromise proposal for lifting the ban, and have also opposed allowing whaling in the Southern Ocean, which was declared a whale sanctuary in 1994. Opponents of the compromise plan want to see an end to all commercial whaling, but are willing to allow subsistence-level catches by indigenous peoples.

Chapter 13

Water Pollution

INTRODUCTION

Water pollution is the contamination of water bodies. Water pollution occurs when pollutants are discharged directly or indirectly into water bodies without adequate treatment to remove harmful compounds. Water pollution affects plants and organisms living in these bodies of water; and, in almost all cases the effect is damaging not only to individual species and populations, but also to the natural biological communities. Water pollution is a major global problem. It has been suggested that it is the leading worldwide cause of deaths and diseases, and that it accounts for the deaths of more than 14,000 people daily.

An estimated 700 million Indians have no access to a proper toilet, and 1,000 Indian children die of diarrheal sickness every day. Some 90% of China's cities suffer from some degree of water pollution, and nearly 500 million people lack access to safe drinking water. In addition to the acute problems of water pollution in developing countries, industrialized countries continue to struggle with pollution problems as well. In the most recent national report on water quality in the United States, 45 per cent of assessed stream miles, 47 per cent of assessed lake acres, and 32 per cent of assessed bay and estuarine square miles were classified as polluted. Water is typically referred to as polluted when it is impaired by anthropogenic contaminants and either does not support a human use, such as drinking water, and/or undergoes a marked shift in its ability to support its constituent biotic communities, such as fish. Natural phenomena such as volcanoes, algae blooms, storms, and earthquakes also cause major changes in water quality and the ecological status of water.

CATEGORIES OF WATER POLLUTION

Surface water and groundwater have often been studied and managed as separate resources, although they are interrelated. Surface water seeps through the soil and becomes groundwater. Conversely, groundwater can also feed surface water sources. Sources of surface water pollution are generally grouped into two categories based on their origin.

POINT SOURCES

Point source water pollution refers to contaminants that enter a waterway from a single, identifiable source, such as a pipe or ditch. Examples of sources in this category include discharges from a sewage treatment plant, a factory, or a city storm drain. The U.S. Clean Water Act defines point source for regulatory enforcement purposes. The CWA definition of point source was amended in 1987 to include municipal storm sewer systems, as well as industrial stormwater, such as from construction sites.

NON–POINT SOURCES

Non–point source pollution refers to diffuse contamination that does not originate from a single discrete source. NPS pollution is often the cumulative effect of small amounts of contaminants gathered from a large area. A common example is the leaching out of nitrogen compounds from fertilized agricultural lands. Nutrient run-off in stormwater from "sheet flow" over an agricultural field or a forest are also cited as examples of NPS pollution. Contaminated storm water washed off of parking lots, roads and highways, called urban run-off, is sometimes included under the category of NPS pollution. However, this run-off is typically channeled into storm drain systems and discharged through pipes to local surface waters, and is a point source. However where such water is not channeled and drains directly to ground it is a non-point source.

GROUNDWATER POLLUTION

Interactions between groundwater and surface water are complex. Consequently, groundwater pollution, sometimes referred to as groundwater contamination, is not as easily classified as surface water pollution. By its very nature, groundwater aquifers are susceptible to contamination from sources that may not directly affect surface water bodies, and the distinction of point vs. non-point source may be irrelevant. A spill or ongoing releases of chemical or radionuclide contaminants into soil may not create point source or non-point source pollution, but can contaminate the aquifer below, defined as a toxin plume. The movement of the plume, called a plume front, may be Analysed through a hydrological transport model or groundwater model. Analysis of groundwater contamination may focus on the soil characteristics and site geology, hydrogeology, hydrology, and the nature of the contaminants.

CAUSES OF WATER POLLUTION

The specific contaminants leading to pollution in water include a wide spectrum of chemicals, pathogens, and physical or sensory changes such as elevated temperature and discolouration. While many of the chemicals and substances that are regulated may be naturally occurring the concentration is often the key in determining what is a natural component of water, and what is a contaminant. High concentrations of naturally–occurring substances can have negative impacts on aquatic flora and fauna. Oxygen–depleting substances may be natural materials, such as plant matter as well as man-made chemicals. Other natural and anthropogenic substances may cause turbidity which blocks light and disrupts plant growth, and clogs the gills of some fish species. Many of the chemical substances are toxic. Pathogens can produce waterborne diseases in either human or animal hosts. Alteration of water's physical chemistry includes acidity, electrical conductivity, temperature, and eutrophication. Eutrophication is an increase in the concentration of chemical nutrients in an ecosystem to an extent that increases in the primary productivity of the ecosystem. Depending on the degree of eutrophication, subsequent negative environmental effects such as anoxia and severe reductions in water quality may occur, affecting fish and other animal populations.

PATHOGENS

Coliform bacteria are a commonly used bacterial indicator of water pollution, although not an actual cause of disease.

Other microorganisms sometimes found in surface waters which have caused human health problems include:

- Burkholderia pseudomallei
- Cryptosporidium parvum
- Giardia lamblia
- Salmonella
- Novovirus and other viruses
- Parasitic worms.

High levels of pathogens may result from inadequately treated sewage discharges. This can be caused by a sewage plant designed with less than secondary treatment. In developed countries, older cities with aging infrastructure may have leaky sewage collection systems, which can cause sanitary sewer overflows. Some cities also have combined sewers, which may discharge untreated sewage during rain storms. Pathogen discharges may also be caused by poorly managed livestock operations.

CHEMICAL AND OTHER CONTAMINANTS

Contaminants may include organic and inorganic substances.

Organic water pollutants include:

- Detergents
- Disinfection by–products found in chemically disinfected drinking water, such as chloroform

- Food processing waste, which can include oxygen–demanding substances, fats and grease
- Insecticides and herbicides, a huge range of organohalides and other chemical compounds
- Petroleum hydrocarbons, including fuels and lubricants and fuel combustion byproducts, from stormwater run-off
- Tree and bush debris from logging operations
- Volatile organic compounds such as industrial solvents, from improper storage. Chlorinated solvents, which are dense non–aqueous phase liquids may fall to the bottom of reservoirs, since they don't mix well with water and are denser.
- Various chemical compounds found in personal hygiene and cosmetic products

Inorganic water pollutants include:

- Acidity caused by industrial discharges
- Ammonia from food processing waste
- Chemical waste as industrial by-products
- Fertilizers containing nutrients—nitrates and phosphates—which are found in stormwater run-off from agriculture, as well as commercial and residential use
- Heavy metals from motor vehicles and acid mine drainage
- Silt in run-off from construction sites, logging, slash and burn practices or land clearing sites
- Macroscopic pollution—large visible items polluting the water—may be termed "floatables" in an urban stormwater context, or marine debris when found on the open seas, and can include such items as:
- Trash or garbage discarded by people on the ground, along with accidental or intentional dumping of rubbish, that are washed by rainfall into storm drains and eventually discharged into surface waters
- Nurdles, small ubiquitous waterborne plastic pellets
- Shipwrecks, large derelict ships

Thermal Pollution

Thermal pollution is the rise or fall in the temperature of a natural body of water caused by human influence. Thermal pollution, unlike chemical pollution, results in a change in the physical properties of water. A common cause of thermal pollution is the use of water as a coolant by power plants and industrial manufacturers.

Elevated water temperatures decreases oxygen levels and affects ecosystem composition, such as invasion by new thermophilic species. Urban run-off may also elevate temperature in surface waters. Thermal pollution can also be caused by the release of very cold water from the base of reservoirs into warmer rivers.

TRANSPORT AND CHEMICAL REACTIONS OF WATER POLLUTANTS

Most water pollutants are eventually carried by rivers into the oceans. In some areas of the world the influence can be traced hundred miles from the mouth by studies using hydrology transport models. Advanced computer models such as SWMM or the DSSAM Model have been used in many locations worldwide to examine the fate of pollutants in aquatic systems. Indicator filter feeding species such as copepods have also been used to study pollutant fates in the New York Bight, for example.

The highest toxin loads are not directly at the mouth of the Hudson River, but 100 kilometers south, since several days are required for incorporation into planktonic tissue. The Hudson discharge flows south along the coast due to coriolis force. Further south then are areas of oxygen depletion, caused by chemicals using up oxygen and by algae blooms, caused by excess nutrients from algal cell death and decomposition. Fish and shellfish kills have been reported, because toxins climb the food chain after small fish consume copepods, then large fish eat smaller fish, etc. Each successive step up the food chain causes a stepwise concentration of pollutants such as heavy metals and persistent organic pollutants such as DDT.

This is known as biomagnification, which is occasionally used interchangeably with bioaccumulation. Large gyres in the oceans trap floating plastic debris. The North Pacific Gyre for example has collected the so-called "Great Pacific Garbage Patch" that is now estimated at 100 times the size of Texas. Many of these long-lasting pieces wind up in the stomachs of marine birds and animals. This results in obstruction of digestive pathways which leads to reduced appetite or even starvation. Many chemicals undergo reactive decay or chemically change especially over long periods of time in groundwater reservoirs. A noteworthy class of such chemicals is the chlorinated hydrocarbons such as trichloroethylene and tetrachloroethylene used in the dry cleaning industry.

Both of these chemicals, which are carcinogens themselves, undergo partial decomposition reactions, leading to new hazardous chemicals. Groundwater pollution is much more difficult to abate than surface pollution because groundwater can move great distances through unseen aquifers. Non–porous aquifers such as clays partially purify water of bacteria by simple filtration, dilution, and, in some cases, chemical reactions and biological activity: however, in some cases, the pollutants merely transform to soil contaminants. Groundwater that moves through cracks and caverns is not filtered and can be transported as easily as surface water.

In fact, this can be aggravated by the human tendency to use natural sinkholes as dumps in areas of Karst topography. There are a variety of secondary effects stemming not from the original pollutant, but a derivative condition. An example is silt-bearing surface run-off, which can inhibit the penetration of sunlight through the water column, hampering photosynthesis in aquatic plants.

MEASUREMENT

Water pollution may be Analysed through several broad categories of methods: physical, chemical and biological. Most involve collection of samples, followed by specialized analytical tests. Some methods may be conducted in situ, without sampling, such as temperature.

Government agencies and research organizations have published standardized, validated analytical test methods to facilitate the comparability of results from disparate testing events.

- *Sampling*: Sampling of water for physical or chemical testing can be done by several methods, depending on the accuracy needed and the characteristics of the contaminant. Many contamination events are sharply restricted in time, most commonly in association with rain events. For this reason "grab" samples are often inadequate for fully quantifying contaminant levels. Scientists gathering this type of data often employ auto-sampler devices that pump increments of water at either time or discharge intervals. Sampling for biological testing involves collection of plants and/or animals from the surface water body. Depending on the type of assessment, the organisms may be identified for biosurveys and returned to the water body, or they may be dissected for bioassays to determine toxicity.
- *Physical testing*: Common physical tests of water include temperature, solids concentration and turbidity.
- *Chemical testing*: Water samples may be examined using the principles of analytical chemistry. Many published test methods are available for both organic and inorganic compounds. Frequently used methods include pH, biochemical oxygen demand, chemical oxygen demand, nutrients, metals, oil and grease, total petroleum hydrocarbons, and pesticides.
- *Biological testing*: Biological testing involves the use of plant, animal, and/or microbial indicators to monitor the health of an aquatic ecosystem.

CONTROL OF WATER POLLUTION

DOMESTIC SEWAGE

Domestic sewage is 99.9 per cent pure water, while the other 0.1 per cent are pollutants. Although found in low concentrations, these pollutants pose risk on a large scale. In urban areas, domestic sewage is typically treated by centralized sewage treatment plants.

In the U.S., most of these plants are operated by local government agencies, frequently referred to as publicly owned treatment works. Municipal treatment plants are designed to control conventional pollutants: BOD and suspended solids.

Well–designed and operated systems can remove 90 per cent or more of these pollutants. Some plants have additional sub-systems to treat nutrients and pathogens. Most municipal plants are not designed to treat toxic pollutants found in industrial wastewater.

Cities with sanitary sewer overflows or combined sewer overflows employ one or more engineering approaches to reduce discharges of untreated sewage, including:

- Utilizing a green infrastructure approach to improve stormwater management capacity throughout the system, and reduce the hydraulic overloading of the treatment plant
- Repair and replacement of leaking and malfunctioning equipment
- Increasing overall hydraulic capacity of the sewage collection system.

A household or business not served by a municipal treatment plant may have an individual septic tank, which treats the wastewater on site and discharges into the soil. Alternatively, domestic wastewater may be sent to a nearby privately owned treatment system.

INDUSTRIAL WASTEWATER

Some industrial facilities generate ordinary domestic sewage that can be treated by municipal facilities. Industries that generate wastewater with high concentrations of conventional pollutants, toxic pollutants or other nonconventional pollutants such as ammonia, need specialized treatment systems. Some of these facilities can install a pre-treatment system to remove the toxic components, and then send the partially-treated wastewater to the municipal system. Industries generating large volumes of wastewater typically operate their own complete on-site treatment systems. Some industries have been successful at redesigning their manufacturing processes to reduce or eliminate pollutants, through a process called pollution prevention. Heated water generated by power plants or manufacturing plants may be controlled with:

- Cooling ponds, man–made bodies of water designed for cooling by evaporation, convection, and radiation
- Cooling towers, which transfer waste heat to the atmosphere through evaporation and/or heat transfer
- Cogeneration, a process where waste heat is recycled for domestic and/or industrial heating purposes.

AGRICULTURAL WASTEWATER

Nonpoint Source Controls

Sediment washed off fields is the largest source of agricultural pollution in the United States. Farmers may utilize erosion controls to reduce run-off flows and retain soil on their fields. Common techniques include contour plowing, crop mulching, crop rotation, planting perennial crops and installing riparian buffers. Nutrients are typically applied to farmland as commercial fertilizer; animal manure; or spraying of municipal or industrial wastewater or sludge. Nutrients may also enter run-off from crop residues, irrigation water, wildlife, and atmospheric deposition. Farmers can develop and implement nutrient management plans to reduce excess application of nutrients. To minimize pesticide impacts, farmers may use Integrated Pest Management techniques to maintain control over pests, reduce reliance on chemical pesticides, and protect water quality.

Point Source Wastewater Treatment

Farms with large livestock and poultry operations, such as factory farms, are called concentrated animal feeding operations or confined animal feeding operations in the U.S. and are being subject to increasing government regulation. Animal slurries are usually treated by containment in lagoons before disposal by spray or trickle application to grassland. Constructed wetlands are sometimes used to facilitate treatment of animal wastes, as are anaerobic lagoons. Some animal slurries are treated by mixing with straw and composted at high temperature to produce a bacteriologically sterile and friable manure for soil improvement.

CONSTRUCTION SITE STORMWATER

Sediment from construction sites is managed by installation of:

- Erosion controls, such as mulching and hydroseeding, and
- Sediment controls, such as sediment basins and silt fences.

Discharge of toxic chemicals such as motor fuels and concrete washout is prevented by use of:

- Spill prevention and control plans, and
- Specially designed containers and structures such as overflow controls and diversion berms.

ENVIRONMENTAL MONITORING

Environmental monitoring describes the processes and activities that need to take place to characterise and monitor the quality of the environment. Environmental monitoring is used in the preparation of environmental impact assessments, as well as in many circumstances in which human activities carry a risk of harmful effects on the natural environment. All monitoring strategies and programmes have reasons and justifications which are often designed to establish the current status of an environment or to establish trends in environmental parameters. In all cases the results of monitoring will be reviewed, analysed statistically and published. The design of a monitoring programme must therefore have regard to the final use of the data before monitoring starts.

DESIGN OF ENVIRONMENTAL MONITORING PROGRAMMES

Monitoring is of little use without a clear and unambiguous definition of the reasons for the monitoring and the objectives that it will satisfy. Almost all monitoring is in some part invasive of the environment under study and extensive and poorly planned monitoring carries a risk of damage to the environment. This may be a critical consideration in wilderness areas or when monitoring very rare organisms or those that are averse to human presence. Some monitoring techniques, such a gill netting fish to estimate populations, can be very damaging, at least to the local population and can also degrade public trust in scientists carrying out the monitoring.

Almost all mainstream environmentalism monitoring projects form part of an overall monitoring strategy or research field, and these field and strategies are themselves derived from the high levels objectives or aspirations of an organisation. Unless individual monitoring projects fit into a wider strategic framework, the results are unlikely to be published and the environmental understanding produced by the monitoring will be lost.

PARAMETERS

Chemical

The range of chemical parameters that have the potential to affect any ecosystem is very large and in all monitoring programmes it is necessary to target a suite of parameters based on local knowledge and past practice for an initial review. The list can be expanded or reduced based on developing knowledge and the outcome of the initial surveys. Freshwater environments have been extensively studied for many years and there is a robust understanding of the interactions between chemistry and the environment across much of the world. However, as new materials are developed and new pressures come to bear, revisions to monitoring programmes will be required. In the last 20 years acid rain, synthetic hormone analogues, halogenated hydrocarbons, greenhouse gases and many others have required changes to monitoring strategies.

Biological

In ecological monitoring, the monitoring strategy and effort is directed at the plants and animals in the environment under review and is specific to each individual study. However in more generalised environmental monitoring, many animals act as robust indicators of the quality of the environment that they are experiencing or have experienced in the recent past. One of the most familiar examples is the monitoring of numbers of Salmonid fish such as Brown trout or Salmon in river systems and lakes to detect slow trends in adverse environmental effects.

The steep decline in salmonid fish populations was one of the early indications of the problem that later became known as acid rain. In recent years much more attention has been given to a more holistic approach in which the ecosystem health is assessed and used as the monitoring tool itself. It is this approach that underpins the monitoring protocols of the Water Framework Directive in the European Union.

Radiological

Radiation monitoring involves the measurement of radiation dose or radionuclide contamination for reasons related to the assessment or control of exposure to ionizing radiation or radioactive substances, and the interpretation of the results. The 'measurement' of dose often means the measurement of a dose equivalent quantity as a proxy for a dose quantity that cannot be measured directly. Also, sampling may be involved as a preliminary step to measurement of the content of radionuclides in environmental media. The methodological and technical details of the design and operation of monitoring programmes and systems for different radionuclides, environmental media and types of facility are given in IAEA Safety Guide RS–G-1.8 and in IAEA Safety Report No. 64. Radiation monitoring is often carried out using networks of fixed and deployable sensors such as the US Environmental Protection Agency's Radnet and the SPEEDI network in Japan. Airborne surveys are also made by organizations like the Nuclear Emergency Support Team.

Microbiological

Bacteria and viruses are the most commonly monitored groups of microbiological organisms monitored and even these are only of great relevance where water in the aquatic environment is subsequently used as drinking water or where water contact recreation such as swimming or canoeing is practised.

Although pathogens are the primary focus of attention, the principal monitoring effort is almost always directed at much more common indicator species such as Escherichia coliý supplemented by overall coliform bacteria counts. The rational behind this monitoring strategy is that most human pathogens originate from other humans via the sewage stream.

Many sewage treatment plants have no sterilisation final stage and therefore discharge an effluent which, although having a clean appearance, still contains many millions of bacteria per litre, the majority of which are relatively harmless coliform bacteria. Counting the number of harmless sewage bacteria allows a judgement to be made about the probability of significant numbers of pathogenic bacteria or viruses being present. Where E. coli or coliform levels exceed pre-set trigger values, more intensive monitoring including specific monitoring for pathogenic species is then initiated.

Populations

Monitoring strategies can produce misleading answers when relaying on counts of species or presence or absence of particular organisms if there is no regard to population size. Understanding the populations dynamics of an organism being monitored is critical. As an example if presence or absence of a particular organism within a 10 km square is the measure adopted by a monitoring strategy, then a reduction of population from 10,000 per square to 10 per square will go unnoticed despite the very significant impact experienced by the organism.

MONITORING PROGRAMMES

All scientifically reliable environmental monitoring is performed in line with a published programme. The programme may include the overall objectives of the organisation, references to the specific strategies that helps deliver the objective and details of specific projects or tasks within those strategies. However the key feature of any programme is the listing of what is being monitored and how that monitoring is to take place and the time-scale over which it should all happen. Typically, a monitoring programme with provide a table of locations, dates and sampling methods that are proposed and which, if undertaken in full, will deliver the published monitoring programme. There are a number of commercial software packages which can assist with the implementation of the programme, monitor its progress and flag up inconsistencies or omissions but none of these can provide the key building block which is the programme itself.

SAMPLING METHODS

There are a wide range of sampling methods which depend on the type of environment, the material being sampled and the subsequent analysis of the sample. At its simplest a sample can be filling a clean bottle with river water and submitting it for conventional chemical analysis. At the more complex end, sample data may be produced by complex electronic sensing devices taking sub-samples over fixed or variable time periods.

Grab Samples

Grab samples are samples taken of a homogeneous material, usually water, in a single vessel. Filling a clean bottle with river water is a very common example. Grab samples provide a good snap-shot view of the quality of the sampled environment at the point of sampling and at the time of sampling. Without additional monitoring, the results cannot be extrapolated to other times or to other parts of the river, lake or ground–water.

In order to enable grab samples or rivers to be treated as representative, repeat transverse and longitudinal transect surveys taken at different times of day and times of year are required to establish that the grab–sample location is as representative as is reasonably possible. For large rivers such surveys should also have regard to the depth of the sample and how to best manage the sampling locations at times of flood and drought. In lakes grab samples are relatively simple to take using depth samplers which can be lowered to a pre-determined depth and then closed trapping a fixed volume of water from the required depth.

In all but the shallowest lakes, there are major changes in the chemical composition of lake water at different depths, especially during the summer months when many lakes stratify into a warm, well oxygenated upper layer and a cool de-oxygenated lower layer. In the open seas marine environment grab samples can establish a wide range of base–line parameters such as salinity and a range of cation and anion concentrations. However, where changing conditions are an issue such as near river or sewage discharges, close to the effects of volcanism or close to areas of freshwater input from melting ice, a grab sample can only give a very partial answer when taken on its own.

Semi–continuous Monitoring and Continuous

There are a wide range of specialist sampling equipment available that can be programmed to take samples at fixed or variable time intervals or in response to an external trigger. For example a sampler can be programmed to start take samples of a river at 8 minute intervals when the rainfall intensity rises above 1 mm/hour. The trigger in this case may be a remote rain gauge communicating with the sampler by using cell phone or meteor burst technology. Samplers can also take individual discrete samples at each sampling occasion or bulk up samples into composite so that in the course of one day, such a sampler might produce 12 composite sample each composed of 6 sub–samples taken at 20 minute intervals. Continuous or quasi–continuous monitoring involves having an automated analytical facility close to the environment being monitored so that results can, if required, be viewed in real time. Such systems are often established to protect important water supplies such as in the River Dee regulation system but may also be part of an overall monitoring strategy on large strategic rivers where early warning of potential problems is essential. Such systems routinely provide data on parameters such as pH, dissolved oxygen, conductivity, turbidity and colour but it is also possible to operate gas liquid chromatography with mass spectrometry technologies to examine a wide range of potential organic pollutants.

In all examples of automated bank-side analysis there is a requirement for water to be pumped from the river into the monitoring station. Choosing a location for the pump inlet is equally as critical as deciding on the location for a river grab sample. The design of the pump and pipework also requires careful design to avoid artefacts being introduced through the action of pumping the water. Dissolved oxygen concentration is difficult to sustain through a pumped system and GLC/MS facilities can detect micro-organic contaminants from the pipework and glands.

Remote Surveillance

Although on–site data collection using electronic measuring equipment is common–place, many monitoring programmes also use remote surveillance and remote access to data in real time. This requires the on-site monitoring equipment to be connected to a base station via either a telemetry network,land–line, cell phone network or other telemetry system such as Meteor burst. The advantage of remote surveillance is that many data feeds can come into a single base station for storing and analysis. It also enable trigger levels or alert levels to be said for individual monitoring sites and/or parameters so that immediate action can be initiated if a trigger level is exceeded. The use of remote surveillance also allows for the installation of very discrete monitoring equipment which can often be buried, camouflaged or tethered at depth in a lake or river with only a short whip aerial protruding. Use of such equipment tends to reduce vandalism and theft when monitoring in locations easily accessible by the public.

Remote Sensing

Environmental remote sensing uses aircraft or satellites to monitor the environment using multi–channel sensors. There are two kinds of remote sensing. Passive sensors detect natural radiation that is emitted or reflected by the object or surrounding area being observed. Reflected sunlight is the most common source of radiation measured by passive sensors and in environmental remote sensing, the sensors used are tuned to specific wavelengths from far infra–red through visible light frequencies through to far ultra violet. The volumes of data that can be collected are very large and require dedicated computational support. The output of data analysis from remote sensing are false colour images which differentiate small differences in the radiation characteristics of the environment being monitored. With a skilful operator choosing specific channels it is possible to amplify differences which are imperceptible to the human eye.

In particular it is possible to discriminate subtle changes in chlorophyll a and chlorophyll b concentrations in plants and show areas of an environment with slightly different nutrient regimes. Active remote sensing emits energy and uses a passive sensor to detect and measure the radiation that is reflected or backscattered from the target.

LIDAR is often used to acquire information about the topography of an area, especially when the area is large and manual surveying would be prohibitively expensive or difficult. Remote sensing makes it possible to collect data on dangerous or inaccessible areas.

Remote sensing applications include monitoring deforestation in areas such as the Amazon Basin, the effects of climate change on glaciers and Arctic and Antarctic regions, and depth sounding of coastal and ocean depths.

Orbital platforms collect and transmit data from different parts of the electromagnetic spectrum, which in conjunction with larger scale aerial or ground-based sensing and analysis, provides information to monitor trends such as El Nino and other natural long and short term phenomena. Other uses include different areas of the earth sciences such as natural resource management, land use planning and conservation.

Bio–monitoring

The use of living organisms as monitoring tools has many advantages. Organisms living in the environment under study are constantly exposed to the physical, biological and chemical influences of that environment.

Organisms that have a tendency to accumulate chemical species can often accumulate significant quantities of material from very low concentrations in the environment. Mosses have been used by many investigators to monitor heavy metal concentrations because of their tendency to selectively adsorb heavy metals. Similarly, eels have been used to study halogenated organic chemicals, as these are adsorbed into the fatty deposits within the eel.

Other Sampling Methods

Ecological sampling requires careful planning to be representative and as non invasive as possible. For grasslands and other low growing habitats the use of a quadrat—a 1 metre square frame—is often used with the numbers and types of organisms growing within each quadrat area counted Sediments and soils require specialist sampling tools to ensure that the material recovered is representative. Such samplers are frequently designed to recover a specified volume of material and may also be designed to recover the sediment or soil living biota as well such as the Ekman grab sampler.

DATA INTERPRETATIONS

The interpretation of environmental data produced from a well designed monitoring programme is a large and complex topic addressed by many publications. Regrettably it is sometimes the case that scientists approach the analysis of results with a pre-conceived outcome in mind and use or misuse statistics to demonstrate that their own particular point of view is correct. This can be clearly seen in the debate about global warming controversy where, for example supporters argue that CO_2 levels have increased by 25% over the last one hundred years whilst opponents argue that CO_2 level have only risen by one percentage point from 3 to 4. Many other scientific debates have less clear–cut positions, but statistics remains a tool that is equally easy to use or to misuse to demonstrate the sessions learnt from environmental monitoring

ENVIRONMENTAL QUALITY INDICES

Since the start of science based environmental monitoring, a number of quality indices have been devised to help classify and clarify the meaning of the considerable volumes of data involved. Stating that a river stretch is in "Class B" is likely to be much more informative than stating that this river stretch has a mean BOD of 4.2, a mean dissolved oxygen of 85%, etc. In the UK the Environment Agency uses a system called GQA—General Quality Assessment which classifies rivers into six quality bands—a, b, c, d, e and f—based on chemical criteria and on biological criteria.

EUTROPHICATION

Eutrophication is the addition of artificial or natural substances, such as nitrates and phosphates, through fertilizers or sewage, to an aquatic system. In other terms, it is the "bloom" or great increase of phytoplankton in a water body. Negative environmental effects include hypoxia, the depletion of oxygen in the water, which induces reductions in specific fish and other animal populations. Other species may experience an increase in population that negatively affects other species.

LAKES AND RIVERS

Eutrophication can be human–caused or natural. Untreated sewage effluent and agricultural run-off carrying fertilizers are examples of human–caused eutrophication. However, it also occurs naturally in situations where nutrients accumulate or where they flow into systems on an ephemeral basis. Eutrophication generally promotes excessive plant growth and decay, favouring simple algae and plankton over other more complicated plants, and causes a severe reduction in water quality. Enhanced growth of aquatic vegetation or phytoplankton and algal blooms disrupts normal functioning of the ecosystem, causing a variety of problems such as a lack of oxygen needed for fish and shellfish to survive. The water becomes cloudy, typically coloured a shade of green, yellow, brown, or red.

Eutrophication also decreases the value of rivers, lakes, and estuaries for recreation, fishing, hunting, and aesthetic enjoyment. Health problems can occur where eutrophic conditions interfere with drinking water treatment. Eutrophication was recognized as a pollution problem in European and North American lakes and reservoirs in the mid–20th century. Since then, it has become more widespread. Surveys showed that 54% of lakes in Asia are eutrophic; in Europe, 53%; in North America, 48%; in South America, 41%; and in Africa, 28%. Although eutrophication is commonly caused by human activities, it can also be a natural process particularly in lakes.

Eutrophy occurs in many lakes in temperate grasslands, for instance. Paleolimnologists now recognise that climate change, geology, and other external influences are critical in regulating the natural productivity of lakes. Some lakes also demonstrate the reverse process, becoming less nutrient rich with time. Eutrophication can also be a natural process in seasonally inundated tropical floodplains. In the Barotse Floodplain of the Zambezi River, the first floodwaters of the rainy season are usually hypoxic because of material such as cattle manure and previous decay of vegetation which grew during the dry season. These so-called "red waters" kill many fish.

The process can be made worse by the use of fertilizers in crops such as maize, rice, and sugarcane grown on the floodplain. Human activities can accelerate the rate at which nutrients enter ecosystems. Run-off from agriculture and development, pollution from septic systems and sewers, and other human-related activities increase the flow of both inorganic nutrients and organic substances into ecosystems. Elevated levels of atmospheric compounds of nitrogen can increase nitrogen availability. Phosphorus is often regarded as the main culprit in cases of eutrophication in lakes subjected to "point source" pollution from sewage pipes. The concentration of algae and the trophic state of lakes correspond well to phosphorus levels in water. Studies conducted in the Experimental Lakes Area in Ontario have shown a relationship between the addition of phosphorus and the rate of eutrophication. Humankind has increased the rate of phosphorus cycling on Earth by four times, mainly due to agricultural fertilizer production and application. Between 1950 and 1995, an estimated 600,000,000 tonnes of phosphorus were applied to Earth's surface, primarily on croplands. Policy changes to control point sources of phosphorus have resulted in rapid control of eutrophication.

OCEAN WATERS

Eutrophication is a common phenomenon in coastal waters. In contrast to freshwater systems, nitrogen is more commonly the key limiting nutrient of marine waters; thus, nitrogen levels have greater importance to understanding eutrophication problems in salt water. Estuaries tend to be naturally eutrophic because land–derived nutrients are concentrated where run-off enters a confined channel. Upwelling in coastal systems also promotes increased productivity by conveying deep, nutrient–rich waters to the surface, where the nutrients can be assimilated by algae.

The World Resources Institute has identified 375 hypoxic coastal zones in the world, concentrated in coastal areas in Western Europe, the Eastern and Southern coasts of the US, and East Asia, particularly Japan. In addition to run-off from land, atmospheric fixed nitrogen can enter the open ocean. A study in 2008 found that this could account for around one third of the ocean's external nitrogen supply, and up to 3% of the annual new marine biological production. It has been suggested that accumulating reactive nitrogen in the environment may prove as serious as putting carbon dioxide in the atmosphere.

TERRESTRIAL ECOSYSTEMS

Terrestrial ecosystems are subject to similarly adverse impacts from eutrophication. Increased nitrates in soil are frequently undesirable for plants. Many terrestrial plant species are endangered as a result of soil eutrophication, such as the majority of orchid species in Europe. Meadows, forests, and bogs are characterized by low nutrient content and slowly growing species adapted to those levels, so they can be overgrown by faster growing and more competitive species. In meadows, tall grasses that can take advantage of higher nitrogen levels may change the area so that natural species may be lost. Species-rich fens can be overtaken by reed or reedgrass species. Forest undergrowth affected by run-off from a nearby fertilized field can be turned into a nettle and bramble thicket. Chemical forms of nitrogen are most often of concern with regard to eutrophication, because plants have high nitrogen requirements so that additions of nitrogen compounds will stimulate plant growth. Nitrogen is not readily available in soil because N_2, a gaseous form of nitrogen, is very stable and unavailable directly to higher plants. Terrestrial ecosystems rely on microbial nitrogen fixation to convert N_2 into other forms such as nitrates.

However, there is a limit to how much nitrogen can be utilized. Ecosystems receiving more nitrogen than the plants require are called nitrogen-saturated. Saturated terrestrial ecosystems then can contribute both inorganic and organic nitrogen to freshwater, coastal, and marine eutrophication, where nitrogen is also typically a limiting nutrient. This is also the case with increased levels of phosphorus. However, because phosphorus is generally much less soluble than nitrogen, it is leached from the soil at a much slower rate than nitrogen. Consequently, phosphorus is much more important as a limiting nutrient in aquatic systems.

ECOLOGICAL EFFECTS

Many ecological effects can arise from stimulating primary production, but there are three particularly troubling ecological impacts: decreased biodiversity, changes in species composition and dominance, and toxicity effects.

- Increased biomass of phytoplankton
- Toxic or inedible phytoplankton species
- Increases in blooms of gelatinous zooplankton
- Decreased biomass of benthic and epiphytic algae
- Changes in macrophyte species composition and biomass
- Decreases in water transparency
- Colour, smell, and water treatment problems
- Dissolved oxygen depletion
- Increased incidences of fish kills
- Loss of desirable fish species
- Reductions in harvestable fish and shellfish
- Decreases in perceived aesthetic value of the water body

Decreased Biodiversity

When an ecosystem experiences an increase in nutrients, primary producers reap the benefits first. In aquatic ecosystems, species such as algae experience a population increase. Algal blooms limit the sunlight available to bottom-dwelling organisms and cause wide swings in the amount of dissolved oxygen in the water.

Oxygen is required by all respiring plants and animals and it is replenished in daylight by photosynthesizing plants and algae. Under eutrophic conditions, dissolved oxygen greatly increases during the day, but is greatly reduced after dark by the respiring algae and by microorganisms that feed on the increasing mass of dead algae.

When dissolved oxygen levels decline to hypoxic levels, fish and other marine animals suffocate. As a result, creatures such as fish, shrimp, and especially immobile bottom dwellers die off. In extreme cases, anaerobic conditions ensue, promoting growth of bacteria such as Clostridium botulinum that produces toxins deadly to birds and mammals. Zones where this occurs are known as dead zones.

New Species Invasion

Eutrophication may cause competitive release by making abundant a normally limiting nutrient. This process causes shifts in the species composition of ecosystems. For instance, an increase in nitrogen might allow new, competitive species to invade and out-compete original inhabitant species. This has been shown to occur in New England salt marshes.

Toxicity

Some algal blooms, otherwise called "nuisance algae" or "harmful algal blooms", are toxic to plants and animals. Toxic compounds they produce can make their way up the food chain, resulting in animal mortality. Freshwater algal blooms can pose a threat to livestock. When the algae die or are eaten, neuro-and hepatotoxins are released which can kill animals and may pose a threat to humans.

An example of algal toxins working their way into humans is the case of shellfish poisoning. Biotoxins created during algal blooms are taken up by shellfish, leading to these human foods acquiring the toxicity and poisoning humans.

Examples include paralytic, neurotoxic, and diarrhoetic shellfish poisoning. Other marine animals can be vectors for such toxins, as in the case of ciguatera, where it is typically a predator fish that accumulates the toxin and then poisons humans.

SOURCES OF HIGH NUTRIENT RUN-OFF

In order to gauge how to best prevent eutrophication from occurring, specific sources that contribute to nutrient loading must be identified. There are two common sources of nutrients and organic matter: point and nonpoint sources.

Point Sources

Point sources are directly attributable to one influence. In point sources the nutrient waste travels directly from source to water. Point sources are relatively easy to regulate.

Nonpoint Sources

Nonpoint source pollution is that which comes from ill-defined and diffuse sources. Nonpoint sources are difficult to regulate and usually vary spatially and temporally. It has been shown that nitrogen transport is correlated with various indices of human activity in watersheds, including the amount of development. Ploughing in Agriculture and development are activities that contribute most to nutrient loading. There are three reasons that nonpoint sources are especially troublesome:

Soil Retention

Nutrients from human activities tend to accumulate in soils and remain there for years. It has been shown that the amount of phosphorus lost to surface waters increases linearly with the amount of phosphorus in the soil. Thus much of the nutrient loading in soil eventually makes its way to water. Nitrogen, similarly, has a turnover time of decades or more.

Run-off to Surface Water and Leaching to Groundwater

Nutrients from human activities tend to travel from land to either surface or ground water. Nitrogen in particular is removed through storm drains, sewage pipes, and other forms of surface run-off. Nutrient losses in run-off and leachate are often associated with agriculture. Modern agriculture often involves the application of nutrients onto fields in order to maximise production.

However, farmers frequently apply more nutrients than are taken up by crops or pastures. Regulations aimed at minimising nutrient exports from agriculture are typically far less stringent than those placed on sewage treatment plants and other point source polluters.

Atmospheric Deposition

Nitrogen is released into the air because of ammonia volatilization and nitrous oxide production. The combustion of fossil fuels is a large human-initiated contributor to atmospheric nitrogen pollution. Atmospheric deposition can also affect nutrient concentration in water, especially in highly industrialized regions.

Other Causes

Any factor that causes increased nutrient concentrations can potentially lead to eutrophication. In modeling eutrophication, the rate of water renewal plays a critical role; stagnant water is allowed to collect more nutrients than bodies with replenished water supplies. It has also been shown that the drying of wetlands causes an increase in nutrient concentration and subsequent eutrophication blooms.

PREVENTION AND REVERSAL

Eutrophication poses a problem not only to ecosystems, but to humans as well. Reducing eutrophication should be a key concern when considering future policy, and a sustainable solution for everyone, including farmers and ranchers, seems feasible.

While eutrophication does pose problems, humans should be aware that natural run-off is common in ecosystems and should thus not reverse nutrient concentrations beyond normal levels.

Effectiveness

Cleanup measures have been mostly, but not completely, successful. Finnish phosphorus removal measures started in the mid-1970s and have targeted rivers and lakes polluted by industrial and municipal discharges. These efforts have had a 90% removal efficiency. Still, some targeted point sources did not show a decrease in run-off despite reduction efforts.

Minimizing Nonpoint Pollution: Future Work

Nonpoint pollution is the most difficult source of nutrients to manage. The literature suggests, though, that when these sources are controlled, eutrophication decreases. The following steps are recommended to minimize the amount of pollution that can enter aquatic ecosystems from ambiguous sources.

Riparian Buffer Zones

Studies show that intercepting non–point pollution between the source and the water is a successful means of prevention. Riparian buffer zones are interfaces between a flowing body of water and land, and have been created near waterways in an attempt to filter pollutants; sediment and nutrients are deposited here instead of in water. Creating buffer zones near farms and roads is another possible way to prevent nutrients from traveling too far.

Still, studies have shown that the effects of atmospheric nitrogen pollution can reach far past the buffer zone. This suggests that the most effective means of prevention is from the primary source.

Prevention Policy

Laws regulating the discharge and treatment of sewage have led to dramatic nutrient reductions to surrounding ecosystems, but it is generally agreed that a policy regulating agricultural use of fertilizer and animal waste must be imposed. In Japan the amount of nitrogen produced by livestock is adequate to serve the fertilizer needs for the agriculture industry. Thus, it is not unreasonable to command livestock owners to clean up animal waste—which when left stagnant will leach into ground water. Policy concerning the prevention and reduction of eutrophication can be broken down into four sectors: Technologies, public participation, economic instruments, and cooperation.

The term technology is used loosely, referring to a more widespread use of existing methods rather than an appropriation of new technologies. Nonpoint sources of pollution are the primary contributors to eutrophication, and their effects can be easily minimized through common agricultural practices. Reducing the amount of pollutants that reach a watershed can be achieved through the protection of its forest cover, reducing the amount of erosion leeching into a watershed.

Also, through the efficient, controlled use of land using sustainable agricultural practices to minimize land degradation, the amount of soil run-off and nitrogen–based fertilizers reaching a watershed can be reduced. Waste disposal technology constitutes another factor in eutrophication prevention. Because a major contributor to the nonpoint source nutrient loading of water bodies is untreated domestic sewage, it is necessary to provide treatment facilities to highly urbanized areas, particularly those in underdeveloped nations, in which treatment of domestic waste water is a scarcity. The technology to safely and efficiently reuse waste water, both from domestic and industrial sources, should be a primary concern for policy regarding eutrophication. The role of the public is a major factor for the effective prevention of eutrophication.

In order for a policy to have any effect, the public must be aware of their contribution to the problem, and ways in which they can reduce their effects. Programmes instituted to promote participation in the recycling and elimination of wastes, as well as education on the issue of rational water use are necessary to protect water quality within urbanized areas and adjacent water bodies. Economic instruments, "which include, among others, property rights, water markets, fiscal and financial instruments, charge systems and liability systems, are gradually becoming a substantive component of the management tool set used for pollution control and water allocation decisions." Incentives for those who practice clean, renewable, water management technologies are an effective means of encouraging pollution prevention.

By internalizing the costs associated with the negative effects on the environment, governments are able to encourage a cleaner water management. Because a body of water can have an effect on a range of people reaching far beyond that of the watershed, cooperation between different organizations is necessary to prevent the intrusion of contaminants that can lead to eutrophication. Agencies ranging from state governments to those of water resource management and non-governmental organizations, going as low as the local population, are responsible for preventing eutrophication of water bodies.

Nitrogen Testing and Modeling

Soil Nitrogen Testing is a technique that helps farmers optimize the amount of fertilizer applied to crops. By testing fields with this method, farmers saw a decrease in fertilizer application costs, a decrease in nitrogen lost to surrounding sources, or both. By testing the soil and modeling the bare minimum amount of fertilizer needed, farmers reap economic benefits while reducing pollution.

Organic Farming

There has been a study that found that organically fertilized fields "significantly reduce harmful nitrate leaching" over conventionally fertilized fields. However, a more recent study found that eutrophication impacts are in some cases higher from organic production than they are from conventional production.

FRESHWATER ENVIRONMENTAL QUALITY PARAMETERS

Freshwater environmental quality parameters are the natural and man-made chemical, biological and microbiological characteristics of rivers, lakes and ground–waters, the ways they are measured and the ways that they change. The values or concentrations attributed to such parameters can be used to describe the pollution status of an environment, its biotic status or to predict the likelihood or otherwise of a particular organisms being present. Monitoring of environmental quality parameters is a key activity in managing the environment, restoring polluted environments and anticipating the effects of man-made changes on the environment.

CHARACTERISATION

The first step in understanding the chemistry of freshwaters is to take samples and analyse them for the chemical constituents that are of interest.

Sampling

Freshwaters are surprisingly difficult to sample because they are rarely homogeneous and their quality varies during the day and during the year. In addition the most representative sampling locations are often at a distance from the shore or bank increasing the logistic complexity.

Rivers

Filling a clean bottle with river water is a very simple task, but a single sample is only representative of that point along the river the sample was taken from and at that point in time. Understanding the chemistry of a whole river, or even a significant tributary, requires prior investigative work to understand how homogeneous or mixed the flow is and to determine if the quality changes during the course of a day and during the course of a year.

Almost all natural rivers will have very significant patterns of change through the day and through the seasons. Many rivers also have a very large flow that is unseen. This flows through underlying gravel and sand layers and is called the hyporheic zone How much mixing there is between the hyporheic zone and the water in the open channel will depend on a variety of factors, some of which relate to flows leaving aquifers which may have been storing water for many years.

Ground-waters

Ground waters by their very nature are often very difficult to access to take a sample. As a consequence the majority of ground–water data comes from samples taken from springs, wells, water supply bore-holes and in natural caves. In recent decades as the need to understand ground water dynamics has increased, an increasing number or monitoring bore–holes have been drilled into aquifers

Lakes

Lakes and ponds can be very large and support a complex eco-system in which environmental parameters vary widely in all three physical dimensions and with time. Large lakes in the temperate zone often stratify in the warmer months into a warmer upper layers rich in oxygen and a colder lower layer with low oxygen levels. In the autumn, falling temperatures and occasional high winds result in the mixing of the two layers into a more homogeneous whole.

When stratification occurs it not only affects oxygen levels but also many related parameters such as iron, phosphate and manganese which are all changed in their chemical form by change in the redox potential of the environment. Lakes also receive waters, often from many different sources with varying qualities. Solids from stream inputs will typically settle near the mouth of the stream and depending on a variety of factors the incoming water may float over the surface of the lake, sink beneath the surface or rapidly mix with the lake water. All of these phenomena can skew the results of any environmental monitoring unless the process are well understood.

MIXING ZONES

Where two rivers meet at a confluence there exists a mixing zone. A mixing zone may be very large and extend for many miles as in the case of the Mississippi and Missouri rivers in the United States and the River Clwyd and River Elwy in North Wales. In a mixing zone water chemistry may be very variable and can be difficult to predict. The chemical interactions are not just simple mixing but may be complicated by biological processes from submerged macrophytes and by water joining the channel from the hyporheic zone or from springs draining an aquifer.

GEOLOGICAL INPUTS

The geology that underlies a river or lake has a major impact on its chemistry. A river flowing across very ancient precambrian schists is likely to have dissolved very little from the rocks and maybe similar to de-ionised water at least in the headwaters. Conversely a river flowing through chalk hills, and especially if its source is in the chalk, will have a high concentration of carbonates and bicarbonates of Calcium and possibly Magnesium. As a river progresses along its course it may pass through a variety of geological types and it may have inputs from aquifers that do not appear on the surface anywhere in the locality. Water chemistry between systems varies tremendously.

ATMOSPHERIC INPUTS

Oxygen is probably the most important chemical constituent of surface water chemistry, as all aerobic organisms require it for survival. It enters the water mostly via diffusion at the water-air interface. Oxygen's solubility in water decreases as water temperature increases. Fast, turbulent streams expose more of the water's surface area to the air and tend to have low temperatures and thus more oxygen than slow, backwaters. Oxygen is a by-product of photosynthesis, so systems with a high abundance of aquatic algae and plants may also have high concentrations of oxygen during the day. These levels can decrease significantly during the night when primary producers switch to respiration. Oxygen can be limiting if circulation between the surface and deeper layers is poor, if the activity of animals is very high, or if there is a large amount of organic decay occurring such as following Autumn leaf-fall. Most other atmospheric inputs come from man-made or anthropogenic sources the most significant of which are the oxides of sulphur produced by burning sulphur rich fuels such as coal and oil which give rise to acid rain. The chemistry of sulphur oxides is complex both in the atmosphere and in river systems. However the effect on the overall chemistry is simple in that it reduces the pH of the water making it more acidic. The pH change is most marked in rivers with very low concentrations of dissolved salts as these cannot buffer the effects of the acid input. Rivers downstream of major industrial conurbations are also at greatest risk. In parts of Scandinavia and West Wales and Scotland many rivers became so acidic from oxides of sulphur that most fish life was destroyed and pHs as low as pH4 were recorded during critical weather conditions.

ANTHROPOGENIC INPUTS

The majority of rivers on the planet and many lakes have received or are receiving inputs from human–kind's activities. In the industrialised world, many rivers have been very seriously polluted, at least during the 19th and the first half of the 20th centuries. Although in general there has been much improvement in the developed world, there is still a great deal of river pollution apparent on the planet.

TOXICITY

In most environmental situations the presence or absence of an organism is determined by a complex web of interactions only some of which will be related to measurable chemical or biological parameters. Flow rate, turbulence, inter and intra specific competition, feeding behaviour, disease, parasatism, commensalism and symbiosis are just a few of the pressures and opportunities facing any organism or population. Most chemical constituents favour some organisms and are less favourable to others. However there are some cases where a chemical constituent exerts a toxic effect. *i.e.* where the concentration can kill or severely inhibit the normal functioning of the organism. Where a toxic effect has been demonstrated this may be noted in the sections below dealing with the individual parameters.

CHEMICAL CONSTITUENTS

Colour and Turbidity

Often it is the colour of freshwater or how clear or hazy the water is that is the most obvious visual characteristic. Unfortunately neither colour nor turbidity are strong indicators of the overall chemical composition of water. However both colour and turbidity reduce the amount of light penetrating the water and can have significant impact on algae and macrophytes. Some algae in particular are highly dependant on water with low colour and turbidity Many rivers draining high moorlands overlain by peat have a very deep yellow brown colour caused by dissolved humic acids.

Organic Constituents

One of the principal sources of elevated concentrations of organic chemical constituents is from treated sewage. Dissolved organic material is most commonly measured using either the Biochemical oxygen demand test of the Chemical oxygen demand test. Organic constituents are significant in river chemistry for the effect that they have on dissolved oxygen concentration and for the impact that individual organic species may have directly on aquatic biota. Any organic and degradable material utilises oxygen as it decomposes.

Where organic concentrations are significantly elevated the effects on oxygen concentrations can be significant and as conditions get extreme the river bed may become anoxic. Some organic constituents such as synthetic hormones, pesticides, pthalates have direct metabolic effects on aquatic biota and even on humans drinking water taken from the river. Understanding such constituents and how they can be identified and quantified is becoming of increasing importance in the understanding of freshwater chemistry.

Metals

A wide range of metals may be found in rivers from natural sources where metal ores are present in the rocks over which the river flows or in the aquifers feeding water into the river. However many rivers have an increased load of metals because of industrial activities which include mining and quarrying and the processing and use of metals.

Iron

Iron, usually as Fe is a common constituent of river waters at very low levels. As concentrations increase visible orange/brown staining appears and any increase in concentrations is likely to create conditions where complex insoluble oxides, hydroxides and carbonates of iron start precipitating out producing a semi–gelatinous and dense floc carpeting the river bed. Such conditions are very deleterious to most organisms and can cause serious damage in a river system. Coal mining is also a very significant source of Iron both in mine–waters and from stocking yards of coal and from coal processing. Long abandoned mines can be a highly intractable source of high concentrations of Iron. Low levels of iron are common in spring waters emanating from deep-seated aquifers and maybe regarding as health giving springs. Such springs are commonly called Chalybeate springs and have given rise to a number of Spa towns in Europe and the United States.

Zinc

Zinc is normally associated with metal mining, especially Lead and Silver mining but is also a component pollutant associated with a variety of other metal mining activities and with Coal mining. Zinc is toxic at relatively low concentrations to many aquatic organisms. Microregma starts to show a toxic reaction at concentrations as low as 0.33 mg/l

Heavy Metals

Lead and silver in river waters are commonly found together and associated with lead mining. Impacts from very old mines can be very long-lived. In the River Ystwyth in Wales for example, the effects of silver and lead mining in the 17th and 18th centuries in the headwaters still causes unacceptably high levels of Zinc and Lead in the river water right down to its confluence with the sea. Silver is very toxic even at very low concentrations but leaves no visible of its contamination. Lead is also highly toxic to freshwater organisms and to humans if the water is used as drinking water.

As with Silver, Lead pollution is not visible to the naked eye. The River Rheidol in west Wales had a major series of lead mines in its headwaters until the end of the 19th century and its mine discharges and waste tips remain to this day. In 1919–1921 only 14 species of invertebrates were found were found in the lower Rheidol when Lead concentrations were between 0.2ppm and 0.5ppm. By 1932 the lead concentration had reduced to 0.02ppm to 0.1ppm because of the abandonment of mining and, at those concentrations, the bottom fauna had stabilized to 103 species including three leeches. Coal mining is also a very significant source of metals, especially Iron, Zinc and Nickel particularly where the coal is rich if pyrites which oxidises on contact with the air producing a very acidic leachate which is able to dissolve metals from the coal. Significant levels of copper are unusual in rivers and where it does it occur the source is most likely to be mining activities, coal stocking, or pig farming. Rarely elevated levels may be of geological origin. Copper is acutely toxic to many freshwater organisms, especially algae, at very low concentrations and significant concentration in river water may have serious adverse effects on the local ecology.

Nitrogen

Nitrogenous compounds have a variety of sources including washout of oxides of nitrogen from the atmosphere, some geological inputs and some from macrophyte and algal nitrogen fixation. However for many rivers in the proximity of humans, the largest input is from sewage whether treated or untreated. The nitrogen derives from breakdown products of proteins found in urine and faeces. These products, being very soluble, often pass through sewage treatment process and are discharged into rivers as a component of sewage treatment effluent. Nitrogen may be in the form of nitrate, nitrite, ammonia or ammonium salts or what is termed albuminoid nitrogen or nitrogen still within an organic proteinoid molecule. The differing forms of nitrogen are relatively stable in most river systems with nitrite slowly transforming into nitrate in well oxygenated rivers and ammonia transforming into nitrite/nitrate.

However, the process are slow in cool rivers and reduction in concentration may more often be attributed to simple dilution. All forms of nitrogen are taken up by macrophytes and algae and elevated levels of nitrogen are often associated with overgrowths of plants or eutrophication. These can have the effect of blocking channels and inhibiting navigation.

However, ecologically, the more significant effect is on dissolved oxygen concentrations which may become super-saturated during daylight due to plant photsynthesis but then drop to very low levels during darkness as plant respiration uses up the dissolved oxygen. Coupled with the release of oxygen in photosynthesis is the creation of bi-carbonate ions which cause a steep rise in pH and this is matched in darkness as carbon dioxide is released through respiration which substantially lowers the pH.

Thus high levels of nitrogenous compounds tends to lead to eutrophication with extreme variations in parameters which in turn can substantially degrade the ecological worth of the watercourse. Ammonium ions also have a toxic effect, especially on fish. The toxicity of ammonia is dependent on both pH and temperature and an added complexity is the buffering effect of the blood/water interface across the gill membrane which masks any additional toxicity over about pH 8.0. The management of river chemistry to avoid ecological damage is particularly difficult in the case of ammonia as a wide range of potential scenarios of concentration, pH and temperature have to be considered and the diurnal pH fluctuation caused by photosynthesis considered. On warm summer days with high-bicarbonate concentrations unexpectedly toxic conditions can be created.

Phosphorus

Phosphorus compounds are usually found as relatively insoluble phosphates in river water and, except in some exceptional circumstances, the origin is from agriculture of human sewage. Phosphorus can encourage excessive growths of plants and algae and contribute to eutrophication. If a river discharges into a lake or reservoir phosphate can be mobilised year after year by natural processes. In the summer time, lakes stratify so that warm oxygen rich water floats on top of cold oxygen poor water. In the warm upper layers—the epilimnion–plants consume the available phosphate. As the plants die in the late summer they fall into the cool water layers underneath–the hypolimnion and decompose. During winter turn–over, when a lake becomes fully mixed through the action of winds on a cooling body of water—the phosphates are spread throughout the lake again to feed a new generation of plants. This process is one of the principal causes of persistent algal blooms at some lakes.

Arsenic

Geological deposits of arsenic may be released into rivers where deep ground-waters are exploited as in parts of Pakistan. Some lead and copper mining also encounters ores containing arsenic which may then be released into local rivers.

Solids

Inert solids are produced in all montane rivers as the energy of the water helps grind away rocks into gravel, sand and finer material. Much of this settles very quickly and provides an important substrate for many aquatic organisms. Many salmonid fish require beds of gravel and sand in which to lay their eggs. Many other types of solids from agriculture, mining, quarrying, urban run-off and sewage may block-out sunlight from the river and may block interstices in gravel beds making them useless for spawning and supporting insect life.

Bacterial, Viral and Parasite Inputs

Both agriculture and sewage treatment produce inputs into rivers with very high concentrations of bateria and viruses including a wide range of pathogenic organisms. Even in areas with little human activity significant levels of bacteria and viruses can be detected originating from fish and aquatic mammals and from animals grazing near rivers such as deer. Upland waters draining areas frequented by sheep, goats or deer may also harbour a variety of opportunistic human parasites such as liver fluke. Consequently there are very few rivers from which the water is safe to drink without some form of sterilisation or disinfection.

In rivers used for contact recreation such as swimming, safe levels of bacteria and viruses can be established based on risk assessment. Under certain conditions bacteria can colonise freshwaters occasionally making large rafts of filamentous mats known as sewage fungus–usually Sphaerotilus natans. The presence of such organisms is almost always an indicator of extreme organic pollution and would be expected to be matched with low dissolved oxygen concentrations and high BOD vales. E. coli bacteria have been commonly found in recreational waters and their presence is used to indicate the presence of recent fecal contamination, but E. coli presence may not be indicative of human waste.

E. coli are harbored in all warm-blooded animals: birds and mammals alike. E. coli bacteria have also been found in fish and turtles. Sand also harbors E. coli bacteria and some strains of E. coli have become naturalized. Some geographic areas may support unique populations of E. coli and conversely, some E. coli strains are cosmopolitan.

pH

pH in rivers is affected by the geology of the water source, atmospheric inputs and a range of other chemical contaminants. pH is only likely to become an issue on very poorly buffered upland rivers where atmospheric sulphur and nitrogen oxides may very significantly depress the pH as low as pH4 or in eutrophic alkaline rivers where photosynthetic bi-carbonate ion production in photosynthesis may drive the pH up above pH10

HYPOXIA

Hypoxia, or oxygen depletion, is a phenomenon that occurs in aquatic environments as dissolved oxygen becomes reduced in concentration to a point where it becomes detrimental to aquatic organisms living in the system. Dissolved oxygen is typically expressed as a percentage of the oxygen that would dissolve in the water at the prevailing temperature and salinity. An aquatic system lacking dissolved oxygen is termed anaerobic, reducing, or anoxic; a system with low concentration—in the range between 1 and 30% saturation—is called hypoxic or dysoxic. Most fish cannot live below 30% saturation. A "healthy" aquatic environment should seldom experience less than 80%. The exaerobic zone is found at the boundary of anoxic and hypoxic zones.

WHERE HYPOXIA OCCURS

Hypoxia can occur throughout the water column and also at high altitudes as well as near sediments on the bottom. It usually extends throughout 20-50% of the water column, but depending on the water depth and location of pycnoclines it can occur in 10–80% of the water column. For example, in a 10–meter water column, it can reach up to 2 meters below the surface. In a 20–meter water column, it can extend up to 8 meters below the surface.

CAUSES OF HYPOXIA

Oxygen depletion can result from a number of natural factors, but is most often a concern as a consequence of pollution and eutrophication in which plant nutrients enter a river, lake, or ocean, and phytoplankton blooms are encouraged. While phytoplankton, through photosynthesis, will raise DO saturation during daylight hours, the dense population of a bloom reduces DO saturation during the night by respiration. When phytoplankton cells die, they sink towards the bottom and are decomposed by bacteria, a process that further reduces DO in the water column.

If oxygen depletion progresses to hypoxia, fish kills can occur and invertebrates like worms and clams on the bottom may be killed as well.Hypoxia may also occur in the absence of pollutants. In estuaries, for example, because freshwater flowing from a river into the sea is less dense than salt water, stratification in the water column can result. Vertical mixing between the water bodies is therefore reduced, restricting the supply of oxygen from the surface waters to the more saline bottom waters.

The oxygen concentration in the bottom layer may then become low enough for hypoxia to occur. Areas particularly prone to this include shallow waters of semi-enclosed water bodies such as the Waddenzee or the Gulf of Mexico, where land run-off is substantial. In these areas a so-called "dead zone" can be created.

The World Resources Institute has identified 375 hypoxic coastal zones around the world, concentrated in coastal areas in Western Europe, the Eastern and Southern coasts of the US, and East Asia, particularly in Japan. Hypoxia may also be the explanation for periodic phenomena such as the Mobile Bay jubilee, where aquatic life suddenly rushes to the shallows, perhaps trying to escape oxygen-depleted water. Recent widespread shellfish kills near the coasts of Oregon and Washington are also blamed on cyclic dead zone ecology.

SOLUTIONS

To combat hypoxia, it is essential to reduce the amount of land-derived nutrients reaching rivers in run-off. Defensively this can be done by improving sewage treatment and by reducing the amount of fertilizers leaching into the rivers. Offensively this can be done by restoring natural environments along a river; marshes are particularly effective in reducing the amount of phosphorus and nitrogen in water. Technological solutions are also possible, such as that used in the redeveloped Salford Docks area of the Manchester Ship Canal in England, where years of run-off from sewers and roads had accumulated in the slow running waters.

In 2001 a compressed air injection system was introduced, which raised the oxygen levels in the water by up to 300%. The resulting improvement in water quality led to an increase in the number of invertebrate species, such as freshwater shrimp, to more than 30. Spawning and growth rates of fish species such as roach and perch also increased to such an extent that they are now amongst the highest in England.

In a very short time the oxygen saturation can drop to zero when offshore blowing winds drive surface water out and anoxic depthwater rises up. At the same time a decline in temperature and a rise in salinity is observed. New approaches of long-term monitoring of oxygen regime in the ocean observe online the behaviour of fish and zooplankton, which changes drastically under reduced oxygen saturations and already at very low levels of water pollution.

BOG CHEMISTRY

In certain northern European sphagnum acidic bogs, a condition of hypoxia arises that prevents tissue decay by impeding micro-organisms in the soil and groundwater. Remarkable preservation of human mummies has occurred in some cases such as the discovery of Haraldskær Woman and Tollund Man in Jutland, Denmark and Lindow man in Cheshire, England.

MARINE DEBRIS

Marine debris, also known as marine litter, is human-created waste that has deliberately or accidentally become afloat in a lake, sea, ocean or waterway. Oceanic debris tends to accumulate at the centre of gyres and on coastlines, frequently washing aground, when it is known as beach litter or tidewrack. Deliberate disposal of wastes at sea is called ocean dumping. Some seeming forms of marine debris, such as driftwood, occur naturally, and human activities have been discharging similar material into the oceans for thousands of years. Recently however, with the increasing use of plastic, human influence has become an issue as many types of plastics do not biodegrade. Waterborne plastic poses a serious threat to fish, seabirds, marine reptiles, and marine mammals, as well as to boats and coastal habitations. Ocean dumping, accidental container spillages, litter washed into storm drains, and wind-blown landfill waste are all contributing to this problem.

TYPES OF DEBRIS

A wide variety of anthropogenic artifacts can become marine debris;

plastic bags, balloons, buoys, rope, medical waste, glass bottles and plastic bottles, cigarette lighters, beverage cans, styrofoam, lost fishing line and nets, and various wastes from cruise ships and oil rigs are among the items commonly found to have washed ashore.

Six pack rings, in particular, are considered a poster child of the damage that garbage can do to the marine environment. Studies have shown that eighty per cent of marine debris is plastic–a component that has been rapidly accumulating since the end of World War II. Plastics accumulate because they don't biodegrade as many other substances do; although they will photodegrade on exposure to sunlight, they do so only under dry conditions, as water inhibits photolysis.

Ghost Nets

Fishing nets left or lost in the ocean by fishermen–ghost nets–can entangle fish, dolphins, sea turtles, sharks, dugongs, crocodiles, seabirds, crabs, and other creatures. Acting as designed, these nets restrict movement, causing starvation, laceration and infection, and, in animals that need to return to the surface to breathe, suffocation.

Nurdles and Plastics Bags

Nurdles, also known as mermaid's tears, are plastic pellets typically under five millimetres in diameter, and are a major component of marine debris. They are used as a raw material in plastics manufacturing, and are thought to enter the natural environment after accidental spillages. Small plastic fragments are also created by the physical weathering of larger plastic debris. Nurdles strongly resemble fish eggs. Plastic waste has reached all oceans around the world. This pollution harms and kills an estimated 100,000 sea turtles and marine mammals and 1,000,000 sea creatures each year.

Pelagic plastic pieces in the center of our ocean's gyres outnumber live marine plankton, and are passed up the food chain to reach all marine life. Plastic shopping bags may clog digestive tracts when consumed and may cause starvation through restricting the movement of food, or by filling the stomach and tricking the animal into thinking it is full. A 1994 study of the seabed using trawl nets in the North-Western Mediterranean around the coasts of Spain, France and Italy reported a particularly high mean concentration of debris; an average of 1,935 items per square kilometre. Plastic debris accounted for 77%, of which 93% was plastic bags.

SOURCE OF DEBRIS

It has been estimated that container ships lose over 10,000 containers at sea each year. One famous spillage occurred in the Pacific Ocean in 1992, when thousands of rubber ducks and other toys went overboard during a storm. The toys have since been found all over the world; Curtis Ebbesmeyer and other scientists have used the incident to gain a better understanding of ocean currents. Similar incidents have happened before, with the same potential to track currents, such as when Hansa Carrier dropped 21 containers. In 2007, MSC Napoli was beached in the English Channel, and dropped hundreds of containers, most of which washed up on the Jurassic Coast, a World Heritage Site. Though it was originally assumed that most oceanic marine waste stemmed directly from ocean dumping, it is now thought that around four fifths of the oceanic debris is from rubbish blown seaward from landfills, and urban run-off washed down storm drains. In the 1987 Syringe Tide, medical waste washed ashore in New Jersey after having been blown from the Fresh Kills Landfill. Even on the remote sub-Antarctic island of South Georgia, fishing related debris made up approximately 80% of plastics have been found washed up and are responsible for the entanglement of large numbers of Antarctic fur seals.

LEGALITY OF OCEAN AND RIVER DUMPING

Ocean dumping is controlled by international law:

- *The London Convention*: A United Nations agreement to control ocean dumping
- *MARPOL 73/78*: An international convention designed to minimize pollution of the seas, including dumping, oil and exhaust pollution

European Law

In 1972 and 1974, conventions were held in Oslo and Paris respectively, and resulted in the passing of the OSPAR Convention, an international treaty controlling marine pollution in the north-east Atlantic Ocean around Europe. A similar Barcelona Convention exists to protect the Mediterranean Sea. The Water Framework Directive of 2000 is a European Union directive committing EU member states to make their inland and coastal waters free from human influence. In the United Kingdom, the Marine and Coastal Access Act is designed to "ensure clean healthy, safe, productive and biologically diverse oceans and seas, by putting in place better systems for delivering sustainable development of marine and coastal environment".

United States Law

In 1972, the United States Congress passed the Ocean Dumping Act, giving the Environmental Protection Agency power to monitor and regulate the dumping of sewage sludge, industrial waste, radioactive waste and biohazardous materials into the nation's territorial waters. The Act was amended sixteen years later to include medical wastes. It is illegal to dispose of any plastic in all US waters. In 2008, the California State Legislature considered several bills aimed at reducing the sources of marine debris, following the recommendations of the California Ocean Protection Council.

Ownership of Debris

Property law, admiralty law, and the law of the sea may be of relevance when lost, mislaid, and abandoned property is found at sea. Salvage law has as a basis that a salvor should be rewarded for risking his life and property to rescue the property of another from peril. On land the distinction between deliberate and accidental loss led to the concept of a "treasure trove". In the United Kingdom, shipwrecked goods should be reported to a Receiver of Wreck, and if identifiable, they should be returned to their rightful owner.

THE GREAT PACIFIC GARBAGE PATCH

Once waterborne, debris is far from immobile. Flotsam can be blown by the wind, or follow the flow of ocean currents, often ending up in the middle of oceanic gyres where currents are weakest. The Great Pacific Garbage Patch is one such example of this, comprising a vast region of the North Pacific Ocean rich with anthropogenic wastes.

Estimated to be double the size of Texas, the area contains more than 3 million tons of plastic. This means that there are approximately six pounds of plastic for every pound of plankton per cubic meter of seawater. The mass of plastic in our oceans may be as high as one hundred million tons. Islands situated within gyres frequently have their coastlines ruined by the waste that inevitably washes ashore; prime examples are Midway and Hawaii. Clean-up teams around the world patrol beaches to clean up this environmental threat.

ENVIRONMENTAL IMPACT

Many animals that live on or in the sea consume flotsam by mistake, as it often looks similar to their natural prey. Plastic debris, when bulky or tangled, is difficult to pass, and may become permanently lodged in the digestive tracts of these animals, blocking the passage of food and causing death through starvation or infection. Tiny floating particles also resemble zooplankton, which can lead filter feeders to consume them and cause them to enter the ocean food chain. In samples taken from the North Pacific Gyre in 1999 by the Algalita Marine Research Foundation, the mass of plastic exceeded that of zooplankton by a factor of six.

Toxic additives used in the manufacture of plastic materials can leach out into their surroundings when exposed to water. Waterborne hydrophobic pollutants collect and magnify on the surface of plastic debris, thus making plastic far more deadly in the ocean than it would be on land. Hydrophobic contaminants are also known to bioaccumulate in fatty tissues, biomagnifying up the food chain and putting great pressure on apex predators. Some plastic additives are known to disrupt the endocrine system when consumed; others can suppress the immune system or decrease reproductive rates. Not all anthropogenic artefacts in the oceans are harmful however.

Iron and concrete do little damage to the environment as they are generally immobile, and can even be used as scaffolding for the creation of artificial reefs, increasing the biodiversity of a coastal region. Entire ships have been deliberately sunk in coastal waters for that purpose. Some organisms have adapted to live on mobile plastic debris, which has allowed the inhabitants to disperse all over the world and become invasive species in remote ecosystems.

DEBRIS REMOVAL

A variety of techniques are used to collect and remove marine debris by concerned jurisdictions or volunteer organizations. Besides collection by hand, some cities operate special beach–cleaning machines that collect trash deposited by the sea along the coast line. Other places arrange for picking debris while it is still floating; such activities are often undertaken regularly where floating debris are perceived to pose danger to navigation. For example, the US Army Corps of Engineers reports removing 90 tons of "drifting material" from San Francisco Bay shipping lanes etc. every month. The Corps has been doing this work since 1942, when a seaplane carrying Admiral Chester W. Nimitz collided with a piece of floating debris and sank, resulting in the death of its pilot. Elsewhere, various kinds of "trash traps" are installed on small rivers flowing into the sea, to capture waterborne debris before it reaches the sea. For example, South Australia's Adelaide operates a number of such traps, known as "trash racks" or "gross pollutant traps" on the Torrens River, which flows into Gulf St Vincent.

MARINE POLLUTION

Marine pollution occurs when harmful effects, or potentially harmful effects, can result from the entry into the ocean of chemicals, particles, industrial, agricultural and residential waste, noise, or the spread of invasive organisms. Most sources of marine pollution are land based. The pollution often comes from nonpoint sources such as agricultural run-off and wind blown debris. Many potentially toxic chemicals adhere to tiny particles which are then taken up by plankton and benthos animals, most of which are either deposit or filter feeders. In this way, the toxins are concentrated upward within ocean food chains.

Many particles combine chemically in a manner highly depletive of oxygen, causing estuaries to become anoxic. When pesticides are incorporated into the marine ecosystem, they quickly become absorbed into marine food webs. Once in the food webs, these pesticides can cause mutations, as well as diseases, which can be harmful to humans as well as the entire food web. Toxic metals can also be introduced into marine food webs.

These can cause a change to tissue matter, biochemistry, behaviour, reproduction, and suppress growth in marine life. Also, many animal feeds have a high fish meal or fish hydrolysate content. In this way, marine toxins can be transferred to land animals, and appear later in meat and dairy products.

HISTORY

Although marine pollution has a long history, significant international laws to counter it were enacted in the twentieth century. Marine pollution was a concern during several United Nations Conferences on the Law of the Sea beginning in the 1950s. Most scientists believed that the oceans were so vast that they had unlimited ability to dilute, and thus render harmless, pollution. In the late 1950s and early 1960s, there were several controversies about dumping radioactive waste off the coasts of the United States by companies licensed by the Atomic Energy Commission, into the Irish Sea from the British reprocessing facility at Windscale, and into the Mediterranean Sea by the French Commissariat à l'Energie Atomique. After the Mediterranean Sea controversy, for example, Jacques Cousteau became a worldwide figure in the campaign to stop marine pollution.

Marine pollution made further international headlines after the 1967 crash of the oil tanker Torrey Canyon, and after the 1969 Santa Barbara oil spill off the coast of California. Marine pollution was a major area of discussion during the 1972 United Nations Conference on the Human Environment, held in Stockholm. That year also saw the signing of the Convention on the Prevention of Marine Pollution by Dumping of Wastes and Other Matter, sometimes called the London Convention. The London Convention did not ban marine pollution, but it established black and gray lists for substances to be banned or regulated by national authorities.

Cyanide and high-level radioactive waste, for example, were put on the black list. The London Convention applied only to waste dumped from ships, and thus did nothing to regulate waste discharged as liquids from pipelines.

PATHWAYS OF POLLUTION

There are many different ways to categorize, and examine the inputs of pollution into our marine ecosystems. Patin notes that generally there are three main types of inputs of pollution into the ocean: direct discharge of waste into the oceans, run-off into the waters due to rain, and pollutants that are released from the atmosphere. One common path of entry by contaminants to the sea are rivers. The evaporation of water from oceans exceeds precipitation.

The balance is restored by rain over the continents entering rivers and then being returned to the sea. The Hudson in New York State and the Raritan in New Jersey, which empty at the northern and southern ends of Staten Island, are a source of mercury contamination of zooplankton in the open ocean. The highest concentration in the filter-feeding copepods is not at the mouths of these rivers but 70 miles south, nearer Atlantic City, because water flows close to the coast.

It takes a few days before toxins are taken up by the plankton. Pollution is often classed as point source or nonpoint source pollution. Point source pollution occurs when there is a single, identifiable, and localized source of the pollution. An example is directly discharging sewage and industrial waste into the ocean. Pollution such as this occurs particularly in developing nations. Nonpoint source pollution occurs when the pollution comes from ill-defined and diffuse sources. These can be difficult to regulate. Agricultural run-off and wind blown debris are prime examples.

Direct Discharge

Pollutants enter rivers and the sea directly from urban sewerage and industrial waste discharges, sometimes in the form of hazardous and toxic wastes. Inland mining for copper, gold. etc., is another source of marine pollution. Most of the pollution is simply soil, which ends up in rivers flowing to the sea. However, some minerals discharged in the course of the mining can cause problems, such as copper, a common industrial pollutant, which can interfere with the life history and development of coral polyps. Mining has a poor environmental track record. For example, just as to the United States Environmental Protection Agency, mining has contaminated portions of the headwaters of over 40% of watersheds in the western continental US. Much of this pollution finishes up in the sea.

Land Run-off

Surface run-off from farming, as well as urban run-off and run-off from the construction of roads, buildings, ports, channels, and harbours, can carry soil and particles laden with carbon, nitrogen, phosphorus, and minerals. This nutrient–rich water can cause fleshy algae and phytoplankton to thrive in coastal areas; known as algal blooms, which have the potential to create hypoxic conditions by using all available oxygen. Polluted run-off from roads and highways can be a significant source of water pollution in coastal areas. About 75 per cent of the toxic chemicals that flow into Puget Sound are carried by stormwater that runs off paved roads and driveways, rooftops, yards and other developed land.

Ship Pollution

Ships can pollute waterways and oceans in many ways. Oil spills can have devastating effects. While being toxic to marine life, polycyclic aromatic hydrocarbons, the components in crude oil, are very difficult to clean up, and last for years in the sediment and marine environment. Discharge of cargo residues from bulk carriers can pollute ports, waterways and oceans. In many instances vessels intentionally discharge illegal wastes despite foreign and domestic regulation prohibiting such actions.

It has been estimated that container ships lose over 10,000 containers at sea each year. Ships also create noise pollution that disturbs natural wildlife, and water from ballast tanks can spread harmful algae and other invasive species. Ballast water taken up at sea and released in port is a major source of unwanted exotic marine life. The invasive freshwater zebra mussels, native to the Black, Caspian and Azov seas, were probably transported to the Great Lakes via ballast water from a transoceanic vessel.

Meinesz believes that one of the worst cases of a single invasive species causing harm to an ecosystem can be attributed to a seemingly harmless jellyfish. Mnemiopsis leidyi, a species of comb jellyfish that spread so it now inhabits estuaries in many parts of the world. It was first introduced in 1982, and thought to have been transported to the Black Sea in a ship's ballast water. The population of the jellyfish shot up exponentially and, by 1988, it was wreaking havoc upon the local fishing industry. "The anchovy catch fell from 204,000 tons in 1984 to 200 tons in 1993; sprat from 24,600 tons in 1984 to 12,000 tons in 1993; horse mackerel from 4,000 tons in 1984 to zero in 1993."

Now that the jellyfish have exhausted the zooplankton, including fish larvae, their numbers have fallen dramatically, yet they continue to maintain a stranglehold on the ecosystem. Invasive species can take over once occupied areas, facilitate the spread of new diseases, introduce new genetic material, alter underwater seascapes and jeopardize the ability of native species to obtain food. Invasive species are responsible for about $138 billion annually in lost revenue and management costs in the US alone.

Atmospheric Pollution

Another pathway of pollution occurs through the atmosphere. Wind blown dust and debris, including plastic bags, are blown seaward from landfills and other areas. Dust from the Sahara moving around the southern periphery of the subtropical ridge moves into the Caribbean and Florida during the warm season as the ridge builds and moves northward through the subtropical Atlantic. Dust can also be attributed to a global transport from the Gobi and Taklamakan deserts across Korea, Japan, and the

Northern Pacific to the Hawaiian Islands. Since 1970, dust outbreaks have worsened due to periods of drought in Africa.

There is a large variability in dust transport to the Caribbean and Florida from year to year; however, the flux is greater during positive phases of the North Atlantic Oscillation. The USGS links dust events to a decline in the health of coral reefs across the Caribbean and Florida, primarily since the 1970s. Climate change is raising ocean temperatures and raising levels of carbon dioxide in the atmosphere. These rising levels of carbon dioxide are acidifying the oceans. This, in turn, is altering aquatic ecosystems and modifying fish distributions, with impacts on the sustainability of fisheries and the livelihoods of the communities that depend on them. Healthy ocean ecosystems are also important for the mitigation of climate change.

Deep Sea Mining

Deep sea mining is a relatively new mineral retrieval process that takes place on the ocean floor. Ocean mining sites are usually around large areas of polymetallic nodules or active and extinct hydrothermal vents at about 1,400–3,700 meters below the ocean's surface. The vents create sulfide deposits, which contain precious metals such as silver, gold, copper, manganese, cobalt, and zinc. The deposits are mined using either hydraulic pumps or bucket systems that take ore to the surface to be processed. As with all mining operations, deep sea mining raises questions about environmental damages to the surrounding areas Because deep sea mining is a relatively new field, the complete consequences of full scale mining operations are unknown.

However, experts are certain that removal of parts of the sea floor will result in disturbances to the benthic layer, increased toxicity of the water column and sediment plumes from tailings. Removing parts of the sea floor disturbs the habitat of benthic organisms, possibly, depending on the type of mining and location, causing permanent disturbances. Aside from direct impact of mining the area, leakage, spills and corrosion would alter the mining area's chemical makeup. Among the impacts of deep sea mining, sediment plumes could have the greatest impact. Plumes are caused when the tailings from mining are dumped back into the ocean, creating a cloud of particles floating in the water. Two types of plumes occur: near bottom plumes and surface plumes. Near bottom plumes occur when the tailings are pumped back down to the mining site. The floating particles increase the turbidity, or cloudiness, of the water, clogging filter–feeding apparatuses used by benthic organisms. Surface plumes cause a more serious problem. Depending on the size of the particles and water currents the plumes could spread over vast areas. The plumes could impact zooplankton and light penetration, in turn affecting the food web of the area.

ACIDIFICATION

The oceans are normally a natural carbon sink, absorbing carbon dioxide from the atmosphere. Because the levels of atmospheric carbon dioxide are increasing, the oceans are becoming more acidic. The potential consequences of ocean acidification are not fully understood, but there are concerns that structures made of calcium carbonate may become vulnerable to dissolution, affecting corals and the ability of shellfish to form shells. Oceans and coastal ecosystems play an important role in the global carbon cycle and have removed about 25% of the carbon dioxide emitted by human activities between 2000 and 2007 and about half the anthropogenic CO_2 released since the start of the industrial revolution. Rising ocean temperatures and ocean acidification means that the capacity of the ocean carbon sink will gradually get weaker, giving rise to global concerns expressed in the Monaco and Manado Declarations.

A report from NOAA scientists published in the journal Science in May 2008 found that large amounts of relatively acidified water are upwelling to within four miles of the Pacific continental shelf area of North America. This area is a critical zone where most local marine life lives or is born. While the paper dealt only with areas from Vancouver to northern California, other continental shelf areas may be experiencing similar effects. A related issue is the methane clathrate reservoirs found under sediments on the ocean floors. These trap large amounts of the greenhouse gas methane, which ocean warming has the potential to release.

In 2004 the global inventory of ocean methane clathrates was estimated to occupy between one and five million cubic kilometres. If all these clathrates were to be spread uniformly across the ocean floor, this would translate to a thickness between three and fourteen metres. This estimate corresponds to 500–2500 gigatonnes carbon and can be compared with the 5000 Gt C estimated for all other fossil fuel reserves.

EUTROPHICATION

Eutrophication is an increase in chemical nutrients, typically compounds containing nitrogen or phosphorus, in an ecosystem. It can result in an increase in the ecosystem's primary productivity and further effects including lack of oxygen and severe reductions in water quality, fish, and other animal populations.

The biggest culprit are rivers that empty into the ocean, and with it the many chemicals used as fertilizers in agriculture as well as waste from livestock and humans. An excess of oxygen depleting chemicals in the water can lead to hypoxia and the creation of a dead zone. Estuaries tend to be naturally eutrophic because land–derived nutrients are concentrated where run-off enters the marine environment in a confined channel. The World

Resources Institute has identified 375 hypoxic coastal zones around the world, concentrated in coastal areas in Western Europe, the Eastern and Southern coasts of the US, and East Asia, particularly in Japan.

In the ocean, there are frequent red tide algae blooms that kill fish and marine mammals and cause respiratory problems in humans and some domestic animals when the blooms reach close to shore. In addition to land run-off, atmospheric anthropogenic fixed nitrogen can enter the open ocean. A study in 2008 found that this could account for around one third of the ocean's external nitrogen supply and up to three per cent of the annual new marine biological production. It has been suggested that accumulating reactive nitrogen in the environment may have consequences as serious as putting carbon dioxide in the atmosphere.

PLASTIC DEBRIS

Marine debris is mainly discarded human rubbish which floats on, or is suspended in the ocean. Eighty per cent of marine debris is plastic—a component that has been rapidly accumulating since the end of World War II. The mass of plastic in the oceans may be as high as one hundred million metric tons. Discarded plastic bags, six pack rings and other forms of plastic waste which finish up in the ocean present dangers to wildlife and fisheries. Aquatic life can be threatened through entanglement, suffocation, and ingestion. Fishing nets, usually made of plastic, can be left or lost in the ocean by fishermen. Known as ghost nets, these entangle fish, dolphins, sea turtles, sharks, dugongs, crocodiles, seabirds, crabs, and other creatures, restricting movement, causing starvation, laceration and infection, and, in those that need to return to the surface to breathe, suffocation.

Many animals that live on or in the sea consume flotsam by mistake, as it often looks similar to their natural prey. Plastic debris, when bulky or tangled, is difficult to pass, and may become permanently lodged in the digestive tracts of these animals, blocking the passage of food and causing death through starvation or infection. Plastics accumulate because they don't biodegrade in the way many other substances do. They will photodegrade on exposure to the sun, but they do so properly only under dry conditions, and water inhibits this process.

In marine environments, photodegraded plastic disintegrates into ever smaller pieces while remaining polymers, even down to the molecular level. When floating plastic particles photodegrade down to zooplankton sizes, jellyfish attempt to consume them, and in this way the plastic enters the ocean food chain. Many of these long-lasting pieces end up in the stomachs of marine birds and animals, including sea turtles, and black–footed albatross. Plastic debris tends to accumulate at the centre of ocean gyres. In particular, the Great Pacific Garbage Patch has a very high level of plastic particulate suspended in the upper water column. In samples taken in 1999, the mass of plastic exceeded that of zooplankton by a factor of six.

Midway Atoll, in common with all the Hawaiian Islands, receives substantial amounts of debris from the garbage patch. Ninety per cent plastic, this debris accumulates on the beaches of Midway where it becomes a hazard to the bird population of the island. Midway Atoll is home to two-thirds of the global population of Laysan Albatross. Nearly all of these albatross have plastic in their digestive system and one-third of their chicks die. Toxic additives used in the manufacture of plastic materials can leach out into their surroundings when exposed to water. Waterborne hydrophobic pollutants collect and magnify on the surface of plastic debris, thus making plastic far more deadly in the ocean than it would be on land. Hydrophobic contaminants are also known to bioaccumulate in fatty tissues, biomagnifying up the food chain and putting pressure on apex predators. Some plastic additives are known to disrupt the endocrine system when consumed, others can suppress the immune system or decrease reproductive rates. Floating debris can also absorb persistent organic pollutants from seawater, including PCBs, DDT and PAHs. Aside from toxic effects, when ingested some of these are mistaken by the animal brain for estradiol, causing hormone disruption in the affected wildlife.

TOXINS

Apart from plastics, there are particular problems with other toxins that do not disintegrate rapidly in the marine environment. Examples of persistent toxins are PCBs, DDT, pesticides, furans, dioxins, phenols and radioactive waste. Heavy metals are metallic chemical elements that have a relatively high density and are toxic or poisonous at low concentrations. Examples are mercury, lead, nickel, arsenic and cadmium. Such toxins can accumulate in the tissues of many species of aquatic life in a process called bioaccumulation. They are also known to accumulate in benthic environments, such as estuaries and bay muds: a geological record of human activities of the last century.

Specific examples:

- Chinese and Russian industrial pollution such as phenols and heavy metals in the Amur River have devastated fish stocks and damaged its estuary soil.
- Wabamun Lake in Alberta, Canada, once the best whitefish lake in the area, now has unacceptable levels of heavy metals in its sediment and fish.
- Acute and chronic pollution events have been shown to impact southern California kelp forests, though the intensity of the impact seems to depend on both the nature of the contaminants and duration of exposure.

- Due to their high position in the food chain and the subsequent accumulation of heavy metals from their diet, mercury levels can be high in larger species such as bluefin and albacore. As a result, in March 2004 the United States FDA issued guidelines recommending that pregnant women, nursing mothers and children limit their intake of tuna and other types of predatory fish.
- Some shellfish and crabs can survive polluted environments, accumulating heavy metals or toxins in their tissues. For example, mitten crabs have a remarkable ability to survive in highly modified aquatic habitats, including polluted waters. The farming and harvesting of such species needs careful management if they are to be used as a food.
- Surface run-off of pesticides can alter the gender of fish species genetically, transforming male into female fish.
- Heavy metals enter the environment through oil spills-such as the Prestige oil spill on the Galician coast-or from other natural or anthropogenic sources.
- In 2005, the 'Ndrangheta, an Italian mafia syndicate, was accused of sinking at least 30 ships loaded with toxic waste, much of it radioactive. This has led to widespread investigations into radioactive waste disposal rackets.
- Since the end of World War II, various nations, including the Soviet Union, the United Kingdom, the United States, and Germany, have disposed of chemical weapons in the Baltic Sea, raising concerns of environmental contamination.

NOISE POLLUTION

Marine life can be susceptible to noise or sound pollution from sources such as passing ships, oil exploration seismic surveys, and naval low-frequency active sonar. Sound travels more rapidly and over larger distances in the sea than in the atmosphere. Marine animals, such as cetaceans, often have weak eyesight, and live in a world largely defined by acoustic information. This applies also to many deeper sea fish, who live in a world of darkness. Between 1950 and 1975, ambient noise in the ocean increased by about ten decibels. Noise also makes species communicate louder, which is called the Lombard vocal response. Whale songs are longer when submarine-detectors are on. If creatures don't "speak" loud enough, their voice can be masked by anthropogenic sounds. These unheard voices might be warnings, finding of prey, or preparations of net-bubbling.

When one species begins speaking louder, it will mask other specie voices, causing the whole ecosystem to eventually speak louder. Sylvia Earle, "Undersea noise pollution is like the death of a thousand cuts. Each sound in itself may not be a matter of critical concern, but taken all together, the noise from shipping, seismic surveys, and military activity is creating a totally different environment than existed even 50 years ago. That high level of noise is bound to have a hard, sweeping impact on life in the sea."

ADAPTATION AND MITIGATION

Much anthropogenic pollution ends up in the ocean. The 2011 edition of the United Nations Environment Programme Year Book identifies as the main emerging environmental issues the loss to the oceans of massive amounts of phosphorus, "a valuable fertilizer needed to feed a growing global population", and the impact billions of pieces of plastic waste are having globally on the health of marine environments. Bjorn Jennssen, "Anthropogenic pollution may reduce biodiversity and productivity of marine ecosystems, resulting in reduction and depletion of human marine food resources".

There are two ways the overall level of this pollution can be mitigated: Either the human population is reduced, or a way is found to reduce the ecological footprint left behind by the average human. If the second way is not adopted, then the first way may be imposed as world ecosystems falter. The second way is for humans, individually, to pollute less. That requires social and political will, together with a shift in awareness so more people respect the environment and are less disposed to abuse it. At an operational level, regulations, and international government participation is needed. It is often very difficult to regulate marine pollution because pollution spreads over international barriers, thus making regulations hard to create as well as enforce.

Without appropriate awareness of marine pollution, there may not be the necessary global will to effectively address the issues. Balanced information on the sources and harmful effects of marine pollution needs to become part of general public awareness, and ongoing research is required to fully establish, and keep current, the scope of the issues.

As expressed in Daoji and Dag's research, one of the reasons why environmental concern is lacking among the Chinese is because the public awareness is low and therefore should be targeted. Likewise, regulation, based upon such in-depth research should be employed. In California, such regulations have already been put in place to protect Californian coastal waters from agricultural run-off. This includes the California Water Code, as well as several voluntary programmes. Similarly, in India, several tactics have been employed that help reduce marine pollution, however, they do not significantly target the problem. In Chennai, sewage has been dumped further into open waters. Due to the mass of waste being deposited, open-ocean is best for diluting, and dispersing pollutants, thus making them less harmful to marine ecosystems.

OIL SPILL

An oil spill is a release of a liquid petroleum hydrocarbon into the environment due to human activity, and is a form of pollution. The term often refers to marine oil spills, where oil is released into the ocean or coastal waters. Oil spills include releases of crude oil from tankers, offshore platforms, drilling rigs and wells, as well as spills of refined petroleum products and their by-products, and heavier fuels used by large ships such as bunker fuel, or the spill of any oily white substance refuse or waste oil. Spills may take months or even years to clean up. Oil also enters the marine environment from natural oil seeps. Public attention and regulation has tended to focus most sharply on seagoing oil tankers.

ENVIRONMENTAL EFFECTS

The oil penetrates into the structure of the plumage of birds, reducing its insulating ability, thus making the birds more vulnerable to temperature fluctuations and much less buoyant in the water. It also impairs or disables bird's flight abilities to forage and escape from predators. As they attempt to preen, birds typically ingest oil that covers their feathers, causing kidney damage, altered liver function, and digestive tract irritation. This and the limited foraging ability quickly causes dehydration and metabolic imbalances.

Hormonal balance alteration including changes in luteinizing protein can also result in some birds exposed to petroleum.

Most birds affected by an oil spill die unless there is human intervention. Marine mammals exposed to oil spills are affected in similar ways as seabirds. Oil coats the fur of Sea otters and seals, reducing its insulation abilities and leading to body temperature fluctuations and hypothermia.

Ingestion of the oil causes dehydration and impaired digestions. Because oil floats on top of water, less sunlight penetrates into the water, limiting the photosynthesis of marine plants and phytoplankton. This, as well as decreasing the fauna populations, affects the food chain in the ecosystem. There are three kinds of oil-consuming bacteria. Sulfate-reducing bacteria and acid-producing bacteria are anaerobic, while general aerobic bacteria are aerobic. These bacteria occur naturally and will act to remove oil from an ecosystem, and their biomass will tend to replace other populations in the food chain.

CLEANUP AND RECOVERY

Cleanup and recovery from an oil spill is difficult and depends upon many factors, including the type of oil spilled, the temperature of the water and the types of shorelines and beaches involved.

Methods for cleaning up include:

- *Bioremediation*: Use of microorganisms or biological agents to break down or remove oil.
- *Bioremediation Accelerator*: Oleophilic, hydrophobic chemical, containing no bacteria, which chemically and physically bonds to both soluble and insoluble hydrocarbons. The bioremedation accelerator acts as a herding agent in water and on the surface, floating molecules to the surface of the water, including solubles such as phenols and BTEX, forming gel-like agglomerations. Undetectable levels of hydrocarbons can be obtained in produced water and manageable water columns. By overspraying sheen with bioremediation accelerator, sheen is eliminated within minutes. Whether applied on land or on water, the nutrient-rich emulsion creates a bloom of local, indigenous, pre-existing, hydrocarbon-consuming bacteria. Those specific bacteria break down the hydrocarbons into water and carbon dioxide, with EPA tests showing 98% of alkanes biodegraded in 28 days; and aromatics being biodegraded 200 times faster than in nature they also sometimes use the hydrofireboom to clean the oil up by taking it away from most of the oil and burning it.
- Controlled burning can effectively reduce the amount of oil in water, if done properly. But it can only be done in low wind, and can cause air pollution.
- Dispersants act as detergents, clustering around oil globules and allowing them to be carried away in the water. This improves the surface aesthetically, and mobilizes the oil. Smaller oil droplets, scattered by currents, may cause less harm and may degrade more easily. But the dispersed oil droplets infiltrate into deeper water and can lethally contaminate coral. Recent research indicates that some dispersants are toxic to corals.
- *Watch and wait*: In some cases, natural attenuation of oil may be most appropriate, due to the invasive nature of facilitated methods of remediation, particularly in ecologically sensitive areas such as wetlands.
- *Dredging*: For oils dispersed with detergents and other oils denser than water.
- *Skimming:* Requires calm waters
- *Solidifying*: Solidifiers are composed of dry hydrophobic polymers that both adsorb and absorb. They clean up oil spills by changing the physical state of spilled oil from liquid to a semi-solid or a rubber-like material that floats on water. Solidifiers are insoluble in water, therefore the removal of the solidified oil is easy and the oil will not leach out. Solidifiers have been proven to be

relatively non-toxic to aquatic and wild life and have been proven to suppress harmful vapours commonly associated with hydrocarbons such as Benzene, Xylene, Methyl Ethyl, Acetone and Naphtha. The reaction time for solidification of oil is controlled by the surf area or size of the polymer as well as the viscosity of the oil. Some solidifier product manufactures claim the solidified oil can be disposed of in landfills, recycled as an additive in asphalt or rubber products, or burned as a low ash fuel. A solidifier called C.I.Agent is being used by BP in granular form as well as in Marine and Sheen Booms on Dauphin Island, AL and Fort Morgan, MS to aid in the Deepwater Horizon oil spill cleanup.

- *Vacuum and centrifuge*: Oil can be sucked up along with the water, and then a centrifuge can be used to separate the oil from the water—allowing a tanker to be filled with near pure oil. Usually, the water is returned to the sea, making the process more efficient, but allowing small amounts of oil to go back as well. This issue has hampered the use of centrifuges due to a United States regulation limiting the amount of oil in water returned to the sea.

Equipment used includes:

- *Booms*: Large floating barriers that round up oil and lift the oil off the water
- *Skimmers*: Skim the oil
- *Sorbents*: Large absorbents that absorb oil
- *Chemical and biological agents:* Helps to break down the oil
- Vacuums: Remove oil from beaches a*nd water surface*
- *Shovels and other road equipments*: Typically used to clean up oil on beaches

Prevention

- Seafood Sensory Training–in an effort to detect oil in seafood, inspectors and regulators are being trained to sniff out seafood tainted by oil and make sure the product reaching consumers is safe to eat.
- Secondary containment—methods to prevent releases of oil or hydrocarbons into environment.
- Oil Spill Prevention Containment and Countermeasures programme by the United States Environmental Protection Agency.
- Double-hulling—build double hulls into vessels, which reduces the risk and severity of a spill in case of a collision or grounding. Existing single–hull vessels can also be rebuilt to have a double hull.

Skimmers–are things that skim the slick off the top of the water. Boomers–are inflatable, rubber blockades that trap the oil, so it is easier to skim.

ENVIRONMENTAL SENSITIVITY INDEX ENVIRONMENTAL SENSITIVITY INDEX MAPPING

Environmental Sensitivity Index maps are used to identify sensitive shoreline resources prior to an oil spill event in order to set priorities for protection and plan cleanup strategies. By planning spill response ahead of time, the impact on the environment can be minimized or prevented. Environmental sensitivity index maps are basically made up of information within the following three categories: shoreline type, and biological and human-use resources.

Shoreline Type

Shoreline type is classified by rank depending on how easy the garet would be to cleanup, how long the oil would persist, and how sensitive the shoreline is. The floating oil slicks put the shoreline at particular risk when they eventually come ashore, covering the substrate with oil. The differing substrates between shoreline types vary in their response to oiling, and influence the type of cleanup that will be required to effectively decontaminate the shoreline. In 1995, the US National Oceanic and Atmospheric Administration extended ESI maps to lakes, rivers, and estuary shoreline types.

The exposure the shoreline has to wave energy and tides, substrate type, and slope of the shoreline are also taken into account–in addition to biological productivity and sensitivity. The productivity of the shoreline habitat is also taken into account when determining ESI ranking. Mangroves and marshes tend to have higher ESI rankings due to the potentially long-lasting and damaging effects of both the oil contamination and cleanup actions. Impermeable and exposed surfaces with high wave action are ranked lower due to the reflecting waves keeping oil from coming onshore, and the speed at which natural processes will remove the oil.

Biological Resources

Habitats of plants and animals that may be at risk from oil spills are referred to as "elements" and are divided by functional group. Further classification divides each element into species groups with similar life histories and behaviours relative to their vulnerability to oil spills. There are eight element groups: Birds, Reptiles Amphibians, Fish, Invertebrates, Habitats and Plants, Wetlands, and Marine Mammals and Terrestrial Mammals. Element groups are further divided into sub-groups, for example, the 'marine mammals' element group is divided into dolphins, manatees, pinnipeds, polar bears, sea otters and whales. Issues taken into consideration when ranking biological resources include the observance of a large number of individuals in a small area, whether special life stages occur ashore and whether there are species present that are threatened, endangered or.rare.

Human-use Resources

Human use resources are divided into four major classifications; archaeological importance or cultural resource site, high-use recreational areas or shoreline access points, important protected management areas, or resource origins. Some examples include airports, diving sites, popular beach sites, marinas, natural reserves or marine sanctuaries.

THERMAL POLLUTION

Thermal pollution is the degradation of water quality by any process that changes ambient water temperature. A common cause of thermal pollution is the use of water as a coolant by power plants and industrial manufacturers. When water used as a coolant is returned to the natural environment at a higher temperature, the change in temperature decreases oxygen supply, and affects ecosystem composition. Urban run-off–stormwater discharged to surface waters from roads and parking lots–can also be a source of elevated water temperatures. When a power plant first opens or shuts down for repair or other causes, fish and other organisms adapted to particular temperature range can be killed by the abrupt rise in water temperature known as "thermal shock."

WARM WATER

Elevated temperature typically decreases the level of dissolved oxygen in water. This can harm aquatic animals such as fish, amphibians and copepods. Thermal pollution may also increase the metabolic rate of aquatic animals, as enzyme activity, resulting in these organisms consuming more food in a shorter time than if their environment were not changed. An increased metabolic rate may result in fewer resources; the more adapted organisms moving in may have an advantage over organisms that are not used to the warmer temperature. As a result, food chains of the old and new environments may be compromised. Some fish species will avoid stream segments or coastal areas adjacent to a thermal discharge. Biodiversity can be decreased as a result. High temperature limits oxygen dispersion into deeper waters, contributing to anaerobic conditions. This can lead to increased bacteria levels when there is ample food supply. Many aquatic species will fail to reproduce at elevated temperatures. Primary producers are affected by warm water because higher water temperature increases plant growth rates, resulting in a shorter lifespan and species overpopulation. This can cause an algae bloom which reduces oxygen levels. Temperature changes of even one to two degrees Celsius can cause significant changes in organism metabolism and other adverse cellular biology effects. Principal adverse changes can include rendering cell walls less permeable to necessary osmosis, coagulation of cell proteins, and alteration of enzyme metabolism. These cellular level effects can adversely affect mortality and reproduction.

A large increase in temperature can lead to the denaturing of life-supporting enzymes by breaking down hydrogen and disulphide bonds within the quaternary structure of the enzymes. Decreased enzyme activity in aquatic organisms can cause problems such as the inability to break down lipids, which leads to malnutrition. In limited cases, warm water has little deleterious effect and may even lead to improved function of the receiving aquatic ecosystem. This phenomenon is seen especially in seasonal waters and is known as thermal enrichment. An extreme case is derived from the aggregational habits of the manatee, which often uses power plant discharge sites during winter. Projections suggest that manatee populations would decline upon the removal of these discharges.

COLD WATER

Releases of unnaturally cold water from reservoirs can dramatically change the fish and macroinvertebrate fauna of rivers, and reduce river productivity. In Australia, where many rivers have warmer temperature regimes, native fish species have been eliminated, and macroinvertebrate fauna have been drastically altered.

CONTROL OF THERMAL POLLUTION

Industrial Wastewater

In the United States, about 75 to 80 per cent of thermal pollution is generated by power plants. The remainder is from industrial sources such as petroleum refineries, pulp and paper mills, chemical plants, steel mills and smelters.

Heated water from these sources may be controlled with:

- Cooling ponds, man–made bodies of water designed for cooling by evaporation, convection, and radiation
- Cooling towers, which transfer waste heat to the atmosphere through evaporation and/or heat transfer
- Cogeneration, a process where waste heat is recycled for domestic and/or industrial heating purposes.

Some facilities use once–through cooling systems which do not reduce temperature as effectively as the above systems. For example, the Potrero Generating Station in San Francisco, which uses OTC, discharges water to San Francisco Bay approximately 10°C above the ambient bay temperature.

Urban Run-off

During warm weather, urban run-off can have significant thermal impacts on small streams, as stormwater passes over hot parking lots, roads and sidewalks. Stormwater management facilities that absorb run-off or direct it into groundwater, such as bioretention systems and infiltration basins, can reduce these thermal effects. Retention basins tend to be less effective at reducing temperature, as the water may be heated by the sun before being discharged to a receiving stream.

SURFACE RUN-OFF

Surface run-off is the water flow that occurs when soil is infiltrated to full capacity and excess water from rain, meltwater, or other sources flows over the land. This is a major component of the water cycle. Run-off that occurs on surfaces before reaching a channel is also called a nonpoint source. If a nonpoint source contains man-made contaminants, the run-off is called nonpoint source pollution. A land area which produces run-off that drains to a common point is called a watershed. When run-off flows along the ground, it can pick up soil contaminants including, but not limited to petroleum, pesticides, or fertilizers that become discharge or nonpoint source pollution.

GENERATION

Surface run-off can be generated either by rainfall or by the melting of snow, or glaciers. Snow and glacier melt occur only in areas cold enough for these to form permanently. Typically snowmelt will peak in the spring and glacier melt in the summer, leading to pronounced flow maxima in rivers affected by them. The determining factor of the rate of melting of snow or glaciers is both air temperature and the duration of sunlight. In high mountain regions, streams frequently rise on sunny days and fall on cloudy ones for this reason. In areas where there is no snow, run-off will come from rainfall.

However, not all rainfall will produce run-off because storage from soils can absorb light showers. On the extremely ancient soils of Australia and Southern Africa, proteoid roots with their extremely dense networks of root hairs can absorb so much rainwater as to prevent run-off even when substantial amounts of rain fall. In these regions, even on less infertile cracking clay soils, high amounts of rainfall and potential evaporation are needed to generate any surface run-off, leading to specialised adaptations to extremely variable streams.

Infiltration Excess Overland Flow

This occurs when the rate of rainfall on a surface exceeds the rate at which water can infiltrate the ground, and any depression storage has already been filled. This is called infiltration excess overland flow, Hortonian overland flow or unsaturated overland flow. This more commonly occurs in arid and semi-arid regions, where rainfall intensities are high and the soil infiltration capacity is reduced because of surface sealing, or in paved areas. This occurs largely in city areas where pavements prevent water infiltration.

Saturation Excess Overland Flow

When the soil is saturated and the depression storage filled, and rain continues to fall, the rainfall will immediately produce surface run-off. The level of antecedent soil moisture is one factor affecting the time until soil becomes saturated. This run-off is called saturation excess overland flow or saturated overland flow.

Antecedent Soil Moisture

Soil retains a degree of moisture after a rainfall. This residual water moisture affects the soil's infiltration capacity. During the next rainfall event, the infiltration capacity will cause the soil to be saturated at a different rate. The higher the level of antecedent soil moisture, the more quickly the soil becomes saturated. Once the soil is saturated, run-off occurs.

Subsurface Return Flow

After water infiltrates the soil on an up-slope portion of a hill, the water may flow laterally through the soil, and exfiltrate closer to a channel. This is called subsurface return flow or throughflow. As it flows, the amount of run-off may be reduced in a number of possible ways: a small portion of it may evapotranspire; water may become temporarily stored in microtopographic depressions; and a portion of it may become run-on, which is the infiltration of run-off as it flows overland. Any remaining surface water eventually flows into a receiving water body such as a river, lake, estuary or ocean.

HUMAN IMPACT ON SURFACE RUN-OFF

Urbanization increases surface run-off, by creating more impervious surfaces such as pavement and buildings, that do not allow percolation of the water down through the soil to the aquifer. It is instead forced directly into streams or storm water run-off drains, where erosion and siltation can be major problems, even when flooding is not. Increased run-off reduces groundwater recharge, thus lowering the water table and making droughts worse, especially for farmers and others who depend on the water wells. When anthropogenic contaminants are dissolved or suspended in run-off, the human impact is expanded to create water pollution. This pollutant load can reach various receiving waters such as streams, rivers, lakes, estuaries and oceans with resultant water chemistry changes to these water systems and their related ecosystems. A 2008 report by the United States National Research Council identified urban stormwater as a leading source of water quality problems in the U.S.

EFFECTS OF SURFACE RUN-OFF

Erosion and Deposition

Surface run-off causes erosion of the Earth's surface; deposition is the depositing of erosion. There are four principal types of erosion: splash erosion, gully erosion, sheet erosion and stream bed erosion. Splash erosion is the result of mechanical collision of raindrops with the soil surface. Dislodged soil particles becoming suspended in the surface run-off and carried into streams and rivers. Gully erosion occurs when the power of run-off is strong enough that it cuts a well defined channel. These channels can be as small as one centimeter wide or as large as several meters. Sheet erosion is the overland transport of run-off without a well defined channel.

In the case of gully erosion, large amounts of material can be transported in a small time period. Stream bed erosion is the attrition of stream banks or bottoms by rapidly flowing rivers or creeks. Reduced crop productivity usually results from erosion, and these effects are studied in the field of soil conservation. The soil particles carried in run-off vary in size from about.001 millimeter to 1.0 millimeter in diameter. Larger particles settle over short transport distances, whereas small particles can be carried over long distances suspended in the water column. Erosion of silty soils that contain smaller particles generates turbidity and diminishes light transmission, which disrupts aquatic ecosystems. Entire sections of countries have been rendered unproductive by erosion. On the high central plateau of Madagascar, approximately ten per cent of that country's land area, virtually the entire landscape is devoid of vegetation, with erosive gully furrows typically in excess of 50 meters deep and one kilometer wide. Shifting cultivation is a farming system which sometimes incorporates the slash and burn method in some regions of the world. Erosion cause loss of the fertile top soil and reduces the its fertility and quality of the agricultural produce. Modern industrial farming is another major cause of erosion. In some areas in the American corn belt, more than 50 per cent of the original topsoil has been carried away within the last 100 years.

Environmental Impacts

The principal environmental issues associated with run-off are the impacts to surface water, groundwater and soil through transport of water pollutants to these systems. Ultimately these consequences translate into human health risk, ecosystem disturbance and aesthetic impact to water resources.

Some of the contaminants that create the greatest impact to surface waters arising from run-off are petroleum substances, herbicides and fertilizers. Quantitative uptake by surface run-off of pesticides and other contaminants has been studied since the 1960s, and early on contact of pesticides with water was known to enhance phytotoxicity.

In the case of surface waters, the impacts translate to water pollution, since the streams and rivers have received run-off carrying various chemicals or sediments. When surface waters are used as potable water supplies, they can be compromised regarding health risks and drinking water aesthetics. Contaminated surface waters risk altering the metabolic processes of the aquatic species that they host; these alterations can lead to death, such as fish kills, or alter the balance of populations present. Other specific impacts are on animal mating, spawning, egg and larvae viability, juvenile survival and plant productivity. Some researches show surface run-off of pesticides, such as DDT, can alter the gender of fish species genetically, which transforms male into female fish. In the case of groundwater, the main issue is contamination of drinking water, if the aquifer is abstracted for human use. Regarding soil contamination, run-off waters can have two important pathways of concern. Firstly, run-off water can extract soil contaminants and carry them in the form of water pollution to even more sensitive aquatic habitats. Secondly, run-off can deposit contaminants on pristine soils, creating health or ecological consequences.

Flooding

Flooding occurs when a watercourse is unable to convey the quantity of run-off flowing downstream. The frequency with which this occurs is described by a return period. Flooding is a natural process, which maintains ecosystem composition and processes, but it can also be altered by land use changes such as river engineering. Floods can be both beneficial to societies or cause damage. Agriculture along the Nile floodplain took advantage of the seasonal flooding that deposited nutrients beneficial for crops. However, as the number and susceptibility of settlements increase, flooding increasingly becomes a natural hazard. Adverse impacts span loss of life, property damage, contamination of water supplies, loss of crops, and social dislocation and temporary homelessness. Floods are among the most devastating of natural disasters.

Agricultural Issues

A common context of run-off deals with agriculture. When farmland is tilled and bare soil is revealed, rainwater carries billions of tons of topsoil into waterways each year, causing loss of valuable topsoil and adding sediment to produce turbidity in surface waters. The other context of agricultural issues involves the transport of agricultural chemicals via surface run-off. This result occurs when chemical use is excessive or poorly timed with respect to high precipitation. The resulting contaminated run-off represents not only a waste of agricultural chemicals, but also an environmental threat to downstream ecosystems. The alternative to conventional farming is organic farming which eliminates chemical usage.

MEASUREMENT AND MATHEMATICAL MODELING

Run-off is Analysed by using mathematical models in combination with various water quality sampling methods. Measurements can be made using continuous automated water quality analysis instruments targeted on pollutants such as specific organic or inorganic chemicals, pH, turbidity etc. or targeted on secondary indicators such as dissolved oxygen. Measurements can also be made in batch form by extracting a single water sample and conducting any number of chemical or physical tests on that sample. In the 1950s or earlier hydrology transport models appeared to calculate quantities of run-off, primarily for flood forecasting. Beginning in the early 1970s computer models were developed to Analyse the transport of run-off carrying water pollutants, which considered dissolution rates of various chemicals, infiltration into soils and ultimate pollutant load delivered to receiving waters. One of the earliest models addressing chemical dissolution in run-off and resulting transport was developed in the early 1970s under contract to the United States Environmental Protection Agency. This computer model formed the basis of much of the mitigation study that led to strategies for land use and chemical handling controls. Other computer models have been developed that allow surface run-off to be tracked through a river course as reactive water pollutants. In this case the surface run-off may be considered to be a line source of water pollution to the receiving waters.

MITIGATION AND TREATMENT

Mitigation of adverse impacts of run-off can take several forms:

- Land use development controls aimed at minimizing impervious surfaces in urban areas
- Erosion controls for farms and construction sites
- Flood control programmes
- Chemical use and handling controls in agriculture, landscape maintenance, industrial use, etc.

Land Use Controls

Many world regulatory agencies have encouraged research on

methods of minimizing total surface run-off by avoiding unnecessary hardscape. Many municipalities have produced guidelines and codes for land developers that encourage minimum width sidewalks, use of pavers set in earth for driveways and walkways and other design techniques to allow maximum water infiltration in urban settings. An example land use control programme can be viewed at seen in the city of Santa Monica, California.

Erosion Controls

Erosion controls have appeared since medieval times when farmers realised the importance of contour farming to protect soil resources. Beginning in the 1950s these agricultural methods became increasingly more sophisticated. In the 1960s some state and local governments began to focus their efforts on mitigation of construction run-off by requiring builders to implement erosion and sediment controls. This included such techniques as: use of straw bales and barriers to slow run-off on slopes, installation of silt fences, programming construction for months that have less rainfall and minimizing extent and duration of exposed graded areas. Montgomery County, Maryland implemented the first local government sediment control programme in 1965, and this was followed by a statewide programme in Maryland in 1970.

Flood Control Programmes

Flood control programmes as early as the first half of the twentieth century became quantitative in predicting peak flows of riverine systems. Progressively strategies have been developed to minimize peak flows and also to reduce channel velocities. Some of the techniques commonly applied are: provision of holding ponds to buffer riverine peak flows, use of energy dissipators in channels to reduce stream velocity and land use controls to minimize run-off.

Chemical Use and Handling

Following enactment of the U.S. Resource Conservation and Recovery Act in 1976, and later the Water Quality Act of 1987, states and cities have become more vigilant in controlling the containment and storage of toxic chemicals, thus preventing releases and leakage. Methods commonly applied are: requirements for double containment of underground storage tanks, registration of hazardous materials usage, reduction in numbers of allowed pesticides and more stringent regulation of fertilizers and herbicides in landscape maintenance. In many industrial cases, pretreatment of wastes is required, to minimize escape of pollutants into sanitary or stormwater sewers. The U.S. Clean Water Act requires that local governments in urbanized areas obtain stormwater discharge permits for their drainage systems. Essentially this means that the locality must operate a stormwater management programme for all surface run-off that enters the municipal separate storm sewer system.

EPA and state regulations and related publications outline six basic components that each local programme must contain:

1. Public education
2. Public involvement
3. Illicit discharge detection and elimination
4. Construction site run-off controls
5. Post-construction stormwater management controls
6. Pollution prevention and "good housekeeping" measures.

Other property owners which operate storm drain systems similar to municipalities, such as state highway systems, universities, military bases and prisons, are also subject to the MS4 permit requirements.

URBAN RUN-OFF

Urban run-off is surface run-off of rainwater created by urbanization. This run-off is a major source of water pollution in many parts of the United States and other urban communities worldwide.Impervious surfaces are constructed during land development. During rain storms and other precipitation events, these surfaces along with rooftops, carry polluted stormwater to storm drains, instead of allowing the water to percolate through soil. This causes lowering of the water table and flooding since the amount of water that remains on the surface is greater. Most municipal storm sewer systems discharge stormwater, untreated, to streams, rivers and bays.

POLLUTANTS IN URBAN RUN-OFF

Water running off these impervious surfaces tends to pick up gasoline, motor oil, heavy metals, trash and other pollutants from roadways and parking lots, as well as fertilizers and pesticides from lawns. Roads and parking lots are major sources of polycyclic aromatic hydrocarbons which are created as combustion byproducts of gasoline and other fossil fuels, as well as of the heavy metals nickel, copper, zinc, cadmium, and lead. Roof run-off contributes high levels of synthetic organic compounds and zinc. Fertilizer use on residential lawns, parks and golf courses is a significant source of nitrates and phosphorus in urban run-off. As stormwater is channeled into storm drains and surface waters, the natural sediment load discharged to receiving waters decreases, but the water flow and velocity increases. In fact, the impervious cover in a typical city creates five times the run-off of a typical woodland of the same size.

EFFECTS OF URBAN RUN-OFF

The run-off also increases temperatures in streams, harming fish and other organisms. Also, road salt used to melt snow on sidewalks and roadways can contaminate streams and groundwater aquifers. One of the most pronounced effects of urban run-off is on watercourses that historically contained little or no water during dry weather periods. When an area around such a stream is urbanized, the resultant run-off creates an unnatural year-round streamflow that hurts the vegetation, wildlife and stream bed of the waterway. Containing little or no sediment relative to the historic ratio of sediment to water, urban run-off rushes down the stream channel, ruining natural features such as meanders and sandbars, and creates severe erosion—increasing sediment loads at the mouth while severely incising the stream bed upstream. As an example, on many Southern California beaches at the mouth of a waterway, urban run-off carries trash, pollutants, excessive silt, and other wastes, and can pose moderate to severe health hazards. Because of fertilizer and organic waste that urban run-off often carries, eutrophication often occurs in waterways affected by this type of run-off. After heavy rains, organic matter in the waterway is relatively high compared with natural levels, spurring growth of algae blooms that soon use up most of the oxygen. Once the naturally occurring oxygen in the water is depleted, the algae blooms die, and in their decomposing process, cause further eutrophication.

Algae blooms mostly occur in areas with still water, such as stream pools and the pools behind dams, weirs, and some drop structures. Eutrophication usually comes with deadly consequences for fish and other aquatic organisms. Excessive stream bank erosion may cause flooding and property damage. For many years governments have often responded to urban stream erosion problems by modifying the streams through construction of hardened embankments and similar control structures using concrete and masonry materials. Use of these hard materials destroys habitat for fish and other animals. Such a project may stabilize the immediate area where flood damage occurred, but often it simply shifts the problem to an upstream or downstream segment of the stream.

PREVENTION AND MITIGATION OF URBAN RUN-OFF

Effective control of urban run-off involves reducing the velocity and flow of stormwater, as well as reducing pollutant discharges. A variety of stormwater management practices and systems may be used to reduce the effects of urban run-off. Some of these techniques, called best management practices in the U.S., focus on water quantity control, while others focus on improving water quality, and some perform both functions. Pollution prevention practices include low impact development techniques, installation of green roofs and improved chemical handling. Run-off mitigation systems include infiltration basins, bioretention systems, constructed wetlands, retention basins and similar devices.

Chapter 14

Recreational Uses of Water

DANGEROUS WATERS

Ever since humans first took to the seas thousands of years ago, sailors have faced numerous dangers. Ancient civilizations tried to explain these dangerous conditions by claiming that they were the work of angry gods or monsters. While scientific explanations have been advanced for dangerous phenomena such as high waves, hurricanes, and treacherous ocean currents (steady flows of water in a prevailing direction), many lives are still lost in the water each year, mainly due to drowning or hypothermia. Hypothermia is a condition where the core body temperature becomes too cold to function properly. Prolonged exposure to waters that may initially seem warm, between 70°–80°F (21°–27° C), can cause death from hypothermia.

WHIRLPOOLS

Some of the earliest written works make references to the dangers of the seas. In the Odyssey, Greek poet Homer mentions a great whirlpool that a group of Greek warriors encountered on their return home from the Trojan War. Many scholars assume that Charybdis, the whirlpool mentioned by Homer, is a whirlpool that still swirls today between mainland Italy and the island of Sicily. Viking poems refer to another famous whirlpool, the Maelstrom, which lies off the rocky coast of Norway. Several factors, working alone or together, can create whirlpools. Ocean currents that converge (come together) can cause a whirlpool. Tides and rock formations can create a whirlpool by forcing ocean currents to flow in a circular motion, as in the Maelstrom.

Also, constant winds on the ocean can create or contribute to a whirlpool, as in the narrow waters between Italy and Sicily. Although movies and literature sometimes refer to people or ships being drawn down into a whirlpool, this rarely happens. Whirlpools can pose a moderate danger to small crafts, as they can experience turbulence or even capsize (turn over) in whirlpools. Modern navigation allows ships to avoid large ocean whirlpools. Today, the greatest danger posed by whirlpools is on rivers, where curious boaters often wander too close to whirlpools and quickly find themselves in their midst.

CAPE HORN AND THE STRAITS OF MAGELLAN

Cape Horn and the Straits of Magellan lie at the southern tip of South America where the Atlantic and Pacific Oceans meet. The Straits of Magellan are a narrow passage between mainland South America and Tierra del Fuego, a large island to the south of the mainland. Portuguese explorer Ferdinand Magellan (1480–1521) discovered the Straits of Magellan in 1520 during his trip around the world. The Straits of Magellan are narrow and often experience rough seas due to high winds. The Atlantic and Pacific Oceans are at different levels, which cause churning currents when their waters meet in the Straits of Magellan. These powerful currents caused numerous ships to sink in the Straits of Magellan.

Isaac Le Maire (1558–1624), a Dutch merchant and explorer, discovered Cape Horn in 1615. Le Maire was looking for a different and safer route between the Atlantic and Pacific Oceans. Le Maire found a different route in what is today called Cape Horn, but it did not prove to be much safer than the Straits of Magellan. Cape Horn has violent weather patterns as a result of the meeting of the Atlantic and Pacific Oceans. Cold air moving north from Antarctica also contributes to the foul weather. Large waves, some over 65 feet (20 meters) tall, often sank ships that tried to round Cape Horn's rough seas. The opening of the Panama Canal in 1914 eliminated the need for most ships to travel through the Straits of Magellan or around Cape Horn in order to pass between the Atlantic and Pacific oceans.

HURRICANES, TYPHOONS, AND CYCLONES

A hurricane is any organized storm with sustained winds of 74 miles per hour (119 kilometers per hour) or greater in the Atlantic Ocean, Gulf of Mexico, Caribbean Sea, or eastern Pacific Ocean. Winds gusts in the strongest hurricanes approach 200 miles per hour (322 kilometers per hour). Hurricanes include circular bands of clouds that slowly swirl around a central core of low atmospheric pressure (the pressure exerted upon Earth's surface by its atmosphere at a given point), called the eye.

A hurricane may be hundreds of miles (kilometers) across, but the eye of the storm is typically only 10–30 miles (16–48 kilometers). Winds are strongest around the eye and weaken further out from the eye. A hurricane that occurs in the Indian Ocean is called a cyclone, and those in the middle and western Pacific are called typhoons. The low pressure of the eye pushes a wall of water in front of the storm called a storm surge. The storm surge is often the most destructive part of a hurricane. Storm surges can sink ships at sea, destroy buildings on the coast, and cause flooding inland. Hurricanes are divided into categories based on the speed their sustained winds.

A category 1 hurricane produces sustained winds of 74–95 miles per hour (119–153 kilometers pr hour) and storm surges 4–5 feet (1.2–1.5 meters) above normal tide levels, enough to flood low-lying coastal roads and buildings. Category 2 storms contain winds 96–110 miles per hour (154–177 km per hour) and produce storm surges 6–8 feet (1.8–2.4 meters) above normal tide levels, enough to flood coastal escape routes (roads and bridges leading away from the coastline) and require some people to evacuate their beachside homes. A category 3 hurricane has sustained winds of 111–130 miles per hour (179–209 km per hour) and storm surges 9–12 feet (2.7–3.6 meters) above tide levels. Storm surges this high can cause major erosion (wearing away) of beaches and destruction of houses and businesses on and near the beach.

A category 4 storm produces winds of 131–155 miles per hour (211–249 km per hour) and storm surges 13–17 feet (4–5.1 meters) above normal tide levels. Wave action from category 4 storms can destroy buildings constructed on land less than 2 feet above sea level, and can cause flooding up to 6 miles (10 kilometers) inland. A category 5 hurricane has sustained winds over 155 miles per hour (249 kilometers per hour) and brings a storm surge 18 feet (5.5 meters) or more above normal tidal levels. Besides massive building damage from wave action and winds, damaging floods occur more than 10 miles (16 kilometers) inland, and large-scale evacuations of coastal communities are necessary.

Only three Category 5 hurricanes have ever hit the United States as of 2004. A storm with sustained winds between 39–74 miles per hour (63–119 kilometers per hour) is called a tropical storm. Tropical storms are known for their ability to produce large amounts of rainfall over a short time. An organized storm with sustained winds below 39 miles per hour (63 km per hour) is called a tropical depression. A tropical depression can become a tropical storm and possibly a hurricane.

NOR'EASTERS

Icebergs are large chunks of ice that break off from glaciers or icepacks (a large expanse of floating ice) and float in the oceans. A glacier is a slow–moving solid pack of ice and snow that forms over thousands of years. Most of the world's glaciers were formed between 10,000 and 15,000 years ago during the last Ice Age. Most glaciers slowly flow towards the sea. When a large piece of a glacier pushes out into the sea, it breaks away from the glacier and becomes an iceberg. Most icebergs break away from glaciers in Greenland or Antarctica. While the majority of icebergs remain far to the north, out of the way of most ships, every year several hundred icebergs drift into areas containing shipping routes. These icebergs pose a major risk to ships. An iceberg can create a large hole in a ship and cause major damage or even sink the ship.

The most famous example of this is the Titanic, which hit an iceberg in the north Atlantic Ocean in 1912. The ship sank within hours, killing more than 1,500 people. Following the sinking of Titanic, several nations formed the International Ice Patrol to search for icebergs and record their positions. Modern technology is also capable of detecting icebergs in shipping lanes during the day and night, and in bad weather as well as clear skies. An instrument called synthetic aperture radar that orbits Earth aboard a satellite (vehicle that orbits Earth) collects and sends pulsed signals back to Earth, where a digital map of icebergs, their size and shape, and their precise location is formed.

REEFS AND ROCKS

Like icebergs, reefs and rocks near the shore can damage the hull of ships, causing them to spill their cargo and even sink within a short time. A reef is an underwater ridge of rock or coral (tiny marine creatures with hard exterior skeletons) that lies just below the surface. Rocks can be difficult to spot with the eye, and it is nearly impossible to see a reef before a collision. Many modern ships rely on sonar (images produced by sound waves) or satellite technology (images produced by light waves) to detect rocks and reefs, but accidents still occur.

In 1989 oil tanker Exxon Valdez ran aground on a reef in Prince William Sound, Alaska, causing an oil spill of 11 million gallons (46.5 million liters) into the Alaskan ecosystem. While the Exxon Valdez was not one of the largest oil spills in history, it did have a major impact on the environment and shipping regulations. The ensuing cleanup cost over $2 billion, and the Prince William Sound ecosystem continues to recover to its former level of biodiversity (range of varying plant and animal species).

ANIMALS IN THE SEAS

Although sharks, jellyfish, and other sea animals do injure people in the ocean every year, the number of these attacks are usually sensationalized. Between 70 and 100 shark attacks on humans occur throughout all the oceans worldwide each year. On average, five to ten people die every year as a result of these attacks. Americans are over 300 times more likely to be killed by a car crash involving a deer than by a shark attack in the ocean.

Many coastal states monitor shark populations in beach areas where sharks and humans mix by regularly counting and mapping shark populations just as to geographic features in their habitat. Areas can use this data to issue shark advisories to beachgoers when shark populations are observed to be greater than the normal number of sharks. Most jellyfish stings cause pain, but they rarely kill humans. One exception is the sea wasp or box jellyfish (Chironex fleckeri) that lives in the waters off northern Australia and Southeast Asia. This species of box jellyfish carries venom (poison) in its tentacles powerful enough that a single sting can cause death without prompt medical treatment. All jellyfish species however, are passive hunters; they do not attack prey for food, but wait until a potential food source (including humans) bump into their tentacles.

RECREATION IN AND ON FRESHWATERS

Freshwater is water that does not contain a high amount of salt or dissolved solids. Examples of freshwater include lakes, river, streams, and creeks. While many Americans do not live within driving distance of the seashore, almost everyone lives close to a freshwater river, lake, or stream, and many people are drawn to water for recreation.

FISHING AND SWIMMING

Fishing is one of the most popular freshwater activities, with over 44 million anglers (people who fish) in the United States. Fish live in almost every lake, river, and stream in the United States, which makes fishing possible for most Americans. There are two main types of freshwater fishing: fly fishing and spin fishing. The form of fishing used depends on location, the type of fish, and the body of water. Fly fishing is most popular on rivers and streams. Popular types of fish for freshwater fly fishing include trout, bass, and salmon. When fly fishing, the weight of the fishing line carries the fly, or lure, out into the stream.

A series of arm motions whip the fishing line overhead like a bullwhip, simulating the movement of the prey. Fly fishers lure fish with artificial flies and other artificial water-loving insects that are the natural prey of river fish. In spin fishing, weights called sinkers are attached to the line and carry the hook and artificial lure out into the water. The hook then sinks in the water and the lure spins as the angler reels in the line impersonating an attractive meal to the fish. Trout, salmon, bass, and pike are popular targets for spin fishers. Swimming is another popular freshwater recreational activity. The principle of buoyancy explains how humans can swim instead of sink in the water. Buoyancy is the ability of an object to float in a liquid. Water exerts an upward force, called buoyant force, on every object that is submerged in it. An object will float if this buoyant force is greater that the downward force of gravity (attraction between all masses).

The object will sink if the weight of the object is greater than the buoyant force. The ancient Greek mathematician and scientist Archimedes (287 B.C.E.–212 B.C.E.) realised that the density of the object determines whether or not an object will float. Density is an expression of the mass of an object within a given volume. A piece of steel has a greater density than a piece of Styrofoam of equal size. Archimedes determined that a solid object would float if its density was less than the density of water. Swimming is possible because the human body is less dense than water.

BOATING

Boating comes in several forms: sailboats; motorboats; and personal watercraft or jet skis. Buoyancy also explains how a ship made of steel can float even though steel is denser than water. The density of the overall shape of an object determines if it will float. A ship is constructed so that most of the interior is filled with air. This makes the overall density of the vessel less than the density of water. A simple experiment involving a piece of modeling clay and a glass of water demonstrate how this principle works.

The clay will sink if it is rolled into a ball and placed in the water, but the clay will float if it is flattened, approximating the shape of a boat. Sailboats harness the energy of the wind in sails and the energy in the water to propel them through the water. When wind blows along the sails it creates aerodynamic lift, much like on an airplane. Trimming (adjusting) the sails harnesses this lift in a manner that moves the boat in the water. Without a keel or centerboard (the structure that protrudes from the bottom, or hull, of a sailboat), the wind would blow the boat sideways.

The keel primarily acts as a stabilizer. Water passing over the keel also provides lift that counteracts the force of the wind. Together, these forces push the boat forward. Speedboats have large engines that propel the boat through the water at high speeds. Speedboats are also known as motorboats or powerboats. These boats are used to zip around on rivers and lakes, pulling water skiers or wakeboarders. Water skiing is where a person holds onto a rope that is attached to the boat while wearing a pair of skis. The boat then pulls the person along the water. Wakeboarding is similar to water skiing, but it involves a single, larger board rather than two skis.

Many fishermen also use motorboats to travel on lakes and rivers in order to reach their favourite fishing spots. Pontoon boats and houseboats are larger forms of motorized boats. A pontoon boat has two long, hollow tubes running the length of the boat. These tubes are called sponsons and help provide buoyancy and reduce rocking. Pontoon boats have a flat deck and have an open, boxy shape, making them stable in calm waters. Pontoon boats have motors, but move much slower than speedboats and are used for leisurely cruising and fishing on lakes and rivers. Houseboats are large, enclosed boats with wide hulls to decrease rocking motion and maximize interior space.

Many people vacation on houseboats, and some people live on houseboats throughout the year. Personal watercraft, or jet skis, are small, motorized boats that usually carry one to three people. Riders straddle a personal watercraft as if riding a horse. Personal watercraft are lightweight and can accelerate quickly. As of late 2003 however, personal watercraft were prohibited in 358 of 379 water recreation areas in the U.S. National Park system because of the noise they generate.

ROWING, CANOEING, KAYAKING, AND RAFTING

Rowing, canoeing, kayaking, and rafting are all forms of transportation that require rowing or paddling to move the craft through the water. A paddle, or oar, is a pole that may have a large, fairly flat end, called a blade. A canoe is a boat that is pointed at both ends and typically has a completely open top, or deck. People sit or kneel in a canoe and use a paddle with a single blade to move the canoe through the water. A canoe usually holds several people. A kayak is a boat that is pointed at both ends and has a closed deck except for a small hole where the paddler sits. A kayak paddle has two blades, with one on each end.

A kayak usually holds only one person, but some models can carry two people. In order to steer the kayak, it is necessary to use the entire body for balancing and leaning along with the paddle. A raft is a flat-bottomed boat, which is usually inflated with air. Several riders use paddles with single blades to move and steer rafts. Rafts are flexible, so they are often used in water that may contains rocks. If the raft hits a rock or goes over a small waterfall, the raft will bend instead of breaking. Kayaks and rafts are often used for riding down river rapids, which are stretches of fast moving water on a river or stream. Rapids form from erosion (wearing away by wind or water), when water erodes rocks in a river at different rates. The soft rocks erode first, creating a steeper gradient (the angle of slope down which a river flows) for the river to flow down among the remaining harder rocks. Whitewater rapids are formed as the water increases speed in order to move along the steeper pathway among the harder rocks. When people travel down rapids, these sports are referred to as whitewater kayaking and whitewater rafting. Specially made canoes may also be used for whitewater canoeing.

THERMAL SPRINGS AND SPAS

Thermal springs (natural flow of groundwater), also commonly called hot springs, were considered healing waters by many ancient cultures and still are by many modern cultures. Thermal springs produce water that has been heated by the earth to a temperature of 70°F (21°C) or above. The ancient Romans constructed elaborate bathhouses, or spas, at the sites of thermal springs, and hot springs continue to be a major attraction in modern times. Modern spas often locate at the source of thermal springs, making hot springs popular destinations. The water that flows from thermal springs becomes heated by geothermal warming in one of two ways.

Geothermal means relating to heat generated from the center of the earth. The presence of underground volcanoes near the surface of the Earth can heat the water. Iceland is famous for its numerous volcanic thermal springs. Thermal springs can also be produced by rainwater seeping deep into the earth and then rising quickly. One example of this method is found in Hot Springs, Arkansas, where rainwater has seeped into the earth for thousands of years.

The water seeps down to a depth of 6,000 to 8,000 feet (1,829 to 2,438 meters) below the surface and warmed by the earth's internal temperature. Cracks in rocks then allow the warmed water to return to the surface in less than a year. Because the water's return trip is quick, the water loses little heat and surfaces at about 147°F (63.8°C).

TOURISM AT NIAGARA FALLS

Sometimes observing water is a recreational activity. Niagara Falls, on the United States–Canada border, became a popular tourist destination in the nineteenth century and has remained a popular destination. Every year, over twelve million people visit Niagara Falls. Niagara Falls actually consists of two main waterfalls. The larger waterfall is Horseshoe Falls, or Canadian Falls. Horseshoe Falls, shaped like a horseshoe, is 167 feet (51 meters) high and 2,600 feet (792 meters) across. Over 600,000 gallons of water flow over Horseshoe Falls every second. On the opposite side of the falls, American Falls is 176 feet (54 meters) high and 1,060 feet (322 meter) across. Over 150,000 gallons flow over American Falls every second. The waterfalls at Niagara were formed nearing the end of the last Ice Age about 12,000 years ago, when melting ice flowed into what is now the Niagara River. The river flowed over the Niagara escarpment (cliff), slowly wearing away the underlying rocks until the falls was carved upstream to its current position.

WINTER SPORTS

Many parts of the country enjoy recreational activities on frozen lakes and ponds. Ice skating and ice hockey are activities that can be enjoyed on frozen bodies of freshwater. Ice fishing is also another popular activity in some parts of the United States. Ice fishing involves cutting a hole in the ice on a lake or river and dropping a fishing line into the water below the ice. Liquid water is denser (heavier per unit) than ice. This explains why ice floats and forms on top of the lake in winter.

RECREATION IN AND ON THE OCEANS

Every year, Americans spend billions of dollars and a large amount of their spare time on recreational activities in and on the oceans. Among others, popular ocean–based activities include swimming, snorkeling, scuba diving, sailing, fishing, and surfing.

IN THE OCEAN

Swimming is one of the most popular forms of ocean recreation. Millions of Americans visit the beach every year to swim in the ocean. While swimming, beachgoers participate in snorkeling. Snorkeling, or skin diving, is a form of diving in which the diver swims at or near the surface of the water. Skin diving is simply holding one's breath underwater for as long as possible. The diver can remain underwater for long periods by breathing through a snorkel, which is a hollow tube attached to a mouthpiece. The snorkel juts out above the surface of the ocean, allowing the diver to breathe surface air through the snorkel like a straw. Snorkeling allows divers to explore ocean animals, plants, and coral reefs (tropical marine ecosystems made up of tiny coral animals and the structures they produce) that lay just below the surface of the ocean. Scuba diving allows divers to fully immerse themselves in the ocean environment. Scuba stands for Self Contained Underwater Breathing Apparatus. Scuba equipment allows divers to go deeper than snorkeling and stay underwater longer. Scuba gear provides oxygen to divers while underwater. Modern scuba equipment is made up of small cylinders of compressed air.

The diver breathes through a mouthpiece, and the air tank provides oxygen with every breath. Recreational scuba divers can explore about 150 feet (46 meters) below the surface and with advanced training they can dive deeper. Dives deeper than 150 feet (46 meters) require gradual rising to the surface and other precautions. Rising too quickly after a deep dive can cause nitrogen to build up in the body, causing a painful, and potentially fatal condition called decompression sickness, or the bends. The world record for a scuba dive set in 2003 is over 1000 feet (313 meters). It took the diver only 12 minutes to reach this depth, but the diver had to rise to the surface of the water over 61D2 hours in order to avoid the bends.

ON THE OCEAN

Sailing involves moving across the water in a boat powered by the wind. Sailing may be done for pleasure or sport. Sailing for sport involves serious competition. Sailboats are divided into numerous classes, or divisions, for competition based on the size and style of the boat. The America's Cup race and the Volvo Ocean Race Round the World are two of the most popular and competitive sailing races. In the Volvo Ocean Race, formerly called the Whitbread Round the World Race, each yacht and its crew receive millions of dollars from corporate sponsors to design and build newer, faster ships. The race also tests the ability and stamina of the crew over the course of nine months. The 2001–2 Volvo Ocean Race Round the World, for instance, was 31,600-nautical-miles long.

A nautical mile is longer than the statutory mile used on highways (1.15 statutory miles). The race, which ran for nine months, consisted of nine legs, or sections. Sailors traveled on the following routes: England to South Africa; South Africa to Australia; Australia to New Zealand; New Zealand to Brazil; Brazil to Miami, Florida; Miami to Baltimore,

Maryland; Baltimore to France; and Sweden to Germany. Recreational fishing on the oceans generally comes in two varieties: shore fishing and deep-sea fishing.

In shore fishing, the angler (one who fishes) casts his or her bait from the shore. This form of fishing catches fish that stay close to land such as redfish, snook, and seatrout. Deep-sea fishing requires boating several miles (kilometers) out to sea in order to catch fish that live far from shore, where sonar (a device that uses sound waves to locate underwater objects) is sometimes used to spot schools of fish. Tuna, marlin, tarpon, and barracuda are examples of deep-sea fish. Some species of deep-sea fish can weigh over 1,000 pounds (454 kilograms). Deep-sea fishing is a large business. Many tourists in popular deep-sea fishing locations pay thousands of dollars to rent boats and equipment for deep-sea fishing trips.

Popular deepsea fishing locations in the United States include Florida, the Gulf of Mexico, and New England. Surfing is the act of riding a board, called a surfboard, on the waves. Surfing requires strength and balance. Recreational surfers typically ride on relatively small waves 3–5 feet (.9–1.5 meters), although some surfers travel worldwide in search of larger waves. Surfing competitions judge competitors on wave size, distance, and quality of performance. Some professional thrill-seeking surfers, called tow surfers, ride out on personal water crafts to ride waves up to 50 feet (15 meters) high.

Chapter 15

Policy and Legislative Landscape for Water Disputes

INTRODUCTION

Growing global concern over the state of the environment and the impact of environmental issues on human security and political stability has, since the 1960s, spurred a series of international conferences to discuss and address environmental concerns. In response to these issues, and culminating from the international conference dynamic of the past three decades, environmental issues have become crucial in political decision making, from the local to the international levels, and have been placed on the political agenda at various levels.

This has resulted in an escalation of environmental policies, legislation and institutions dealing with environmental issues at the international, the regional and the local levels. The importance of fresh water on international, regional and national policy agendas has likewise been duly acknowledged since the 1970s. The unprecedented increase in water consumption, the realities of growing water scarcity in the face of population pressure, and the resultant socio-political tensions associated with competition over this scarce resource have all served to focus increased attention on both water scarcity and the issues surrounding this concern.

To understand the current situation with regard to water scarcity, the capacity of the existing policy and the institutional framework to deal with water issues, it is necessary to devote some attention to an historical analysis of the main policy and the legal and institutional developments that are currently shaping how water issues are dealt with at both the international and the regional policy levels. Furthermore, the international conference dynamic that has shaped international environmental policy and institutions in general, and water policy and institutions specifically need to be outlined and analysed. A legislative framework has furthermore emerged concomitantly with the policy and institutional frameworks. Moreover, since legislative considerations are gaining increased importance in the context of international cooperation over water, attention is thus also devoted to the development of an international legislative framework to deal with water issues. Lastly, developments at the international level have filtered down to inform regional policy and institutional framework on water in the Southern African region. This framework developed within the bounds of two larger developments. Firstly, a growing awareness of water scarcity in the region provided impetus to the development of a regional instrument, the SADC Protocol on Shared Watercourses, to govern relations in respect of water in the region. Secondly, within the larger regional political and economic framework, emphasis is placed on regional cooperation and integration at various levels.

The process by means of which the regional framework for cooperation over freshwater resources developed is discussed and critically evaluated in terms of its efficacy to prevent and intervene in conflicts over freshwater resources in the region. This chapter therefore focuses distinctly on two aspects, namely the international policy and legislative framework and the regional policy and legislative framework for dealing with water issues. In addressing the international and regional policy and institutional landscape, this chapter also includes references to data gathered by means of in-depth interviews with key informants and thus also comprises an empirical component. Attention is first devoted to the historical context of the development of water policy.

WATER POLICY DEVELOPMENT IN THE HISTORICAL CONTEXT: SHIFTING EMPHASIS

The political importance of water was recognised very early in the history of humankind, and since early historical times water has played a major role in political decision making. Prominent agricultural and urban developments in history all took place near large sources of water, such as along major rivers and on the banks of lakes. Wong *et al.* even go as far as to state that human civilization was born on a river bank.

By developing near water sources, societies took advantage of the ecological services and benefits that these water bodies provided, such as the possibility of flood-recession agriculture. However, owing to the need to use freshwater resources more effectively for societal needs, the need to manage and control these natural systems by means of political decision making also arose.

Political leaders accordingly used the manipulation and control of water to prove their power and win the favour of citizens. The ruler of Assyria during the 9th century BC was reported to have had her tombstone inscribed with a reference to her constraining the 'mighty river' to flow according to her will, thereby providing water to fertilize previously barren uninhabited lands. The mindset of controlling and managing water, which today still influences water policy and water-management practices, is therefore deeply imbedded in societal memory. Over time, technological advances have made the control and management of large water bodies increasingly possible.

Society has progressed far beyond the building of aqueducts to relay water to the large cities in ancient history, for example, to the engineering feats of dam-building and water transfer schemes of which society is today capable. Traditionally, though, water was viewed as an infinite substance that should be made readily available to all competing sectors. During the 20th century, emphasis in policy was thus largely placed on satisfying increasing human demand for fresh water, without taking cognizance of the adverse effects of this practice on water availability and on the natural environmental systems that support such water resources. Epitomizing this view, Winston Churchill in 1908 envisaged that every drop of water draining into the Nile River Valley would one day be divided among the people and that the river would one day 'gloriously' perish and not reach the sea.

Until the 1960s, there was no clear conception in the minds of people, or in the minds of those in decision-making positions, of the gravity of society's actions on the state of the planet in general, and more particularly on water resources. Water scarcity was dealt with by increasing the supply of water to the area of scarcity, and the entire policy and institutional framework that developed until late in the 20th century was geared to meeting demand and managing the allocation of water to the various user sectors. Emphasis was therefore placed on increasing supply and the policy environment was characterised by a supply-orientated approach.

Water policies were thus geared towards advancing water supplies to the various sectors of society–agriculture, domestic and industrialisation. As a result of such supply orientation, water policy became foremost centered on satisfying humanity's rising demand for irrigation, domestic and industrial water supply, as well as on flood reduction and hydropower generation. This aligns with the technocentric spirit that characterized environmental thinking during much of the 20th century. With regard to the currently dominant worldview, Petrella states that, humankind's views of society have increasingly been permeated by a 'techno-economist' culture. The emphasis placed on large-scale infrastructure further fits into the realist political mindset that dominated the political arena during the 20th century and in which the protection of national interests over the interests of other parties took a central position. During the Cold War era there was an increase in dam building that in effect started with the building of the Hoover Dam on the Colorado River in the 1930s. The building of large infrastructural developments in this era are, according to Turton linked to larger political and economic advantages over other nations.

These developments provided countries not only with water security but also with other political and economic benefits derived from these infrastructural developments. With the building of the Hoover Dam, for example, America increased its strategic advantage over German forces in World War II. The dam provided cheap hydroelectricity to produce aluminium, a key component in aircraft production. Since the USA could outpace German production of aircraft as a result of the hydroelectricity generated by Hoover Dam, the Allied Forces won some crucial battles that turned the tide of the war in the favour of the Allied Forces.

This societal emphasis on technological ability and economic advancement is reflected strongly in many policy frameworks on water developed until the 1970s. However, the rising demand for water, exacerbated by the focus on supply management, in combination with the environmental impacts of ecosystem alteration, caused the depletion of water supplies, triggered large-scale damage to ecosystems and placed a large share of freshwater life at risk of extinction. Furthermore, during the last 50 years of the 20th century, significant emphasis in the political arena was placed on national sovereignty and the protection of and control over resources within geopolitical boundaries. In the case of Southern Africa, water was firmly linked firmly with political interests and issues such as sovereignty, even during colonial times.

Collaboration between countries was limited, especially over the sharing of natural resources. Within this orientation, waterresource developments usually took place at a localised level and did not involve a wide array of stakeholders, let alone the catchment as a whole. During the 1940s, when Swaziland was still a British Protectorate, the Commonwealth Development Corporation built a large weir, the Issys Canal, on the Komati River to supply irrigation water to farmers in Swaziland, but there was no consultation with downstream farmers in South Africa. The building of this weir led to decreased water flows to farmers in South Africa to the extent that the South African farmers took the CDC to court to restore some of the water flows. Eventually an agreement to release 45 cubic seconds of water from the Komati River to South Africa was reached in the 1940s. In the face of growing water scarcity, it has become increasingly important to view the development, utilisation and management of water resources interdependently–as has indeed been the case with policy developments in recent years.

Dlamini links the above conflict between South African farmers and the CDC with later developments in water cooperation between Swaziland, Mozambique and South Africa. This reflects the growing emphasis on interdependence in water resource management. He relates it as follows: "So this is the key thing that started the whole cooperation on the Komati. It started in 1945 when these small farmers [in South Africa] started complaining to the CDC, 'We don't even have drinking water', But as that happened there were more developments... there was major developments of commercial farmers in this area [South Africa].... But that resulted in a major increase in demand for water, because suddenly these people here [needed water] not only for domestic but even for participation in agriculture. As a result of encouragement by the Department of Agriculture, they started demanding more water and then there was conflict now and the whole issue went back to square one.

'You are taking all the water'. That was the biggest problem. As years go by the colonial era passed - it ended in 1970–and the [South African] government was in a predicament. They were no longer talking to Britain, but Swaziland Government and Swaziland Government look at this as a major asset... It employed a large number of people....It became very difficult, until this whole thing came to a point where South Africa and Swaziland agreed to talk about it." Since the end of the Cold War at the end of the 1980s, the world has undergone some dramatic social, political and economic changes that have challenged the policy frameworks and institutional setups developed during and resulting from the Cold War.

Those governing in the post-Cold War context must have the adaptive capacity ability to function in situations of rapid change. Thus, the rapidly changing social context demands that institutions and governing bodies adapt to change quickly and also have second order resources–social adaptive capacity and social ingenuity - in place that so that they could be positive agents for change and cooperation. On the political front, the Cold War was the dominant force that divided the world into distinct zones during those decades. Political relations revolved around a struggle for power and dominance between the world's superpowers, and the ability to protect a nation's political independence and territorial integrity became paramount. In this context, Kegley and Wittkopf state that most states rejected measures such as reformation or integration of "… governmental procedures for democratically making national security decisions…" and the building of institutions for world law.

Some political changes, amongst others increasing internationali-sation and an emphasis on closer economic collaboration in the context of globalisation, have emphasised the need to collaborate on the sharing of increasingly scarce natural resources. In response to the emerging recognition of socio-economic, socio-political and environmental challenges, the United Nations responded by initiating and sponsoring a number of cross-cutting conferences starting in 1972, which attempted to deal with the multifaceted nature of the world's environmental and population issues. This conference dynamic will now be considered.

INTERNATIONAL CONFERENCE DYNAMIC AND THE IMPACT ON WATER RESOURCE MANAGEMENT

The policy developments that took place within the international conference dynamic can broadly be divided into decadal time periods. The first significant set of developments took place during the 1970s, starting with the UN Conference on the Human Environment in 1972. Further developments took place in the 1980s, the 1990s and since 2000.

1970–1980: LINKING HUMAN ACTIONS WITH ENVIRONMENTAL CONSEQUENCES

The first international conference to highlight the interrelatedness of social and environmental factors in achieving a sustainable world, was the landmark UNCHE in Stockholm organised by UNESCO. This conference "...considered the need for a common outlook and for common principles to inspire and guide the peoples of the world in the preservation and enhancement of the human environment". Emphasis was placed on humankind taking better cognisance of the environmental consequences of its actions, while the growing interconnectedness of environmental problems formed the point of departure for this conference. An important focus of the conference was international çooperation on the environment and environmental issues. This conference led to the establishment of environmental agencies and ministries in more than a 100 countries, while growth in the number of NGOs dealing with environmental issues also skyrocketed. On the international front, the conference led to the establishment of the UN Environmental Programme for promoting the results of the conference. Thus, this conference facilitated the development of a general institutional framework to tackle environmental issues at the international and the local level. Furthermore, the Declaration and Action Plan resulting from the Stockholm Conference were also instrumental in the development of international environmental law after this conference. To a large extent, the technocratic worldview dominant in this era also infused the conference content with the notion that technological solutions could be found to all environmental problems facing the world at that stage.

1980–1990: FOCUS ON SUSTAINABLE DEVELOPMENT

The Stockholm +10 Conference in Nairobi, Kenya, was held as a follow-up to the 1972 Conference. During the 1980s, addressing environmental issues became more comprehensive and the social and economic drivers of environmental problems began to receive attention. In 1980, the World Commission on Environment and Development was established and proceeded with unlocking the relationship between the environment and the economy. The conclusions of this commission were included in the 1987 Our Common Future report.

A part from entrenching the concept of sustainable development 65 in the environmental policy framework, the commission also argued strongly that the major ecological problems to be addressed by sustainable development were global-scale problems Humphrey *et al.* see a significant contribution of the Commission's work being that it led to a measure of compromise among representatives of world governments, environmental organisations, international NGOs and development agencies on the apparent contradiction between economic growth, development and environmental sustainability. Therefore, through the Commission's work, a certain level of agreement was reached that environmental issues could be dealt with more effectively by cooperation and agreement between parties, and that the nature of environmental issues necessitated the involvement of higher levels of governance and multiple disciplines.

1990–2000: THE DECADE DOMINATED BY THE EARTH SUMMIT

The 1990s were characterised by a search to increasingly understand what sustainable development entailed, both conceptually and in terms of its significance. Sustainable development became the central idea steering the United Nations Conference on Environment and Development held in Rio de Janeiro in 1992. The Rio Conference, or Earth Summit, culminated both in the Rio Declaration on Environment and Development and in Agenda.

The former expressly emphasised establishing a new and equitable global partnership through creating new levels of cooperation between states and people. Brynard and Stone noted the latter, commonly regarded as the blueprint for all environmental policy and implementation, emphasised the holistic management of fresh water and the integration of water plans and programmes into national economic and social policy. Agenda, specifically called for the holistic management of fresh water and the integration of water programmes within national economic and social policy, thereby entrenching the linkages between development and water issues. Water resources are given specific attention and the overall goal is to ensure that the supply and quality of water is sufficient to meet both human and ecological needs across the world.

According to Petrella: The Rio conference did indeed help to reaffirm, within the framework of Agenda, the urgent need for a world water policy". At an institutional level, the United Nations Commission on Sustainable Development was created to see to it that the decisions, resolutions and agreements of the Rio Conference were translated into practice. This commission became an important centre for discussions and meetings on water issues, and two influential world conferences on water were a direct result of discussions preceding the sixth session of the CSD in 1998, organised by the German and French governments respectively. The International Conference on Population and Development of 1994 was markedly more population centered.

One of the outcomes of the ICPD was the Programme of Action that underpinned many of the principles expressed at the UNCED. The ICPD reiterated that a healthy environment was crucial towards meeting the basic human needs of a growing world population, and that access to water and sanitation was fundamental in achieving a healthier environment for the world population. Linking with sustainable development, this conference further emphasised that population, environmental and poverty eradication factors should be integrated in sustainable development policies.

Environmental Factors into Policy Frameworks

In 2000, governmental heads of state, through collaboration between the World Bank and the United Nations in an effort to focus development assistance more effectively, negotiated a Millennium Declaration. This Declaration encompassed eight MDGs in the areas of life expectancy, education, housing, gender equality, trade and environmental protection to be achieved by 2015. Significantly, these goals reinforced one another and an integrated approach was therefore followed in achieving the targets set. For example, water security holds implications for reaching all of the eight MDGs as a result of the direct and indirect impact of water resource availability on dealing with developmental challenges such as poverty, hunger and disease. Water scarcity was thus considered to be a factor that constrains prospects for attaining human security as envisioned by the international policy framework, and emphasis was placed on providing adequate access to fresh water for all people.

A further characteristic of the MDGs was the recognition by the Summit of the sovereignty of nations to determine their own needs based on their culture and history, while past experience with international cooperation was important in informing and shaping action. The World Summit on Sustainable Development, in Johannesburg built on the above and made it clear that freshwater availability was vital to sustainable development and further had implications for all areas of concern identified by the Summit. The WSSD focused intently on the identified gaps in implementing the agenda set in Rio for the achievement of sustainable development.

Among these gaps, it was recognised in the PrepComs preceding the WSSD that, amongst others, there was a fragmented approach to sustainable development and core items such as water, energy, health, agriculture, and that biodiversity had received insufficient attention. Therefore, the issue was not the ideas inherent in the quest for sustainable development, but rather the implementation of said ideas. Two outcomes of the Summit included the Johannesburg Declaration on Sustainable Development and the WSSD Plan of Implementation. The former recognised that the eradication of poverty, the changing of consumption and production patterns and the protection and management of the natural resource base formed the foundation of socioeconomic development and were essential to achieving sustainable development.

This document further identified the most important instruments for a sustainable development policy and these included capacity building, forming new partnerships and good governance, among others. The latter document is regarded as the central document produced by the Summit and provides a list of actions to be implemented to enforce Agenda. With regard to water, it is again emphasised that clean drinking water and adequate sanitation are necessary to protect the health of people and the environment. The goal of halving the proportion of people without access to safe water and adequate sanitation by 2015–as set in the Millennium Summit–is again agreed upon again in the Plan of Implementation.

The overarching outcome of these and other conferences was that environmental, population and developmental issues were recognised as the foundations of sustainable development and human well-being. More significantly, the premise that population factors and environmental resources cannot be detached from social develop-mental factors was widely acknowledged and imbedded in the international policy framework governing environmental sustainability. With regard specifically to water, the international policy framework emanating from the above recognises water security as integral in achieving the well-being of current and future generations.

INTERNATIONAL POLICY DEVELOPMENTS PERTAINING SPECIFICALLY TO WATER

Concurrently with the above international conference dynamic, some specific international policy developments pertaining to water also occurred since the 1970s.

1970–1990: WATER IS PLACED ON THE INTERNATIONAL POLICY AGENDA

The 1970s saw the first international conference specifically dedicated to water, namely the United Nations Conference on Water, Mar del Plata. An outlook similar to that in Stockholm characterised this conference with regard to a technocentric mindset and emphasis was largely placed on the assessment of water resources, while the processing and compilation of data were seen thus far to have been neglected.

This conference defined water as a common good and declared the right of access to basic drinking water for all people. The main issues dealt with in this conference were the assessment of water resources, as well as water use and efficiency. Petrella notes that Mar del Plata managed to outline the basic facts on water and made water a top issue on the international political agenda.

Philips *et al.* summarise the contribution of Mar del Plata as essentially arguing for "...better informed and more flexible water plans; viable institutions for the implementation of such plans; comprehensive and updated water laws to create an adequate enabling environment; and participation by all stakeholders". During the next 15 years, until the Earth Summit in 1992, no major conferences were held that specifically focused on water issues. However, the decade from 1981 to 1990 was declared the International Drinking Water Supply and Sanitation Decade by the UN. Efforts during this decade were aimed at providing every person with access to an adequate and safe supply of water and a satisfactory means of excreta and sullage disposal by 1990.

In spite of ambitious efforts worldwide, this goal was not achieved, and in 1990 the Global Consultation on Safe Water and Sanitation for the 1990s took place in New Delhi. An important contribution made by the Consultation was putting across the idea that supplying safe water and proper waste disposal should be a central aspect of integrated water resource management. The IDWSSD, as well as the Safe Water and Sanitation Consultation, brought to the fore the issue of human development as a central aspect of water policy.

THE DUBLIN DECLARATION AND ITS IMPACT

In 1992 the International Conference on Water and the Environment, a preparatory session to the Earth Summit, was held in Dublin. While the Earth Summit was largely successful in addressing many environmental issues on the international policy agenda, it was generally deemed a failure by water specialists. However, the Dublin meeting did succeed in placing water on the global development agenda. This conference dealt with a number of issues, amongst others, the economic value of water, women, poverty, conflict resolution and natural disaster awareness, and it resulted in the Dublin Statement on Water and Sustainable Development that encompasses four basic principles on water.

These principles were:

- *Principle* 1: Fresh water is finite and vulnerable, essential to life, development and the environment.
- *Principle* 2: Water development and management should be based on a participatory approach, involving users, planners and policy makers at all levels.
- *Principle* 3: Women play a central part in the provision, management and safeguarding of water.
- *Principle* 4: Water has economic value in all its competing uses and should be recognised as an economic good.

The contributions of the Dublin meeting came to play a central role in guiding policy developments by modernising thinking on how to deal with the emerging global scarcity of water. It further paved the way for better water management underpinned by participation and recognition of the holistic nature of water resources. In the aftermath of Rio/Dublin, the societal perception of water scarcity became so strong that Philips *et al.* note that this perception led to a scramble to secure supply of water at all cost and water resource management was "...elevated to the ...realm of national security".

It is against this background that the water sector increasingly came to realise the importance of finding common ground in dealing with issues of growing scarcity and potential conflict over water resources. Particularly in view of the importance of international cooperation in addressing the world's water issues jointly, the UN proclaimed 2005-2015 as The International Decade for Action,"Water for Life".

THE WORLD WATER FORUMS

During 1996, two NGOs working in the field of water were formed, namely the World Water Council and the Global Water Partnership. Both the WWC and the GWP were formed through the collaboration of various UN agencies, some countries and some private water corporations. The WWC's purpose was to act as a policy think tank on water issues and to create and promote a common world vision on water-related issues, while the GWP established itself with the aim to support countries in sustainably managing their water resources by getting public institutions and private companies to work together on water policy.

In 1997, the WWC organised the First World Water Forum in Marrakech and undertook to organise similar forums every three years from then on. At this first forum it was explicitly recognised that water might be at risk of becoming a marketable and expensive resource that could become an object of conflict in the same way as oil had. Prioritised areas of discussion included drinking water and sanitation, preservation of ecosystems, gender equity, water-use efficiency and the management of shared water resources.

Through the Declaration of Marrakech, this Forum linked back to previous international efforts to guide the way forward in addressing the world's water issues. To this effect, the WWC was charged with the task of producing a global Vision for Water, Life and the Environment. Emphasis was placed on providing "... policy relevant conclusions and recommendations for action to be taken by the world's leaders to meet the needs of future generations". The Council set up the World Commission on Water for the Twenty First Century to oversee the drafting of this Vision under the chairmanship of Ismail Serageldin, World Bank vice-president and chair of both the GWP and The Consultative Group on International Agricultural Research. The Commission in turn formed a Vision Unit to achieve the objective of drafting this vision. The Vision Document was presented and discussed at the Second World Water Forum in The Hague, 2000.

Five recommendations for action were outlined in the document:

- Involvement of all stakeholders in integrated water management.
- Moving towards full-cost pricing of all water services.
- Increasing public funding for research and innovation in the public interest.
- Increasing cooperation in international water basins.
- Increasing investments in water.

The World Water Vision Document highlighted the scarcity of the resource and that it was a vital social and economic asset. Water thus had to be brought under the market laws governing the use of other natural resources such as oil. Flowing from this, it was further stated that the rational and efficient management of water resources required scrupulous economic culture and practice. Furthermore, since water was a primary factor in health, rational and efficient water policy should aim towards the best possible quality by investing in infrastructure and maintenance. This entailed that water policy became a financial issue, guided by access to investment and profitability. The Vision Document did not however provide a unified vision on dealing with water issues, the whole process being wrought with disagreement amongst those involved and by conflicting positions on various issues.

Eventually the document, subtitled "Making Water Everybody's Business", lacked clarity. The vagueness of the document led the Commission to draft its own report advocating clear actions under the categories of water pricing, institutions, research, and data and investments. Another document, "Towards Water Security, A Framework for Action", was produced by the Global Water Partnership, another prominent NGO, to be presented at the Second World Water Forum. This NGO's goal was to help developing countries in the sustainable management of their national water resources. Two other significant developments took place at the Second World Water Forum. Firstly, conflicting viewpoints were brought to the fore when at an ad hoc meeting of NGOs, primarily working in the fields of environment and labour, took a strong position against the pro-corporate stance of the Commission, the GWP and the WWC.

This group of NGOs strongly advocated the position that access to water and sanitation and a healthy environment were basic human rights, while food and water insecurity were linked to the current global trade system embodied by the World Trade Organisation. No resolution of viewpoints was accomplished, even after a meeting of the NGOs, the Director of the World Water Vision Unit, leaders of the GWP and Serageldin, chair of the Commission. Instead, the GWP and the WWC belittled the statement of the NGOs and accused them of being a divided group.

What transpired from this process was the fact that the task was given to the WWC of developing a unified perspective on the conflicting issue of how to approach water scarcity in the 21st century and of ensuring water security for all. Although this institution attempted to fulfil the objective of drafting a common vision, the process lacked participation and eventually the result was that the document strongly reflected the values of the WWC, but lacked definite recommendations for future action. It is also significant of the conflicting views between the Vision Unit and the Commission that the Commission thought it necessary to draft another document with much clearer recommendations for action.

A further indication of the lack of agreement at the international level on the way forward at that stage was the fact that the GWP chose to draft yet another document to guide actions for water security in the 21st century. However, in spite of these apparently conflicting positions, there are a number of similar positions held in each of these documents as outlined in Table. The second development to take place was the Ministerial Conference that ran concurrently with the Second World Water Forum. This conference, in which 158 delegations, representing 130 countries participated, was organised by the Netherlands government and produced The Ministerial Declaration of The Hague: Water Security in the Twenty-First Century.

In compiling this document, the conference drew on inputs provided by the Commission, the WWC, the GWP and the NGO caucus in addition to other sources and experiences from delegates. The Ministerial Declaration emphasised that neither the threats to water security, nor the attempts to address these threats were new, that discussions and actions to this end had continued since Mar del Plata through to the Second World Water Forum and that this process would continue in the future. Seven challenges and concordant actions to meet these challenges were outlined in the Ministerial Declaration.

In summary the challenges included:

- Access to safe and sufficient water and sanitation were basic human needs, essential to health and well being;
- Enhancing food security was needed, with specific reference to the poor and the vulnerable;
- Ensuring ecosystem integrity through sustainable water resource management was crucial in achieving water security;
- Promoting peaceful cooperation and developing synergies between water users at all levels, also in the case of boundary and transboundary resources weree necessary;
- Providing security from water related hazards was an important aspect of water security;
- Water needed to be managed with regard to its economic, social, environmental and cultural value, with emphasis on the pricing of water services to reflect the cost of its provision, while taking

the needs of the poor and vulnerable into account;

- Ensuring good governance of water resources was essential.

The Declaration also strongly emphasised the Ministerial Conference's commitment to IWRM, taking into account the social, economic and environmental factors in managing water resources. To achieve IWRM, the need was expressed for coherent national, regional and international policies that would overcome fragmentation. To this end, the Declaration recognised the importance of increasing coherence in international water-related activities. The United Nations, multilateral institutions, international financial institutions and bodies established through intergovernmental treaties were specifically mentioned in the Declaration to work towards strengthening water-related policies and programmes to achieve water security and address the challenges identified by the declaration.

Also significant in this Declaration was the recognition of collaboration and partnerships ranging from the individual citizen through to international organisations. However, although the Ministerial Declaration included a reference to water sharing between states and users as a challenge for the 21st century, the importance of water sharing between countries was not elevated above the other six challenges identified.

Table. Contributions to Water Policy Development (Second World Water Forum)

The World Commission on Water for the 21st Century	**Global Water Patnership**	**Vision Unit**	**Ministerial Conference**
	Towards water security: A frame work for action	World water vision: Making water every-body's business	Minsterial dec-laration of the Hague on water security in the 21stcentury
Water pricing–full cost-pricing of water services.	Water wisdom. Expanding and deepening dialo-gue between stakeholders.	Involvement of all stakeholders in integrated water mange-ment.	Meeting basic needs: access to safe and sufficient water and sanita-tion.
Institutions–broader role of private sector and smaller role for public sector.	Strengthening capacities of organisations in water manage-ment.	Moving towards full-costpricing of all water services.	Securing food supply, with particular refer-ence to the poor and the vulner-able.

The World Commission on Water for the 21st Century	Global Water Patnership	Vision Unit	Ministerial Conference
Research and data– funding for agricultural research.	Ensuring adequate financial resources to pay for actions required.	Increasing public funding for research and and innovation in the public interest.	Ensuring ecosystem integrity through sustainable water resource mangement.
Investment– subsidies to faciliate entry of private operations		Increasing cooperation internatonal water basins. Increasing investment in water	Promoting peaceful cooperation and devloping synergies between water user at all levels.

The Third World Water Forum was held in Kyoto in 2003. According to the report World Water Actions released by the WWC at the forum, significant progress was made since the second forum and the optimistic view was expressed that, through the continuation of current efforts, it was possible to meet the water challenges that the world faced.

Priorities set at this forum were: governance, integrated water resources, gender, pro-poor policies, financing, cooperation, capacity-building, water-use efficiency, waterpollution prevention and disaster mitigation.

The most recent Fourth World Water Forum was held in Mexico in 2006, during which emphasis shifted strongly towards implementing local actions to confront global water problems.

At this Forum, a Ministerial Declaration was again signed by the ministers participating. This declaration reaffirmed the critical nature of fresh water for all aspects of sustainable development and it again stressed the importance of including water and sanitation as national priorities. The declaration also recognises the continued role of the CSD process in keeping water on the sustainable development agenda. Despite the above tension-ridden process, the GWP has been playing a significant role in creating a neutral multi-stakeholder partnership that aims at bridging the gap between the various stakeholders in the water sector.

In the regional context, the GWP-Southern Africa links stakeholders from the regional to the local level by creating a platform in which needs and challenges are identified and addressed within the framework of IWRM.

Takawira thus highlights the significance of the GWP in the regional context: "[GWP] goes beyond intergovernmental [cooperation]. We cut across. We bring up right from the local level right up to the regional level. We are kind of bridging that gap. We are trying to see how this vertical integration is functioning. And what we do is capacity development of institutions.

What we are trying to do is talking up and making sure that for example Lower Manyane a sub catchment is integrated within the Zambezi River Basin."

The GWP-SA has also played an important role in helping the SADC in developing a regional SADC Vision for Water Life and Environment. Beukman indicates that this process formed part of the global vision process, but Southern African stakeholders were very active in the regional vision process that ran concurrently with the global process.

CONTRIBUTIONS OUTSIDE THE WORLD WATER FORUM PROCESS

Since Rio, a number of influential international conferences on water also took place outside of the dynamic of the World Water Forums. First among the former, the International Conference on Water and Sustainable Development was hosted in Paris in 1998. The second, the International Conference on International River Basin Management, an initiative of the German government, was also held in 1998. Both these conferences were organised in the run-up to the Sixth Session of the CSD in that same year, which focused on applying Agenda in the protection of water resources.

The first major international conference on water since the Millennium, apart from the Second World Water Forum, was the International Conference on Freshwater in Bonn, where, 118 national governments, 47 international organisations and 73 organisations from civil society convened to discuss and plan recommended actions on how to cope with the issue of water as a key to sustainable development.

Two outcomes of the conference were the Ministerial Declaration and the Recommendations for Action. Main issues discussed at the conference included governance, mobilising financial resources, capacity building and sharing knowledge.

The Ministerial Declaration stressed the importance of combating poverty to achieve equitable and sustainable development, while acknowledging that water plays a central role in human health, livelihood, economic growth and in sustaining ecosystems.

Bonn also specifically raised the issues of transboundary water management and virtual water. With regard to transboundary water resources management, it was accepted that water may be a source of conflict among different water users in a water basin, but it was emphasised that water can promote regional cooperation when watersheds, lakes, river basins and aquifers become the primary frame of reference for water resources management. Therefore, Bonn had a strong focus on IWRM with the water basin as the main unit of analysis.

Table. Overview of Main Water Policy Development Milestones (1977 – 2005).

Year	Conference/Event	Outcome(Policy)
1977	UN Conference on Water, Mardel Plata	Mardel Plata Action Plan
1981-1990	International Drinking Water and Sanitation Decade	New Delhi Statement
1990	Global Consultation on safe and Sanitation for the 1990's, New delhi	New Delhi Statement
1992	International Conference on Water and Environment, Dublin	Dublin Statement on Water and Sustainable Development
1994	Ministerial Conference on Drinking Water and Supply and Environment Sanitation, Noordwijk	Programme of Action
1997	First World Water Forum, Marrakech	Marrakech Declaration
1998	International Conference on Water and Sustainable Development, Paris	Paris Declaration
1998	International Conference on International River Basin Management, Bonn	
2000	Ministerial Conference on Water Security in the 21st century	Ministerial Declaration
2001	International Conference on Freshwater, Bonn	Ministerial Declaration
2003	International year of Freshwater	
2003	Third World Water Forcum, Kyoto	Portfolio of Water Actions

Since 1977, there has been considerable international consultation on the many dimensions and challenges that water represents. Petrella notes that between the dawning awareness of a water crisis from the 1970s to the late 1990s, numerous action programmes, projects and resolutions were developed, which, apart from increasing awareness, also defined new concepts such as the right to water and new solutions. However, judging both from historical evidence since the Mar del Plata Conference and from current developments in the field of water, humankind is no closer to solving the problem of growing scarcity and potential conflict over water.

In this regard, Petrella emphatically asks: "Why haven't we been able to lessen the scale of the water crisis in the world, despite the number of major national and international initiatives taken over the past twenty years, with considerable investment and the involvement of thousands of NGOs?" Postel remarks that despite a steady stream of global initiatives in the form of commissions, conferences and networks emphasising water's importance to food production, human health, poverty alleviation, ecosystem protection and political stability, the world's water problems have worsened markedly. While a number of reasons are put forth to explain the seeming lack of progress in dealing with water scarcity and conflict, this issue is partly imbedded in a policy and institutional environment that still lacks coherent focus and that is characterised by many subtle tensions that prevent real and practical solutions.

Currently the world is faced with the same issues resurfacing at every international platform created specifically to provide a focused and agreed-upon framework for dealing with water issues. There is increasingly better information, assessment and monitoring of water resources and related issues, but still the social challenges of providing access to sanitation and clean water remain, while complete international cooperation in which the needs of the different sectors are balanced perfectly with the available water resources is still an idealistic and unattainable goal. Against this historical background, the focus now shifts to an evaluation of the current legal and policy framework that serves both to prevent and intervene in water-related conflicts at the international level.

THE DEVELOPMENT OF AN INTERNATIONAL LEGISLATIVE FRAMEWORK

International relations over water are influenced by the legislative framework within which decisions are made. The International Court of Justice points to three sources of international law that govern relations between states and that also apply to relations over water resources, namely international conventions, international custom and general principles of law. Although treaties or conventions have been a common method for governing international environmental issues, these agreements are not without problems. While political states may adopt a treaty, it does not imply that the state agrees to be bound by it. As Hunt emphasises, states may support the treaty in principle, but hold back on final approval to see whether critical allies or opponents also adopt the treaty.

Furthermore, in an effort to attract the maximum number of signatories, conventions or treaties may be agreed on the basis of the lowest common denominator, or tend to be very unspecific in their terms. A third problem is that compliance to the obligations of the treaty is often neither enforced nor monitored effectively. Since the 1950s there have been a few attempts in international legislation to codify a set of rules or principles into international water law, but in many respects the development of international law is still in a formative phase; in fact, the lack of a codified body of rules acceptable to all states as customary international law has restricted the governance of fresh water. Adam emphasises that "... regional cooperation is...a stumbling block and you can see it in a number of shared rivers.

Many efforts have been made to ease things. The International Law Association, an NGO, and the International Law Commission, a subsidiary organ of the General Assembly of the UN, worked on the international rivers for many years". International water law has developed alongside the developments taking place in environmental law, while also simultaneously constituting a sub-set of international environmental law. Modern environmental and water laws trace their development back to the UNCHE in Stockholm in 1972 and the UNCED in Rio de Janeiro in 1992. Principle of the Stockholm Declaration and Principle of the Rio Declaration form the basis of many current international environmental law conventions. In summary, these principles state that nations have the right to exploit their own resources in accordance with their own environmental and developmental policies, but must ensure that their activities do not damage the environment of other states beyond the limits of their national jurisdictions. Since the 1800s several doctrines of water rights have been developed to guide the use and distribution of water among users. In that they are incorporated into international customary water law, these doctrines fulfil an important role in the relations between sovereign states over water resources.

INTERNATIONAL CUSTOMARY WATER LAW: DOCTRINES OF WATER USE

The doctrines are divided into the theory of absolute territorial sovereignty, absolute territorial integrity, limited territorial integrity, community of interests, and, equitable utilisation theory. Another legal position widely adopted in earlier centuries is the doctrine of prior appropriation. This doctrine is based on the assumption that the earliest or first user has the right to the full amount of water, regardless of the needs of any subsequent users.

This view, if put into practice, favours neither the upstream nor the downstream user, since the first user has the strongest claim. In the international arena this view is often taken by downstream nations, since downstream nations are often the first developers of water resources. Because of the perception of unjust water use associated with this doctrine, it is now not widely supported in international water law. In contrast to the doctrine of prior appropriation, the Harmon Doctrine, also known as the territorial sovereignty theory, asserts a riparian state's exclusive, sovereign rights over the water flowing through its territory.

This implies that a nation can use the water flowing through their borders in any chosen way without regard for other riparian states. Shiva notes that this doctrine, too, has also never won complete acceptance since it violates the concept of justice and even nations that could benefit from this principle choose to concede rights to other riparians. The absolute territorial integrity or natural flow theory, on the other hand, prevents states from altering the environment within their own territory in such a way that it disadvantages neighbouring states. Therefore, states may not affect the quantity or quality of water to the disadvantage of neighbouring or downstream states. Lower riparian states are thus entitled to the natural flow of the river, and upper riparians must allow water to flow its natural course by adhering to reasonable use. This view places certain duties on upstream states, but no reciprocal actions from downstream states are provided for, with the result that this position has not received wide support in the international law fraternity. Closely related to this doctrine is the doctrine of limited territorial integrity. According to this doctrine no one state can be the sole appropriator of a shared water source.

This principle ensures that all states can utilise the waters of a shared source within reasonable limits. Applying this principle may however cause conflict, since issues such as determining reasonable use and equitable sharing of water are vague. Two theories built on the twin premises of the equitable and just sharing of water resources and of cooperation among states over shared water resources are the theories of community of interests and of equitable utilisation. The community-of-interests theory holds that water should be regarded as common property, and that consultation and cooperation should therefore govern relations over water.

States have, according to this theory, equal rights to the waters flowing through the system. Theories of equitable utilisation/equitable apportionment are closely related to the community-ofinterests theory and state that international watercourses should be used equitably by different states. This theory is one of the principles on which the 1966 Helsinki Rules on the Uses of International Rivers were built. These last two theories, according to Petrella, are a considerable step forward in helping to resolve certain conflicts over water. If these principles were respected more definitively, it would certainly serve to equalise some conflicting power relations over water.

However, in practice it appears that if states are not compelled by legislative arrangements, such states tend to maximise their own interests, often at the expense of other states. From these doctrines, the following general principles of law guide the use of shared water resources.

These principles were incorporated into the 1997 Convention on the Law of Non-navigational Uses of International Watercourses:

- The harmonious application of national laws in the case of conflict between states;
- The obligation to promote cooperative management of international rivers, implying that it is the duty of all riparians to participate in the development, use and conservation of the water system;
- The obligation to settle disputes peacefully;
- The obligation to share data relating to the shared water system.
- The obligation to share information on activities that may affect other riparian states;
- The principle of equitable apportionment, entitling every basin state to an equitable and reasonable share of an international watercourse. Equitability does not imply equal use, but rather that various factors be taken into consideration in allocating water rights;
- The principle of equitable utilisation, according to which international law does not act in favour of any group or state;
- The principle of reciprocity, which implies that every nation that acts within its rights and fulfils its obligations can expect the same conduct form other states;
- The principle of Sic utero tuo ut aleneum non laedes, according to which a state's right to use shared waters is limited by the rights of co-basin states using the same resource, without being harmed in a significant manner;

One specific set of principles drafted in an effort to bring uniformity to international water law was the Helsinki Rules, mentioned above. They were formulated by the International Law Association in 1966. Adam states that "the Helsinki Rules are widely adopted as a basis for international negotiation and collaboration on river basin development. The Rules include also the sovereignty of each state within its territory to a reasonable and equitable share of the beneficial use of the international waters".

Among the most notable contributions of the Helsinki Rules to international water law were the provisions in Article IV and Article V regarding the equitable and reasonable apportionment of water in transboundary waters. According to these provisions, states are entitled to a reasonable and equitable share of the uses of water in an international basin; but, as Beaumont and Shiva both point out, it is often difficult to determine or measure precisely how to share the available resources equitably. The Helsinki Rules did come to play a crucial role in the development and codification of international water law and have also formed the basis for negotiations among riparian states over shared water. However, these rules never received recognition as official codifications of international water law, since the ILA operates as a private NGO without official status to develop international law. During 1970, the International Law Commission, an organisation with considerably more authority in the international law arena, were entrusted by the UN with the task of drafting a set of articles to govern the use of transboundary waters. Gradually an in-depth legal framework for transboundary waters began to emerge following this assignment. In 1992, a Draft Convention on the Protection and Use of Transboundary Watercourses and International Lakes was tabled by the ILC. This document formed the basis from which the UN tasked a working group in 1996 to draw up a framework convention on the uses of transboundary water resources.

The process of drafting an acceptable agreement brought the tensions and debates between upstream/downstream nations surrounding equitable use to the fore yet again. The apparently irreconcilable views of these nations were, however, eventually reconciled to a certain extent, and the rights of and obligations on all watercourse states were accommodated in the 1996 Helsinki Convention for the Protection and Use of Transboundary Watercourses and International Lakes.

This agreement was the first legal document to reformulate the 1966 Helsinki Rules. When this agreement was put to the vote in the Working Group, forty-three states voted in favour and three states–China, France and Turkey–voted against the agreement. The Helsinki Convention obligates states to prevent, control and reduce water pollution and also provides for the use of water in a reasonable, equitable way. This agreement also deals very specifically with the issue of dispute settlement. However, this document does not apply to African countries.

THE 1997 UNITED NATIONS CONVENTION ON THE LAW OF NON-NAVIGATIONAL USES OF INTERNATIONAL WATERCOURSES

In 1997, the United Nations Convention on the Law of Non-Navigational Uses of International Watercourses was adopted by the UN General Assembly after 27 years of discussion and negotiations. This is a potentially influential convention in terms of transboundary water issues. UNESCO hails this convention as "one post-Rio accomplishment that specifically focuses on transboundary water resources"

The UN Watercourse Convention emanated directly from the perception of an emerging water crisis and the likelihood that water scarcity could lead to regional conflicts over water. The Convention thus emphasised establishing collaborative relationships between countries sharing international watercourses. When this treaty was brought to a vote, 103 states voted in favour and three against the treaty. Burundi, China and Turkey opposed the treaty–all three being upper riparian states and significant actors in the world's major water basins, namely the Nile, the Mekong and the Tigris/Euphrates Rivers respectively. In the case of China and Turkey, both were then in the process of developing their parts of the water resources in a way that might threaten the use by downstream countries, leading to considerable tension in these river basins. Only 16 countries signed, and nine ratified the Convention.

This fell below the required 35 countries needed to bring the Convention into force. The Convention thus currently has no legal status. Similarly, in Southern Africa, at present Namibia and South Africa are the only two signatories to the Convention, making it impossible for the Convention to have any legally binding status on states in various shared basins. Eckstein emphasises that although the document is far from entering into force, the mere fact that the Convention was adopted signals that there is at least broad agreement in the international community on the basic principles that govern transboundary waters and that this Convention is indeed a positive step forward in laying down commonly accepted principles in water governance at the international level.

Three significant provisions of the Convention are the obligation not to cause significant harm, the need both for prior notification and for equitable and reasonable use of water resources. Equitable and reasonable use of resources, together with participation between states on the management of shared water sources are outlined in Article of this Convention. Article also outlines the right to utilise watercourses and the obligation to cooperate in the development and protection of such watercourses. Thus, the principle of equitable apportionment and that of cooperative management are contained in this article of the Convention.

Article further elaborates on the obligation to participate, stating that: "Watercourse states shall cooperate on the basis of sovereign equality, territorial integrity, mutual benefits and good faith in order to attain optimal utilisation and adequate protection of an international watercourse". Another important provision of the treaty is that any watercourse state is entitled to participate in and become party to any watercourse agreement applicable to the whole international watercourse.

Vital human needs are explicitly protected by the stipulation that, in a situation of competing water use, highest priority is given to the human needs requirement. The Convention also deals clearly with mechanisms for dispute settlement. Article refers to the provisions made in earlier articles that deal with equitable use and the obligation not to cause significant harm in resolving conflicts between users. In Article, specific stipulations for dealing with conflict situations are outlined. It states, amongst others, that parties in dispute must seek a settlement by peaceful means. This Convention would, if ratified, provide a legally binding framework for managing international watercourses. However, as UNESCO indicates "... the Convention ... does not entirely resolve many legal questions concerning the management of internationally shared waters". Beaumont points out that article states that "... an international watercourse shall be used and developed by watercourse states with a view of attaining optimal and sustainable utilisation thereof and benefits therefrom ...".

According to Beaumont, the word 'optimal' can be interpreted differently by different countries–some might say that optimal use of water does not allow for irrigation under conditions of scarcity, while others may feel that irrigation should in this case only be practised in parts of the basin where evapotranspiration losses are minimal. Furthermore, the ability of the Convention successfully to prevent and provide a legal framework for intervening in conflicts over water is severely hampered by the fact that those nations appearing not to benefit from the Convention's stipulations are not willing to sign it. Hunt emphasises that "there appears to be an interesting but disturbing pattern in the countries... that have/have not signed and ratified the Convention" in that countries that may gain from dominating the water resources for their own needs and developments are not prepared to become signatories to such Conventions.

In this regard, Petrella observes: "... in most of the basins in question, the warlords are located in the upriver states. Asserting a principle of absolute territorial sovereignty, they claim to have exclusive ownership of water resources on their territory... and the right to use them in any way they see fit". In spite of the promising developments in the international law arena in providing a solid legal framework for dealing with water issues, it appears that the impacts of these developments have been diluted by existing tensions between nations sharing watercourses.

The fact that nations standing to benefit from the signing and ratification of conventions have indeed signed, while those who would be disadvantaged by the enforcement of a legal instrument have not, is an indication that there exist some very powerful underlying tensions that may not easily be resolved through legal sanctions alone. Therefore, while there appears–at least in theory–to have emerged a legal framework for cooperation, conflict prevention and intervention between states, the lack of ratification has rendered many of these instruments ineffective in preventing and intervening in potential conflict situations. One cannot enforce any country to be bound by an agreement to which it has in principle agreed, but has not actually ratified formally. However, UNEP emphasises that these international policy and legislative developments are key components in achieving cooperation between stakeholders. While the above outlines certain policy and institutional developments at the international level, it is also important to explore the regional policy and institutional landscape.

Developments at the international level have influenced regional and local developments with regard to policy and institutional frameworks, but the environmental and social realities prevalent in the Southern African region have also called for more specifically tailored policy and institutional developments to deal particularly with the unique socio-economic, political and environmental realities of this region. These developments are now considered.

REGIONAL POLICY AND LEGISLATIVE DEVELOPMENTS

In line with the above international policy developments of the past 30 years, there is now general agreement among African governments that socio-economic well-being and a healthy natural environment are intertwined. Therefore, emphasis must be placed on policy options that safeguard vital natural resources such as water, while simultaneously striving to meet the development needs of the continent. Referring specifically to Southern Africa, Ashton remarks likewise that "Southern Africa's pressing need for social and economic development has prompted governments of the SADC countries to focus on broader issues of social equity, and resource stewardship". Therefore, the development of a regional legal and policy framework to deal with water issues is firmly imbedded in the international policy developments of the past 30 years, while wider political and socio-economic developments taking place in the Southern African region, particularly since the middle of the 20th century have indeed had a significant impact on the development of a regional water policy and legal framework. To understand the impact of the larger socio-political and economic forces that shaped developments in the water sector, it is useful briefly to outline the most significant historical political and social developments in the region.

LINKING REGIONAL POLICY DEVELOPMENTS IN THE WATER SECTOR WITH A WIDER SOCIAL CONTEXT

The current socio-economic and political context in Southern Africa has been markedly influenced by colonisation from Europe, which occurred since the 1600s. As a result of this process, Southern African states have firstly been artificially divided by arbitrary borders; secondly, across the region indigenous peoples have fought for their independence from colonial powers in a number of wars of independence. Southern Africa has, thus, historically been characterised by prolonged periods of civil and liberation wars that have left a legacy of weakened economies, fragmented governance systems, inequities created by past political dispensations and shattered infrastructure.

These challenges place pressure on governments in the region. Furthermore, specifically with regard to water and water policy developments, the colonial legacy has created a political environment in which water has for long been treated as a geo-political security concern and thus dealt with mainly by national states, without regard for the needs of other states sharing such water resources. In the light of this tumultuous past and its ongoing legacy, Southern African leaders have attempted to ensure the political security of the region's nations through measures taken at the regional level. These measures include:, the formation of the Front-Line States in 1979; the establishment of the Southern African Development Co-ordination Conference in 1980; and, the formation of the South African Development Community in 1992.

The purpose of the FLS was to assist in struggles for liberation from white-ruled states, and was formed by the United Republic of Tanzania, Mozambique and Zambia. As more countries in the region gained their independence, they joined the FLS. Angola joined in 1976, Zimbabwe in 1980, Namibia in 1990 and South Africa in 1994. These developments emanated from a growing concern over the strong position of South Africa within the region's economy, the subsequent dependence of the rest of the region on South Africa, and the fear of regional political domination by the Apartheid Regime that was in power in this country from 1948 to 1994.

Following the demise of Apartheid in South Africa in 1994, the original liberation focus of the FLS inevitably changed to reflect the political changes in the region. These political developments also gave rise to the transformation of the SADCC into SADC at the Windhoek Summit in 1992. Since its inception, the SADC has established itself as the collective political voice for the region's nations. Flowing from the changes in the political context, particularly the end of Apartheid in South Africa, the focus of the SADC changed from that of working towards greater economic independence from an Apartheid South Africa, to a drive towards cooperation and integration at various institutional levels in the region. The signing of the SADC Treaty on 27 August 1992 in Windhoek, formally established the SADC, its goal being regional integration based on balance, equity and mutual benefit. This treaty made provision for member states to develop, negotiate, and agree upon a range of protocols that would give effect, through their objectives and institutional mechanisms, to achieving regional integration.

Tagawira thus sums up the mission of the SADC: "I think in SADC the main thing is the drive towards regional integration and what the region is doing is supporting integration, which is in the SADC Treaty". Within this policy context, one of the first protocols to be developed within the provisions of the SADC Treaty was the SADC Protocol on Shared Watercourses.

THE IMPACT OF NATURAL ENVIRONMENTAL REALITIES ON POLICY DEVELOPMENTS IN THE REGION

Coinciding with, and resulting from, these wider socio-political and economic developments were specific political and institutional developments in the water sector. In addition, since this region's early history, the realities of aridity, recurrent droughts, fluctuating rainfall patterns and water scarcity had to be factored into economic and political decision-making processes. Turton remarks that water has always been of strategic significance in the region in the context of national economies that are constrained by water. Since early times, the arid climatic conditions of the region led to the adoption of a nomadic huntingand- gathering lifestyle by the indigenous Khoisan people of Southern Africa. This lifestyle became increasingly difficult to follow as settlers from Europe took control of land and consequently also the water supply flowing through these lands.

Between 1800 and the early 1900s, a process of water, land and institutional reform in South Africa increasingly linked water rights to land rights, while also recognising the importance of irrigation and agriculture. During this process, the issue of water scarcity was always a major precursor to important institutional, legislative and policy developments. In the case of South Africa, major institutional and legislative reforms not only coincided with larger changes in the political sphere, but environmental challenges also impacted upon these changes. New water legislation in 1912, 1956 and 1998 were preceded by political changes–Unification in 1910, the election of the National Party in 1948 and the election of the African National Congress in 1994. Some of the major water infrastructure developments in the region were also influenced by political developments. The rapid decolonisation of the African continent in the 1960s led to a closer collaboration between South Africa and Portugal in an effort to hold onto their positions of power in Southern Africa. One result of the stronger ties between South Africa and Portugal at that stage was the Agreement between the Republic of South Africa and the Government of Portugal in Regard to Rivers of Mutual Interest and the Cunene River Scheme. which laid the foundation for some major infrastructure developments, among which the construction of the Cahora Bassa Dam on the Zambezi River and the Ruacana Hydropower Scheme on the Kunene River.

Water's strategic importance in the region is exemplified by the fact that both these rivers became important focal points for the armed struggle later on in the region's history. Socio-economic and socio-political shifts are, therefore, closely linked to the importance of water in the region. Access to water at the national level has been a main driver of policy and institutional developments at the regional level, with nations continuously seeking and attempting to maintain collaborative relations with neighbours to secure access to water. To this effect, many formal agreements over water have since 1926 been reached between Southern African states over the course of the past century. As a result of the environmental realities of the region, there has thus been–at least in principle–an aspiration to pursue opportunities for cooperation over water, both at the regional and at the national levels. The examples of Southern Africa and also the Middle East illustrate that, while demographic pressure, socio-economic inequality and political tension may form part of the realities confronting these nations, a solid institutional environment can decrease the likelihood of tension over scarce resources. Even during the times of severe political conflict in the Middle East, Israel and Jordan embarked on secret 'picnic table talks' to discuss issues surrounding the sharing of the Jordan River.

Likewise, in Southern Africa, countries signed a number of river basin agreements in the midst of the wars of the 1970s and 1980s. South Africa, Mozambique and Swaziland managed to establish a Tripartite Permanent Technical Commission in 1983 to discuss the use of the Incomati River even in the face of tension over South Africa's involvement in the Mozambican Civil War and its Apartheid policy.

THE SADC AND POLICY DEVELOPMENTS PERTAINING TO WATER

Legislative and policy reforms in the water sector have gained considerable momentum in Southern Africa since the 1990s. According to Beukman, all countries in the SADC region are "in some process to reform their [water] legislation or their policies" to make their countries IWRM aligned. She also emphasises that SADC is "doing incredibly in terms of supporting the countries to make the transition, so that is a good thing". The importance of SADC in water policy reform cannot thus be underestimated.

Underpinning these reforms at the national level and at the regional level are principles taken from years of consultation and discussion in international forums and conferences discussed above. Among these, legislative developments in Southern Africa adhere to the Dublin Principles, while other notable factors taken into consideration in the process of reform have been equity, efficiency, sustainability, political and public acceptability, fiscal impact and health.

Furthermore, at the regional level within the SADC institution, reforms have not only taken place to facilitate cooperation and integration in general, but also more specifically in the water sector. Transpiring from regional policy developments, the SADC has recognised the most serious security problems in the region as being political, social, economic and environmental issues that are solved more effectively through socio-economic development and democratisation rather than through military force. Therefore, the SADC focuses strongly on human security, recognising various socio-economic and sociopolitical drivers of insecurity. Tawana maintains that the adoption of a wider security focus in the Southern African region should over time remove the core sources of human insecurity, such as access to food and clean water, health services, energy and economic opportunities. Environmental degradation in general and water scarcity particularly are, therefore, acknowledged in the Southern African political context through the SADC regional policy framework. Given the central importance of water scarcity in the context of human security in the region, it was imperative that the SADC find ways to facilitate cooperation between states over water resources.

With 70% of freshwater resources being shared by two or more countries in the SADC, a coordinated approach to the use and preservation of water had to be acknowledged at the regional level. It is thus not coincidental that the first protocol to be drafted by the SADC in 1995, namely the SADC Protocol on the Non-Navigational Uses of Shared Watercourses, was a protocol to deal with the water issues of the region. This Protocol, in principle, is regarded as the foundation for cooperation over the region's scarce water resources and fits into the wider regional goal of integration and cooperation at various levels in the region and this is therefore now discussed in more depth.

THE SADC PROTOCOL ON THE NON-NAVIGATIONAL USES OF SHARED WATERCOURSES

The SADC Protocol had its origins in the implementation of the Zambezi River Basin System Action Plan to facilitate the management of the Zambezi River by the SADC in 1993. During the negotiations taking place to establish a Zambezi River Basin Commission within the context of this plan, the SADC felt that instead of developing a single legal instrument applicable only to this river basin, it would be more beneficial to develop a regional legal framework on which all cooperative relationships in the region could be based. This decision set in motion a process of negotiation that led to the adoption of the SADC Protocol on the Non-navigational Uses of Shared Watercourses in 1995. Although the Protocol was met with some reservation by some member states, it was subsequently adopted by eleven of the fourteen member states, namely Botswana, Lesotho, Malawi, Mauritius, Mozambique, Namibia, South Africa, Swaziland, Tanzania, Zambia and Zimbabwe, allowing it to enter into force in 1998.

Two-thirds of the States had ratified the Protocol by 2000. The concerns expressed by member states at the signing of the document in 1995 led to a process of consultation and negotiation that culminated in the adoption, in 2000, of the SADC Revised Protocol on Shared Watercourses. The SADC Protocol would remain in force until twelve months after the Revised

Protocol came into force. The Revised Protocol was signed by all member states and has subsequently been ratified. As was mentioned before, not only the acute awareness of water scarcity, but also the unequal distribution of water across time and space in the region have played a major role in bringing countries in the region to the point of establishing a regional instrument to facilitate cooperation over shared water resources.

The Protocol was developed within the international policy framework guided by international water law instruments such as the Helsinki Rules, the Dublin Principles and Agenda, and thus follows principles of international rules and conventions in the rights and obligations set out. As Ramoeli points out in reference to the SADC Protocol, " it recognises international consensus on a number of concepts and principles related to water resource development and management in an environmentally sound manner".

The Revised Protocol also takes into account the international legal and policy framework and recognises the progress made in international water law through the development of instruments such as the Helsinki Rules and the UN Convention on the Uses of International Watercourses. Recognition is also given to the provisions of Agenda, and specific reference is made to the concepts of environmentally sound management, sustainable development and equitable utilisation of shared watercourses in the region. In line with the regional policy framework, the SADC Protocol and Revised Protocol have as their primary objective the development of close cooperation for the sensible and coordinated use of shared watercourses and they also encompass harmonisation of sectoral policies and laws with the wider regional goals of integration and cooperation in SADC.

In referring to the issue of regional cooperation, Green Cross International emphasises that the SADC Protocol has offered states the opportunity for cooperation created by the natural partnerships existing through shared watercourses, and that such cooperation can then be transferred to other areas of mutual interest. The SADC Protocol has established an "enabling environment" for the development of water resource management to the extent that it could even become a powerful driver of regional integration. However, for this to transpire there needs to be a high degree of political stability and a commitment to negotiating outcomes beneficial to all stakeholders–positive-sum outcomes. Principles specifically laid out in both the Protocol and the Revised Protocol are: the acceptance of the sovereignty of member states in utilising the water resources within their territories; the application of the doctrines of community of interests and equitable utilisation; the imperative of balancing development and environment needs; cooperation in joint projects and studies; sharing of data and information; the obligation of notification of emergencies and the establishment of an institutional framework.

In applying the doctrine of community of interests, the Revised Protocol has also included the concept of benefitsharing. Benefit-sharing takes cooperation on water resources to a higher level in the sense that the water alone is not the main focus of negotiation and cooperation, but benefits derived from the water source are also additionally included into agreements.

The Protocol and Revised Protocol both make specific mention of establishing access to water through water-sharing agreements, while simultaneously maintaining a balance between the need for resource development to attain higher standards of living and the need for environmental protection and sustainable development and thus contributing to the positive-sum outcomes mentioned above. To the region's credit, many agreements over the sharing of freshwater resources exist, although it is not certain whether these agreements will prove resilient in the face of high levels of human insecurity and rising tensions over access to scarce water resources. With regard to institutional developments, an SADC Water Sector was established in 1995, and Ashton envisions that the SADC as an institution will become a strong regional force in preventing water conflicts.

The Revised Protocol makes provision for the establishment of SADC Water Sector Organs that include a Committee of Water Ministers, a Committee of Water Senior Officials, A Water Sector Coordinating Unit and a Water Resources Technical Committee. At present, these institutions are established and are functioning to various degrees. Yet as Beukman points out: "Everyone knows there are capacity constraints within SADC. It is only small sectors for large regions. [However] SADC tries to work through their main partners and through their member states as well as those other structures like the Country Water Partnership that exists....So I think the will is there, it is that these things are still in [a process of] formulating and streamlining how in a SADC framework [the partners] can add value". Provision is also made for shared watercourse institutions.

These institutions are to be established by watercourse states and include watercourse commissions, water authorities and water boards. A number of such watercourse commissions are currently functioning within the framework of the Protocol, for example the Tripartite Technical Commission that incorporates Mozambique, Swaziland and South Africa. This institution is "in line with the SADC Protocol which encourages the development of river commissions". Another example is ORESACOM that was also established between Botswana, Lesotho, Namibia and South Africa and functions within the framework provided by the SADC Protocol.

Referring to prevention of and intervention in conflict over water resources, Article of the Revised Protocol is specifically devoted to shared watercourse agreements, while Article deals with settlement of disputes. Article states that all watercourse states are entitled to negotiate and become party to any watercourse agreement, while the rights and obligations of states not party to a particular agreement shall not be affected under the provisions of this Protocol. Settlement of disputes in this Protocol encompasses the obligation to resolve all disputes regarding implementation, interpretation or application of the provisions in this Protocol amicably, and where this is not possible, that the case shall be referred to a Tribunal.

While the legal and policy framework provided by the SADC Protocol and Revised Protocol has certainly made the intentions of the region's states with regard to cooperation over water resources clear, some very important challenges persist. Heyns states that the Protocol is difficult to enforce and remains essentially a gentlemen's agreement, since it is not able to prevent civil war, destruction of infrastructure, or to avoid natural water-related disasters such as droughts and floods. Such crisis situations will ultimately test the ability of this regional instrument to intervene effectively in and prevent conflict over freshwater resources. However, at the regional level, there is a definite need for transboundary institutional cooperation to deal with the management of transboundary water resources and to prevent and deal with conflict over shared water resources. The Protocol at least provides a mechanism that is mutually sustainable and acceptable for negotiating the peaceful use of shared water.

For this to happen, it is important that regional institutions have a mutually agreed framework of criteria and agreements as a basis for decisions regarding shared water resources. Although instruments such as the SADC Protocol are a necessary step towards cooperation and prevention of conflict, the region is hampered by weak economies, a lack of technical and human resources, and vast differences between countries in terms of management systems–all factors that make the implementation of such an instrument difficult. With regard to these issues in the wider, continental context, but with specific relevance to Southern Africa, The Commission of African Union emphasises that "needless to say, Africa's future depends on how countries in the region address issues relating to economic and political governance, and the extent to which countries avoid civil conflicts, and undertake effective economic and social reforms to effectively address the poverty issue and bring about sustainable development".

Therefore, the ability of the regional policy and institutional framework effectively to prevent and intervene in conflicts over water would rest in large part on the institutional capacity to govern the region's stakeholders where water is concerned. Good governance of transboundary regional resources should encompass institutional structures–such as those created through the provisions of the SADC Protocol and Revised Protocol at regional level–to govern shared resources. In this regard, Ashton states that it is in the interest of both societies and individuals that appropriate national and international institutions act jointly in developing management plans for shared river basins and draw up workable protocols to prevent conflicts over water. While individual governments are increasing their institutional strength, the same is not necessarily true of regional institutional structures. The SADC as a regional institution did not, for example, manage to resolve the Sedudu/Kasikili dispute between Namibia and Botswana, despite provisions for dispute resolution contained in the SADC Protocol. Ramoeli reiterates that although some efforts have been made towards implementing the Protocol, a number of challenges persist: establishing institutions at the basin level, dispute-resolution mechanisms and institutions, as well how to harmonise national legislation with the Protocol and other international water laws. Since the 1970s considerable and regular international consultation has taken place on the many dimensions and challenges that water represents.

Numerous action programmes, projects and resolutions have been developed that, apart from increasing awareness, also defined new concepts and new solutions. However, judging from historical evidence, the problem of the growing scarcity of fresh water and the continuous threat of conflict is far from being effectively addressed in practice by the existing policy and legislative frameworks governing relations over water. This apparent lack of progress can be attributed to a number of reasons, although the issue is partly imbedded in a policy and institutional environment that still lacks coherent focus and is characterised by many subtle tensions that prevents real and practical solutions to be found.

Since the 1970s it has been increasingly recognised that environmental, population and development issues are linked reciprocally and, that water availability is thus central to achieving population- and development-related goals for the future. More specifically, the international policy framework has recognised water security as being integral towards achieving the well-being of current and future generations. Furthermore, particularly with regard to the development of a legal framework, the impacts of these developments have been weakened by existing tensions between nations sharing watercourses.

These tensions centre on issues of national socio-economic development and on the advantages gained from use of water resources outside of the legal provisions and also on sovereignty issues. In theory at least, a legal framework for cooperation, conflict prevention and intervention between states have emerged, although the lack of ratification has rendered many of these instruments ineffective in preventing and intervening in potential conflict situations. Therefore, although there seem to be sufficient policy and legislative frameworks on paper, there is still a wide gap between the intended purposes and outcomes of such frameworks and any real action towards dealing with the impending impacts of water scarcity, both international, and also specifically in Southern Africa. In Southern Africa itself a regional legal and policy framework has likewise developed over the past two decades. This policy and legislative framework has as its primary objective the facilitation of regional cooperation and integration, not only in terms of water, but also with regard to other sectors of shared interests and benefits. The importance of water in the water-scarce Southern African region is highlighted by the fact that the regional SADC Protocol on Shared Watercourses was one of the first legal and policy documents to be drafted by the SADC in 1995. In principle, the development of this document and its successor indicated a definite desire to cooperate over joint water issues in the region.

However, it remains to be seen whether this document will prove to achieve the goal of cooperation in the face of potential crisis situations such as prolonged droughts–especially in the absence of both truly effective governance and of adequate institutional capacity to enforce the requirements of the protocol. From these developments at the international level some recurring dilemmas may in future hamper both the prevention of water conflicts and intervention in water conflicts, both internationally and at the regional level. Some such dilemmas that emerged from the foregoing discussion are the issue of sovereignty vs. regional cooperation; and the issue of the equitable distribution of water taking into consideration the competing demands of water as an economic asset or a common good and the tension between socio-economic development and ecosystem.

Therefore, in order to explore the tensions between intentions and actual actions, one needs to deconstruct the recurrent dilemmas in the above international policy and institutional developments that are currently hampering progress in implementing practical and agreed-upon policy solutions to the world's water issues. An analysis of these challenges becomes crucial when we consider that such dilemmas may serve to hamper effective prevention of and intervention in potential conflicts over water.

Chapter 16

Treatment of Water

Water treatment describes those processes used to make water more acceptable for a desired end-use. These can include use as drinking water, industrial processes, medical and many other uses. The goal of all water treatment process is to remove existing contaminants in the water, or reduce the concentration of such contaminants so the water becomes fit for its desired end-use. One such use is returning water that has been used back into the natural environment without adverse ecological impact.

The processes involved in treating water for drinking purpose may be solids separation using physical such as settling and filtration, chemical such as disinfection and coagulation. Biological processes are also employed in the treatment of wastewater and these processes may include, for example, aerated lagoons, activated sludge or slow sand filters.

WATER PURIFICATION

Water purification is the process of removing undesirable chemicals, materials, and biological contaminants from raw water. The goal is to produce water fit for a specific purpose. Most water is purified for human consumption but water purification may also be designed for a variety of other purposes, including to meet the requirements of medical, pharmacology, chemical and industrial applications.

In general the methods used include physical process such as filtration and sedimentation, biological processes such as slow sand filters or activated sludge, chemical process such as flocculation and chlorination and the use of electromagnetic radiation such as ultraviolet light. The purification process of water may reduce the concentration of particulate matter including suspended particles, parasites, bacteria, algae, viruses, fungi; and a range of dissolved and particulate material derived from the surfaces that water may have made contact with after falling as rain. The standards for drinking water quality are typically set by governments or by international standards. These standards will typically set minimum and maximum concentrations of contaminants for the use that is to be made of the water. It is not possible to tell whether water is of an appropriate quality by visual examination. Simple procedures such as boiling or the use of a household activated carbon filter are not sufficient for treating all the possible contaminants that may be present in water from an unknown source.

Even natural spring water - considered safe for all practical purposes in the 1800s - must now be tested before determining what kind of treatment, if any, is needed. Chemical analysis, while expensive, is the only way to obtain the information necessary for deciding on the appropriate method of purification. According to a 2007 World Health Organization report, 1.1 billion people lack access to an improved drinking water supply, 88% of the 4 billion annual cases of diarrheal disease are attributed to unsafe water and inadequate sanitation and hygiene, and 1.8 million people die from diarrheal diseases each year.

The WHO estimates that 94% of these diarrheal cases are preventable through modifications to the environment, including access to safe water. Simple techniques for treating water at home, such as chlorination, filters, and solar disinfection, and storing it in safe containers could save a huge number of lives each year.

SOURCES OF WATER

- *Groundwater*: The water emerging from some deep ground water may have fallen as rain many decades, hundreds, thousands or in some cases millions of years ago. Soil and rock layers naturally filter the ground water to a high degree of clarity before it is pumped to the treatment plant. Such water may emerge as springs, artesian springs, or may be extracted from boreholes or wells.

Deep ground water is generally of very high bacteriological quality but the water typically is rich in dissolved solids, especially carbonates and sulfates of calcium and magnesium. Depending on the strata through which the water has flowed, other ions may also be present including chloride, and bicarbonate. There may be a requirement to reduce the iron or manganese content of this water to make it pleasant for drinking, cooking, and laundry use. Disinfection may also be required. Where groundwater recharge is practised; a process in which river water is injected into an aquifer to store the water in times of plenty so that it is available in times of drought; it is equivalent to lowland surface waters for treatment purposes.

- *Upland lakes and reservoirs*: Typically located in the headwaters of river systems, upland reservoirs are usually sited above any human habitation and may be surrounded by a protective zone to restrict the opportunities for contamination. Bacteria and pathogen levels are usually low, but some bacteria, protozoa or algae will be present. Where uplands are forested or peaty, humic acids can colour the water. Many upland sources have low pH which require adjustment.
- *Rivers, canals and low land reservoirs*: Low land surface waters

will have a significant bacterial load and may also contain algae, suspended solids and a variety of dissolved constituents.

- Atmospheric water generation is a new technology that can provide high quality drinking water by extracting water from the air by cooling the air and thus condensing water vapor.
- Rainwater harvesting or fog collection which collects water from the atmosphere can be used especially in areas with significant dry seasons and in areas which experience fog even when there is little rain.
- Desalination of seawater by distillation or reverse osmosis.

Treatment

The processes below are the ones commonly used in water purification plants. Some or most may not be used depending on the scale of the plant and quality of the water.

Pre-treatment

- *Pumping and containment*: The majority of water must be pumped from its source or directed into pipes or holding tanks. To avoid adding contaminants to the water, this physical infrastructure must be made from appropriate materials and constructed so that accidental contamination does not occur.
- *Screening*: The first step in purifying surface water is to remove large debris such as sticks, leaves, trash and other large particles which may interfere with subsequent purification steps. Most deep groundwater does not need screening before other purification steps.
- *Storage*: Water from rivers may also be stored in bankside reservoirs for periods between a few days and many months to allow natural biological purification to take place. This is especially important if treatment is by slow sand filters. Storage reservoirs also provide a buffer against short periods of drought or to allow water supply to be maintained during transitory pollution incidents in the source river.
- Pre-conditioning: Many waters rich in hardness salts are treated with soda-ash to precipitate calcium carbonate out utilising the common-ion effect.
- *Pre-chlorination*: In many plants the incoming water was chlorinated to minimise the growth of fouling organisms on the pipe-work and tanks. Because of the potential adverse quality effects, this has largely been discontinued.

Widely varied techniques are available to remove the fine solids, micro-organisms and some dissolved inorganic and organic materials. The choice of method will depend on the quality of the water being treated, the cost of the treatment process and the quality standards expected of the processed water.

pH Adjustment

Distilled water has a pH of 7 and sea water has an average pH of 8.3 . If the water is acidic , lime, soda ash, or sodium hydroxide is added to raise the pH. For somewhat acidic, alkaline waters , forced draft degassifiers are the cheapest way to lower the pH, as the process raises the pH by stripping dissolved carbon dioxide from the water.

Lime is commonly used for pH adjustment for municipal water, or at the start of a treatment plant for process water, as it is cheap, but it also increases the ionic load by raising the water hardness. Making the water slightly alkaline ensures that coagulation and flocculation processes work effectively and also helps to minimize the risk of lead being dissolved from lead pipes and lead solder in pipe fittings.

Acid may be added to alkaline waters in some circumstances to lower the pH. Having an alkaline water does not necessarily mean that lead or copper from the plumbing system will not be dissolved into the water but as a generality, water with a pH above 7 is much less likely to dissolve heavy metals than a water with a pH below 7.

FLOCCULATION

Flocculation is a process which clarifies the water. Clarifying means removing any turbidity or colour so that the water is clear and colourless. Clarification is done by causing a precipitate to form in the water which can be removed using simple physical methods. Initially the precipitate forms as very small particles but as the water is gently stirred, these particles stick together to form bigger particles - this process is sometimes called flocculation. Many of the small particles that were originally present in the raw water absorb onto the surface of these small precipitate particles and so get incorporated into the larger particles that coagulation produces.

In this way the coagulated precipitate takes most of the suspended matter out of the water and is then filtered off, generally by passing the mixture through a coarse sand filter or sometimes through a mixture of sand and granulated anthracite .

Coagulants / flocculating agents that may be used include:

- *Iron (III) hydroxide*: This is formed by adding a solution of an iron (III) compound such as iron(III) chloride to pre-treated water with a pH of 7 or greater. Iron (III) hydroxide is extremely insoluble and forms even at a pH as low as 7. Commercial formulations of iron salts were traditionally marketed in the UK under the name Cuprus.

- Aluminium hydroxide is also widely used as the flocculating precipitate although there have been concerns about possible health impacts and mis-handling led to a severe poisoning incident in 1988 at Camelford in south-west UK when the coagulant was introduced directly into the holding reservoir of final treated water.
- PolyDADMAC is an artificially produced polymer and is one of a class of synthetic polymers that are now widely used. These polymers have a high molecular weight and form very stable and readily removed flocs, but tend to be more expensive in use compared to inorganic materials.

SEDIMENTATION

Water exiting the flocculation basin may enter the sedimentation basin, also called a clarifier or settling basin. It is a large tank with slow flow, allowing floc to settle to the bottom. The sedimentation basin is best located close to the flocculation basin so the transit between does not permit settlement or floc break up.

Sedimentation basins can be in the shape of a rectangle, where water flows from end to end, or circular where flow is from the centre outward. Sedimentation basin outflow is typically over a weir so only a thin top layer - furthest from the sediment - exits.The amount of floc that settles out of the water is dependent on the time the water spends in the basin and the depth of the basin. The retention time of the water must therefore be balanced against the cost of a larger basin. The minimum clarifier retention time is normally 4 hours. A deep basin will allow more floc to settle out than a shallow basin. This is because large particles settle faster than smaller ones, so large particles bump into and integrate smaller particles as they settle. In effect, large particles sweep vertically through the basin and clean out smaller particles on their way to the bottom.

As particles settle to the bottom of the basin, a layer of sludge is formed on the floor of the tank. This layer of sludge must be removed and treated. The amount of sludge that is generated is significant, often 3%-5% of the total volume of water that is treated. The cost of treating and disposing of the sludge can be a significant part of the operating cost of a water treatment plant. The tank may be equipped with mechanical cleaning devices that continually clean the bottom of the tank or the tank can be taken out of service when the bottom needs to be cleaned.

FILTRATION

After separating most floc, the water is filtered as the final step to remove remaining suspended particles and unsettled floc. The most common type of filter is a rapid sand filter. Water moves vertically through sand which often has a layer of activated carbon or anthracite coal above the sand. The top layer removes organic compounds, which contribute to taste and odor. The space between sand particles is larger than the smallest suspended particles, so simple filtration is not enough. Most particles pass through surface layers but are trapped in pore spaces or adhere to sand particles. Effective filtration extends into the depth of the filter. This property of the filter is key to its operation: if the top layer of sand were to block all the particles, the filter would quickly clog.

To clean the filter, water is passed quickly upward through the filter, opposite the normal direction to remove embedded particles. Prior to this, compressed air may be blown up through the bottom of the filter to break up the compacted filter media to aid the backwashing process; this is known as *air scouring*. This contaminated water can be disposed of, along with the sludge from the sedimentation basin, or it can be recycled by mixing with the raw water entering the plant. Some water treatment plants employ pressure filters. These work on the same principle as rapid gravity filters, differing in that the filter medium is enclosed in a steel vessel and the water is forced through it under pressure.

Advantages:

- Filters out much smaller particles than paper and sand filters can.
- Filters out virtually all particles larger than their specified pore sizes.
- They are quite thin and so liquids flow through them fairly rapidly.
- They are reasonably strong and so can withstand pressure differences across them of typically 2-5 atmospheres.
- They can be cleaned and reused.

Membrane filters are widely used for filtering both drinking water and sewage . For drinking water, membrane filters can remove virtually all particles larger than 0.2 um—including Giardia and cryptosporidium. Membrane filters are an effective form of tertiary treatment when it is desired to reuse the water for industry, for limited domestic purposes, or before discharging the water into a river that is used by towns further downstream. They are widely used in industry, particularly for beverage preparation . However no filtration can remove substances that are actually dissolved in the water such as phosphorus, nitrates and heavy metal ions.

SLOW SAND FILTERS

Slow sand filters may be used where there is sufficient land and space as the water must be passed very slowly through the filters. These filters rely on biological treatment processes for their action rather than physical filtration. The filters are carefully constructed using graded layers of sand with the coarsest sand, along with some gravel, at the bottom and finest sand at the top. Drains at the base convey treated water away for disinfection.

Filtration depends on the development of a thin biological layer, called the zoogleal layer or Schmutzdecke, on the surface of the filter. An effective slow sand filter may remain in service for many weeks or even months if the pre-treatment is well designed and produces water with a very low available nutrient level which physical methods of treatment rarely achieve. Very low nutrient levels allow water to be safely sent through distribution system with very low disinfectant levels thereby reducing consumer irritation over offensive levels of chlorine and chlorine by-products. Slow sand filters are not backwashed; they are maintained by having the top layer of sand scraped off when flow is eventually obstructed by biological growth.

A specific 'large-scale' form of slow sand filter is the process of bank filtration, in which natural sediments in a riverbank are used to provide a first stage of contaminant filtration. While typically not sufficiently clean enough to be used directly for drinking water, the water gained from the associated extraction wells is much less problematic than river water taken directly from the major streams where bank filtration is often used.

LAVA FILTERS

Lava filters are similar to sand filters and may also only be used where there is sufficient land and space. Like sand filters, the filters rely on biological treatment processes for their action rather than physical filtration. Unlike slow sand filters however, they are constructed out of 2 layers of lava pebbles and a top layer of nutrient-free soil . On top, water-purifying plants are placed. Usually, around 1/4 of the dimension of lavastone is required to purify the water and just like slow sand filters, a series of herringbone drains are placed .

Removal of Ions and Other Dissolved Substances

Ultrafiltration membranes use polymer membranes with chemically formed microscopic pores that can be used to filter out dissolved substances avoiding the use of coagulants. The type of membrane media determines how much pressure is needed to drive the water through and what sizes of micro-organisms can be filtered out. Ion exchange: Ion exchange systems use ion exchange resin- or zeolite-packed columns to replace unwanted ions. The most common case is water softening consisting of removal of Ca and Mg ions replacing them with benign Na or K ions. Ion exchange resins also used to remove toxic ions such as nitrate, nitrite, lead, mercury, arsenic and many others.

Electrodeionization: Water is passed between a positive electrode and a negative electrode. Ion exchange membranes allow only positive ions to migrate from the treated water toward the negative electrode and only negative ions toward the positive electrode. High purity deionized water is produced with a little worse degree of purification in comparison with ion exchange treatment. Complete removal of ions from water is regarded as electrodialysis. The water is often pre-treated with a reverse osmosis unit to remove non-ionic organic contaminants.

OTHER MECHANICAL AND BIOLOGICAL TECHNIQUES

In addition to the many techniques used in large-scale water treatment, several small-scale, less polluting techniques are also being used to treat polluted water. These techniques include those based on mechanical and biological processes.

An overview:

- *Mechanical systems*: Sand filtration, lava filter systems and systems based on UV-radiation)
- *Biological systems*: Plant systems as constructed wetlands and treatment ponds and compact systems as activated sludge systems, biorotors, aerobic and anaerobic biofilters, submerged aerated filters, and biorolls

In order to purify the water adequately, several of these systems are usually combined to work as a whole. Combination of the systems is done in two to three stages, namely primary and secondary purification. Sometimes tertiary purification is also added.

Disinfection

Disinfection is accomplished both by filtering out harmful microbes and also by adding disinfectant chemicals in the last step in purifying drinking water. Water is disinfected to kill any pathogens which pass through the filters. Possible pathogens include viruses, bacteria, including *Escherichia coli*, *Campylobacter* and *Shigella*, and protozoans, including *Giardia lamblia* and other cryptosporidia. In most developed countries, public water supplies are required to maintain a residual disinfecting agent throughout the distribution system, in which water may remain for days before reaching the consumer. Following the introduction of any chemical disinfecting agent, the water is usually held in temporary storage - often called a contact tank or clear well to allow the disinfecting action to complete.

- *Chlorination*: The most common disinfection method is some form of chlorine or its compounds such as chloramine or chlorine dioxide. Chlorine is a strong oxidant that rapidly kills many harmful micro-organisms. Because chlorine is a toxic gas, there is a danger of a release associated with its use. This problem is avoided by the use of sodium hypochlorite, which is a relatively inexpensive solution that releases free chlorine when dissolved in water. Chlorine solutions can be generated on site by electrolyzing common salt solutions. A solid form, calcium hypochlorite exists that releases chlorine on contact with water. Handling the solid, however, requires greater routine human contact through opening bags and pouring than the use of gas cylinders or bleach which are more easily automated. The generation of liquid sodium hypochlorite is both inexpensive and safer than the use of gas or solid chlorine. All forms of chlorine are widely used despite their respective drawbacks. One drawback is that chlorine from any source reacts with natural organic compounds in the water to form potentially harmful chemical by-products trihalomethanes and haloacetic acids, both of which are carcinogenic in large quantities and regulated by the United States Environmental Protection Agency . The formation of THMs and haloacetic acids may be minimized by effective removal of as many organics from the water as possible prior to chlorine addition. Although chlorine is effective in killing bacteria, it has limited effectiveness against protozoans that form cysts in water .

- Chlorine dioxide is another faster-acting disinfectant. It is, however, relatively rarely used, because in some circumstances it may create excessive amounts of chlorite, which is a by-product regulated to low allowable levels in the United States. Chlorine dioxide is made in water and added/used in water to avoid gas handling problems; chlorine dioxide gas accumulations may spontaneously detonate.
- Chloramines are another chlorine-based disinfectant. Although chloramine is not as strong of an oxidant it does provide a longer-lasting residual than free chlorine, and it won't form THMs or haloacetic acids. It is possible to convert chlorine to chloramine by adding ammonia to the water after addition of chlorine: The chlorine and ammonia react to form chloramine. Water distribution systems disinfected with chloramines may experience nitrification, wherein ammonia is used a nutrient for bacterial growth, with nitrates being generated as a byproduct.
- Ozone is an unstable molecule, a "free radical" of oxygen which readily gives up one atom of oxygen providing a powerful oxidising agent which is toxic to most waterborne organisms. It is a very strong, broad spectrum disinfectant that is widely used in Europe. It is an effective method to inactivate harmful protozoans that form cysts. It also works well against almost all other pathogens. Ozone is made by passing oxygen through ultraviolet light or a "cold" electrical discharge. To use ozone as a disinfectant, it must be created on-site and added to the water

by bubble contact. Some of the advantages of ozone include the production of fewer dangerous by-products and the lack of taste and odour produced by ozonation. Although fewer by-products are formed by ozonation, it has been discovered that the use of ozone produces a small amount of the suspected carcinogen bromate, although little bromine should be present in treated water. Another of the main disadvantages of ozone is that it leaves no disinfectant residual in the water. Ozone has been used in drinking water plants since 1906 where the first industrial ozonation plant was built in Nice, France. The U.S. Food and Drug Administration has accepted ozone as being safe; and it is applied as an anti-microbiological agent for the treatment, storage, and processing of foods.

- UV radiation is very effective at inactivating cysts, as long as the water has a low level of colour so the UV can pass through without being absorbed. The main disadvantage to the use of UV radiation is that, like ozone treatment, it leaves no residual disinfectant in the water. Because neither ozone nor UV radiation leaves a residual disinfectant in the water, it is sometimes necessary to add a residual disinfectant after they are used. This is often done through the addition of chloramines, discussed above as a primary disinfectant. When used in this manner, chloramines provide an effective residual disinfectant with very little of the negative aspects of chlorination.
- Hydrogen peroxide works in a similar way to ozone. Activators such as formic acid are often added to increase the efficacy of disinfection. It has the disadvantages that it is slow-working, phytotoxic in high dosage, and decreases the pH of the water it purifies.
- Various portable methods are available for disinfection in emergencies or in remote locations. Disinfection is the primary goal, since aesthetic considerations such as taste, odour, appearance, and trace chemical contamination do not affect the short-term safety of drinking water.
- Solar water disinfection is a low-cost method of disinfecting water that can often be implemented with locally available materials. Unlike methods that rely on firewood, it has low impact on the environment.

ADDITIONAL TREATMENT OPTIONS

- *Water fluoridation*: In many areas fluoride is added to water for the purpose of preventing tooth decay. Fluoride is usually added after the disinfection process. In the U.S., fluoridation is usually accomplished by the addition of hexafluorosilicic acid, which decomposes in water, yielding fluoride ions.

- *Water conditioning*: This is a method of reducing the effects of hard water. Hardness salts are deposited in water systems subject to heating because the decomposition of bicarbonate ions creates carbonate ions that crystallise out of the saturated solution of calcium or magnesium carbonate. Water with high concentrations of hardness salts can be treated with soda ash which precipitates out the excess salts, through the common-ion effect, producing calcium carbonate of very high purity. The precipitated calcium carbonate is traditionally sold to the manufacturers of toothpaste. Several other methods of industrial and residential water treatment are claimed to include the use of magnetic or/and electrical fields reducing the effects of hard water.
- *Plumbosolvency reduction*: In areas with naturally acidic waters of low conductivity, the water may be capable of dissolving lead from any lead pipes that it is carried in. The addition of small quantities of phosphate ion and increasing the pH slightly both assist in greatly reducing plumbo-solvency by creating insoluble lead salts on the inner surfaces of the pipes.
- *Radium Removal*: Some groundwater sources contain radium, a radioactive chemical element. Typical sources include many groundwater sources north of the Illinois River in Illinois. Radium can be removed by ion exchange, or by water conditioning. The back flush or sludge that is produced is, however, a low-level radioactive waste.
- *Fluoride Removal*: Although fluoride is added to water in many areas, some areas of the world have excessive levels of natural fluoride in the source water. Excessive levels can be toxic or cause undesirable cosmetic effects such as staining of teeth. One method of reducing fluoride levels is through treatment with activated alumina.

OTHER WATER PURIFICATION TECHNIQUES

Other popular methods for purifying water, especially for local private supplies are listed below.

In some countries some of these methods are also used for large scale municipal supplies. Particularly important are distillation and reverse osmosis.

1. *Boiling*: Water is heated hot enough and long enough to inactivate or kill micro-organisms that normally live in water at room temperature. Near sea level, a vigorous rolling boil for at least one minute is sufficient. At high altitudes three minutes is recommended. In areas where the water is "hard" , boiling decomposes the bicarbonate ions, resulting in partial

precipitation as calcium carbonate. This is the "fur" that builds up on kettle elements, etc., in hard water areas. With the exception of calcium, boiling does not remove solutes of higher boiling point than water and in fact increases their concentration . Boiling does not leave a residual disinfectant in the water. Therefore, water that has been boiled and then stored for any length of time may have acquired new pathogens.

2. *Granular Activated Carbon filtering*: a form of activated carbon with a high surface area, adsorbs many compounds including many toxic compounds. Water passing through activated carbon is commonly used in municipal regions with organic contamination, taste or odors. Many household water filters and fish tanks use activated carbon filters to further purify the water. Household filters for drinking water sometimes contain silver to release silver ions which have an anti-bacterial effect.
3. Distillation involves boiling the water to produce water vapour. The vapour contacts a cool surface where it condenses as a liquid. Because the solutes are not normally vaporised, they remain in the boiling solution. Even distillation does not completely purify water, because of contaminants with similar boiling points and droplets of unvaporised liquid carried with the steam. However, 99.9% pure water can be obtained by distillation.
4. *Reverse osmosis*: Mechanical pressure is applied to an impure solution to force pure water through a semi-permeable membrane. Reverse osmosis is theoretically the most thorough method of large scale water purification available, although perfect semi-permeable membranes are difficult to create. Unless membranes are well-maintained, algae and other life forms can colonize the membranes.
5. The use of iron in removing arsenic from water.
6. *Direct contact membrane distillation*: Applicable to desalination. Heated seawater is passed along the surface of a hydrophobic polymer membrane. Evaporated water passes from the hot side through pores in the membrane into a stream of cold pure water on the other side. The difference in vapour pressure between the hot and cold side helps to push water molecules through.
7. *Gas hydrate crystals centrifuge method*: If carbon dioxide gas is mixed with contaminated water at high pressure and low temperature, gas hydrate crystals will contain only clean water. This is because the water molecules bind to the gas molecules at molecule level. The contaminated water is in liquid form. A centrifuge may be used to separate the crystals and the concentrated contaminated water.

WATER PURIFICATION FOR HYDROGEN PRODUCTION

For the small scale production of hydrogen, water purifiers are installed to prevent formation of minerals on the surface of the electrodes and to remove organics and chlorine from utility water. First, the water passes through a 20 micrometre interference filter to remove sand and dust particles, then a charcoal filter using activated carbon to remove organics and chlorine and finally a de-ionizing filter to remove metallic ions. Testing can be done before and after the filter to verify the proper removal of barium, calcium, potassium, magnesium, sodium and silica. Another method that is used is reverse osmosis.

SAFETY AND CONTROVERSIES

Accidents have also been known to happen. In April, 2007, the water supply of Spencer, Massachusetts became contaminated with excess sodium hydroxide when its treatment equipment malfunctioned. Many municipalities have moved from free chlorine to chloramine as a disinfection agent. However, chloramine in some water systems, appears to be a corrosive agent. Chlormaine can dissolve the "protective" film inside older service line, with the leaching of lead into residential spigots. This can result in harmful exposure to lead, with elevated blood levels of lead the outcome. Lead is a known neurotoxin.

DEMINERALIZED WATER

Distillation removes all minerals from water, and the membrane methods of reverse osmosis and nanofiltration remove most to all minerals. This results in demineralized water which is not considered ideal drinking water. The World Health Organization has investigated the health effects of demineralized water since 1980. Experiments in humans found that demineralized water increased diuresis and the elimination of electrolytes, with decreased blood serum potassium concentration.

Magnesium, calcium, and other minerals in water can help to protect against nutritional deficiency. Demineralized water may also increase the risk from toxic metals because it more readily leaches materials from piping like lead and cadmium, which is prevented by dissolved minerals such as calcium and magnesium. Low-mineral water has been implicated in specific cases of lead poisoning in infants, when lead from pipes leached at especially high rates into the water. Recommendations for magnesium have been put at a minimum of 10 mg/L with 20–30 mg/L optimum; for calcium a 20 mg/L minimum and a 40–80 mg/L optimum, and a total water hardness of 2 to 4 mmol/L. At water hardness above 5 mmol/L, higher incidence of gallstones, kidney stones, urinary stones, arthrosis, and arthropathies have been observed. Additionally, desalination processes can increase the risk of bacterial contamination. Manufacturers of home water distillers, of course, claim the opposite — that minerals in water are the cause of many diseases, and that most beneficial minerals come from food, not water. They quote the American Medical Association as saying "The body's need for minerals is largely met through foods, not drinking water." The WHO report agrees that "drinking water, with some rare exceptions, is not the major source of essential elements for humans" and is "not the major source of our calcium and magnesium intake", yet states that demineralized water is harmful anyway. "Additional evidence comes from animal experiments and clinical observations in several countries. Animals given zinc or magnesium dosed in their drinking water had a significantly higher concentration of these elements in the serum than animals given the same elements in much higher amounts with food and provided with low-mineral water to drink."

AGRICULTURAL WASTEWATER TREATMENT

Agricultural wastewater treatment relates to the treatment of wastewaters produced in the course of agricultural activities. Agriculture is a highly intensified industry in many parts of the world, producing a range of wastewaters requiring a variety of treatment technologies and management practices.

NONPOINT SOURCE POLLUTION

Nonpoint source pollution from farms is caused by surface runoff from fields during rain storms. Agricultural runoff is a major source of pollution, in some cases the only source, in many watersheds.

Sediment Run-off

Soil washed off fields is the largest source of agricultural pollution in the United States. Excess sediment causes high levels of turbidity in water bodies, which can inhibit growth of aquatic plants, clog fish gills and smother animal larvae. Farmers may utilize erosion controls to reduce runoff flows and retain soil on their fields.

Common techniques include:

- Contour plowing
- Crop mulching
- Crop rotation
- Planting perennial crops
- Installing riparian buffers.

Nutrient Run-off

Nitrogen and phosphorus are key pollutants found in runoff, and they are applied to farmland in several ways:

- Commercial fertilizer

- Animal manure
- Municipal or industrial wastewater or sludge.

These chemicals may also enter runoff from crop residues, irrigation water, wildlife, and atmospheric deposition.

Farmers can develop and implement nutrient management plans to mitigate impacts on water quality:

- Map and document fields, crop types, soil types, water bodies
- Develop realistic crop yield projections
- Conduct soil tests and nutrient analyses of manures and/or sludges applied
- Identify other significant nutrient sources
- Evaluate significant field features such as highly erodible soils, subsurface drains, and shallow aquifers
- Apply fertilizers, manures, and/or sludges based on realistic yield goals and using precision agriculture techniques.

PESTICIDES

Pesticides are widely used by farmers to control plant pests and enhance production, but chemical pesticides can also cause water quality problems.

Pesticides may appear in surface water due to:

- Direct application
- Runoff during rain storms
- Aerial drift .

Some pesticides have also been detected in groundwater. Farmers may use Integrated Pest Management techniques to maintain control over pests, reduce reliance on chemical pesticides, and protect water quality. There are few safe ways of disposing of pesticide surpluses other than through containment in well managed landfills or by incineration. In some parts of the world, spraying on land is a permitted method of disposal.

POINT SOURCE POLLUTION

Farms with large livestock and poultry operations, such as factory farms, can be a major source of point source wastewater. In the United States, these facilities are called concentrated animal feeding operations or confined animal feeding operations and are being subject to increasing government regulation.

Animal Wastes

The constituents of animal wastewater typically contain:

- Strong organic content — much stronger than human sewage
- High solids concentration
- High nitrate and phosphorus content
- Antibiotics

- Synthetic hormones
- Often high concentrations of parasites and their eggs
- Spores of *Cryptosporidum* resistant to drinking water treatment processes
- Spores of *Giardia*
- Human pathogenic bacteria such as *Brucella* and *Salmonella*

Animal wastes from cattle can be produced as solid or semisolid manure or as a liquid slurry. The production of slurry is especially common in housed dairy cattle.

Treatment

Whilst solid manure heaps outdoors can give rise to polluting wastewaters from runoff, this type of waste is usually relatively easy to treat by containment and/or covering of the heap. Animal slurries require special handling and are usually treated by containment in lagoons before disposal by spray or trickle application to grassland. Constructed wetlands are sometimes used to facilitate treatment of animal wastes, as are anaerobic lagoons. Excessive application or application to sodden land or insufficient land area can result in direct runoff to watercourses, with the potential for causing severe pollution. Application of slurries to land overlying aquifers can result in direct contamination or, more commonly, elevation of nitrogen levels as nitrite or nitrate. The disposal of any wastewater containing animal waste upstream of a drinking water intake can pose serious health problems to those drinking the water because of the highly resistant spores present in many animals that are capable of causing disabling in humans. This risk exists even for very low-level seepage via shallow surface drains or from rainfall run-off. Some animal slurries are treated by mixing with straws and composted at high temperature to produce a bacteriologically sterile and friable manure for soil improvement.

PIGGERY WASTE

Piggery waste is comparable to other animal wastes except that many piggery wastes contain elevated levels of copper that can be toxic in the natural environment. Ascarid worms and their eggs are also common and can infect humans if wastewater treatment is ineffective.

Treatment

As for general animal waste, although the liquid fraction of the waste is frequently separated off and re-used in the piggery to avoid the prohibitively expensive costs of disposing of a copper-rich liquor.

SILAGE LIQUOR

Fresh or wilted grass or other green crops can be made into the semi-fermented product called silage which can be stored and used as winter forage for cattle and sheep. The production of silage often involves the use of an acid conditioner such as sulfuric acid or formic acid. The process of silage making frequently produces a yellow-brown strongly smelling liquid which is very rich in simple sugars, alcohol, short-chain organic acids and silage conditioner. This liquor is one of the most polluting organic substances known. The volume of silage liquor produced is generally in proportion to the moisture content of the ensiled material.

Treatment

Silage liquor is best treated through prevention by wilting crops well before silage making. Any silage liquor that is produced can be used as part of the food for pigs. The most effective treatment is by containment in a slurry lagoon and by subsequent spreading on land following substantial dilution with slurry. Containment of silage liquor on its own can cause structural problems in concrete pits because of the acidic nature of silage liquor.

MILKING PARLOUR (DAIRY FARMING) WASTES

Although milk has a deserved reputation as an important and valuable food product, its presence in wastewaters is highly polluting because of its organic strength, which can lead to very rapid de-oxygenation of receiving waters. Milking parlour wastes also contain large volumes of wash-down water, some animal waste together with cleaning and disinfection chemicals.

Treatment

Milking parlour wastes are often treated in admixture with human sewage in a local sewage treatment plant. This ensures that disinfectants and cleaning agents are sufficiently diluted and amenable to treatment. Running milking wastewaters into a farm slurry lagoon is a possible option although this tends to consume lagoon capacity very quickly. Land spreading is also a treatment option.

SLAUGHTERING WASTE

Wastewater from slaughtering activities is similar to milking parlour waste although considerably stronger in its organic composition and therefore potentially much more polluting.

Treatment

Milking parlour wastes are often treated in admixture with human sewage in a local sewage treatment plant. This ensures that disinfectants and cleaning agents are sufficiently diluted and amenable to treatment. Running milking wastewaters into a farm slurry lagoon is a possible option although this tends to consume lagoon capacity very quickly. Land spreading is also a treatment option.

VEGETABLE WASHING WATER

Washing of vegetables produces large volumes of water contaminated by soil and vegetable pieces. Low levels of pesticides used to treat the vegetables may also be present together with moderate levels of disinfectants such as chlorine.

Treatment

Most vegetable washing waters are extensively recycled with the solids removed by settlement and filtration. The recovered soil can be returned to the land.

FIREWATER

Although few farms plan for fires, fires are nevertheless more common on farms than on many other industrial premises. Stores of pesticides, herbicides, fuel oil for farm machinery and fertilizers can all help promote fire and can all be present in environmentally lethal quantities in firewater from fire fighting at farms.

Treatment

All farm environmental management plans should allow for containment of substantial quantities of firewater and for its subsequent recovery and disposal by specialist disposal companies. The concentration and mixture of contaminants in firewater make them unsuited to any treatment method available on the farm. Even land spreading has produced severe taste and odour problems for downstream water supply companies in the past.

INDUSTRIAL WASTEWATER TREATMENT

Industrial wastewater treatment covers the mechanisms and processes used to treat waters that have been contaminated in some way by anthropogenic industrial or commercial activities prior to its release into the environment or its re-use. Most industries produce some wet waste although recent trends in the developed world have been to minimise such production or recycle such waste within the production process. However, many industries remain dependent on processes that produce wastewaters.

SOURCES OF INDUSTRIAL WASTEWATER

Iron and Steel Industry

The production of iron from its ores involves powerful reduction reactions in blast furnaces. Cooling waters are inevitably contaminated with products especially ammonia and cyanide. Production of coke from coal in coking plants also requires water cooling and the use of water in by-products separation.

Contamination of waste streams includes gasification products such as benzene, naphthalene, anthracene, cyanide, ammonia, phenols, cresols together with a range of more complex organic compounds known collectively as polycyclic aromatic hydrocarbons . The conversion of iron or steel into sheet, wire or rods requires hot and cold mechanical transformation stages frequently employing water as a lubricant and coolant.

Contaminants include hydraulic oils, tallow and particulate solids. Final treatment of iron and steel products before onward sale into manufacturing includes *pickling* in strong mineral acid to remove rust and prepare the surface for tin or chromium plating or for other surface treatments such as galvanisation or painting.

The two acids commonly used are hydrochloric acid and sulfuric acid. Wastewaters include acidic rinse waters together with waste acid. Although many plants operate acid recovery plants, where the mineral acid is boiled away from the iron salts, there remains a large volume of highly acid ferrous sulfate or ferrous chloride to be disposed of. Many steel industry wastewaters are contaminated by hydraulic oil also known as soluble oil.

MINES AND QUARRIES

The principal waste-waters associated with mines and quarries are slurries of rock particles in water. These arise from rainfall washing exposed surfaces and haul roads and also from rock washing and grading processes.

Volumes of water can be very high, especially rainfall related arisings on large sites. Some specialized separation operations, such as coal washing to separate coal from native rock using density gradients, can produce wastewater contaminated by fine particulate haematite and surfactants.

Oils and hydraulic oils are also common contaminants. Wastewater from metal mines and ore recovery plants are inevitably contaminated by the minerals present in the native rock formations. Following crushing and extraction of the desirable materials, undesirable materials may become contaminated in the wastewater.

For metal mines, this can include unwanted metals such as zinc and other materials such as arsenic. Extraction of high value metals such as gold and silver may generate slimes containing very fine particles in where physical removal of contaminants becomes particularly difficult.

FOOD INDUSTRY

Wastewater generated from agricultural and food operations has distinctive characteristics that set it apart from common municipal wastewater managed by public or private wastewater treatment plants throughout the world: it is biodegradable and nontoxic, but that has high concentrations of biochemical oxygen demand and suspended solids . The constituents of food and agriculture wastewater are often complex to predict due to the differences in BOD and pH in effluents from vegetable, fruit, and meat products and due to the seasonal nature of food processing and postharvesting. Processing of food from raw materials requires large volumes of high grade water. Vegetable washing generates waters with high loads of particulate matter and some dissolved organics. It may also contain surfactants.

Animal slaughter and processing produces very strong organic waste from body fluids, such as blood, and gut contents. This wastewater is frequently contaminated by significant levels of antibiotics and growth hormones from the animals and by a variety of pesticides used to control external parasites.

Insecticide residues in fleeces is a particular problem in treating waters generated in wool processing. Processing food for sale produces wastes generated from cooking which are often rich in plant organic material and may also contain salt, flavourings, colouring material and acids or alkali. Very significant quantities of oil or fats may also be present.

COMPLEX ORGANIC CHEMICALS INDUSTRY

A range of industries manufacture or use complex organic chemicals. These include pesticides, pharmaceuticals, paints and dyes, petro-chemicals, detergents, plastics, paper pollution, etc. Waste waters can be contaminated by feed-stock materials, by-products, product material in soluble or particulate form, washing and cleaning agents, solvents and added value products such as plasticisers.

Nuclear Industry

The waste production from the nuclear and radio-chemicals industry is dealt with at *Radioactive waste.*

Water Treatment

Water treatment for the production of drinking water is dealt with elsewhere. Many industries have a need to treat water to obtain very high quality water for demanding purposes. Water treatment produces organic and mineral sludges from filtration and sedimentation. Ion exchange using natural or synthetic resins removes calcium, magnesium and carbonate ions from water, replacing them with hydrogen and hydroxyl ions. Regeneration of ion exchange columns with strong acids and alkalis produces a wastewater rich in hardness ions which are readily precipitated out, especially when in admixture with other wastewaters.

Treatment of Industrial Wastewater

The different types of contamination of wastewater require a variety of strategies to remove the contamination.

SOLIDS REMOVAL

Most solids can be removed using simple sedimentation techniques with the solids recovered as slurry or sludge. Very fine solids and solids with densities close to the density of water pose special problems. In such case filtration or ultrafiltration may be required. Although, flocculation may be used, using alum salts or the addition of polyelectrolytes.

OILS AND GREASE REMOVAL

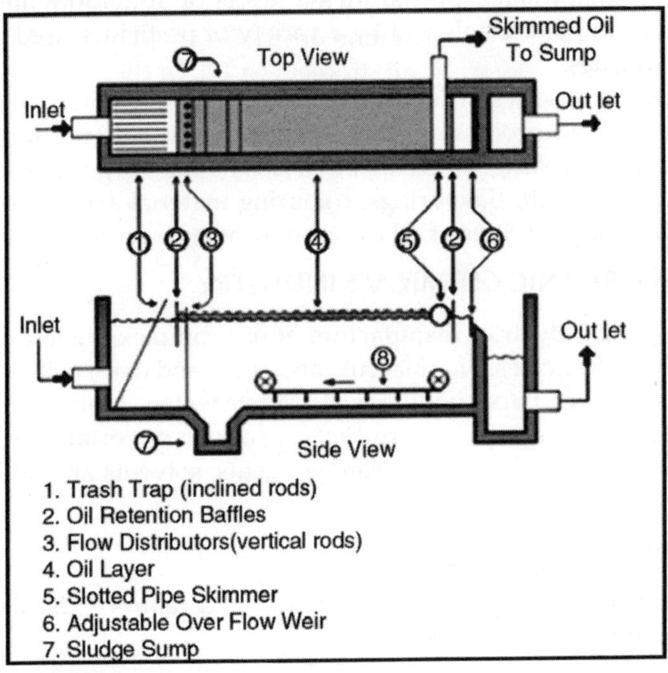

Fig. A Typical API Oil-water Separator Used in Many Industries.

Many oils can be recovered from open water surfaces by skimming devices. Considered a dependable and cheap way to remove oil, grease and other hydrocarbons from water, oil skimmers can sometimes achieve the desired level of water purity. At other times, skimming is also a cost-efficient method to remove most of the oil before using membrane filters and chemical processes. Skimmers will prevent filters from blinding prematurely and keep chemical costs down because there is less oil to process. Because grease skimming involves higher viscosity hydrocarbons, skimmers must be equipped with heaters powerful enough to keep grease fluid for discharge. If floating grease forms into solid clumps or mats, a spray bar, aerator or mechanical apparatus can be used to facilitate removal. However, hydraulic oils and the majority of oils that have degraded to any extent will also have a soluble or emulsified component that will require further treatment to eliminate. Dissolving or emulsifying oil using surfactants or solvents usually exacerbates the problem rather than solving it, producing wastewater that is more difficult to treat. The wastewaters from large-scale industries such as oil refineries, petrochemical plants, chemical plants, and natural gas processing plants commonly contain gross amounts of oil and suspended solids. Those industries use a device known as an API oil-water separator which is designed to separate the oil and suspended solids from their wastewater effluents.

The name is derived from the fact that such separators are designed according to standards published by the American Petroleum Institute. The API separator is a gravity separation device designed by using Stokes Law to define the rise velocity of oil droplets based on their density and size. The design is based on the specific gravity difference between the oil and the wastewater because that difference is much smaller than the specific gravity difference between the suspended solids and water.

The suspended solids settles to the bottom of the separator as a sediment layer, the oil rises to top of the separator and the cleansed wastewater is the middle layer between the oil layer and the solids. Typically, the oil layer is skimmed off and subsequently re-processed or disposed of, and the bottom sediment layer is removed by a chain and flight scraper and a sludge pump.

The water layer is sent to further treatment consisting usually of a Electroflotation module for additional removal of any residual oil and then to some type of biological treatment unit for removal of undesirable dissolved chemical compounds.

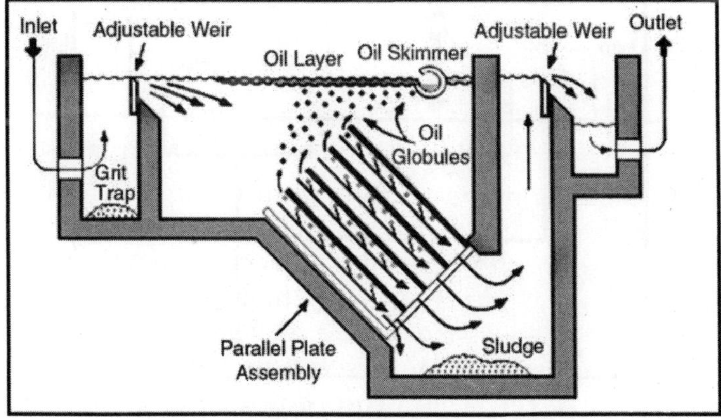

Fig. A Typical Parallel Plate Separator.

Parallel plate separators are similar to API separators but they include tilted parallel plate assemblies. The parallel plates provide more surface for suspended oil droplets to coalesce into larger globules. Such separators still depend upon the specific gravity between the suspended oil and the water. However, the parallel plates enhance the degree of oil-water separation. The result is that a parallel plate separator requires significantly less space than a conventional API separator to achieve the same degree of separation.

REMOVAL OF BIODEGRADABLE ORGANICS

Biodegradable organic material of plant or animal origin is usually possible to treat using extended conventional wastewater treatment processes such as activated sludge or trickling filter. Problems can arise if the wastewater is excessively diluted with washing water or is highly concentrated such as neat blood or milk. The presence of cleaning agents, disinfectants, pesticides, or antibiotics can have detrimental impacts on treatment processes.

Activated Sludge Process

Activated sludge is a biochemical process for treating sewage and industrial wastewater that uses air and microorganisms to biologically oxidize organic pollutants, producing a waste sludge containing the oxidized material.

In general, an activated sludge process includes:

- An aeration tank where air is injected and thoroughly mixed into the wastewater.
- A settling tank to allow the waste sludge to settle. Part of the waste sludge is recycled to the aeration tank and the remaining waste sludge is removed for further treatment and ultimate disposal.

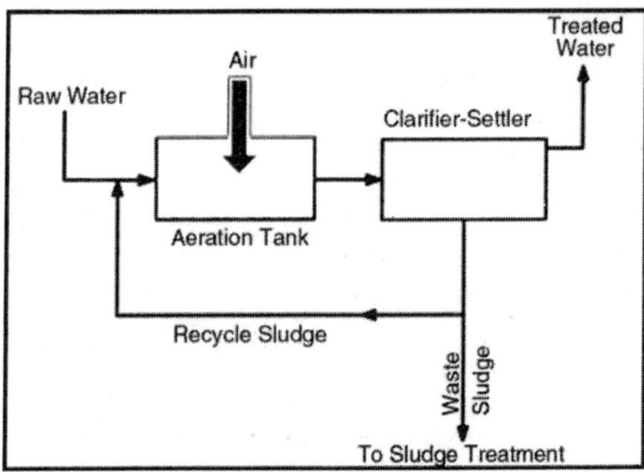

Fig. A Generalized, Schematic Diagram of an Activated Sludge Process.

Trickling Filter Process

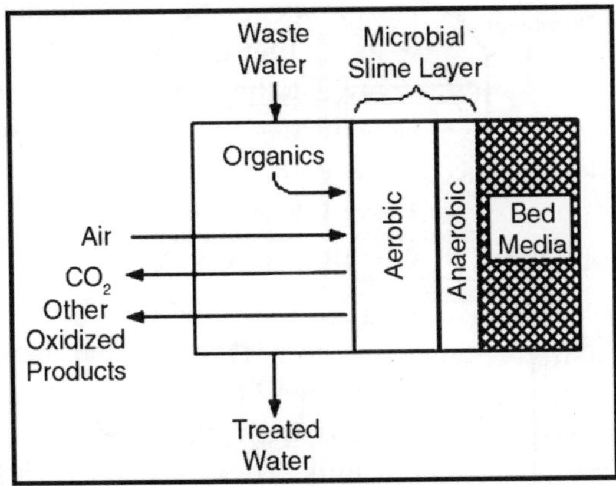

Fig. A Schematic Cross-section of the Contact Face of the Bed Media in a Trickling Filter.

A trickling filter consists of a bed of rocks, gravel, slag, peat moss, or plastic media over which wastewater flows downward and contacts a layer of microbial slime covering the bed media. Aerobic conditions are maintained by forced air flowing through the bed or by natural convection of air.

The process involves adsorption of organic compounds in the wastewater by the microbial slime layer, diffusion of air into the slime layer to provide the oxygen required for the biochemical oxidation of the organic compounds.

The end products include carbon dioxide gas, water and other products of the oxidation. As the slime layer thickens, it becomes difficult for the air to penetrate the layer and an inner anaerobic layer is formed.

The components of a complete trickling filter system are: fundamental components:

- A bed of filter medium upon which a layer of microbial slime is promoted and developed.
- An enclosure or a container which houses the bed of filter medium.
- A system for distributing the flow of wastewater over the filter medium.
- A system for removing and disposing of any sludge from the treated effluent.

The treatment of sewage or other wastewater with trickling filters is among the oldest and most well characterized treatment technologies. A trickling filter is also often called a trickle filter, trickling biofilter, biofilter, biological filter or biological trickling filter.

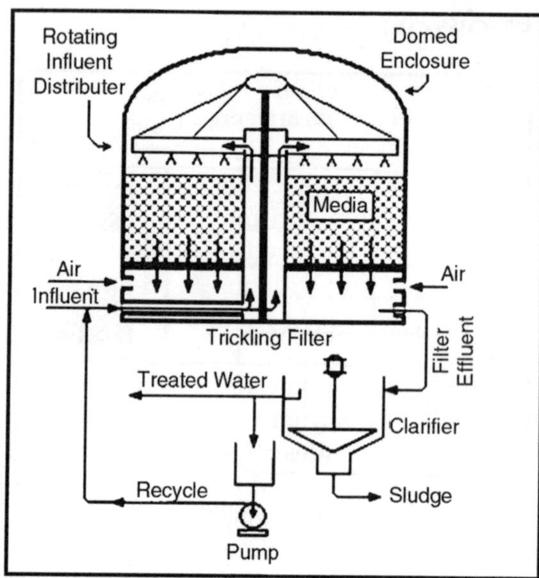

Fig. A Typical Complete Trickling Filter System.

TREATMENT OF OTHER ORGANICS

Synthetic organic materials including solvents, paints, pharmaceuticals, pesticides, coking products and so forth can be very difficult to treat. Treatment methods are often specific to the material being treated. Methods include Advanced Oxidation Processing, distillation, adsorption, vitrification, incineration, chemical immobilisation or landfill disposal. Some materials such as some detergents may be capable of biological degradation and in such cases, a modified form of wastewater treatment can be used.

TREATMENT OF ACIDS AND ALKALIS

Acids and alkalis can usually be neutralised under controlled conditions. Neutralisation frequently produces a precipitate that will require treatment as a solid residue that may also be toxic. In some cases, gasses may be evolved requiring treatment for the gas stream. Some other forms of treatment are usually required following neutralisation.

Waste streams rich in hardness ions as from de-ionisation processes can readily lose the hardness ions in a buildup of precipitated calcium and magnesium salts. This precipitation process can cause severe *furring* of pipes and can, in extreme cases, cause the blockage of disposal pipes. A 1 metre diameter industrial marine discharge pipe serving a major chemicals complex was blocked by such salts in the 1970s. Treatment is by concentration of de-ionisation waste waters and disposal to landfill or by careful pH management of the released wastewater.

TREATMENT OF TOXIC MATERIALS

Toxic materials including many organic materials, metals acids, alkalis, non-metallic elements are generally resistant to biological processes unless very dilute. Metals can often be precipitated out by changing the pH or by treatment with other chemicals. Many, however, are resistant to treatment or mitigation and may require concentration followed by landfilling or recycling. Disolved organics can be *incinerated* within the wastewater by Advanced Oxidation Processes.

SEWAGE TREATMENT

Sewage treatment, or domestic wastewater treatment, is the process of removing contaminants from wastewater and household sewage, both runoff and domestic. It includes physical, chemical, and biological processes to remove physical, chemical and biological contaminants. Its objective is to produce a waste stream and a solid waste or sludge suitable for discharge or reuse back into the environment. This material is often inadvertently contaminated with many toxic organic and inorganic compounds.

ORIGINS OF SEWAGE

Sewage is created by residences, institutions, hospitals and commercial and industrial establishments. Raw influent includes household waste liquid from toilets, baths, showers, kitchens, sinks, and so forth that is disposed of via sewers. In many areas, sewage also includes liquid waste from industry and commerce.

The separation and draining of household waste into greywater and blackwater is becoming more common in the developed world, with greywater being permitted to be used for watering plants or recycled for flushing toilets. A lot of sewage also includes some surface water from roofs or hard-standing areas.

Municipal wastewater therefore includes residential, commercial, and industrial liquid waste discharges, and may include stormwater runoff. Sewage systems capable of handling stormwater are known as combined systems or combined sewers. Such systems are usually avoided since they complicate and thereby reduce the efficiency of sewage treatment plants owing to their seasonality.

The variability in flow also leads to often larger than necessary, and subsequently more expensive, treatment facilities. In addition, heavy storms that contribute more flows than the treatment plant can handle may overwhelm the sewage treatment system, causing a spill or overflow. It is preferable to have a separate storm drain system for stormwater in areas that are developed with sewer systems.

As rainfall runs over the surface of roofs and the ground, it may pick up various contaminants including soil particles and other sediment, heavy metals, organic compounds, animal waste, and oil and grease. Some jurisdictions require stormwater to receive some level of treatment before being discharged directly into waterways. Examples of treatment processes used for stormwater include sedimentation basins, wetlands, buried concrete vaults with various kinds of filters, and vortex separators.

PROCESS OVERVIEW

Sewage can be treated close to where it is created or collected and transported via a network of pipes and pump stations to a municipal treatment plant. Sewage collection and treatment is typically subject to local, state and federal regulations and standards. Industrial sources of wastewater often require specialized treatment processes. Conventional sewage treatment involves three stages, called primary, secondary and tertiary treatment. First, the solids are separated from the wastewater stream. Then dissolved biological matter is progressively converted into a solid mass by using indigenous, water-borne micro-organisms.

Finally, the biological solids are neutralized then disposed of or re-used, and the treated water may be disinfected chemically or physically . The final effluent can be discharged into a stream, river, bay, lagoon or wetland, or it can be used for the irrigation of a golf course, green way or park. If it is sufficiently clean, it can also be used for groundwater recharge or agricultural purposes.

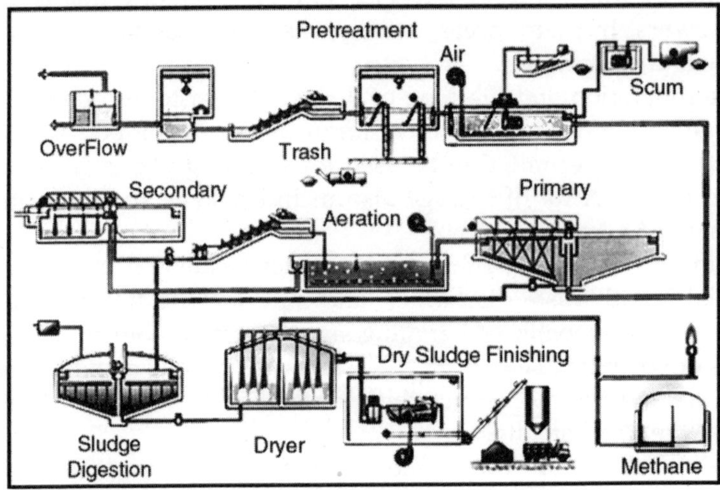

PRE-TREATMENT

Pre-treatment removes the materials that can be easily collected from the raw wastewater and disposed of. The typical materials that are removed during pre treatment include fats, oils, and greases , sand, gravels and rocks , larger settleable solids and floating materials. In modern plants serving large populations, sophisticated equipment with remote operation and control are employed whilst in smaller or less modern plants manually cleaned screen may be used.

SCREENING

The influent sewage water is strained to remove all large objects carried in the sewage stream, such as rags, sticks, tampons, cans, fruit, etc. This is most commonly done with a manual or automated mechanically raked bar screen. The raking action of a mechanical bar screen is typically paced according to the accumulation on the bar screens and/or flow rate. The bar screen is used because large solids can damage or clog the equipment used later in the sewage treatment plant. The large solids can also hinder the biological process. The solids are collected and later disposed in a landfill or incinerated. Pre treatment also typically includes a sand or grit channel or chamber where the velocity of the incoming wastewater is carefully controlled to allow sand grit and stones to settle, while keeping the majority of the suspended organic material in the water column. This equipment is called a de-gritter or sand catcher.

Sand, grit, and stones need to be removed early in the process to avoid damage to pumps and other equipment in the remaining treatment stages. Sometimes there is a sand washer followed by a conveyor that transports the sand to a container for disposal. The contents from the sand catcher may be fed into the incinerator in a sludge processing plant, but in many cases, the sand and grit is sent to a landfill.

PRIMARY TREATMENT

Sedimentation

In the primary sedimentation stage, sewage flows through large tanks, commonly called "primary clarifiers" or "primary sedimentation tanks". The tanks are large enough that sludge can settle and floating material such as grease and oils can rise to the surface and be skimmed off. The main purpose of the primary sedimentation stage is to produce both a generally homogeneous liquid capable of being treated biologically and a sludge that can be separately treated or processed. Primary settling tanks are usually equipped with mechanically driven scrapers that continually drive the collected sludge towards a hopper in the base of the tank from where it can be pumped to further sludge treatment stages.

Secondary Treatment

Secondary treatment is designed to substantially degrade the biological content of the sewage such as are derived from human waste, food waste, soaps and detergent. The majority of municipal plants treat the settled sewage liquor using aerobic biological processes. For this to be effective, the biota require both oxygen and a substrate on which to live. There are a number of ways in which this is done. In all these methods, the bacteria and protozoa consume biodegradable soluble organic contaminants and bind much of the less soluble fractions into floc.

Secondary treatment systems are classified as:

- Fixed-film or
- Suspended-growth.

Fixed-film treatment process including trickling filter and rotating biological contactors where the biomass grows on media and the sewage passes over its surface. In suspended-growth systems, such as activated sludge, the biomass is well mixed with the sewage and can be operated in a smaller space than fixed-film systems that treat the same amount of water. However, fixed-film systems are more able to cope with drastic changes in the amount of biological material and can provide higher removal rates for organic material and suspended solids than suspended growth systems. Roughing filters are intended to treat particularly strong or variable organic loads, typically industrial, to allow them to then be treated by conventional secondary treatment processes. Characteristics include typically tall, circular filters filled with open synthetic filter media to which wastewater is applied at a relatively high rate. They are designed to allow high hydraulic loading and a high flow-through of air. On larger installations, air is forced through the media using blowers. The resultant wastewater is usually within the normal range for conventional treatment processes.

SURFACE-AERATED BASINS

Most biological oxidation processes for treating industrial wastewaters have in common the use of oxygen and microbial action. Surface-aerated basins achieve 80 to 90% removal of Biochemical Oxygen Demand with retention times of 1 to 10 days. The basins may range in depth from 1.5 to 5.0 metres and use motor-driven aerators floating on the surface of the wastewater.

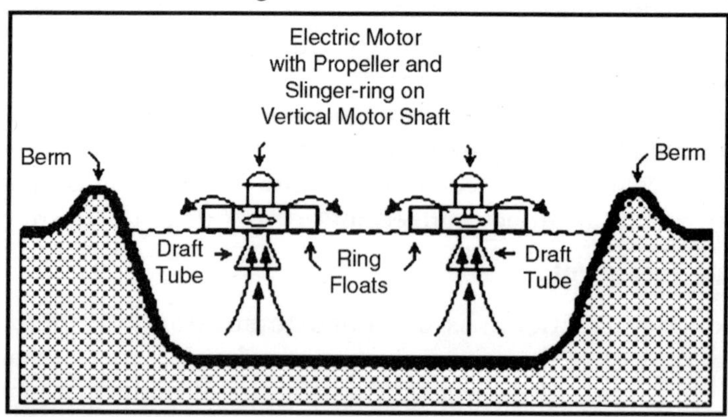

Fig. A Typical Surface-Aerated Basin

In an aerated basin system, the aerators provide two functions: they transfer air into the basins required by the biological oxidation reactions, and they provide the mixing required for dispersing the air and for contacting the reactants. Typically, the floating surface aerators are rated to deliver the amount of air equivalent to 1.8 to 2.7 kg O_2/kW·h. However, they do not provide as good mixing as is normally achieved in activated sludge systems and therefore aerated basins do not achieve the same performance level as activated sludge units. Biological oxidation processes are sensitive to temperature and, between 0 °C and 40 °C, the rate of biological reactions increase with temperature. Most surface aerated vessels operate at between 4 °C and 32 °C.

Filter Beds

In older plants and plants receiving more variable loads, trickling filter beds are used where the settled sewage liquor is spread onto the surface of a deep bed made up of coke, limestone chips or specially fabricated plastic media. Such media must have high surface areas to support the biofilms that form. The liquor is distributed through perforated rotating arms radiating from a central pivot. The distributed liquor trickles through this bed and is collected in drains at the base.

These drains also provide a source of air which percolates up through the bed, keeping it aerobic. Biological films of bacteria, protozoa and fungi form on the media's surfaces and eat or otherwise reduce the organic content. This biofilm is grazed by insect larvae and worms which help maintain an optimal thickness. Overloading of beds increases the thickness of the film leading to clogging of the filter media and ponding on the surface.

Biological Aerated Filters

Biological Aerated Filter or Biofilters combine filtration with biological carbon reduction, nitrification or denitrification. BAF usually includes a reactor filled with a filter media. The media is either in suspension or supported by a gravel layer at the foot of the filter.

The dual purpose of this media is to support highly active biomass that is attached to it and to filter suspended solids. Carbon reduction and ammonia conversion occurs in aerobic mode and sometime achieved in a single reactor while nitrate conversion occurs in anoxic mode. BAF is operated either in upflow or downflow configuration depending on design specified by manufacturer.

Membrane Bioreactors

Membrane bioreactors combines activated sludge treatment with a membrane liquid-solid separation process. The membrane component uses low pressure microfiltration or ultra filtration membranes and eliminates the need for clarification and tertiary filtration. The membranes are typically immersed in the aeration tank. One of the key benefits of a membrane bioreactor system is that it effectively overcomes the limitations associated with poor settling of sludge in conventional activated sludge processes. The technology permits bioreactor operation with considerably higher mixed liquor suspended solids concentration than CAS systems, which are limited by sludge settling.

The process is typically operated at MLSS in the range of 8,000–12,000 mg/L, while CAS are operated in the range of 2,000–3,000 mg/L. The elevated biomass concentration in the membrane bioreactor process allows for very effective removal of both soluble and particulate biodegradable materials at higher loading rates. Thus increased Sludge Retention Times—usually exceeding 15 days—ensure complete nitrification even in extremely cold weather. The cost of building and operating a MBR is usually higher than conventional wastewater treatment, however, as the technology has become increasingly popular and has gained wider acceptance throughout the industry, the life-cycle costs have been steadily decreasing. The small footprint of MBR systems, and the high quality effluent produced, makes them particularly useful for water reuse applications.

Secondary Sedimentation

The final step in the secondary treatment stage is to settle out the biological floc or filter material and produce sewage water containing very low levels of organic material and suspended matter.

ROTATING BIOLOGICAL CONTACTORS

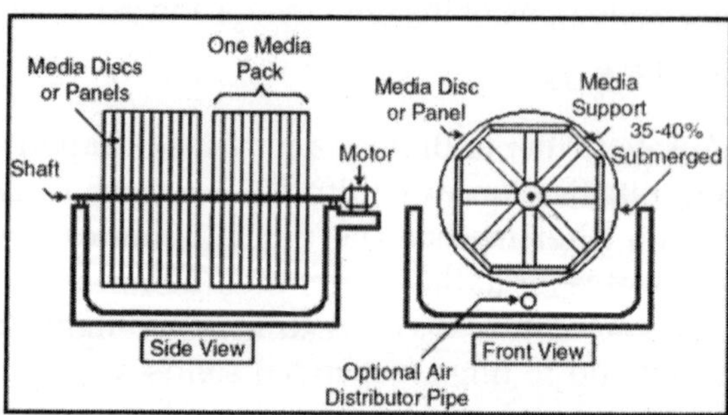

Fig. Schematic Diagram of a Typical Rotating Biological Contactor.

Rotating biological contactors are mechanical secondary treatment systems, which are robust and capable of withstanding surges in organic load. RBCs were first installed in Germany in 1960 and have since been developed and refined into a reliable operating unit. The rotating disks support the growth of bacteria and micro-organisms present in the sewage, which breakdown and stabilise organic pollutants.

To be successful, micro-organisms need both oxygen to live and food to grow. Oxygen is obtained from the atmosphere as the disks rotate. As the micro-organisms grow, they build up on the media until they are sloughed off due to shear forces provided by the rotating discs in the sewage. Effluent from the RBC is then passed through final clarifiers where the micro-organisms in suspension settle as a sludge. The sludge is withdrawn from the clarifier for further treatment.

A functionally similar biological filtering system has become popular as part of home aquarium filtration and purification. The aquarium water is drawn up out of the tank and then cascaded over a freely spinning corrugated fiber-mesh wheel before passing through a media filter and back into the aquarium. The spinning mesh wheel develops a biofilm coating of microorganisms that feed on the suspended wastes in the aquarium water and are also exposed to the atmosphere as the wheel rotates. This is especially good at removing waste urea and ammonia urinated into the aquarium water by the fish and other animals.

TERTIARY TREATMENT

The purpose of tertiary treatment is to provide a final treatment stage to raise the effluent quality before it is discharged to the receiving environment. More than one tertiary treatment process may be used at any treatment plant. If disinfection is practiced, it is always the final process. It is also called "effluent polishing".

Filtration

Sand filtration removes much of the residual suspended matter. Filtration over activated carbon removes residual toxins.

Lagooning

Lagooning provides settlement and further biological improvement through storage in large man-made ponds or lagoons. These lagoons are highly aerobic and colonization by native macrophytes, especially reeds, is often encouraged. Small filter feeding invertebrates such as Daphnia and species of Rotifera greatly assist in treatment by removing fine particulates.

CONSTRUCTED WETLANDS

Constructed wetlands include engineered reedbeds and a range of similar methodologies, all of which provide a high degree of aerobic biological improvement and can often be used instead of secondary treatment for small communities. One example is a small reedbed used to clean the drainage from the elephants' enclosure at Chester Zoo in England.

NUTRIENT REMOVAL

Wastewater may contain high levels of the nutrients nitrogen and phosphorus. Excessive release to the environment can lead to a build up of nutrients, called eutrophication, which can in turn encourage the overgrowth of weeds, algae, and cyanobacteria. This may cause an algal bloom, a rapid growth in the population of algae. The algae numbers are unsustainable and eventually most of them die.

The decomposition of the algae by bacteria uses up so much of oxygen in the water that most or all of the animals die, which creates more organic matter for the bacteria to decompose. In addition to causing deoxygenation, some algal species produce toxins that contaminate drinking water supplies. Different treatment processes are required to remove nitrogen and phosphorus.

Nitrogen removal

The removal of nitrogen is effected through the biological oxidation of nitrogen from ammonia (nitrification) to nitrate, followed by denitrification, the reduction of nitrate to nitrogen gas. Nitrogen gas is released to the atmosphere and thus removed from the water. Nitrification itself is a two-step aerobic process, each step facilitated by a different type of bacteria.

The oxidation of ammonia to nitrite is most often facilitated by Nitrosomonas spp. Nitrite oxidation to nitrate, though traditionally believed to be facilitated by Nitrobacter spp. is now known to be facilitated in the environment almost exclusively by Nitrospira spp.

Denitrification requires anoxic conditions to encourage the appropriate biological communities to form. It is facilitated by a wide diversity of bacteria. Sand filters, lagooning and reed beds can all be used to reduce nitrogen, but the activated sludge process can do the job the most easily. Since denitrification is the reduction of nitrate to dinitrogen gas, an electron donor is needed.

This can be, depending on the wastewater, organic matter, sulfide, or an added donor like methanol. Sometimes the conversion of toxic ammonia to nitrate alone is referred to as tertiary treatment.

Phosphorus Removal

Phosphorus removal is important as it is a limiting nutrient for algae growth in many fresh water systems . It is also particularly important for water reuse systems where high phosphorus concentrations may lead to fouling of downstream equipment such as reverse osmosis.

Phosphorus can be removed biologically in a process called enhanced biological phosphorus removal. In this process, specific bacteria, called polyphosphate accumulating organisms, are selectively enriched and accumulate large quantities of phosphorus within their cells . When the biomass enriched in these bacteria is separated from the treated water, these biosolids have a high fertilizer value.

Phosphorus removal can also be achieved by chemical precipitation, usually with salts of iron , aluminum, or lime. This may lead to excessive sludge productions as hydroxides precipitates and the added chemicals can be expensive. Despite this, chemical phosphorus removal requires significantly smaller equipment footprint than biological removal, is easier to operate and is often more reliable than biological phosphorus removal. Once removed, phosphorus, in the form of a phosphate rich sludge, may be land filled or, if in suitable condition, resold for use in fertilizer.

DISINFECTION

The purpose of disinfection in the treatment of wastewater is to substantially reduce the number of microorganisms in the water to be discharged back into the environment. The effectiveness of disinfection depends on the quality of the water being treated , the type of disinfection being used, the disinfectant dosage , and other environmental variables.

Cloudy water will be treated less successfully since solid matter can shield organisms, especially from ultraviolet light or if contact times are low. Generally, short contact times, low doses and high flows all militate against effective disinfection. Common methods of disinfection include ozone, chlorine, or ultraviolet light. Chloramine, which is used for drinking water, is not used in wastewater treatment because of its persistence. Chlorination remains the most common form of wastewater disinfection in North America due to its low cost and long-term history of effectiveness. One disadvantage is that chlorination of residual organic material can generate chlorinated-organic compounds that may be carcinogenic or harmful to the environment. Residual chlorine or chloramines may also be capable of chlorinating organic material in the natural aquatic environment. Further, because residual chlorine is toxic to aquatic species, the treated effluent must also be chemically dechlorinated, adding to the complexity and cost of treatment.

Ultraviolet light can be used instead of chlorine, iodine, or other chemicals. Because no chemicals are used, the treated water has no adverse effect on organisms that later consume it, as may be the case with other methods. UV radiation causes damage to the genetic structure of bacteria, viruses, and other pathogens, making them incapable of reproduction.

The key disadvantages of UV disinfection are the need for frequent lamp maintenance and replacement and the need for a highly treated effluent to ensure that the target microorganisms are not shielded from the UV radiation . In the United Kingdom, light is becoming the most common means of disinfection because of the concerns about the impacts of chlorine in chlorinating residual organics in the wastewater and in chlorinating organics in the receiving water. Edmonton and Calgary, Alberta, Canada also use UV light for their effluent water disinfection. Ozone O_3 is generated by passing oxygen O_2 through a high voltage potential resulting in a third oxygen atom becoming attached and forming O_3. Ozone is very unstable and reactive and oxidizes most organic material it comes in contact with, thereby destroying many pathogenic microorganisms.

Ozone is considered to be safer than chlorine because, unlike chlorine which has to be stored on site, ozone is generated onsite as needed. Ozonation also produces fewer disinfection by-products than chlorination. A disadvantage of ozone disinfection is the high cost of the ozone generation equipment and the requirements for special operators.

Package Plants and Batch Reactors

In order to use less space, treat difficult waste, deal with intermittent flow or achieve higher environmental standards, a number of designs of hybrid treatment plants have been produced. Such plants often combine all or at least two stages of the three main treatment stages into one combined stage. In the UK, where a large number of sewage treatment plants serve small populations, package plants are a viable alternative to building discrete structures for each process stage. One type of system that combines secondary treatment and settlement is the sequencing batch reactor. Typically, activated sludge is mixed with raw incoming sewage and mixed and aerated. The resultant mixture is then allowed to settle producing a high quality effluent. The settled sludge is run off and re-aerated before a proportion is returned to the headworks. SBR plants are now being deployed in many parts of the world including North Liberty, Iowa, and Llanasa, North Wales, Southwestern United States.

The disadvantage of such processes is that precise control of timing, mixing and aeration is required. This precision is usually achieved by computer controls linked to many sensors in the plant. Such a complex, fragile system is unsuited to places where such controls may be unreliable, or poorly maintained, or where the power supply may be intermittent. Package plants may be referred to as high charged or low charged. This refers to the way the biological load is processed. In high charged systems, the biological stage is presented with a high organic load and the combined floc and organic material is then oxygenated for a few hours before being charged again with a new load. In the low charged system the biological stage contains a low organic load and is combined with flocculate for a relatively long time.

Sludge Treatment and Disposal

The sludges accumulated in a wastewater treatment process must be treated and disposed of in a safe and effective manner. The purpose of digestion is to reduce the amount of organic matter and the number of disease-causing microorganisms present in the solids. The most common treatment options include anaerobic digestion, aerobic digestion, and composting. Choice of a wastewater solid treatment method depends on the amount of solids generated and other site-specific conditions. However, in general, composting is most often applied to smaller-scale applications followed by aerobic digestion and then lastly anaerobic digestion for the larger-scale municipal applications.

ANAEROBIC DIGESTION

Anaerobic digestion is a bacterial process that is carried out in the absence of oxygen. The process can either be *thermophilic* digestion, in which sludge is fermented in tanks at a temperature of 55°C, or *mesophilic*, at a temperature of around 36°C. Though allowing shorter retention time, thermophilic digestion is more expensive in terms of energy consumption for heating the sludge. One major feature of anaerobic digestion is the production of biogas, which can be used in generators for electricity production and/or in boilers for heating purposes.

AEROBIC DIGESTION

Aerobic digestion is a bacterial process occurring in the presence of oxygen. Under aerobic conditions, bacteria rapidly consume organic matter and convert it into carbon dioxide. The operating costs are characteristically much greater for aerobic digestion because of the energy costs needed to add oxygen to the process.

SLUDGE DISPOSAL

When a liquid sludge is produced, further treatment may be required to make it suitable for final disposal. Typically, sludges are thickened to reduce the volumes transported off-site for disposal. There is no process which completely eliminates the need to dispose of biosolids. There is, however, an additional step some cities are taking to superheat the wastewater sludge and convert it into small pelletized granules that are high in nitrogen and other organic materials.

In New York City, for example, several sewage treatment plants have dewatering facilities that use large centrifuges along with the addition of chemicals such as polymer to further remove liquid from the sludge. The removed fluid, called centrate, is typically reintroduced into the wastewater process. The product which is left is called "cake" and that is picked up by companies which turn it into fertilizer pellets. This product is then sold to local farmers and turf farms as a soil amendment or fertilizer, reducing the amount of space required to dispose of sludge in landfills.

TREATMENT IN THE RECEIVING ENVIRONMENT

Many processes in a wastewater treatment plant are designed to mimic the natural treatment processes that occur in the environment, whether that environment is a natural water body or the ground. If not overloaded, bacteria in the environment will consume organic contaminants, although this will reduce the levels of oxygen in the water and may significantly change the overall ecology of the receiving water. Native bacterial populations feed on the organic contaminants, and the numbers of disease-causing microorganisms are reduced by natural environmental conditions such as predation or exposure to ultraviolet radiation. Consequently, in cases where the receiving environment provides a high level of dilution, a high degree of wastewater treatment may not be required.

However, recent evidence has demonstrated that very low levels of certain contaminants in wastewater, including hormones and synthetic materials such as phthalates that mimic hormones in their action, can have an unpredictable adverse impact on the natural biota and potentially on humans if the water is re-used for drinking water.

In the US and EU, uncontrolled discharges of wastewater to the environment are not permitted under law, and strict water quality requirements are to be met. A significant threat in the coming decades will be the increasing uncontrolled discharges of wastewater within rapidly developing countries.

SEWAGE TREATMENT IN DEVELOPING COUNTRIES

There are few reliable figures on the share of the wastewater collected in sewers that is being treated in the world. In many developing countries the bulk of domestic and industrial wastewater is discharged without any treatment or after primary treatment only. In Latin America about 15% of collected wastewater passes through treatment plants. In Venezuela, a below average country in South America with respect to wastewater treatment, 97 percent of the country's sewage is discharged raw into the environment.

In a relatively developed Middle Eastern country such as Iran, Tehran's majority of population has totally untreated sewage injected to the city's groundwater. Israel has also aggressively pursued the use of treated sewer water for irrigation. In 2008, agriculture in Israel consumed 500 million cubic metres of potable water and an equal amount of treated sewer water. The country plans to provide a further 200 million cubic metres of recycled sewer water and build more desalination plants to supply even more water. chal Most of sub-Saharan Africa is without wastewater treatment.

Water utilities in developing countries are chronically underfunded because of low water tariffs, the inexistence of sanitation tariffs in many cases, low billing efficiency and poor operational efficiency. In addition, wastewater treatment typically is the process within the utility that receives the least attention, partly because enforcement of environmental standards is poor. As a result of all these factors, operation and maintenance of many wastewater treatment plants is poor. This is evidenced by the frequent breakdown of equipment, shutdown of electrically operated equipment due to power outages or to reduce costs, and sedimentation due to lack of sludge removal.

Developing countries as diverse as Egypt, Algeria, China or Colombia have invested substantial sums in wastewater treatment without achieving a significant impact in terms of environmental improvement. Even if wastewater treatment plants are properly operating, it can be argued that the environmental impact is limited in cases where the assimilative capacity of the receiving waters is high, as it is often the case.

IN DEVELOPING COUNTRIES

As of 2006, waterborne diseases are estimated to cause 1.8 million deaths each year. These deaths are attributable to inadequate public sanitation systems and it is clear that proper sewerage need to be installed. Appropriate technology options in water treatment include both community-scale and household-scale point-of-use designs. Military surplus water treatment units like the ERDLator are still seen in developing countries. Newer military style Reverse Osmosis Water Purification Units are portable, self-contained water treatment plants are becoming more available for public use. In order for the decrease of waterborne diseases to have long term effects, water treatment programs implemented by research and development groups in developing countries must be sustainable by its own residents. This can ensure the efficiency of such programs after the departure of the research team as monitoring is difficult because of the remoteness of many locations.

Index

D

E

F

G

H

I

J

K

L

M

N